Advanced Engineering Mathematics

Second Edition

Advanced Engineering Mathematics

Second Edition

A. C. BAJPAI

L. R. MUSTOE

D. WALKER

Department of Mathematical Sciences
Loughborough University of Technology

JOHN WILEY & SONS
Chichester . New York . Brisbane . Toronto . Singapore

Wiley Editorial Offices

John Wiley & Sons Ltd, Baffins Lane, Chichester, West Sussex, PO19 1UD, England

John Wiley & Sons, Inc., 605 Third Avenue, New York, NY 10158-0012, USA

Jacaranda Wiley Ltd, G.P.O. Box 859, Brisbane, Queensland 4001, Australia

John Wiley & Sons (Canada) Ltd, 22 Worcester Road, Rexdale, Ontario, M9W 1L1, Canada

John Wiley & Sons (SEA) Pte Ltd, 37 Jalan Pemimpin #05-04, Block B, Union Industrial Building, Singapore 2057

Copyright © 1990 by John Wiley & Sons Ltd.

Library of Congress Cataloging-in-Publication Data:

Bajpai, A. C. (Avinash Chandra)
    Advanced engineering mathematics / A.C. Bajpai, L.R. Mustoe, D. Walker. – 2nd ed.
      p . cm.
    Includes bibliographical references and index.
    ISBN 0 471 92595 0
      1.Engineering mathematics. I. Mustoe, L. R. II.
Walker, Dennis, Ph. D. III. Title
TA330.B33 1990
510′ .2462 – dc20   90-37523
                                   CIP

British Library Cataloguing in Publication Data:

Bajpai, A. C. (Avinash Chandra) *1925-*
  Advanced engineering mathematics. –2nd ed.
  1. Mathematics
  I. Title II. Mustoe, L. R. (Leslie R.) III. Walker, D.
  1932-
  510

ISBN 0 471 92595 0

Printed and bound in Great Britain by
Courier International Ltd, Tiptree, Essex

PREFACE

It is well accepted that a good mathematical grounding is essential for all engineers and scientists. This book is the second in a series and is aimed at second year undergraduate science and engineering students in universities, polytechnics and colleges in all parts of the world. It would also be useful for students preparing for the Engineering Council examinations in mathematics at Part 2 standard. The authors have continued the 'integrated' approach of their first book *Engineering Mathematics* in the present volume. However, students who have covered the earlier material from an alternative approach will still benefit from using this book.

The basic concept of the book is that it should provide a motivation for the student. Thus, wherever possible, a topic is introduced by considering a real example and formulating the mathematical model for the problem; its solution is considered by both analytical and numerical techniques. In this way, it is hoped to integrate the two approaches, whereas many texts have regarded the analytical and numerical methods as separate entities. As a consequence, students have failed to realise the possibilities of the different methods or that on occasions a combination of both analytical and numerical techniques is needed. Indeed, in most practical cases met by the engineer and scientist the desired answer is a set of numbers; even if the solution can be obtained completely by analytical methods, the final process is often to obtain discrete values from the analytical expression.

The authors believe that some proofs are necessary where basic principles are involved. However, in other cases where it is thought that the proof is too difficult for students at this stage, it has been omitted or only outlined. For the numerical techniques, the approach has been to form a heuristically derived algorithm to illustrate how it is used and a formal justification is given only in the simpler cases.

In parts of the text the running of an appropriate computer program will aid understanding of certain techniques. Many books are devoted solely to the teaching of computer programming; we therefore do not attempt to cover this topic in detail. It is hoped that students will either have prior knowledge of programming or will be studying this in a parallel course.

Throughout the text there is a generous supply of worked examples which illustrate both the theory and its application. Supplementary problems are provided at the end of each chapter. Some harder problems have been included for which the student may have to seek help from his teacher.

Since the first edition appeared the authors have received much helpful comment and criticism. The overwhelming reaction, fortunately, was favourable. However, the advent of the pocket calculator and the microcomputer and the shifting in emphasis of parts of the syllabus meant that a new edition of this book was needed. Accordingly, taking the view of our readers, both staff and students, into account, the text material has been revised. The opportunity has also been taken to remove some mathematical 'dead wood'; this made room for some new ideas.

Changes in the revised edition
The present edition differs from its predecessor in the following respects.

1 *Chapters* The text material has been reorganised into eleven chapters instead of the fourteen of the first edition. This reflects the revisions to the material.

2 *Problem sets* These now appear at the end of each chapter instead of at the end of each section. The problems have been modified and have included more up to date examples from Engineering Council examinations.

3 *Transform methods* The chapter on integral transforms has been extended to meet the needs of Electrical Engineering applications. There is more material on Fourier Transforms and the Z-transform and Discrete Fourier Transform are introduced.

4 *Vector Field Theory* Some material on Vector Field Theory has been removed and the surviving subject-matter has been amalgamated with some of the material on integration.

5 *Statistical methods* The material from the two chapters in the first edition has been condensed. The application of the χ^2 distribution to work on frequencies has been taken out and multiple regression has been restricted to an introductory account.

The authors would like to thank all staff and students who have made helpful suggestions for the improvement of the text. In addition they acknowledge with pleasure a debt of gratitude to the following:

Staff and students of Loughborough University of Technology and other institutions who have participated in the development of the original text and the revisions incorporated in this edition.

John Wiley & Sons Ltd for their help and cooperation.

The University of London and The Engineering Council for permission to use questions from their past examination papers. (These are denoted by LU and EC respectively.)

Miss H A Wyatt for typing this book.

Mr Alex Ford for preparing the diagrams.

CONTENTS

1

LINEAR ALGEBRA

1.1 Introduction

Much of the application of mathematics to engineering problems involves linear models. A system is said to be **linear** if the following property obtains. Suppose the input x_1 causes an output y_1 and the input x_2 causes an output y_2 then the input $\alpha x_1 + \beta x_2$, where α and β are scalars, causes an output $\alpha y_1 + \beta y_2$. Great advantages occur if a system is linear. The output from a combination of inputs can be predicted if the outputs from each separate input are known. In this way, a great simplification of the analysis of the system is possible.

But not all systems are linear. We are then faced with the choice of attempting a non-linear analysis or making a linear approximation to the system; do the powerful methods of linear analysis that we can bring to bear outweigh the possible crudeness of approximation? Many of the theoretical models which are widespread in engineering *are* linear and these models have served engineers well. In this chapter we intend to show how linear algebra can act as a unifying link between many branches of mathematics. Sometimes we shall be able to apply our theory directly to practical problems; sometimes we shall not, and you will have to accept the theory as a stepping-stone to further theory which does have direct practical application. Your understanding of some techniques will be enriched by a deeper appreciation of the underlying theory.

1.2 Vector Spaces

Consider the differential equation $\qquad \dfrac{d^2y}{dx^2} - 4\dfrac{dy}{dx} + 3y = 0 \qquad\qquad (1.1)$

We find its general solution by forming the auxiliary equation $m^2 - 4m + 3 = 0$ which has roots $m = 3$ and $m = 1$ and we state that two basic solutions of (1.1) are e^{3x} and e^x. We then claim that the **general solution** of (1.1) is found by taking a **linear combination** of the basic solutions, viz.

$$y = Ae^{3x} + Be^x \qquad\qquad (1.2)$$

where A and B are real constants. What do we mean by a *general* solution? We mean that *any* solution of (1.1) can be expressed in the form (1.2). Now it is easy to show that (1.2) is *a* solution of (1.1), but its generality is more difficult to establish. For the moment, note that we could represent (1.2) as a pair of coordinates (A, B). The particular values of A and B will be determined by the initial conditions attached to (1.1). Each **particular** solution corresponds to a particular coordinate pair (A, B) which represents in some sense a point in some plane; but, which plane? Also note that if $y_1(x)$ is a solution of (1.1) then so is $\alpha y_1(x)$, where α is any scalar. Further, if $y_2(x)$ is another solution of (1.1) then so is the sum $y_1(x) + y_2(x)$.

Now let us turn to an apparently different example. Consider all polynomials
$$a + bx + cx^2 + dx^3 \tag{1.3}$$
where a, b, c and d are real constants. If we multiply the polynomial (1.3) by a scalar, we obtain a polynomial of the same form; likewise, if we add together two such polynomials we again obtain one of the same form. Then the set of polynomials (1.3) is said to be **closed under scalar multiplication** and **closed under addition**. Furthermore we can represent a polynomial of the form (1.3) as the coordinate quartet (a, b, c, d). The structure underlying both examples is seen to be similar.

Axioms of a Vector Space

Let $v_1, v_2, v_3, \dots$ be elements (vectors) of a set V and α, β any real scalars.

(i)   To every pair of vectors $v_1, v_2 \in V$ there is a unique vector $(v_1 + v_2) \in V$.

(ii) Addition is associative: $v_1 + (v_2 + v_3) = (v_1 + v_2) + v_3$ for any three vectors $v_1, v_2, v_3 \in V$.

(iii) Addition is commutative: $v_1 + v_2 = v_2 + v_1$.

(iv) There is a zero vector, $0 \in V$, such that $v_1 + 0 = v_1$ for all $v_1 \in V$.

(v)   For each vector $v_1 \in V$ there is a unique vector $-v_1 \in V$ such that $v_1 + (-v_1) = 0$.

(vi) To each vector $v_1 \in V$ and to each scalar $\alpha \in R$ there is a unique product $\alpha v_1 \in V$.

(vii) $(\alpha\beta)v_1 = \alpha(\beta v_1)$.

(viii) $1\, v_1 = v_1$.

(ix) $\alpha(v_1 + v_2) = \alpha v_1 + \alpha v_2$ } Distributive laws.

(x) $(\alpha + \beta)v_1 = \alpha v_1 + \beta v_1$ }

Subspaces

A subset of vectors from a vector space may form a new vector space by themselves; in such a case we shall call the subset a **subspace** of V. In order to check whether a subset of vectors does form a subspace it is sufficient to demonstrate that the subset is closed under addition and scalar multiplication. For example, if we consider the subset of solutions of (1.1) for which $y(0) = 0$ then the condition on A and B of (1.2) is that $A + B = 0$. Hence, we are dealing with solutions of the form $y = A(e^{3x} - e^x)$. We can show that these solutions are closed under addition and scalar multiplication in one fell swoop: we need merely to show that if $y_1(x)$ and $y_2(x)$ are two such solutions, then so is $[\alpha y_1(x) + \beta y_2(x)]$ for any scalars α and β (i.e. any **linear combination** of y_1 and y_2). Let $y_1 = A_1(e^{3x} - e^x)$ and $y_2 = A_2(e^{3x} - e^x)$ and consider $[\alpha y_1(x) + \beta y_2(x)] = \alpha A_1(e^{3x} - e^x) + \beta A_2(e^{3x} - e^x)$; this can be shown to be a solution of the equation (1.1) which has value zero when $x = 0$. Hence the subset of solutions for which $y(0) = 0$ is a subspace. What about the subset of solutions for which $y(0) = 1$?

Example 1

The subset of polynomials of degree ≤ 3 for which the x term is absent forms a subspace.

In vector terms, the subset consists of all vectors of the form $(a, 0, c, d)$. Consider two vectors: $(a_1, 0, c_1, d_1)$ and $(a_2, 0, c_2, d_2)$; the linear combination $\alpha(a_1, 0, c_1, d_1) + \beta(a_2, 0, c_2, d_2) = (\alpha a_1 + \beta a_2, 0, \alpha c_1 + \beta c_2, \alpha d_1 + \beta d_2)$ satisfies the criterion of a zero second coordinate. The subset is closed under addition and scalar multiplication and hence forms a subspace.

Example 2

The subset of polynomials of degree ≤ 3 for which the constant term is 3 do *not* form a subspace. Why not?

Linear Independence

When all the vectors in a vector space can be expressed as linear combinations of a subset of those vectors, the subset is said to **span** the space. For example, the polynomials 1, x, x^2 and x^3 span the space of polynomials expressed in (1.3). The unit vectors $\mathbf{i}, \mathbf{j}, \mathbf{k}$ span the space of geometric vectors with three components since any vector (x, y, z) can be written as $x\mathbf{i} + y\mathbf{j} + z\mathbf{k}$.

The functions $3e^{3x} - e^x$ and $2e^{3x} + e^x$ span the space (1.2). Consider a typical solution $4e^{3x} + 3e^x$; this can be written as $\alpha(3e^{3x} - e^x) + \beta(2e^{3x} + e^x) = (3\alpha + 2\beta)e^{3x} + (-\alpha + \beta)e^x$, provided that $3\alpha + 2\beta = 4$ and $-\alpha + \beta = 3$; in this case $\alpha = -2/5$ and $\beta = 13/5$. Another subset of solutions which span the space is $\{e^{3x}, e^x, 3e^{3x} - e^x\}$; try to show this. The function $3e^{3x} - e^x$ is *itself* a linear combination of e^{3x} and e^x and is in effect redundant. However, neither e^{3x} nor e^x can be expressed as a multiple (special case of a linear combination) of each other. A set of vectors $\{\mathbf{v}_1, \mathbf{v}_2, \mathbf{v}_3,..., \mathbf{v}_n\}$ is said to be **linearly dependent** if and only if there is a set of real numbers $\{\alpha_1, \alpha_2, \alpha_3,... , \alpha_n\}$ *not all zero* such that

$$\alpha_1\mathbf{v}_1 + \alpha_2\mathbf{v}_2 + \alpha_3\mathbf{v}_3 + ... + \alpha_n\mathbf{v}_n = \mathbf{0} \qquad (1.4)$$

Note that (1.4) can be rearranged to give a formula for any one of the vectors in terms of the rest; in fact, as a linear combination of the rest.

A set of vectors which is not linearly dependent is **linearly independent**; the condition (1.4) implies in this case that $\alpha_1 = \alpha_2 = \alpha_3 = ... = \alpha_n = 0$.

Example 1
The set of polynomials $\{1, x+2, x^2+x, x^3\}$ is linearly independent. We write the functions in vector form as $(1,0,0,0)$, $(2,1,0,0)$, $(0,1,1,0)$, $(0,0,0,1)$. Consider the equation $\alpha(1,0,0,0) + \beta(2,1,0,0) + \gamma(0,1,1,0) + \delta(0,0,0,1) = \mathbf{0} = (0,0,0,0)$. Comparing components we obtain the equations $\alpha + 2\beta = 0$, $\beta + \gamma = 0$, $\gamma = 0$, $\delta = 0$. Hence $\alpha = \beta = \gamma = \delta = 0$ and therefore the set is linearly independent.

Example 2
The set of polynomials $\{1, x+2, x^2+x, x^3, x^3-x\}$ is linearly dependent. *Prove this.*

Example 3
The set of polynomials $\{x+2, x^3-2x, x^3+4\}$ is linearly dependent, since (resorting to vector notation) $\alpha(2,1,0,0) + \beta(0,-2,0,1) + \gamma(4,0,0,1) = (0,0,0,0)$ $\Rightarrow 2\alpha + 4\gamma = 0$, $\alpha - 2\beta = 0$, $\beta + \gamma = 0$ and these equations have infinitely many solutions of the form $(\alpha, \beta, \gamma) = (-2c, -c, c)$ for any value c.

We can show that a subset of a linearly independent set of vectors is also linearly independent and that a linearly dependent set will remain linearly dependent if one or more vectors from V are added to it.

Basis of a Vector Space
We have seen that we can produce subsets of vectors from a vector space which span the space. If we restrict our attention to those spanning sets which are linearly independent we can summarise the information about the vector space by listing the vectors in a linearly independent spanning set. Since the set spans the space we can produce any vector in the space by a linear combination of the elements of the set. Since the set is linearly independent, we cannot afford to lose any vectors from it or else it will cease to span the space. In other words a spanning set which is linearly independent contains *precisely* the right amount of information to describe the space. Such a set is called a **basis** for the vector space.

Example 1
We can show that the set $\{(1,0,0,0), (0,1,0,0,), (0,0,1,0), (0,0,0,1)\}$ is linearly independent, and furthermore it spans the space of vectors with four real components. Hence it forms a basis for the space. So does the subset $\{(1,0,0,0), (2,1,0,0), (0,1,1,0), (0,0,0,1)\}$, for example.

Example 2

For the space of solutions of $\dfrac{d^2y}{dx^2} - 4\dfrac{dy}{dx} + 3y = 0$, each of the following sets

form a basis: $\{(1,0), (0,1)\}$, $\{(1,1), (0,1)\}$, $\{(1,1), (1,0)\}$. There is one thing that *all* bases for a given space have in common: they contain the same number of vectors. The number of vectors in a basis for a given space is called the **dimension** of the space. It is important to realise that any set of n linearly independent vectors will be a basis for a space of dimension n.

Example 3

The set of polynomials of degree ≤ 3 with no term in x form a vector space. A possible basis is $\{(1,0,0,0), (0,0,1,0), (0,0,0,1)\}$. Since the basis consists of three vectors, the dimension of this space is 3. Notice that we have taken a subset of the four unit vectors which comprise a basis for the space of all polynomials of degree ≤ 3. By imposing the one restriction that the x term is absent we have reduced the dimension of the space by 1. When we express a vector of the space as a linear combination of the vectors in the basis, the coefficients are called the **coordinates** of the vector *relative* to the particular basis.

Orthonormal Bases

One advantage of using a basis consisting of the unit vectors is that they are **mutually orthogonal**, i.e. the scalar product of any two of them is zero. Furthermore, they are unit vectors so that the scalar product of any of them with itself is unity; we say that the vectors are **normalised**. A set of normalised, mutually orthogonal vectors is called an **orthonormal set**. If we introduce the **Kronecker delta** symbol δ_{ij} which follows the rules that $\delta_{ij} = 0$ for $i \neq j$ and $\delta_{ij} = 1$, $i = j$, then we may succinctly define an orthonormal set of vectors $\{\mathbf{v}_1, \mathbf{v}_2, ..., ..., \mathbf{v}_n\}$ by requiring that $\mathbf{v}_i \cdot \mathbf{v}_j = \delta_{ij}$.

An important result is that an **orthonormal set is linearly independent**. *Prove this.*

In geometry we are used to Cartesian axes being mutually orthogonal. If we project a point in space (3 dimensional) onto the x-y plane then the coordinates of the point reduce in a simple way. Thus, the point $(3,4,-2)$ when projected onto the x-y plane, becomes in the new axes the point $(3,4)$.

Gram-Schmidt Orthogonalisation Process

This method takes a linearly independent set of vectors from a space V and produces an orthonormal set of vectors from V. In particular it will convert a basis for V into an orthonormal basis. The idea is to start with a linearly independent set $\{\mathbf{v}_1, \mathbf{v}_2, ..., \mathbf{v}_s\}$ and produce an orthonormal set

$$\{\xi_1, \xi_2, ..., \xi_s\}, \quad \text{where each } \xi_k = \sum_{i=1}^{k} a_{ik} \mathbf{v}_i .$$

We know that $\mathbf{v}_1 \neq 0$ and hence $|\mathbf{v}_1| > 0$; therefore we can sensibly define $\xi_1 = \mathbf{v}_1/|\mathbf{v}_1|$ so that $|\xi_1| = 1$.

We now assume that we have found the first r vectors of the orthonormal set and show how to find ξ_{r+1}. We assume that the set $\{\xi_1, \xi_2, ..., \xi_r\}$ is orthonormal and that each of the vectors ξ_k in it is a linear combination of the set $\{\mathbf{v}_1, \mathbf{v}_2, ..., \mathbf{v}_k\}$. Now consider

$$\mathbf{v}'_{r+1} = \mathbf{v}_{r+1} - (\xi_1 \cdot \mathbf{v}_{r+1})\xi_1 - (\xi_2 \cdot \mathbf{v}_{r+1})\xi_2 - ... - (\xi_r \cdot \mathbf{v}_{r+1})\xi_r \qquad (1.5)$$

By taking the scalar product of (1.5) in turn with $\xi_1, \xi_2, ..., \xi_r$ we can show that the vector $\mathbf{v}'_{r+1}$ is orthogonal to the vectors $\xi_1, \xi_2, ..., \xi_r$ and hence $\{\xi_1, \xi_2, ..., \xi_r, \mathbf{v}'_{r+1}\}$ is a mutually orthogonal set. Is $\mathbf{v}'_{r+1}$ of unit length? We can ensure that the additional vector is of unit length by choosing it to be $\xi_{r+1} = \mathbf{v}'_{r+1}/|\mathbf{v}'_{r+1}|$. Hence $\{\xi_1, \xi_2, ..., \xi_r, \xi_{r+1}\}$ is an orthonormal set. In addition, $\mathbf{v}'_{r+1}$ (and therefore ξ_{r+1}) must be a linear combination of the vectors $\mathbf{v}_1, \mathbf{v}_2, ..., \mathbf{v}_{r+1}$; this follows from (1.5) and because we assumed that $\xi_1, \xi_2, ..., \xi_r$ were linear combinations of the vectors $\mathbf{v}_1, \mathbf{v}_2, ..., \mathbf{v}_r$.

If we proceed to replace each $\mathbf{v}_k$ by ξ_k as prescribed above, we shall obtain an orthonormal set $\{\xi_1, \xi_2, ..., \xi_s\}$.

Example

We start with the set $\{\mathbf{v}_1, \mathbf{v}_2, \mathbf{v}_3\} = \{(1,0,1), (1,-2,1), (0,1,1)\}$ which we can show is a basis for the space of vectors with three components. (Note that we merely need to demonstrate linear independence.)

Then $|\mathbf{v}_1| = \sqrt{1^2 + 0^2 + 1^2} = \sqrt{2}$ and $\xi_1 = \dfrac{1}{\sqrt{2}}(1,0,1)$

Now $\mathbf{v}'_2 = \mathbf{v}_2 - (\xi_1 \cdot \mathbf{v}_2)\xi_1 = (1,-2,1) - \left[\dfrac{1}{\sqrt{2}}(1,0,1).(1,-2,1) \right] \dfrac{1}{\sqrt{2}}(1,0,1)$

$= (1,-2,1) - \dfrac{1}{2}(2)(1,0,1) = (0,-2,0)$

Hence $\xi_2 = \mathbf{v}'_2/|\mathbf{v}'_2| = \dfrac{1}{2}(0,-2,0) = (0,-1,0)$

Then $\mathbf{v}'_3 = \mathbf{v}_3 - (\xi_1 \cdot \mathbf{v}_3)\xi_1 - (\xi_2 \cdot \mathbf{v}_3)\xi_2$

$= (0,1,1) - \left[\dfrac{1}{\sqrt{2}}(1,0,1).(0,1,1) \right] \dfrac{1}{\sqrt{2}}(1,0,1)$

$- [(0,-1,0) . (0,1,1)] (0,-1,0)$

$= (0,1,1) - \dfrac{1}{2}(1,0,1) - (-1)(0,-1,0) = (-\dfrac{1}{2}, 0, \dfrac{1}{2})$

Finally, $\xi_3 = \dfrac{1}{\sqrt{2}}(-1,0,1)$.

We can easily check that $\xi_i . \xi_j = \delta_{ij}$ and hence that $\{\xi_1, \xi_2, \xi_3\}$ is an orthonormal set. Furthermore these three vectors are linearly independent. We have therefore constructed an orthonormal basis.

1.3 Linear Transformations

This section is concerned with mappings from one vector space to another.

The particular class of mappings which interests us is that class which preserves linear structure. Let T be a transformation mapping vectors from a space U, known as the **domain**, to another space V, known as the **co-domain**, in such a way that a vector $\mathbf{u}$ from U is associated uniquely with a vector $\mathbf{v}$ from V i.e. $T(\mathbf{u}) = \mathbf{v}$. If we impose the requirement that addition is preserved i.e.

$$T(\mathbf{u}_1 + \mathbf{u}_2) = T(\mathbf{u}_1) + T(\mathbf{u}_2) \tag{1.6a}$$

for all $\mathbf{u}_1, \mathbf{u}_2 \in U$, and if we impose the requirement that multiplication by a scalar is preserved i.e.

$$T(\alpha\mathbf{u}) = \alpha T(\mathbf{u}) \tag{1.6b}$$

then T is called a **linear transformation**. Note that (1.6a) and (1.6b) can be replaced by

$$T(\alpha\mathbf{u}_1 + \beta\mathbf{u}_2) = \alpha T(\mathbf{u}_1) + \beta T(\mathbf{u}_2) \tag{1.7}$$

Example 1

Let U and V both be the space of polynomials of degree ≤ 3 and let T be defined by the rule that a polynomial is mapped onto its derivative. Resorting to a more customary notation, we denote elements of U by $p(x)$ so that

$$T[p(x)] = \frac{d}{dx}p(x) = p'(x)$$

Now $\dfrac{d}{dx}\left[\alpha p_1(x) + \beta p_2(x) \right] = \alpha p'_1(x) + \beta p'_2(x)$ so that

$$T(\alpha\mathbf{u}_1 + \beta\mathbf{u}_2) = \alpha T(\mathbf{u}_1) + \beta T(\mathbf{u}_2)$$

which proves the linearity of T. One point to note is that the polynomials $p'(x)$ can never be of degree greater than 2. In other words the polynomial $a + bx + cx^2 + dx^3 \equiv (a,b,c,d)$ is mapped onto the polynomial $(b + 2cx + 3dx^2) \equiv (b, 2c, 3d, 0)$. This indicates that we have lost something by the mapping operation.

Example 2

Suppose we extend the above ideas and consider U comprising all functions which can be differentiated twice. Let the mapping T associate the function

$f(x)$ from U with the function $\left[\dfrac{d^2}{dx^2} - 4 \dfrac{d}{dx} + 3 \right] f(x)$.   It is left to you to

show that T is a linear transformation. Has anything been lost this time?

Example 3
   Consider U = all points (x,y) in the plane. Then let T be the operation rotating the plane through an anticlockwise angle θ about the origin so that

$$T[(x,y)] = (x',y') \quad \text{where} \quad \begin{cases} x' = x \cos \theta - y \sin \theta \\[2mm] y' = x \sin \theta + y \cos \theta \end{cases} \tag{1.8}$$

   However, in this example, it appears that points have merely changed the values of their coordinates and nothing has been lost.
   The subset of vectors of V which are the images under T of vectors from U, i.e. the set of all vectors $T(\mathbf{u})$ where $\mathbf{u}$ is a vector from U, forms a subspace of V. This is called the **image space** of T and is sometimes denoted by $T(U)$. If it is a subspace, its dimension must be less than or equal to the dimension of V. Under what circumstances is equality achieved? The dimension of the image space is called the **rank** of the linear transformation.

The kernel of a linear transformation
   Suppose we return to Example 2. We know that there are solutions of the

differential equation $\dfrac{d^2y}{dx^2} - 4 \dfrac{dy}{dx} + 3y = 0$ i.e. that there are several functions

$f(x)$ mapped to the polynomial 0 by the linear transformation $\left[\dfrac{d^2}{dx^2} - 4 \dfrac{d}{dx} + 3 \right]$.

On the other hand, from Example 3 we find that only the point $(0,0)$ is mapped to the point $(0,0)$.
   The subset of vectors in the domain which map onto the zero vector in the co-domain is called the **kernel** or the **null-space** of the linear transformation. We can show that the kernel is itself a vector space and that, moreover, the dimension of the kernel is the dimension of the space V less the dimension of the image space. We call the dimension of the kernel the **nullity** of the linear transformation and we state the result that if the linear transformation T maps vectors from U into vectors from V then

$$\text{rank of } T + \text{nullity of } T = \text{dimension of } U \tag{1.9}$$

For the projection mapping, the dimension of U is 3, the nullity is 1 since the vector $(0,0,1)$ is a basis for the kernel, and the rank is 2.
   Let us look back at our three other examples. In Example 1, the domain U had dimension 4 and the image space had dimension 3. The kernel consisted of all polynomials with zero derivative, i.e. all constants, which may be represented

as $(a,0,0,0)$; the dimension of this space is 1. In Example 2 it is not possible to state the dimension of the domain U, but we have already seen that the dimension of the kernel is 2. In Example 3 only the point $(0,0)$ is mapped to the point $(0,0)$ and the kernel, which merely consists of the zero vector, has zero dimension; the dimensions of the domain and the image space are both equal to 2.

Properties of the kernel

Let T map vectors from space U to vectors from space V, and let the kernel of T be written K.

(i) If $\mathbf{u}_0 \in U$ and $\mathbf{k} \in K$ then $\mathbf{u}_0 + \mathbf{k}$ is mapped to the same vector as $\mathbf{u}_0$. This follows since $T(\mathbf{u}_0 + \mathbf{k}) = T(\mathbf{u}_0) + T(\mathbf{k}) = T(\mathbf{u}_0) + \mathbf{0} = T(\mathbf{u}_0)$.

(ii) If $\mathbf{u}_0$ maps to the vector $\mathbf{v}_0$, then all vectors which map to $\mathbf{v}_0$ are of the form $\mathbf{u}_0 + \mathbf{k}$ where $\mathbf{k}$ is some vector from K.

This second property is important since it allows us to produce general solutions to linear ordinary differential equations and simultaneous linear algebraic equations. Before we demonstrate this, consider the operation of differentiating polynomials of degree less than or equal to 3. We have seen that this is an example of a linear transformation. The kernel is the set of constant polynomials. Therefore, the problem of integration of polynomials of degree <3 can be seen as aiming to find the polynomial which maps to a given polynomial under the operation of differentiation, i.e. we seek $p(x)$ such that $p'(x) = p_1(x)$, which is given. If we can find any polynomial $q(x)$ whose derivative is $p_1(x)$, then by property (ii) we know that all solutions to the problem are of the form $q(x) + c$ where c is a constant.

With regard to the solution of the non-homogeneous differential equation

$$\frac{d^2y}{dx^2} - 4\frac{dy}{dx} + 3y = g(x) \tag{1.10}$$

where $g(x)$ is a given function, we can extend our results to show that its most general solution may be written in the form $y = y_{PI} + y_{CF}$, where y_{PI} is *any* solution of (1.10) and y_{CF} is the general solution of the associated homogeneous equation

$$\frac{d^2y}{dx^2} - 4\frac{dy}{dx} + 3y = 0$$

Example

We require to find the solutions of the equations

$$\left.\begin{array}{l} 3x + 2y - z = 4 \\ x + y + z = 3\frac{1}{2} \end{array}\right\} \tag{1.11}$$

Here we have three unknowns restricted by two equations and therefore we must expect an infinite set of solutions. One *particular* solution is $(x,y,z) = (0, 2\frac{1}{2}, 1)$. How can we find other solutions? If we now think in terms of a vector space, the equations are imposing a mapping which associates (x,y,z) with $(3x + 2y - z, x + y + z)$ and we seek those vectors which are mapped to $(4, 3\frac{1}{2})$. The mapping is a linear transformation, and therefore by property (ii) we need only now find the kernel to obtain the general solution. The kernel is the set of (x,y,z) which map to $(0,0)$ and which therefore satisfy the equations

$$\left.\begin{array}{r} 3x + 2y - z = 0 \\ x + y + z = 0 \end{array}\right\} \tag{1.12}$$

We obtain $(x,y,z) = (3\lambda, -4\lambda, \lambda)$. By varying λ we obtain all elements in the kernel. The general solution of (1.11) can then be expressed as the sum of the particular solution and the general solution of (1.12) viz. $(x,y,z) = (3\lambda, 2\frac{1}{2} - 4\lambda, 1 + \lambda)$.

You might be tempted to ask what would have happened if we had taken a different particular solution, for example, $(-3, 6\frac{1}{2}, 0)$. We should then have expressed the solution set as $(x,y,z) = (-3 + 3\mu, 6\frac{1}{2} - 4\mu, \mu)$ where μ can be varied. Any specific solution will require different values of μ; for example the solution $(x,y,z) = (1, \frac{7}{6}, \frac{4}{3})$ requires $\lambda = \frac{1}{3}$ or $\mu = \frac{4}{3}$. Further, if the equations (1.11) are replaced by ones with different right-hand sides then the kernel will remain unaltered, and all that is needed is a new particular solution.

Matrices and Linear Transformations

We generalise to m equations in n unknowns. The equations can be regarded as defining a linear transformation from the space of vectors with n real components, denoted R^n, to the vector space R^m. It can be shown that *any* linear transformation mapping R^n to R^m can be defined by a set of m linear equations in n unknowns.

Such a set of equations can be summarised as $\mathbf{A}\mathbf{x} = \mathbf{b}$ where

$$\mathbf{A} = \begin{bmatrix} a_{11} & a_{12} & \cdots & a_{1n} \\ a_{21} & a_{22} & \cdots & a_{2n} \\ \vdots & \vdots & & \vdots \\ a_{m1} & a_{m2} & \cdots & a_{mn} \end{bmatrix}, \quad \mathbf{x} = \begin{bmatrix} x_1 \\ x_2 \\ \vdots \\ x_n \end{bmatrix} \quad \text{and} \quad \mathbf{b} = \begin{bmatrix} b_1 \\ b_2 \\ \vdots \\ b_m \end{bmatrix} \tag{1.13}$$

$\mathbf{A}$ is called the **associated matrix** of the linear transformation.

Suppose we take the vectors $\mathbf{e}_1 = (1,0,0,...,0)$, $\mathbf{e}_2 = (0,1,0,...,0)$, ..., $\mathbf{e}_n = (0,0,0,...,1)$ as a basis for R^n. Then let $\mathbf{x}$ be any vector of R^n so that $\mathbf{x} = x_1\mathbf{e}_1 + x_2\mathbf{e}_2 + ... + x_n\mathbf{e}_n$, i.e. with respect to this basis, $\mathbf{x} = (x_1,..., x_n)$. It follows that $T(\mathbf{x}) = x_1T(\mathbf{e}_1) + x_2T(\mathbf{e}_2) + ... + x_nT(\mathbf{e}_n)$.

Suppose $T(e_1) = (a_{11}, a_{21}, a_{31}, ..., a_{m1})$, $T(e_2) = (a_{12}, a_{22}, a_{32}, ..., a_{m2})$ and so on; then

$$T(x) = (a_{11}x_1 + a_{12}x_2 + ... + a_{1n}x_n, a_{21}x_1 + a_{22}x_2 + ... + a_{2n}x_n, ...,$$
$$a_{m1}x_1 + a_{m2}x_2 + ... + a_{mn}x_n)$$
$$= (b_1, b_2, ..., b_m)$$

Hence the transformation can be represented by equations (1.13).

Notice that the columns of the associated matrix are merely $T(e_1)$, $T(e_2)$, ..., $T(e_n)$. We can follow one transformation T_1 mapping vectors from R^n to vectors from R^m by a second transformation T_2 mapping vectors from R^m to vectors from R^p. The combined transformation is denoted by $T_2 \circ T_1$ so that if x is a vector from R^n, then $(T_2 \circ T_1)(x) \equiv T_2[T_1(x)]$.

Example 1

Consider T_1 and T_2 both mapping vectors from R^2 to vectors from R^2, defined by the set of equations

$$\left. \begin{array}{l} x'_1 = a_{11}x_1 + a_{12}x_2 \\ x'_2 = a_{21}x_1 + a_{22}x_2 \end{array} \right\} \text{ and } \left. \begin{array}{l} x''_1 = a'_{11}x'_1 + a'_{12}x'_2 \\ x''_2 = a'_{21}x'_1 + a'_{22}x'_2 \end{array} \right\} \text{ respectively.}$$

Substituting for x'_1, x'_2 from the first set into the second set, we obtain

$$\left. \begin{array}{l} x''_1 = (a'_{11} a_{11} + a'_{12} a_{21}) x_1 + (a'_{11} a_{12} + a'_{12} a_{22}) x_2 \\ x''_2 = (a'_{21} a_{11} + a'_{22} a_{21}) x_1 + (a'_{21} a_{12} + a'_{22} a_{22}) x_2 \end{array} \right\}$$

The matrix of this combined transformation is

$$\begin{bmatrix} a'_{11} a_{11} + a'_{12} a_{21} & a'_{11} a_{12} + a'_{12} a_{22} \\ a'_{21} a_{11} + a'_{22} a_{21} & a'_{21} a_{12} + a'_{22} a_{22} \end{bmatrix} = \begin{bmatrix} a'_{11} & a'_{12} \\ a'_{21} & a'_{22} \end{bmatrix} \begin{bmatrix} a_{11} & a_{12} \\ a_{21} & a_{22} \end{bmatrix}$$

In this instance the *associated matrix of the combined transformation is the product of the associated matrices of the separate transformations in reverse order of application*. The result applies quite generally.

Example 2

The matrices of the anti-clockwise rotation of the coordinate axes through angles α and β are, respectively $\begin{bmatrix} \cos \alpha & -\sin \alpha \\ \sin \alpha & \cos \alpha \end{bmatrix}$ and $\begin{bmatrix} \cos \beta & -\sin \beta \\ \sin \beta & \cos \beta \end{bmatrix}$.

The combined transformation has matrix $\begin{bmatrix} \cos \beta & -\sin \beta \\ \sin \beta & \cos \beta \end{bmatrix} \begin{bmatrix} \cos \alpha & -\sin \alpha \\ \sin \alpha & \cos \alpha \end{bmatrix}$, i.e.

$$\begin{bmatrix} \cos\beta\cos\alpha - \sin\beta\sin\alpha & -\cos\beta\sin\alpha - \sin\beta\cos\alpha \\ \sin\beta\cos\alpha + \cos\beta\sin\alpha & -\sin\beta\sin\alpha + \cos\beta\cos\alpha \end{bmatrix} = \begin{bmatrix} \cos[\beta+\alpha] & -\sin[\beta+\alpha] \\ \sin[\beta+\alpha] & \cos[\beta+\alpha] \end{bmatrix}$$

This suggests that the combined transformation is a rotation through an angle $\beta + \alpha$, as we should have expected. Note that, *in this instance,* the end product is independent of the order of the separate transformations.

1.4 The Solution of Simultaneous Linear Algebraic Equations

A system of m simultaneous linear equations in n unknowns may have a unique solution, an infinite number of solutions, or no solution at all. Let us investigate the underlying reasons. We denote the system by $\mathbf{Ax} = \mathbf{b}$. If a solution does exist then we first find *one* solution of the system and then find the *general* solution of the system $\mathbf{Ax} = \mathbf{0}$; the sum of these is the general solution of the full system. If the kernel of $\mathbf{A}$ comprises only the zero vector then the solution of the full system will be unique. We denote the columns of $\mathbf{A}$ as $\mathbf{a}_1, \mathbf{a}_2, ..., \mathbf{a}_n$; in other words, we are regarding the matrix as an aggregation of column vectors. We shall choose the usual basis for R^n, viz. $\mathbf{e}_1, \mathbf{e}_2, ..., \mathbf{e}_n$. It follows that if $\mathbf{x} = x_1\mathbf{e}_1 + x_2\mathbf{e}_2 + ... + x_n\mathbf{e}_n$ then $\mathbf{Ax} = x_1\mathbf{a}_1 + x_2\mathbf{a}_2 + ... + x_n\mathbf{a}_n$. For example, if we have the system of equations

$$\left. \begin{array}{c} x_1 + 2x_2 = 1 \\ 4x_1 + x_2 = 3 \end{array} \right\} \tag{1.14}$$

then this may be written as $\mathbf{Ax} \equiv x_1 \begin{bmatrix} 1 \\ 4 \end{bmatrix} + x_2 \begin{bmatrix} 2 \\ 1 \end{bmatrix} = \begin{bmatrix} 1 \\ 3 \end{bmatrix} \equiv \mathbf{b}$

Existence of a solution

Any vector $\mathbf{Ax}$ must be a linear combination of the column vectors $\mathbf{a}_1, \mathbf{a}_2, ..., \mathbf{a}_n$. If a solution of $\mathbf{Ax} = \mathbf{b}$ is to exist, $\mathbf{b}$ must itself be a linear combination of $\mathbf{a}_1, \mathbf{a}_2, ..., \mathbf{a}_n$.

In the example considered, $\mathbf{b}$ is a linear combination of $\begin{bmatrix} 1 \\ 4 \end{bmatrix}$ and $\begin{bmatrix} 2 \\ 1 \end{bmatrix}$ since we can take $x_1 = 5/7$ and $x_2 = 1/7$; this is of course the solution of the system. If we consider the system

$$\left. \begin{array}{c} x_1 + 2x_2 = 1 \\ 4x_1 + 8x_2 = 3 \end{array} \right\} \tag{1.15}$$

we require that $\begin{bmatrix} 1 \\ 3 \end{bmatrix}$ is a linear combination of $\begin{bmatrix} 1 \\ 4 \end{bmatrix}$ and $\begin{bmatrix} 2 \\ 8 \end{bmatrix}$; since $\begin{bmatrix} 2 \\ 8 \end{bmatrix}$ is a

multiple of $\begin{bmatrix} 1 \\ 4 \end{bmatrix}$ we merely require that $\begin{bmatrix} 1 \\ 3 \end{bmatrix}$ is a multiple of $\begin{bmatrix} 1 \\ 4 \end{bmatrix}$. It is *not* and

therefore the system has no solutions.

How can we extend these ideas? We first define the **rank** of a matrix as the minimum number of linearly independent vectors from the set of its column vectors. In the system (1.14), the rank of **A** was 2 (the maximum possible for a matrix with two columns) but in the system (1.15) the rank of **A** was 1.

Since **b** has to be a linear combination of $a_1, a_2, ..., a_n$ for a solution to exist, then the sets $\{a_1, a_2, ..., a_n\}$ and $\{a_1, a_2, ..., a_n, b\}$ have the same number of linearly independent vectors. Put another way, the matrices **A** = $(a_1, a_2, ..., a_n)$ and $(A | b) = (a_1, a_2, ..., a_n, b)$ have the same rank. The matrix $(A | b)$, which is formed by adding the column **b** to the matrix **A**, is called the **augmented matrix**.

Example 1

The system (1.14) has augmented matrix $\begin{bmatrix} 1 & 2 & \vdots & 1 \\ 4 & 1 & \vdots & 3 \end{bmatrix}$ which has rank 2.

Note that since column rank and row rank are equal, the augmented matrix cannot have rank >2.

Example 2

The system (1.15) with augmented matrix $\begin{bmatrix} 1 & 2 & \vdots & 1 \\ 4 & 8 & \vdots & 3 \end{bmatrix}$ has no solution and

the augmented matrix has rank 2 which is greater than the rank of **A**. You can check the rank from a consideration of row vectors.

Uniqueness of a solution

We have said earlier that the sum of the rank of a linear transformation and the dimension of its kernel is equal to the dimension of the domain space. If the rank of the linear transformation equals the dimension of its domain space then the kernel contains only the zero vector. With reference to the matrix of the transformation, if its rank equals the number of columns then there is a unique solution of the system of equations of which it is the associated matrix.

If a system of m equations in n unknowns can be written as $Ax = b$ (where **A** is an m by n matrix) then, *provided* a solution can be found, it will be a unique solution if and only if the rank of **A** is n. In the special case where $m = n$ then a unique solution will exist if and only if the rank of $A = n$.

But suppose that the rank of **A** is less than n and more than one solution exists. We quote the general result that if the rank of **A** is equal to $r < n$ then all solutions of $Ax = b$ can be expressed in terms of $(n - r)$ independent parameters.

Example

Consider the system of equations

$$3x_1 + 2x_2 - x_3 + 4x_4 = 4$$
$$4x_1 + 3x_2 - x_3 + 5x_4 = 4$$
$$-2x_1 - x_2 + 2x_3 - 2x_4 = -3$$
$$3x_1 + 2x_2 + 2x_3 + 7x_4 = 7$$

$$(1.16)$$

which has augmented matrix $\begin{bmatrix} 3 & 2 & -1 & 4 & . & 4 \\ 4 & 3 & -1 & 5 & . & 4 \\ -2 & -1 & 2 & -2 & . & -3 \\ 3 & 2 & 2 & 7 & . & 7 \end{bmatrix}$. Performing a

Gauss-Jordan elimination we finally produce $\begin{bmatrix} 1 & 0 & 0 & 3 & . & 5 \\ 0 & 1 & 0 & -2 & . & -5 \\ 0 & 0 & 1 & 1 & . & 1 \\ 0 & 0 & 0 & 0 & . & 0 \end{bmatrix}$.

This has a 3×3 identity submatrix in its top left-hand corner. The row of zeros indicates that there is no unique solution. Now the rank of **A** = rank of (**A**|**b**) = $3 < 4$. We should expect to express all solutions in terms of $4 - 3 = 1$ independent parameter, for example x_4.

We have $x_1 = 5 - 3x_4$, $x_2 = -5 + 2x_4$, $x_3 = 1 - x_4$. Hence $(x_1, x_2, x_3, x_4) = (5,-5,1,0) + x_4(-3,2,-1,1)$ and you can check that $(-3,2,-1,1)$ is a basis for the kernel of the transformation whose matrix is **A**.

The Gauss procedure can be regarded as a sequence of pre-multiplications by certain elementary matrices. It can be shown that the rank of the matrix at each stage of the procedure is the same as that of its predecessor.

1.5 Schemes for Solution of Linear Equations

In this section we shall examine a number of variations on Gaussian Elimination which are specially designed to take advantage of particular features of the matrix.

Tri-diagonal Matrices

In problems in which finite difference approximations are used to solve differential equations, matrices occur with non-zero elements confined to the leading diagonal, the first sub-diagonal and the first super-diagonal. For an $m \times n$ matrix **A** this means that $a_{ij} = 0$ if $i > j+1$ or if $i < j-1$. An example is shown on the next page.

$$A = \begin{bmatrix} 1 & 4 & 0 & 0 & 0 \\ -2 & 2 & 2 & 0 & 0 \\ 0 & 6 & 1 & 3 & 0 \\ 0 & 0 & 4 & 4 & 2 \\ 0 & 0 & 0 & 2 & 7 \end{bmatrix}$$

We are able to develop an efficient algorithm for Gauss Elimination when the coefficient matrix is tri-diagonal. The popular notation for a tri-diagonal matrix is not that used for general matrices. The notation is

$$\begin{bmatrix} b_1 & c_1 & 0 & 0 & & & \\ a_2 & b_2 & c_2 & 0 & & O & \\ 0 & a_3 & b_3 & c_3 & & & \\ & & & & \ddots & & \\ & O & & a_{n-1} & b_{n-1} & c_{n-1} \\ & & & 0 & a_n & b_n \end{bmatrix} \qquad (1.17)$$

Consider a simple case. We wish to solve the equations

$$\begin{bmatrix} b_1 & c_1 & 0 & 0 \\ a_2 & b_2 & c_2 & 0 \\ 0 & a_3 & b_3 & c_3 \\ 0 & 0 & a_4 & b_4 \end{bmatrix} \begin{bmatrix} x_1 \\ x_2 \\ x_3 \\ x_4 \end{bmatrix} = \begin{bmatrix} d_1 \\ d_2 \\ d_3 \\ d_4 \end{bmatrix}$$

We can eliminate x_1 from the second equation by the operation Row $2 - (a_2/b_1)$ Row 1; this will reduce the element a_2 to zero and leave the elements c_2 and 0 unchanged. The only operation which need concern us is that b_2 is replaced by $b'_2 = b_2 - (a_2/b_1)c_1$. Similar operations on rows 2 and 3, viz $b'_3 = b_3 - (a_3/b'_2)c_2$ and $b'_4 = b_4 - (a_4/b'_3)c_3$ will reduce the matrix to

$$\begin{bmatrix} b_1 & c_1 & 0 & 0 \\ 0 & b'_2 & c_2 & 0 \\ 0 & 0 & b'_3 & c_3 \\ 0 & 0 & 0 & b'_4 \end{bmatrix} \qquad (1.18)$$

If we write at the outset $b'_1 = b_1$ then these three operations can be summarised as
$$b'_r = b_r - (a_r/b'_{r-1})c_{r-1}, \qquad r = 2, 3, 4.$$
We may then back substitute via
$$x_4 = d_4/b'_4 \quad \text{and} \quad x_r = (d_r - c_r x_{r+1})/b'_r, \qquad r = 3, 2, 1.$$

These operations constitute the **Thomas algorithm**. The operations count on multiplication and division for n equations is $5n - 4$, compared with $\frac{1}{3}n^3 + n^2 - \frac{1}{3}n$ for Gauss elimination.

Triangular decomposition

A matrix which is either upper-triangular (zeros below the leading diagonal) or lower-triangular (zeros above the leading diagonal) is relatively easy to invert.

We first assume that a square matrix A can be written as the product of a lower-triangular matrix L and an upper-triangular matrix U in the order

$$A = LU \tag{1.19}$$

This is usually possible when A is non-singular.

Example

Let $A = \begin{bmatrix} 1 & 5 & 3 \\ 3 & 19 & 17 \\ 8 & 36 & 25 \end{bmatrix}$ and suppose that $A = LU$

where $L = \begin{bmatrix} l_{11} & 0 & 0 \\ l_{21} & l_{22} & 0 \\ l_{31} & l_{32} & l_{33} \end{bmatrix}$ and $U = \begin{bmatrix} u_{11} & u_{12} & u_{13} \\ 0 & u_{22} & u_{23} \\ 0 & 0 & u_{33} \end{bmatrix}$. It follows that

$$LU = \begin{bmatrix} l_{11}u_{11} & l_{11}u_{12} & l_{11}u_{13} \\ l_{21}u_{11} & l_{21}u_{12} + l_{22}u_{22} & l_{21}u_{13} + l_{22}u_{23} \\ l_{31}u_{11} & l_{31}u_{12} + l_{32}u_{22} & l_{31}u_{13} + l_{32}u_{23} + l_{33}u_{33} \end{bmatrix} = A$$

We can set about solving these equations for l_{ij}, u_{ij} in a systematic fashion. However, there are 6 unknown l_{ij} and 6 unknown u_{ij} with only 9 equations governing them. We can put the uncertainty in the products $l_{11}u_{11}$, $l_{22}u_{22}$ and $l_{33}u_{33}$. As long as we specify the values of these three products we can choose the values of the components freely. One method takes $u_{ii} = 1$; this is the **Crout** decomposition. A second method takes $l_{ii} = 1$; this is the **Doolittle** method. We shall work with the Crout method.

We can immediately see that $l_{11} = 1$, $l_{21} = 3$, $l_{31} = 8$.

Then $u_{12} = 5/l_{11} = 5$, $u_{13} = 3/l_{11} = 3$, $l_{22} = 19 - l_{21}u_{12} = 4$, $l_{32} = 36 - l_{31}u_{12} = -4$. Further, $u_{23} = (17 - l_{21}u_{13})/l_{22} = 2$ and, finally, $l_{33} = 25 - l_{31}u_{13} - l_{32}u_{23} = 9$.

Hence $A = \begin{bmatrix} 1 & 0 & 0 \\ 3 & 4 & 0 \\ 8 & -4 & 9 \end{bmatrix}\begin{bmatrix} 1 & 5 & 3 \\ 0 & 1 & 2 \\ 0 & 0 & 1 \end{bmatrix}$

General Crout Algorithm

Let $\mathbf{A}$ be an $n \times n$ matrix with elements a_{ij}. Then the steps are

(i) $l_{i1} = a_{i1}, \quad i = 1, \quad n$

(ii) $u_{1j} = a_{1j}/l_{11}, \quad j = 2, \quad n$

(iii) $l_{ik} = a_{ik} - \displaystyle\sum_{m=1}^{k-1} l_{im} u_{mk}, \quad i = k, \quad n$

(iv) $u_{kj} = \left(a_{kj} - \displaystyle\sum_{m=1}^{k-1} l_{km} u_{mj} \right) \bigg/ l_{kk}, \quad j = k+1, \quad n$

$\left. \vphantom{\sum_{m=1}^{k-1}} \right\}$ Repeat for $k = 2, \quad n$

Try and write the equations out in full for a 4×4 matrix.

For a 4×4 matrix the elements are evaluated in the following order:
$$l_{11}, l_{21}, l_{31}, l_{41}, u_{12}, u_{13}, u_{14}, l_{22}, l_{32}, l_{42}, u_{23}, u_{24}, l_{33}, l_{43}, u_{34}, l_{44}.$$

We can store the non-zero elements of $\mathbf{L}$ and $\mathbf{U}$ in the space occupied by $\mathbf{A}$; a_{11} is replaced by l_{11}, a_{12} by u_{12} and so on.

If some $l_{ii} = 0$ then either the decomposition is not possible or, if $\mathbf{A}$ is singular, there is an infinite number of possible decompositions.

Solution of linear equations

To solve the system $\mathbf{Ax} = \mathbf{b}$ we first factorise $\mathbf{A}$ into $\mathbf{LU}$ so that $(\mathbf{LU})\mathbf{x} = \mathbf{b}$ i.e. $\mathbf{L}(\mathbf{Ux}) = \mathbf{b}$. If we write $\mathbf{Ux} = \mathbf{y}$ (1.20a)
then we need to solve $\mathbf{Ly} = \mathbf{b}$ (1.20b)
Having found $\mathbf{y}$ we then use (1.20a) to find $\mathbf{x}$. Each of these two eliminations is easier then the general Gauss Elimination. We work through an example using the decomposition already obtained. Suppose we wish to solve the system

$$\left. \begin{aligned} x_1 + 5x_2 + 3x_3 &= 22 \\ 3x_1 + 19x_2 + 17x_3 &= 94 \\ 8x_1 + 36x_2 + 25x_3 &= 166 \end{aligned} \right\} \qquad (1.21)$$

Then we have the matrix of coefficients as $\mathbf{A}$ of page 16. First we solve $\mathbf{Ly} = \mathbf{b}$ for $\mathbf{y}$ i.e. we solve

$$\begin{bmatrix} 1 & 0 & 0 \\ 3 & 4 & 0 \\ 8 & -4 & 9 \end{bmatrix} \begin{bmatrix} y_1 \\ y_2 \\ y_3 \end{bmatrix} = \begin{bmatrix} 22 \\ 94 \\ 166 \end{bmatrix}$$

Clearly $y_1 = 22$; then $y_2 = 7$ and $y_3 = 2$. We next solve $\mathbf{Ux} = \mathbf{y}$, i.e.

$$
\begin{bmatrix} 1 & 5 & 3 \\ 0 & 1 & 2 \\ 0 & 0 & 1 \end{bmatrix} \begin{bmatrix} x_1 \\ x_2 \\ x_3 \end{bmatrix} = \begin{bmatrix} 22 \\ 7 \\ 2 \end{bmatrix}
$$

and find $x_3 = 2$, $x_2 = 3$, $x_1 = 1$. We can check that this is the solution of the original equations.

General Algorithm

Having found L and U we may summarise the solutions for y and x by the equations

$$
\left.
\begin{aligned}
y_i &= \left(b_i - \sum_{j=1}^{i-1} l_{ij} b_j\right)/l_{ii} \qquad i = 1, 2, \ldots, n \\[2em]
x_i &= y_i - \sum_{j=i+1}^{n} u_{ij} x_j \qquad j = n, (n-1), \ldots, 1
\end{aligned}
\right\} \qquad (1.22)
$$

Choleski's Scheme

When the matrix A is real and symmetric, and positive definite† we can choose the diagonal elements of L so that L is real and $U = L^T$, the transpose of L. The requirement of positive definiteness for A is for the purpose of greater numerical stability.

$$
\text{If } A = \begin{bmatrix} a_{11} & a_{21} & \cdots & a_{n1} \\ a_{21} & a_{22} & \cdots & a_{n2} \\ \vdots & \vdots & & \vdots \\ a_{n1} & a_{n2} & \cdots & a_{nn} \end{bmatrix} = \begin{bmatrix} l_{11} & 0 & \cdots & 0 \\ l_{21} & l_{22} & \cdots & 0 \\ \vdots & \vdots & & \vdots \\ l_{n1} & l_{n2} & \cdots & l_{nn} \end{bmatrix} \begin{bmatrix} l_{11} & l_{21} & \cdots & l_{n1} \\ 0 & l_{22} & \cdots & l_{n2} \\ \vdots & \vdots & & \vdots \\ 0 & 0 & \cdots & l_{nn} \end{bmatrix} = LU
$$

it follows that $a_{ii} = \sum_{k=1}^{i} l^2_{ik}$ and $a_{ij} = \sum_{k=1}^{i} l_{ik} l_{jk}$. Therefore we obtain the scheme shown in (1.23) on the next page.

The order of the determination of the elements follows the Crout pattern.

To solve a set of equations $Ax = b$ where A is real, symmetric and positive definite, we first solve $Ly = b$ for y and then solve $L^Tx = y$ for x. The process uses the formulae (1.24) on the next page.

---

† A is **positive definite** if and only if $x^TAx > 0$ for all $x \neq 0$

$$l_{ii} = \sqrt{a_{ii} - \sum_{k=1}^{i-1} l_{ik}^2} \qquad 1 \le i \le n$$

$$l_{ji} = \left(a_{ij} - \sum_{k=1}^{i-1} l_{ik}\, l_{jk}\right)\!\Big/ l_{ii} \qquad 1 \le i < j \qquad\qquad (1.23)$$

$$y_i = \left(b_i - \sum_{k=1}^{i-1} l_{ik}\, y_k\right)\!\Big/ l_{ii} \qquad i = 1, 2, ..., n$$

$$x_i = \left(y_i - \sum_{j=i+1}^{n} l_{ij}\, x_j\right)\!\Big/ l_{ii} \qquad i = n, (n-1), ..., 1 \qquad (1.24)$$

1.6 Partitioned Matrices

In some problems in structural analysis the equations relating imposed forces and resulting deflections can be cast in matrix form. The so-called **flexibility matrix** of coefficients can often be split or **partitioned** into rectangular blocks. The advantage of partitioning is that we can treat the blocks as individual elements and this makes certain operations, for example inversion, more straightforward. In this section, we indicate briefly some of the basic ideas using simple examples for ease of understanding.

(i) **Addition**

$$\text{Let } A = \begin{bmatrix} 1 & 2 & 5 \\ 3 & 4 & 6 \\ 7 & 8 & 9 \end{bmatrix} = \begin{bmatrix} A_1 & A_2 \\ A_3 & A_4 \end{bmatrix} \text{ and } B = \begin{bmatrix} 6 & 7 & 8 \\ 4 & 5 & 9 \\ 1 & 2 & 3 \end{bmatrix} = \begin{bmatrix} B_1 & B_2 \\ B_3 & B_4 \end{bmatrix}$$

$$\text{Then } A + B = \begin{bmatrix} A_1 + B_1 & A_2 + B_2 \\ A_3 + B_3 & A_4 + B_4 \end{bmatrix} = \begin{bmatrix} 7 & 9 & 13 \\ 7 & 9 & 15 \\ 8 & 10 & 12 \end{bmatrix}$$

(ii) **Multiplication**
Provided the partitioning has produced compatible blocks we can carry out multiplication in two stages.
Let A and B be as above. Then, as you can show,

$$AB = \begin{bmatrix} A_1 B_1 + A_2 B_3 & A_1 B_2 + A_2 B_4 \\ A_3 B_1 + A_4 B_3 & A_3 B_2 + A_4 B_4 \end{bmatrix} = \begin{bmatrix} 19 & 27 & 41 \\ 40 & 53 & 78 \\ 83 & 107 & 155 \end{bmatrix}$$

Let $C = \begin{bmatrix} 6 & 3 & | & 4 \\ 3 & 5 & | & 2 \\ \hline 1 & 2 & | & 3 \end{bmatrix} = \begin{bmatrix} C_1 & | & C_2 \\ \hline C_3 & | & C_4 \end{bmatrix}$. Then, in theory,

$$AC = \begin{bmatrix} A_1 & | & A_2 \\ \hline A_3 & | & A_4 \end{bmatrix} \begin{bmatrix} C_1 & | & C_2 \\ \hline C_3 & | & C_4 \end{bmatrix} = \begin{bmatrix} A_1\,C_1 + A_2\,C_3 & | & A_1\,C_2 + A_2\,C_4 \\ \hline A_3\,C_1 + A_4\,C_3 & | & A_3\,C_2 + A_4\,C_4 \end{bmatrix}$$

However, for example, A_1 is 2×2 and C_1 is 1×2 so that multiplication is not possible.

(iii) **Inverse**

We consider the special case of a 4×4 matrix partitioned into four equal blocks of 2×2 submatrices.

Let $A = \begin{bmatrix} 4 & 2 & | & 1 & 2 \\ 2 & 2 & | & 2 & 3 \\ \hline 1 & 2 & | & 1 & 0 \\ 0 & 1 & | & 2 & 1 \end{bmatrix} = \begin{bmatrix} A_1 & | & A_2 \\ \hline A_3 & | & A_4 \end{bmatrix}$ and let $C = \begin{bmatrix} C_1 & | & C_2 \\ \hline C_3 & | & C_4 \end{bmatrix} = A^{-1}$

so that $AC = \begin{bmatrix} A_1\,C_1 + A_2\,C_3 & | & A_1\,C_2 + A_2\,C_4 \\ \hline A_3\,C_1 + A_4\,C_3 & | & A_3\,C_2 + A_4\,C_4 \end{bmatrix} = I = \begin{bmatrix} 1 & 0 & | & 0 & 0 \\ 0 & 1 & | & 0 & 0 \\ \hline 0 & 0 & | & 1 & 0 \\ 0 & 0 & | & 0 & 1 \end{bmatrix}$

It follows that

$$
\left.
\begin{aligned}
A_1\,C_1 + A_2\,C_3 &= I & \text{(a)} \\
A_1\,C_2 + A_2\,C_4 &= 0 & \text{(b)} \\
A_3\,C_1 + A_4\,C_3 &= 0 & \text{(c)} \\
A_3\,C_2 + A_4\,C_4 &= I & \text{(d)}
\end{aligned}
\right\} \qquad (1.25)
$$

If we substitute $C_3 = -A_4^{-1}\,A_3\,C_1$ from (1.25c) into (1.25a) we obtain $A_1\,C_1 - A_2\,A_4^{-1}\,A_3\,C_1 = I$ and hence $C_1 = (A_1 - A_2\,A_4^{-1}\,A_3)^{-1}$. We therefore may determine C_1 and C_3. Similarly, we obtain $C_2 = -A_1^{-1}\,A_2\,C_4$, $C_4 = (A_4 - A_3\,A_1^{-1}\,A_2)^{-1}$ and hence determine C_4 and C_2. Of course A_1^{-1} and A_4^{-1} are found in the usual way for 2×2 matrices.

Problems

Section 1.2

1　Show that the vectors $(1,2,2,1)$, $(3,4,4,3)$ and $(1,0,0,1)$ are linearly dependent and span a vector space of dimension 2; find a basis for this space. Repeat for the vectors $(1,1,1,0)$, $(4,3,2,-1)$, $(2,1,0,-1)$ and $(4,2,0,-2)$.

2 Show that the vector space spanned by the vectors (1,2,1), (1,2,3) and (3,6,5) is identical with that spanned by (1,2,5) and (0,0,1).

3 Find the dimension of the vector space spanned by each of the following sets of vectors; choose a basis for each space.
(a) (1,2,3,4,5), (5,4,3,2,1) and (1,1,1,1,1) (b) (1,1,0,–1), (1,2,3,4) and (2,3,3,3)
(c) (1,1,1,1), (3,4,5,6), (1,2,3,4) and (1,0,–1,–2)

4 Which of the following sets are subspaces of the vector space of which a typical member is (a,b,c,d), all components being real numbers.
(i)   all vectors with $a = b = c = d$ (ii) all vectors with $a = 2$
(iii) all vectors with $a = b$ and $c = 2d$ (iv) all vectors with integer components
(v)   all vectors with $d = 0$

5 Find a vector orthogonal to
(i) (1,–2,–2) and (2,–1,2) (ii) (1,2,1) and (2,1,2) (iii) (1,2,1) and (2,1,4).

6 Construct an orthonormal basis for the space of vectors with three real components, $V_3(R)$, given the basis
(i)   (2,1,3), (1,2,3), (1,1,1) (ii) (1,–1,0), (2,–1,–2), (1,–1,–2)
(iii) (1,0,1), (1,3,1), (3,2,1) (vi) (2,–1,0), (4,–1,0)(4,0,–1).

7 Find an orthonormal basis for $V_3(R)$ given (i) (1,1,–1) and (2,1,0) (ii) (7,–1,–1).

8 Given the basis {(1,0,1,0), (1,1,0,0), (0,1,1,1), (0,1,1,0)}, use the Gram-Schmidt process to obtain an orthonormal basis.

9 Let V be a subspace of U spanned by {(0,1,1,0), (0,5,–3,–2), (–3,–3,5,–7)}. Find an orthonormal basis for V.

Section 1.3
10 Find a linear transformation which maps (1,0,0) to (1,2,3); (0,1,0) to (3,1,2) and (0,0,1) to (2,1,3).

11 Find the images under the transformation in Problem 10 of (1,1,1), (3,–1,4), (4,0,5).

12 Which of the following are linear transformations?
(i) $\sigma(x_1, x_2) = (x_1 + 2, 2x_2)$ (ii) $\sigma(x_1, x_2) = (x_1 - x_2, x_2)$
(iii) $\sigma(x_1, x_2) = (x_1 \cos 2a - x_2 \sin 2a, \ x_1 \sin 2a + x_2 \cos 2a)$
(iv) $\sigma(x_1, x_2, x_3) = (2x_1, \ x_2 + x_3, \ 3x_3, \ x_1 + x_2)$
A linear transformation maps (1,0,1) to (2,3,–1); (1,–1,1) to (3,0,–2) and (1,2,–1) to (–2,7,–1). Find the images under this transformation of (1,0,0), (0,1,0), (0,0,1).

13 Find vectors which span the kernel of the transformations whose matrices are

(i) $\begin{bmatrix} 1 & 1 & 0 \\ 2 & 3 & 1 \\ -2 & 3 & 5 \end{bmatrix}$ (ii) $\begin{bmatrix} 1 & 1 & 3 \\ 1 & 2 & 4 \\ 1 & 1 & 3 \end{bmatrix}$ (iii) $\begin{bmatrix} 1 & 2 & 3 \\ 2 & 4 & 6 \\ 3 & 6 & 9 \end{bmatrix}$

14 Find the rank and nullity of the linear transformations whose associated matrices are

(i) $\begin{bmatrix} 1 & 2 & 3 \\ 1 & 3 & 5 \end{bmatrix}$ (ii) $\begin{bmatrix} 1 & 3 & 7 \\ 2 & 7 & 16 \end{bmatrix}$ (iii) $\begin{bmatrix} 1 & 2 & 4 & -1 \\ 1 & 3 & 5 & -2 \end{bmatrix}$

(iv) $\begin{bmatrix} 0 & 1 & -1 \\ -1 & 0 & 1 \\ 1 & -1 & 0 \end{bmatrix}$ (v) $\begin{bmatrix} 0 & 1 & 2 \\ 1 & 2 & 3 \\ 2 & 3 & 4 \end{bmatrix}$ (vi) $\begin{bmatrix} 1 & 1 & 1 & 1 \\ 1 & 2 & 3 & 4 \\ 0 & 1 & 2 & 3 \end{bmatrix}$

15 If $\sigma_1(x_1, x_2) = (x_2, -x_1)$ and $\sigma_2(x_1, x_2) = (x_1, -x_2)$ find $\sigma_1\sigma_2$ and $\sigma_2\sigma_1$.

16 Find the kernel of the transformation $\sigma(x_1, x_2, x_3, x_4) = (3x_1 - 2x_2 - x_3 - 4x_4, x_1 + x_2 - 2x_3 - 3x_4)$.

17 Find the solutions of the equations

(i) $\begin{cases} x_1 + x_2 + x_3 = 3 \\ 2x_1 + 5x_2 - 2x_3 = 3 \end{cases}$ (ii) $\begin{cases} x_1 - 2x_2 + 3x_3 = 0 \\ 2x_1 + 5x_2 + 6x_3 = 0 \end{cases}$

18 Find the dimension of the vector space of solutions of the following systems of linear equations. Find a basis for this space of solutions.

(i) $x + y - z = 0$ (ii) $\begin{cases} 2x + y - z = 0 \\ y + z = 0 \end{cases}$ (iii) $\begin{cases} 2x - 3y + z = 0 \\ x + y - z = 0 \end{cases}$

(iv) $\begin{cases} x + y + z = 0 \\ x - y = 0 \\ y + z = 0 \end{cases}$ (v) $\begin{cases} 2x - 3y + z = 0 \\ x + y - z = 0 \\ 3x + 4y = 0 \\ 5x + y + z = 0 \end{cases}$ (vi) $\begin{cases} x + y + z = 0 \\ 2x + 2y + 2z = 0 \end{cases}$

19 Find the dimension of the set of solutions of the following systems. Find a basis of the space of solutions of the associated homogeneous system and find one solution of the non-homogeneous system. Hence produce general solutions.

(i) $2x + 3y - z = 1$ (ii) $\begin{cases} 2x - y + z = 1 \\ 2x + y + z = 1 \end{cases}$ (iii) $\begin{cases} -x + 4y + z = 2 \\ 3x + y - z = 0 \end{cases}$

(iv) $\begin{cases} x - y + z = 1 \\ 2x - 3y + z = 0 \\ x + y - z = 5 \end{cases}$

Section 1.4

20 Determine the rank of the matrix $A = \begin{bmatrix} 1 & 2 & 3 \\ 2 & 4 & 7 \\ -1 & -2 & -2 \end{bmatrix}$ and reduce A to normal form by

a sequence of elementary transformations.   (EC)

21 Find the rank of the following matrices:

$$\begin{bmatrix} 1 & 5 & -7 \\ 2 & 3 & 1 \end{bmatrix}, \begin{bmatrix} 2 & 1 & 1 \\ 0 & 1 & -1 \end{bmatrix}, \begin{bmatrix} 2 & 1 & 3 \\ 7 & 2 & 0 \end{bmatrix}, \begin{bmatrix} -1 & 2 & -2 \\ 3 & 4 & -5 \end{bmatrix}, \begin{bmatrix} 1 & 2 & -3 \\ -1 & 2 & 3 \\ 4 & 8 & 12 \\ 0 & 0 & 0 \end{bmatrix}.$$

22 Prove that A and A^T have the same rank.

23 A system of m linear equations in n unknowns is said to be *consistent* if it has one or more solutions; otherwise it is *inconsistent*. Show that the equations $x_1 + 2x_2 = 3$, $x_2 - x_3 = 2$, $x_1 + x_2 + x_3 = 1$ are consistent, but that if the third equation is replaced by $x_1 + x_2 + x_3 = 2$ the new set is inconsistent. Interpret the two sets of equations geometrically.

24 Prove that if A and B are $m \times n$ matrices, then the rank of $A + B$ is less than or equal to the sum of the ranks of A and of B.

25 A matrix is said to be *regular by rows* if its row-vectors are linearly independent and *regular by columns* if its column-vectors are linearly independent. A matrix A which is regular either by rows or by columns is called a *regular* matrix. Show that an $m \times n$ matrix has rank r if and only if it contains at least one regular $k \times k$ submatrix for each value of k, $1 \leq k \leq r$, but none for $k > r$. A submatrix of A is obtained by deleting complete rows and/or complete columns.

Section 1.5

26 Write a computer program to solve a set of linear equations based on the Thomas algorithm. Use suitable test data.

27 Write a computer program to decompose a matrix by the Crout method. Then incorporate routines to invert a matrix so decomposed and to solve a set of linear equations.

28 Repeat Problem 27 for the Doolittle method and for Choleski's method.

29 Consider the following matrices, where only non-zero elements are shown:

$$A = \begin{bmatrix} 1 & & \\ a_2 & 1 & \\ & a_3 & 1 \end{bmatrix} \qquad B = \begin{bmatrix} b_1 & c_1 & \\ & b_2 & c_2 \\ & & b_3 \end{bmatrix}$$

Show that $E = AB$ has a special form and give the formulae for its elements. Deduce the general result when A and B are matrices of the same form but of larger order.
If a matrix E having the above structure is given, describe briefly how the matrix factors A and B can be determined, and how they can be used to solve equations of the form $Ex = d$ when d is given. Illustrate this in the following case:

$$E = \begin{bmatrix} 4 & 2 & & \\ 2 & 5 & 2 & \\ & 2 & 5 & 2 \\ & & 2 & 5 \end{bmatrix} \qquad d = \begin{bmatrix} 6 \\ 1 \\ 7 \\ 12 \end{bmatrix} \qquad \text{(EC)}$$

30 Define upper (U) and lower (L) triangular matrices of order 3. Evaluate matrices U and L

explicitly if a matrix $\mathbf{M} = \mathbf{LU}$, where $\mathbf{M} = \begin{bmatrix} 2 & -2 & 3 \\ 1 & -4 & 7 \\ -1 & 1 & 2 \end{bmatrix}$ and the diagonal elements in U

are all unity. Deduce the value of the determinant of **M**.   (LU)

31 Apply the method of factorising **A** to determine the solution of the equations:

(i) $4x_1 - x_2 + 3x_3 = 15$ (ii) $x_1 + 4x_2 + 7x_3 = 7$

            $8x_1 - x_2 + 11x_3 = 43$ $2x_1 + 11x_2 + 5x_3 = 8$

            $-12x_1 + 7x_2 + 18x_3 = 28$ $8x_1 + 6x_2 + 9x_3 = 9$

(iii) $4x_1 + 2x_2 + x_3 + x_4 = 25$

            $8x_1 + 6x_2 + 4x_3 + 3x_4 = 61$

            $4x_1 - 2x_2 + 3x_3 + x_4 = 17$

            $8x_1 + 8x_2 + 12x_3 + 9x_4 = 79$

(iv) $\mathbf{A} = \begin{bmatrix} 2 & 1 & 6 & 3 \\ 4 & -1 & 14 & 8 \\ 8 & -5 & 26 & 19 \\ 10 & -1 & 30 & 25 \end{bmatrix}$, $\mathbf{b} = \begin{bmatrix} 1 \\ -4 \\ 1 \\ 23 \end{bmatrix}$, $\mathbf{x} = \begin{bmatrix} x_1 \\ x_2 \\ x_3 \\ x_4 \end{bmatrix}$

(v) $x + 3y + 6z = 17$

            $2x + 8y + 16z = 42$

            $5x + 21y + 45z = 91$

Show any necessary checks.   (LU)

32 Show that the inverse of $\mathbf{A} = \mathbf{LU}$ can be written $\mathbf{A}^{-1} = \mathbf{U}^{-1} \mathbf{L}^{-1}$, if it exists. Hence find

the inverse of $\mathbf{A} = \begin{bmatrix} 4 & 6 & 8 \\ 6 & 10 & 17 \\ 8 & 17 & 25 \end{bmatrix}$ and hence solve the set of equations $\mathbf{Ax} = \mathbf{b}$ where

$\mathbf{b} = (18, 33, 50)^T$.

33 (a)   Find lower and upper triangular matrices **L** and **U** such that the product **LU** equals the
            matrix

$$\mathbf{A} = \begin{bmatrix} 1 & 1 & 2 & 1 \\ 0 & 1 & 1 & 1 \\ 1 & 2 & 1 & 1 \\ 2 & 2 & 4 & 0 \end{bmatrix}$$

    (b)   Using the values of L and U obtained in part (a), solve the simultaneous equations

$$x_2 + x_3 + x_4 = 2$$
$$x_1 + x_2 + 2x_3 + x_4 = 1$$
$$x_1 + 2x_2 + x_3 + x_4 = 3$$
$$2x_1 + 2x_2 + 4x_3 = 4$$

(c) Find the inverse matrices L^{-1} and U^{-1} and hence determine the inverse matrix A^{-1}.

(EC)

34 If L is a lower triangular matrix and a matrix A is such that $A = LL^T$, then $A^{-1} = [L^{-1}]^T[L^{-1}]$. Use this result to find the inverse of the matrix

$$A = \begin{bmatrix} 4 & 6 & 8 \\ 6 & 10 & 17 \\ 8 & 17 & 25 \end{bmatrix} \qquad \text{(EC)}$$

35 (a) Use the method of Choleski to solve the simultaneous equations

$$8x_1 + x_2 = 4$$
$$2x_1 - \frac{3}{4}x_2 + x_3 = 1$$
$$x_2 + 2x_3 + x_4 = 2$$
$$x_3 - 3x_4 = -1$$

(b) Repeat using the Thomas algorithm.

(c) For this problem, show that $(L^T)^{-1} = (L^{-1})^T$.

36 Express the Gauss-Seidel method of solution of simultaneous linear equations as an iterative method in terms of an upper triangular matrix U and a lower triangular matrix L. (Use the example below to illustrate the formation of these matrices.) If the (column) matrix $e^{(n)}$ stands for the difference between the exact solution and the nth iteration show that $e^{(n+1)} = (I - L)^{-1} U e^{(n)}$.

Give the explicit forms for L, U corresponding to the equations

$$25x + 2y + z = 70$$
$$2x + 10y + z = 60$$
$$x + y + 4z = 40$$

Starting from the matrix $x^{(0)} = (1,1,1)$ find $x^{(2)}$.

(LU)

Section 1.6

37 Prove for the matrix $A = \begin{bmatrix} 2 & 1 & 3 & 4 \\ 5 & 3 & 2 & 3 \\ 2 & 1 & 2 & 3 \\ 9 & 5 & 1 & 2 \end{bmatrix} = \begin{bmatrix} A_1 & A_2 \\ A_3 & A_4 \end{bmatrix}$ that $\det A \neq \det \begin{bmatrix} \det A_1 & \det A_2 \\ \det A_3 & \det A_4 \end{bmatrix}$

38 Find the products of the partitioned matrices

(i) $\begin{bmatrix} 1 & 2 \\ 3 & 4 \\ \hline 5 & 6 \end{bmatrix} \begin{bmatrix} 9 & 8 \\ 7 & 6 \end{bmatrix}$

(ii) $\begin{bmatrix} 1 & 0 & 5 \\ 0 & 1 & 6 \\ \hline 1 & 0 & 7 \\ 0 & 1 & 8 \end{bmatrix} \begin{bmatrix} 2 & 0 \\ 0 & 2 \\ \hline 9 & 10 \end{bmatrix}$

(iii) $\begin{bmatrix} 1 & 2 & 3 \\ 4 & 5 & 6 \\ 7 & 8 & 9 \end{bmatrix} \begin{bmatrix} 3 \\ 2 \\ 1 \end{bmatrix}$

(iv) $\begin{bmatrix} a & b \\ c & d \end{bmatrix} \begin{bmatrix} x \\ y \end{bmatrix}$

(v) $\begin{bmatrix} a & h & g \\ h & b & f \\ \hline g & f & c \end{bmatrix} \begin{bmatrix} a & h & g \\ h & b & f \\ \hline g & f & c \end{bmatrix}$

(vi) $\begin{bmatrix} \mathbf{A}_1 & \mathbf{A}_2 \\ \hline \mathbf{A}_3 & \mathbf{A}_4 \end{bmatrix} \begin{bmatrix} \mathbf{I} & \mathbf{0} \\ \hline \mathbf{0} & \mathbf{I} \end{bmatrix}$

(vii) $\begin{bmatrix} \mathbf{A}_1 & \mathbf{A}_2 \\ \hline \mathbf{A}_3 & \mathbf{A}_4 \end{bmatrix} \begin{bmatrix} \mathbf{I} & -\mathbf{A}_1^{-1}\mathbf{A}_2 \\ \hline \mathbf{0} & \mathbf{I} \end{bmatrix}$

39   Prove that if the submatrices are compatible for multiplication,

(i) $\begin{bmatrix} \mathbf{A}_1 & \mathbf{I} \\ \hline \mathbf{I} & \mathbf{A}_1^{-1} \end{bmatrix} \begin{bmatrix} \mathbf{A}_1^{-1} & \mathbf{I} \\ \hline \mathbf{I} & \mathbf{A}_1 \end{bmatrix} = 2 \begin{bmatrix} \mathbf{I} & \mathbf{A}_1 \\ \hline \mathbf{A}_1^{-1} & \mathbf{I} \end{bmatrix}$

(ii) $\begin{bmatrix} \mathbf{A}_1 & \mathbf{0} \\ \hline \mathbf{I} & \mathbf{A}_1^{-1} \end{bmatrix} \begin{bmatrix} \mathbf{A}_1^{-1} & \mathbf{I} \\ \hline \mathbf{0} & \mathbf{A}_1 \end{bmatrix} = \begin{bmatrix} \mathbf{I} & \mathbf{A}_1 \\ \hline \mathbf{A}_1^{-1} & 2\mathbf{I} \end{bmatrix}$

40   Prove that $\det \begin{bmatrix} \mathbf{A}_1 & \mathbf{A}_2 \\ \hline \mathbf{A}_3 & \mathbf{A}_4 \end{bmatrix} = \det\left[(\mathbf{A}_1\,\mathbf{A}_4 - \mathbf{A}_3\,\mathbf{A}_2)\right]$

41   By partitioning, find the inverses of the following matrices and verify your results.

(i) $\begin{bmatrix} 2 & 4 & 3 & 2 \\ 3 & 6 & 5 & 2 \\ 2 & 5 & 2 & -3 \\ 4 & 5 & 14 & 14 \end{bmatrix}$

(ii) $\begin{bmatrix} 1 & 2 & 3 & 1 \\ 1 & 3 & 3 & 2 \\ 2 & 4 & 3 & 3 \\ 1 & 1 & 1 & 1 \end{bmatrix}$

(iii) $\begin{bmatrix} 1 & -2 & 1 & 0 \\ 1 & -2 & 2 & -3 \\ 0 & 1 & -1 & 1 \\ -2 & 3 & -2 & 3 \end{bmatrix}$

(iv) $\begin{bmatrix} 2 & 1 & -1 & 2 \\ 1 & 3 & 2 & -3 \\ -1 & 2 & 1 & -1 \\ 2 & -3 & -1 & 4 \end{bmatrix}$

(v) $\begin{bmatrix} \cos\alpha & \sin\alpha & \cos\beta & \sin\beta \\ -\sin\alpha & \cos\alpha & -\sin\beta & \cos\beta \\ 0 & 0 & \cos\gamma & \sin\gamma \\ 0 & 0 & -\sin\gamma & \cos\gamma \end{bmatrix}$

(vi) $\begin{bmatrix} 4 & 2 & 1 & 2 \\ 2 & 2 & 2 & 3 \\ 1 & 2 & 1 & 0 \\ 0 & 1 & 2 & 1 \end{bmatrix}$

2

EIGENVALUE PROBLEMS

2.1 Introduction

An important set of problems is the class of **boundary value problems**. In these, the values of the dependent variable and its derivative are specified at end-points, or more generally on the boundaries of the domain of the problem. For example, we may be interested in the deflected profile of a beam simply-supported at its ends, where the boundary conditions are the specified values of the deflection and slope of the beam at its two ends. A second example would be the propagation of heat in a rectangular slab with insulated faces, where the temperature along the four edges was specified.

These are examples of **continuous systems**. We shall be concerned in this chapter, as in the previous one, with problems where the system is **discrete**. We shall assume that the system possesses a finite number (> 1) of degrees of freedom. Alternative names for these systems are **distributed-parameter** and **lumped-parameter** respectively. In the latter class, we make such assumptions as shafts having negligible masses. The reason for these names will become clear as we study the examples in this chapter.

A special class of problems is known as the class of **characteristic-value** or **eigenvalue problems**. The main contexts in which we meet such cases are in the study of vibration problems or in certain problems in structures. The distinction that these problems share is that a solution can exist for only a set of values of a parameter of the problem: the eigenvalues. As examples, a system may be able to vibrate only at certain frequencies or a structure may only be in equilibrium in certain modes of deflection.

Vibration problems

Let us pursue a little further the problem of vibrating systems. A system which is subject to an imposed periodic force will vibrate in a way which partly reflects the nature of the imposed force and partly reflects something inherent in the system: the so-called **natural vibration**. The squares of the natural frequencies of oscillation are related to the eigenvalues and the descriptions of the possible vibrations (possibly in terms of displacements of the mass-centres of the system) are the normal modes. If the frequency of the imposed periodic force is close to that of one of these normal modes then the amplitude of the resulting oscillations may build up in an alarming way. It is important therefore for a designer to know the natural frequencies of the system under consideration. The kinds of problem that he wishes to avoid are typified by the collapse of the Ferrybridge cooling towers, the collapse of the Tacoma Narrows bridge, the flutter of aircraft wings and the transverse vibrations of turbine shafts.

2.2 Algebraic Determination of Eigenvalues

In this section we concentrate on systems with 3 degrees of freedom; these give rise to 3×3 matrices. The system shown in Figure 2.1 represents a physical model of a compressor for a jet engine; each disk corresponds to a set of rotor blades. We have ignored any damping effects either between blades and machine housing or between blades. There will be an oscillatory motion in addition to any rigid body rotational motion.

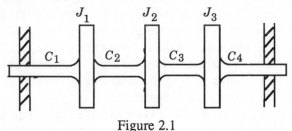

Figure 2.1

We consider the case where the central shaft is free to rotate at either end, the torsional stiffnesses C_i are equal and the moments of inertia J_i of the disks are equal. We shall choose as our coordinates θ_1, θ_2 and θ_3, the angular displacements of the disks in the oscillatory motion. We may write the equations of motion for the oscillatory motion of the disks as follows

$$\left. \begin{array}{l} J\ddot{\theta}_1 = -C(\theta_1 - \theta_2) \\ J\ddot{\theta}_2 = C(\theta_1 - \theta_2) - C(\theta_2 - \theta_3) \\ J\ddot{\theta}_3 = C(\theta_2 - \theta_3) \end{array} \right\} \tag{2.1}$$

For an oscillatory motion we may expect $\ddot{\theta}_i = -\omega^2 \theta_i$; substituting for $\ddot{\theta}_i$, writing $\lambda = J\omega^2/C$ and rearranging the equations we obtain

$$\left. \begin{array}{l} \theta_1 - \theta_2 = \lambda\theta_1 \\ -\theta_1 + 2\theta_2 - \theta_3 = \lambda\theta_2 \\ -\theta_2 + \theta_3 = \lambda\theta_3 \end{array} \right\} \tag{2.2}$$

In matrix form, equations (2.2) can be written

$$\begin{bmatrix} 1 & -1 & 0 \\ -1 & 2 & -1 \\ 0 & -1 & 1 \end{bmatrix} \begin{bmatrix} \theta_1 \\ \theta_2 \\ \theta_3 \end{bmatrix} = \lambda \begin{bmatrix} \theta_1 \\ \theta_2 \\ \theta_3 \end{bmatrix} \tag{2.3}$$

or $\quad A\theta = \lambda\theta$

Alternatively, we may write the equations as

$$\begin{bmatrix} 1-\lambda & -1 & 0 \\ -1 & 2-\lambda & -1 \\ 0 & -1 & 1-\lambda \end{bmatrix} \begin{bmatrix} \theta_1 \\ \theta_2 \\ \theta_3 \end{bmatrix} = \begin{bmatrix} 0 \\ 0 \\ 0 \end{bmatrix} \tag{2.4}$$

or $\quad (A - \lambda I)\theta = 0$

These equations have a non-zero solution if $|A - \lambda I| = 0$. This last condition becomes $(1-\lambda)[(2-\lambda)(1-\lambda) - 1] - (1-\lambda) = 0$ or $(1-\lambda)\lambda(3-\lambda) = 0$. This is called the **characteristic equation** for the system. Its roots are called **characteristic roots** or **eigenvalues** of the system. The eigenvalues of matrix A are therefore 0, 1 and 3.

Definition

The values of λ for which the equations $A\theta = \lambda\theta$ have other than the zero solution $\theta = 0$ are called the **eigenvalues** of the matrix A. To each eigenvalue there is a non-zero solution θ called the associated **eigenvector**.

Example

Find the eigenvalues of the matrix $D = \begin{bmatrix} 4 & 2 & -2 \\ 1 & 3 & 1 \\ -1 & -1 & 5 \end{bmatrix}$

We first form the characteristic equation by expanding the condition $|D - \lambda I| = 0$,

i.e. $\quad \begin{vmatrix} 4-\lambda & 2 & -2 \\ 1 & 3-\lambda & 1 \\ -1 & -1 & 5-\lambda \end{vmatrix} = 0.$ We obtain the equation

$$(4-\lambda)[(3-\lambda)(5-\lambda) + 1] - 2[1(5-\lambda) + 1] - 2[-1 + (3-\lambda)] = 0$$

which factorises into $(4-\lambda)(2-\lambda)(6-\lambda) = 0$. Let us be honest; we deliberately chose an example which would factorise relatively easily and, of course, we will not always be so lucky.

Determination of Eigenvectors

We shall now find the eigenvectors corresponding to each eigenvalue of a matrix. Returning to equations (2.2), we shall substitute $\lambda = 0, 1$ and 3 in turn.

$\underline{\lambda = 0}$. The equations become, with $\lambda = 0$, when written out in full

$$\theta_1 - \theta_2 = 0 \qquad -\theta_1 + 2\theta_2 - \theta_3 = 0 \qquad -\theta_2 + \theta_3 = 0$$

If we add the first and third of these equations we obtain $\theta_1 - 2\theta_2 + \theta_3 = 0$ which is effectively the same as the second equation of the set. We will not therefore be able to obtain a unique solution (as we must have expected) but will have to content ourselves with merely obtaining the ratio $\theta_1 : \theta_2 : \theta_3$. From the first equation we see that $\theta_1 = \theta_2$ and from the third we find $\theta_2 = \theta_3$. Hence the ratio $\theta_1 : \theta_2 : \theta_3$ is $1 : 1 : 1$. Typical eigenvectors satisfying this one condition are $(1,1,1)$, $(-8, -8, -8)$, $(1/\sqrt{3}, 1/\sqrt{3}, 1/\sqrt{3})$ and $(12.3, 12.3, 12.3)$. For simplicity's sake we shall choose the form $(1,1,1)$.

$\underline{\lambda = 1}$. The equations become

$$-\theta_2 = 0 \qquad -\theta_1 + \theta_2 - \theta_3 = 0 \qquad -\theta_2 = 0$$

Again we see that we have only two independent equations. If we substitute $\theta_2 = 0$ into the second we obtain $\theta_3 = -\theta_1$. Therefore the ratio $\theta_1 : \theta_2 : \theta_3$ is $1 : 0 : -1$. Therefore typical eigenvectors are $(1, 0, -1)$, $(-3, 0, 3)$, $(1/\sqrt{2}, 0, -1/\sqrt{2})$. We shall choose the first of these to represent its class.

$\underline{\lambda = 3}$. Now the equations can be written as

$$-2\theta_1 - \theta_2 = 0 \qquad -\theta_1 - \theta_2 - \theta_3 = 0 \qquad -\theta_2 - 2\theta_3 = 0$$

We see that the second equation is redundant (why?) so that we need only consider the first and the last, which give $\theta_1 : \theta_2 : \theta_3 = 1 : -2 : 1$. Two typical eigenvectors are $(1, -2, 1)$ and $(1/\sqrt{6}, -2/\sqrt{6}, 1/\sqrt{6})$.

Interpretation of eigenvectors

We have already remarked that the eigenvectors represent the modes of vibration. Let us investigate the situation further. Corresponding to $\lambda = 0$ we have the eigenvector $(1,1,1)$. In other words there is no relative motion of the disks and the disks and the shaft rotate freely in the bearings as a single rigid body. In the case $\lambda = 1$, we choose the eigenvector $(1,0,-1)$. In this mode the central disk remains motionless whilst the outer two disks rotate with equal amplitudes in opposite senses. In the mode corresponding to $\lambda = 3$ we chose as eigenvector $(1,-2,1)$ and this may be interpreted as the outer two disks rotating in the same sense, keeping pace with each other whereas the centre disk rotates in the opposite sense with twice the amplitude. In Figure 2.2 we depict these modes; diagrams (a), (b) and (c) schematically represent the modes, while diagrams (d), (e) and (f) attempt to show the motion by considering the positions of markers on each disk at a certain time, given that these markers were initially in a line.

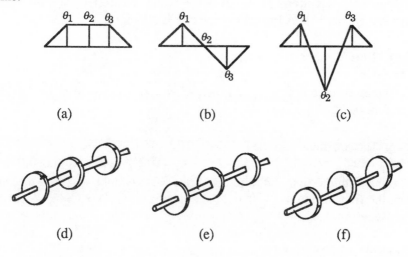

(a)   (b)   (c)

(d)   (e)   (f)

Figure 2.2

In general, the free oscillatory motion will be a combination of these normal modes. Hence

$$\theta_i = A_i \cos \omega_1 t + B_i \sin \omega_1 t + E_i \cos \omega_2 t + F_i \sin \omega_2 t + G_i \cos \omega_3 t + H_i \sin \omega_3 t$$

where ω_1, ω_2 and ω_3 are the three frequencies. Now $\lambda_1 = 0$ and hence $\omega_1 = 0$; $\lambda_2 = 1$ and so $\omega_2 = \sqrt{C/J}$ and $\lambda_3 = 3$ so that $\omega_3 = \sqrt{3C/J}$. Therefore we may write

$$\theta_i = A_i + B_i t + E_i \cos \sqrt{C/J}\, t + F_i \sin \sqrt{C/J}\, t + G_i \cos \sqrt{3C/J}\, t + H_i \sin \sqrt{3C/J}\, t$$

where A_i, B_i, E_i, F_i, G_i and H_i are constants to be determined by the initial displacements and velocities of each disk. (Explain the presence of the B_i term.)

Example

For the matrix $\mathbf{D} = \begin{bmatrix} 4 & 2 & -2 \\ 1 & 3 & 1 \\ -1 & -1 & 5 \end{bmatrix}$ we found eigenvalues 2, 4, 6.

(i) $\underline{\lambda = 2}$. We solve the equations $(\mathbf{D} - \lambda \mathbf{I})\mathbf{x} = \mathbf{0}$ i.e.

$$\begin{bmatrix} 2 & 2 & -2 \\ 1 & 1 & 1 \\ -1 & -1 & 3 \end{bmatrix} \begin{bmatrix} x_1 \\ x_2 \\ x_3 \end{bmatrix} = \begin{bmatrix} 0 \\ 0 \\ 0 \end{bmatrix}$$

The first row of $(\mathbf{D} - \lambda\mathbf{I})$ is equal to the third row subtracted from the second. Let us concentrate on rows 2 and 3 and work with the equations
$$x_1 + x_2 + x_3 = 0 \qquad\qquad -x_1 - x_2 + 3x_3 = 0$$
Adding these equations produces $4x_3 = 0$ and we are left with the single equation $x_1 + x_2 = 0$. A typical eigenvector is $(1, -1, 0)$.

(ii) $\underline{\lambda = 4}$. The equations to be solved are
$$2x_2 - 2x_3 = 0 \qquad\quad x_1 - x_2 + x_3 = 0 \qquad\quad -x_1 - x_2 + x_3 = 0$$
We therefore obtain $x_2 = x_3$ and, hence, $x_1 = 0$. A typical eigenvector is therefore $(0, 1, 1)$.

(iii) $\underline{\lambda = 6}$. The equations become
$$-2x_1 + 2x_2 - 2x_3 = 0 \qquad\quad x_1 - 3x_2 + x_3 = 0 \qquad\quad -x_1 - x_2 - x_3 = 0$$
Working with the first and third equations we may subtract twice the third from the first to find $4x_2 = 0$ and then deduce that $x_1 + x_3 = 0$. A typical eigenvector is $(1, 0, -1)$.

Comparison of systems

The governing equations for the system of Figure 2.1 were found to be $\mathbf{A}\theta = \lambda\theta$ where θ was the vector of angular displacements, $\mathbf{A}$ was the matrix
$$\begin{bmatrix} 1 & -1 & 0 \\ -1 & 2 & -1 \\ 0 & -1 & 1 \end{bmatrix} \quad \text{and } \lambda = J\omega^2/C.$$

In this system the parameters were moments of inertia J and torsional stiffness C. The coordinates were angles θ_i and the critical angular frequencies ω_i had to be determined. We now examine two related systems.

(i) **Linear train**   (Figure 2.3(a))
If we choose as coordinates x_1, x_2, and x_3 – the displacements of the masses from their equilibrium positions – then the governing equation of motion can be written as $\mathbf{A}\mathbf{x} = \lambda\mathbf{x}$ where $\mathbf{x}$ is the vector of linear displacements (x_1, x_2, x_3), $\lambda = M\omega^2/k$ and $\mathbf{A}$ is as before. The system parameters are masses M and spring stiffness k. This time it is frequencies ω which are critical.

(ii) **Electrical Network**
For the electrical circuit shown in Figure 2.3(b) we may write the governing equation $\mathbf{A}\mathbf{I} = \lambda\mathbf{I}$ where $d^2I_1/dt^2 = -\omega^2I_1$, etc., $\lambda = \omega^2CL$ and $\mathbf{I} = (I_1, I_2, I_3)$. The system parameters are inductances L and capacitances C.

All the systems are lumped-parameter approximations of physical situations.

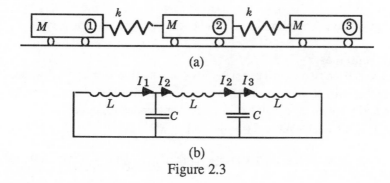

(a)

(b)

Figure 2.3

2.3 Further Results on Eigenvalues

We now develop some results concerning eigenvalues and eigenvectors which will be useful in later work.

Miscellaneous results
(i) It is obvious that A must be square in order that it possesses eigenvalues.
(ii) If the matrix A is singular then $|A| = 0$ and $\lambda = 0$ is one eigenvalue of A.
(iii) If A is a lower triangular matrix, then the eigenvalues of A are simply the diagonal elements; a similar result holds for an upper triangular matrix.
(iv) The eigenvalues of A^T are those of A: this follows by considering the quantity $|A^T - \lambda I|$.
(v) The product of the eigenvalues of A is $|A|$.
(vi) The sum of the eigenvalues of A is the *trace* of A, i.e. the sum of the diagonal elements.
(These last two results form useful checks on numerical estimates of the eigenvalues of a matrix.)

Eigenvalues of a symmetric matrix

Let A be a **Hermitian** matrix so that $\bar{A} = A^T$, where $\bar{A}$ is the matrix formed by replacing each element of A by its complex conjugate. Then let λ be an eigenvalue of A and x the corresponding eigenvector so that $Ax = \lambda x$. Taking transposes we obtain $x^T A = \lambda x^T$. We then take complex conjugates to produce the equation $\bar{x}^T \bar{A} = \bar{\lambda} \bar{x}^T$. Then we take the scalar product of both sides of this equation with x to give $\bar{x}^T \bar{A} x = \bar{\lambda} \bar{x}^T x$. But $Ax = \lambda x$ so that $\bar{x}^T \bar{A}^T x = \bar{x}^T A x$ $= \lambda \bar{x}^T x$. Hence $\lambda \bar{x}^T x = \bar{\lambda} \bar{x}^T x$. Now $\bar{x}^T x \neq 0$ unless $x = 0$ which is not a valid eigenvector. Therefore $\lambda = \bar{\lambda}$ and hence the eigenvalues are real.

Similar matrices

Let A and B be two matrices related by the equation $B = P^{-1} AP$ where P is a non-singular matrix, then A and B are said to be **similar**. Further let $Ax = \lambda x$ where λ and x are an eigenvalue of A and its associated eigenvector.

Now $|B - \lambda I| = |P^{-1} AP - \lambda I| = |P^{-1} AP - \lambda P^{-1} IP|$

$$= |P^{-1}(A - \lambda I)P| = |P^{-1}| \cdot |A - \lambda I| \cdot |P| = \frac{1}{|P|} \cdot |A - \lambda I| \cdot |P|$$

$$= |A - \lambda I|.$$

It therefore follows that two similar matrices have the same eigenvalues. If we can find a simpler matrix B, especially a diagonal matrix, which is similar to A, then the determination of the eigenvalues should become easier.

Example

We saw that the eigenvalues of $A = \begin{bmatrix} 1 & -1 & 0 \\ -1 & 2 & -1 \\ 0 & -1 & 1 \end{bmatrix}$ were 0, 1 and 3. Let

$P = \begin{bmatrix} 1 & 1 & 1 \\ 1 & 0 & -2 \\ 1 & -1 & 1 \end{bmatrix}$; we can show that $P^{-1} = \frac{1}{6}\begin{bmatrix} 2 & 2 & 2 \\ 3 & 0 & -3 \\ 1 & -2 & 1 \end{bmatrix}$ and that $P^{-1} AP =$

$\begin{bmatrix} 0 & 0 & 0 \\ 0 & 1 & 0 \\ 0 & 0 & 3 \end{bmatrix} = B$. From result (iii), the eigenvalues of B are 0, 1 and 3.

The columns of P are the eigenvectors of A; P is called the **modal matrix**. Note the order of the columns.

Powers of a matrix

If a matrix A has an eigenvalue λ then A^m has an eigenvalue λ^m, for integer m. The proof starts by considering $Ax = \lambda x$. Then $A^2x = A\lambda x = \lambda Ax = \lambda^2 x$ and the result for positive integers m follows by induction.

Note that if A is non-singular then the eigenvalues of A^{-1} are the reciprocals of the eigenvalues of A.

In all cases, corresponding eigenvalues share the same eigenvector.

Example

The eigenvalues of $A = \begin{bmatrix} -5 & 2 \\ 2 & -2 \end{bmatrix}$ are $\lambda = -6$ and -1. For $\lambda = -6$ we have an eigenvector $(-2,1)^T$ and for $\lambda = -1$ we have an eigenvector $(1,2)^T$.

Now $A^2 = \begin{bmatrix} 29 & -14 \\ -14 & 8 \end{bmatrix}$ (note that A^2 is also symmetric). Then the equation

$$|A^2 - \lambda I| = \begin{vmatrix} 29-\lambda & -14 \\ -14 & 8-\lambda \end{vmatrix} = 0 \quad \text{produces} \quad 232 - 37\lambda + \lambda^2 - 196 = 0$$

i.e. $36 - 37\lambda + \lambda^2 = 0$ which has roots $\lambda = 1 = (-1)^2$ and $\lambda = 36 = (-6)^2$.

Now $(A^2 - I)x = 0$ becomes the equations $28x_1 - 14x_2 = 0$, $-14x_1 + 7x_2 = 0$, so that a typical eigenvector is $(1,2)^T$. Similarly, we can show that a typical eigenvector corresponding to $\lambda = 36$ is $(-2,1)^T$. Note that these eigenvectors are orthogonal to each other, i.e. their scalar product is zero. We now prove this last result generally.

Theorem

If x_1 and x_2 are two eigenvectors of a symmetric matrix A corresponding to distinct eigenvalues λ_1, λ_2 respectively then $x_1^T . x_2 = 0$.

Proof

Let $Ax_1 = \lambda_1 x_1$ and $Ax_2 = \lambda_2 x_2$. Then $x_1^T Ax_2 = \lambda_2 x_1^T x_2$

i.e. $x_1^T A^T x_2 = \lambda_2 x_1^T x_2$ (since A is symmetric)   i.e. $(Ax_1)^T x_2 = \lambda_2 x_1^T x_2$

i.e. $(\lambda_1 x_1)^T x_2 = \lambda_2 x_1^T x_2$ or $\lambda_1 x_1^T x_2 = \lambda_2 x_1^T x_2$

Now if $\lambda_1 \neq \lambda_2$ it follows that $x_1^T x_2 = 0$ as required.

Equal eigenvalues

Consider the matrix $C = \begin{bmatrix} 1 & 0 & -2 \\ 0 & 0 & 1 \\ -2 & 0 & 4 \end{bmatrix}$

Then $|C - \lambda I| = \begin{vmatrix} 1-\lambda & 0 & -2 \\ 0 & -\lambda & 1 \\ -2 & 0 & 4-\lambda \end{vmatrix} = -(1 - \lambda)\lambda(4 - \lambda) - 2(-2\lambda) = -\lambda^3 + 5\lambda^2.$

The eigenvalues are $\lambda = 0$ (twice) and $\lambda = 5$.

For $\lambda = 5$ we solve the equations $-4x_1 - 2x_3 = 0$, $-5x_2 + x_3 = 0$, $-2x_1 - x_3 = 0$ to obtain a typical eigenvector $(5, -2, -10)^T$.

For $\lambda = 0$ we have the equations $x_1 - 2x_3 = 0$, $x_3 = 0$, $-2x_1 + 4x_3 = 0$ and we obtain a typical eigenvector $(0,1,0)^T$.

Now consider the matrix $D = \begin{bmatrix} 1 & 0 & -2 \\ 0 & 0 & 0 \\ -2 & 0 & 4 \end{bmatrix}$. It should be clear that the

eigenvalues of D are the same as those of C. However for $\lambda = 5$ we now have to solve the equations $-4x_1 - 2x_3 = 0$, $-5x_2 = 0$, $-2x_1 - x_3 = 0$. Hence a typical eigenvector is $(1,0,-2)^T$.

For $\lambda = 0$ we have the equations $x_1 - 2x_3 = 0$, $-2x_1 + 4x_3 = 0$. We now have only one equation for three unknowns: this leaves us two degrees of freedom. Any vector of the form $(2\alpha,\beta,\alpha)^T$ will do for the eigenvector. We shall choose two vectors from this system so that as far as possible we copy the unit vectors $(1,0,0)^T$, $(0,1,0)^T$ and $(0,0,1)^T$. The eigenvector $(1,0,-2)^T$ is close to the first unit vector. As it happens $(0,1,0)^T$ will do as a second eigenvector. Now we cannot quite copy $(0,0,1)^T$ but $\alpha = 1$ and $\beta = 0$ gives a third eigenvector $(2,0,1)$. Note that the three eigenvectors $(1,0,-2)^T$, $(0,1,0)^T$, $(2,0,1)^T$ are **mutually orthogonal**.

Why do you think that in the case of matrix **C** there was only one degree of freedom for the set of eigenvectors associated with the repeated eigenvalue $\lambda = 0$, whereas in the case of matrix **D** the eigenvalue $\lambda = 0$ has a class of eigenvectors with two degrees of freedom?

Recall that when solving linear second order ordinary differential equations, for the complementary function we sometimes encountered repeated roots. For example with the equation

$$\frac{d^2y}{dt^2} - 6\frac{dy}{dt} + 9y = 0 \tag{2.5}$$

when we try $y = \exp(mt)$ we find that $m = 3$ is the only root, yielding $y = \exp(3t)$ as one basic solution. If we rewrite (2.5) as

$$(D - 3)^2 y = 0 \tag{2.6}$$

where D is the differential operator d/dt then we notice that the basic solution we obtained satisfies $(D - 3)y = 0$; operating on both sides with $(D - 3)$ produces equation (2.6). To find a second basic solution we can solve

$$(D - 3)y = \exp(3t) \tag{2.7}$$

Again note that operating on (2.7) with $(D - 3)$ gives equation (2.6). Now (2.7) can be solved by integrating factor to give $y = (c + t)\exp(3t)$, c constant, from which we see that $y = t\exp(3t)$ is the required second solution.

We take the ideas forward into the search for a third eigenvector for the matrix **C**. For $\lambda = 0$ we have found an eigenvector $y = (0,\alpha,0)^T$. This satisfies the equation $(C - \lambda I)x = 0$ when $\lambda = 0$. To find a second eigenvector for this λ we solve the equation $(C - \lambda I)x = y$

i.e. $\begin{bmatrix} 1 & 0 & -2 \\ 0 & 0 & 1 \\ -2 & 0 & 4 \end{bmatrix}\begin{bmatrix} x_1 \\ x_2 \\ x_3 \end{bmatrix} = \begin{bmatrix} 0 \\ \alpha \\ 0 \end{bmatrix}$

from which we find that $x_2 = \alpha$ and $x_1 = 2x_3$. Hence $(x_1, x_2, x_3) = (2\beta,\alpha,\beta)$ is the general solution. A suitable choice is $\beta = 1$, $\alpha = 0$ so that we obtain (2,0,1) which satisfies the equation

$$(C - \lambda I)^2 \, x = 0$$

for $\lambda = 0$, as you can verify directly. We call the new vector a **generalised eigenvector** of **C**.

Transformation of coordinates

In Figure 2.4 we show a unit square before and after a shear transformation applied in the x-direction.

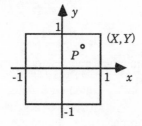 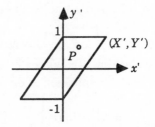

Figure 2.4

The point P has original coordinates (X,Y) and new coordinates (X', Y').

These coordinates are related by $\begin{bmatrix} X' \\ Y' \end{bmatrix} = \begin{bmatrix} 1 & 1 \\ 0 & 1 \end{bmatrix}\begin{bmatrix} X \\ Y \end{bmatrix}$. Now the eigenvalues of

the matrix $\begin{bmatrix} 1 & 1 \\ 0 & 1 \end{bmatrix}$ are $\lambda = 1$ (twice). A typical eigenvector, found in the usual

way, is $(1,0)^T$. The interpretation of this result is that points on the x-axis are the only points for which the vectors connecting them to the origin remain unchanged in direction. These points keep the same distance from the origin since $\lambda = 1$.

Now consider the matrix $D = \begin{bmatrix} 2 & 0 \\ 0 & 1/3 \end{bmatrix}$. You should be able to show that this

represents the transformation shown in Figure 2.5.

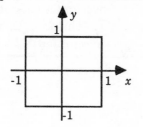 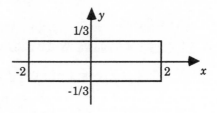

Figure 2.5

The eigenvalues of **D** are $\lambda = 2$ and $\lambda = 1/3$. Corresponding to $\lambda = 2$ we have an eigenvector $(1,0)^T$ and corresponding to $\lambda = 1/3$ we have $(0,1)^T$. Geometrically, this means that points on the x-axis have their distance from the origin doubled, whereas points on the y-axis are brought closer to the origin (by a factor of 3). It is interesting to note that $|D| = 2/3$ which is the ratio of the new area to the area of the original square.

Note that the orthogonal matrix $\begin{bmatrix} \cos \alpha & \sin \alpha \\ -\sin \alpha & \cos \alpha \end{bmatrix}$ represents a rotation about

the origin through an angle α. The eigenvalues are given by

$$\cos^2 \alpha - 2 \cos \alpha \, \lambda + \lambda^2 + \sin^2 \alpha = 0$$

i.e. $\lambda = \cos \alpha \pm \sqrt{\cos^2 \alpha - 1} = \cos \alpha \pm i \sin \alpha = e^{i\alpha} \text{ or } e^{-i\alpha}$

Both eigenvalues are of modulus 1, so that points stay the same distance from the origin after the transformation as they were before it.

Note that $\begin{vmatrix} \cos \alpha & \sin \alpha \\ -\sin \alpha & \cos \alpha \end{vmatrix} = 1$ so that the rotation has not changed any areas:

it is a **rigid-body transformation**.

By applying a suitable transformation, we may be able to simplify the solution

of a given problem. Consider the vibration problem $\ddot{x} = A x$ where

$A = \begin{bmatrix} -2 & 1 \\ 1 & -2 \end{bmatrix}$. If we take $P = \begin{bmatrix} 1 & -1 \\ 1 & 1 \end{bmatrix}$ then $P^{-1} = \frac{1}{2} \begin{bmatrix} 1 & 1 \\ -1 & 1 \end{bmatrix}$. (Note that

P^{-1} is a multiple of P^T.) It follows that $P^{-1} A P = \begin{bmatrix} -1 & 0 \\ 0 & -3 \end{bmatrix}$. We transform to

coordinates $y = P^{-1} x$ so that becomes $\ddot{y} = P^{-1} \ddot{x} = P^{-1} A x = (P^{-1} A P)y$, i.e.

$\ddot{y} = \begin{bmatrix} -1 & 0 \\ 0 & -3 \end{bmatrix} y$ giving $\ddot{y}_1 = -y_1, \; \ddot{y}_2 = -3y_2$.

The solution of these equations is

$y_1 = A \cos t + B \sin t$ and $y_2 = C \cos \sqrt{3}t + D \sin \sqrt{3}t$.

In terms of the old coordinates we have, by using $x = Py$,

$$x_1 = A \cos t + B \sin t - (C \cos \sqrt{3}t + D \sin \sqrt{3}t)$$

$$x_2 = A \cos t + B \sin t + (C \cos \sqrt{3}t + D \sin \sqrt{3}t)$$

which are both linear combinations of the normal modes. The **normal** or **natural coordinates** y_1 and y_2 reflect the normal modes of vibration.

2.4 Quadratic Forms and their Reduction

An expression such as $3x_1^2 + 4x_1 x_2 - x_2^2$ is known as a **quadratic form**; in general, a quadratic form in $x_1, x_2, ..., x_n$ is a polynomial in these variables where every term is of the form $a_{ij} x_i x_j$ and the a_{ij} are constants. Quadratic forms occur in many branches of mathematics as the following examples show.

Equation of a conic. The general equation of a conic is

$$a_{11} x_1^2 + 2a_{12} x_1 x_2 + a_{22} x_2^2 + b_1 x_1 + b_2 x_2 + c_2 = 0$$

The last three terms may be removed by translating the origin of coordinates to the centre of the conic, which will then have the form

$$a_{11} x_1^2 + 2a_{12} x_1 x_2 + a_{22} x_2^2 = 0$$

The nature of the conic will be determined by the nature of this quadratic form.

Kinetic and potential energies. With reference to the linear train of Figure 2.3(a), the kinetic energy of the system is $T = \frac{1}{2} M (\dot{x}_1^2 + \dot{x}_2^2 + \dot{x}_3^2)$ and the potential energy of the system is

$$V = \frac{1}{2} k (x_2 - x_1)^2 + \frac{1}{2} k (x_3 - x_2)^2 = \frac{1}{2} k x_1^2 + k x_2^2 + \frac{1}{2} k x_3^2 - k x_1 x_2 - k x_2 x_3.$$

We see that V is a quadratic form in x_1, x_2 and x_3 while T is a quadratic form in $\dot{x}_1$, $\dot{x}_2$ and $\dot{x}_3$.

Approximation of a function near a stationary point. Near a stationary point a function $f(x,y)$ may be approximated by

$$f(x_0+h, y_0+k) = f(x_0,y_0) + \frac{1}{2} h^2 f_{xx}(x_0,y_0) + hk f_{xy}(x_0,y_0) + \frac{1}{2} k^2 f_{yy}(x_0,y_0)$$

The nature of the stationary point may be determined by

$$\delta f \equiv f(x_0+h, y_0+k) - f(x_0,y_0) \cong \frac{1}{2} h^2 f_{xx}(x_0,y_0) + hk f_{xy}(x_0,y_0) + \frac{1}{2} k^2 f_{yy}(x_0,y_0)$$

Since the second partial derivatives are evaluated at (x_0,y_0), δf is given approximately by a quadratic form in h and k. If this quadratic form is **positive definite** (i.e. it takes zero value if h and k are both zero, otherwise it takes positive values), then the stationary point is a local minimum (why?).

Associated symmetric matrix

We can abstract the essential information about a quadratic form into an associated matrix. For example the quadratic form

$$3x_1^2 + 4x_1 x_2 - x_2^2 \quad \text{may be written as} \quad (x_1, x_2) \begin{bmatrix} 3 & 2 \\ 2 & -1 \end{bmatrix} \begin{bmatrix} x_1 \\ x_2 \end{bmatrix}$$

as you can verify by multiplication.

However, the representation is not unique; for example, the representations

$$(x_1, x_2)\begin{bmatrix} 3 & 4 \\ 0 & -1 \end{bmatrix}\begin{bmatrix} x_1 \\ x_2 \end{bmatrix}, \quad (x_1, x_2)\begin{bmatrix} 3 & 0 \\ 4 & -1 \end{bmatrix}\begin{bmatrix} x_1 \\ x_2 \end{bmatrix} \text{ and } (x_1, x_2)\begin{bmatrix} 3 & 3 \\ 1 & -1 \end{bmatrix}\begin{bmatrix} x_1 \\ x_2 \end{bmatrix}$$

are equally valid. We usually choose that matrix which is symmetric. In this

example, the **associated symmetric matrix** is $\begin{bmatrix} 3 & 2 \\ 2 & -1 \end{bmatrix}$.

The quadratic form $x_1{}^2 - 2x_2{}^2 + 3x_3{}^2 - 6x_1 x_2 - 2x_1 x_3 + 4x_2 x_3$ has

associated symmetric matrix $\begin{bmatrix} 1 & -3 & -1 \\ -3 & -2 & 2 \\ -1 & 2 & 3 \end{bmatrix}$.

Nature of a quadratic form

In determining the nature of a quadratic form, the two particular forms in which we are specially interested are positive definite and negative definite forms. The formal definitions are as follows; it is assumed obvious that any quadratic form takes a zero value if all its constituent variables are zero.

The form $Q = \mathbf{x}^T \mathbf{A} \mathbf{x}$ is **positive definite** if $Q > 0$ for all $\mathbf{x} \neq \mathbf{0}$.

The form $Q = \mathbf{x}^T \mathbf{A} \mathbf{x}$ is **negative definite** if $Q < 0$ for all $\mathbf{x} \neq \mathbf{0}$.

Now it is easy to see that the form $3x_1{}^2 + 4x_2{}^2 + x_3{}^2$ is positive definite, that the form $-x_1{}^2 - 2x_2{}^2 - 5x_3{}^2$ is negative definite and that the form $2x_1{}^2 - x_2{}^2 + 3x_3{}^2$ is neither since it can take both positive and negative values (for example $x_1 = 1, x_2 = 2, x_3 = 1$ gives a positive value whereas $x_1 = 1, x_2 = 14, x_3 = 1$ gives a negative value.) We are able to determine their nature easily because each of these forms comprises only squared terms, i.e. the associated matrix is **diagonal**. If we could somehow reduce a quadratic form to a sum of squares or, equivalently, find a diagonal matrix related to the original symmetric matrix in such a way that the nature of the quadratic form is unchanged then we shall be able to determine that nature easily.

Reduction of a quadratic form via eigenvalues

We have already seen that a real symmetric matrix has eigenvectors which are orthogonal, provided they correspond to distinct eigenvalues.

We now develop a systematic method, using eigenvalues, for reducing a quadratic form to a sum of squares.

First, we state and prove an important result.

Theorem

Let A and P be $n \times n$ matrices and let P^{-1} exist. Then $D = P^{-1} AP$ is a diagonal matrix if and only if the columns of P are the eigenvectors of A corresponding to the eigenvalues $\lambda_1, \lambda_2, ..., \lambda_n$.

D is then the matrix $\quad\quad\quad \operatorname{diag}(\lambda_1, \lambda_2, ..., \lambda_n)$

Proof

If the columns of P, viz $p_1, p_2, ..., p_n$, are the eigenvectors of A corresponding to eigenvalues $\lambda_1, \lambda_2, ..., \lambda_n$ then

$$AP = (Ap_1, Ap_2, ..., Ap_n) = (\lambda_1 p_1, \lambda_2 p_2, ..., \lambda_n p_n)$$

$$= \begin{bmatrix} \lambda_1 p_{11} & \lambda_2 p_{12} & \cdots & \lambda_n p_{1n} \\ \lambda_1 p_{21} & \lambda_2 p_{22} & \cdots & \lambda_n p_{2n} \\ \vdots & \vdots & \vdots & \vdots \\ \lambda_1 p_{n1} & \lambda_2 p_{n2} & \cdots & \lambda_n p_{nn} \end{bmatrix} = \begin{bmatrix} p_{11} & p_{12} & \cdots & p_{1n} \\ p_{21} & p_{22} & \cdots & p_{2n} \\ \vdots & \vdots & \vdots & \vdots \\ p_{n1} & p_{n2} & \cdots & p_{nn} \end{bmatrix} \begin{bmatrix} \lambda_1 & & & 0 \\ & \lambda_2 & & \\ & & \ddots & \\ 0 & & & \lambda_n \end{bmatrix}$$

$$= PD$$

Hence $D = P^{-1} AP$.

The argument can be reversed, completing the proof.

Now when A has n linearly independent eigenvectors it follows that the columns of P are linearly independent and hence P^{-1} exists.

We have already seen that the eigenvalues of an orthogonal matrix are of modulus 1 and, since an orthogonal matrix represents rotation about an axis, we would expect to be able to multiply a matrix representing a conic by a suitable orthogonal matrix to realign the coordinate axes along the axes of symmetry of the conic. Let C be an orthogonal matrix. Then the following properties hold.

(i) $C^{-1} = C^T$.

(ii)   The columns of C are unit vectors which are mutually orthogonal.

(iii) C^T is also an orthogonal matrix.

(iv) $|C| = 1$.

Note that since the distinct eigenvalues of a real symmetric matrix A are associated with eigenvectors which are mutually orthogonal then the matrix P which has these eigenvectors as its columns will be orthogonal. Hence we may use the theorem to see that $P^{-1} AP$ will be diagonal, with the eigenvalues of A as its diagonal elements.

Example

The matrix $A = \begin{bmatrix} -2 & 1 \\ 1 & -2 \end{bmatrix}$ is real and symmetric.

The eigenvalues of A are $\lambda = -1$ and -3; corresponding to $\lambda = -1$ we have an eigenvector $(1,1)^T$ and corresponding to $\lambda = -3$ we have an eigenvector $(-1,1)^T$.

Unit eigenvectors are $(1/\sqrt{2}, \ 1/\sqrt{2})^T$ and $(-1/\sqrt{2}, \ 1/\sqrt{2})^T$. The matrix

$$\mathbf{P} = \begin{bmatrix} 1/\sqrt{2} & -1/\sqrt{2} \\ 1/\sqrt{2} & 1/\sqrt{2} \end{bmatrix} \text{ is orthogonal and, therefore, } \mathbf{P}^{-1} = \mathbf{P}^T = \begin{bmatrix} 1/\sqrt{2} & 1/\sqrt{2} \\ -1/\sqrt{2} & 1/\sqrt{2} \end{bmatrix}.$$

You can show directly that $\mathbf{P}^{-1}\mathbf{AP} = \begin{bmatrix} -1 & 0 \\ 0 & -3 \end{bmatrix}$.

Now A is the matrix associated with the quadratic form
$Q = -2x_1^2 + 2x_1x_2 - 2x_2^2$. Consider then the transformation $y = \mathbf{P}^{-1}x$.

i.e.
$$\begin{bmatrix} y_1 \\ y_2 \end{bmatrix} = \begin{bmatrix} 1/\sqrt{2} & 1/\sqrt{2} \\ -1/\sqrt{2} & 1/\sqrt{2} \end{bmatrix}\begin{bmatrix} x_1 \\ x_2 \end{bmatrix}$$

If we express Q in terms of y_1 and y_2 we use $\mathbf{x} = \mathbf{P}y$ so that

$$x_1 = \frac{1}{\sqrt{2}}y_1 - \frac{1}{\sqrt{2}}y_2 \ \text{ and } \ x_2 = \frac{1}{\sqrt{2}}y_1 + \frac{1}{\sqrt{2}}y_2. \text{ Then}$$

$$Q = -2(\tfrac{1}{2}y_1^2 - y_1y_2 + \tfrac{1}{2}y_2^2) + 2(\tfrac{1}{2}y_1^2 - \tfrac{1}{2}y_2^2) - 2(\tfrac{1}{2}y_1^2 + y_1y_2 + \tfrac{1}{2}y_2^2)$$

$$= -y_1^2 - 3y_2^2 = \lambda_1 y_1^2 + \lambda_2 y_2^2$$

This last result holds generally.

If **A** is a real symmetric matrix associated with a quadratic form Q in the variables $x_1, x_2,..., x_n$ then we can find an orthogonal matrix **P** such that $\mathbf{P}^{-1}\mathbf{AP}$ is a diagonal matrix. The coordinate transformation $y = \mathbf{P}^{-1}x$ allows Q to be expressed in terms of $y_1, y_2,..., y_n$. If all the eigenvalues of **A** are positive the reduced quadratic form is $Q = z_1^2 + z_2^2 + ... + z_n^2$. This result follows because we can apply a transformation **P** as demonstrated in the last example to reduce Q to $\lambda_1 y_1^2 + \lambda_2 y_2^2 + ... + \lambda_n y_n^2$. A further transformation

$$y_1 = (1/\sqrt{\lambda_1})z_1, \ y_2 = (1/\sqrt{\lambda_2})z_2 \text{ etc. will produce } Q = z_1^2 + z_2^2 + ... + z_n^2.$$

The second transformation represents contractions in each of the coordinate directions y_i.

Example

The quadratic form $Q = 2x_1^2 - x_2^2 - x_3^2 - 4x_1x_2 + 4x_1x_3 + 8x_2x_3$ has

associated matrix $\mathbf{A} = \begin{bmatrix} 2 & -2 & 2 \\ -2 & -1 & 4 \\ 2 & 4 & -1 \end{bmatrix}$.

The eigenvalues are given by $\begin{vmatrix} 2-\lambda & -2 & 2 \\ -2 & -1-\lambda & 4 \\ 2 & 4 & -1-\lambda \end{vmatrix} = 0$ and are $\lambda = 3$ (twice) and

$\lambda = -6$. For $\lambda = -6$ we need to solve the equations $8x_1 - 2x_2 + 2x_3 = 0$, $-2x_1 + 5x_2 + 4x_3 = 0$, $2x_1 + 4x_2 + 5x_3 = 0$. We obtain as eigenvectors $(\alpha, 2\alpha, -2\alpha)^T$. We take the unit vector $(1/3, 2/3, -2/3)^T$. For $\lambda = 3$ we solve $-x_1 - 2x_2 + 2x_3 = 0$ (the other equations are really the same). We shall choose two eigenvectors which are orthogonal to $(1/3, 2/3, -2/3)^T$ and orthogonal to each other. Suitable vectors are $(2/3, 1/3, 2/3)^T$ and $(-2/3, 2/3, 1/3)^T$. Verify that these three vectors are mutually orthogonal and that the last two satisfy the appropriate equation.

Let $P = 1/3 \begin{bmatrix} 1 & -2 & 2 \\ 2 & 2 & 1 \\ -2 & 1 & 2 \end{bmatrix}$. Then P is orthogonal. Furthermore,

$P^{-1} AP = \begin{bmatrix} -6 & 0 & 0 \\ 0 & 3 & 0 \\ 0 & 0 & 3 \end{bmatrix}$. Now consider $\begin{bmatrix} x_1 \\ x_2 \\ x_3 \end{bmatrix} = P \begin{bmatrix} y_1 \\ y_2 \\ y_3 \end{bmatrix}$. You can check that this

transformation allows us to write $Q = -6y_1^2 + 3y_2^2 + 3y_3^2$.

Theorem
A quadratic form is positive definite if and only if all the eigenvalues of the associated symmetric matrix are positive.

Proof
We know that the transformation $x = Py$, as used above, will transform Q to

the expression $\sum_{i=1}^{n} \lambda_i y_i^2$. Now if all $\lambda_i > 0$, then $Q > 0$ unless $y_i = 0$ for all

i, i.e. $x_i = 0$ for all i. Conversely, suppose an eigenvalue, say λ_1, is ≤ 0. Then by choosing $y_1 = 1$ and all other $y_i = 0$, Q becomes ≤ 0 for a non-zero y and hence for a non-zero x, violating the positive definite quality.

Example 1
For the quadratic form of the last example one eigenvalue is negative and so the form is not positive definite.

Example 2

The quadratic form $Q = x_1^2 + 2x_2^2 + 7x_3^2 - 2x_1 x_2 + 4x_1 x_3 - 2x_2 x_3$

has associated symmetric matrix $\begin{bmatrix} 1 & -1 & 2 \\ -1 & 2 & -1 \\ 2 & -1 & 7 \end{bmatrix}$ which has eigenvalues satisfying

$\lambda^3 - 10\lambda^2 + 17\lambda - 2 = 0$. To find the eigenvalues would require numerical techniques. However we may write

$Q = (x_1 - x_2 + 2x_3)^2 + 2x_2^2 + 7x_3^2 - 2x_2 x_3 - x_2^2 - 4x_3^2 + 4x_2 x_3$

$= (x_1 - x_2 + 2x_3)^2 + x_2^2 + 3x_3^2 + 2x_2 x_3$

$= (x_1 - x_2 + 2x_3)^2 + (x_2 + x_3)^2 + 2x_3^2$

Therefore we see that Q is positive definite.

2.5 Boundary-Value Problems

We consider as a first example the bending of a uniform strut of length l under an axial compressive force P at either end. In Figure 2.6 we demonstrate an approximate method of solution. We aim to estimate the deflections at five equally spaced internal points and use these as a guide to the deflected profile. The strut is fixed at either end.

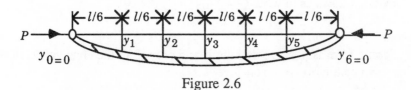

Figure 2.6

The governing differential equation is the bending moment equation

$$EI \frac{d^2 y}{dx^2} = -Py \qquad (2.8)$$

where EI is the (constant) flexural rigidity of the strut and y is the deflection of a point on the strut which is distant x from the left-hand end of the strut. Since both ends are fixed, $y(0) = 0 = y(l)$; in our approximate method this implies that $y_0 = 0 = y_6$. Since the problem is symmetrical, it follows that $y_1 = y_5$ and $y_2 = y_4$ and we need only find y_1, y_2 and y_3.

We approximate $\dfrac{d^2 y}{dx^2}$ at the point x_r (where the deflection is y_r) by the central difference formula $\dfrac{y_{r+1} - 2y_r + y_{r-1}}{(l/6)^2}$ which has error $O(l^2/36)$. At x_2, therefore,

the differential equation may be approximated by the finite difference equation

$$EI\left[\frac{y_3 - 2y_2 + y_1}{l^2/36}\right] = -Py_2 \quad \text{or} \quad y_3 - 2y_2 + y_1 = \left[\frac{-Pl^2}{36EI}\right]y_2. \quad \text{Now at } x_1$$

the finite difference approximation to $\dfrac{d^2y}{dx^2}$ is $\dfrac{y_2 - 2y_1 + y_0}{l^2/36}$; but $y_0 = 0$ and the

equation at x_1 reduces to $y_2 - 2y_1 = \dfrac{-Pl^2}{36EI}y_1$. Find the equation at x_3. Put

$\lambda = Pl^2/(36EI)$, and rearrange the three finite difference equations to

$$\left.\begin{array}{c} 2y_1 - y_2 = \lambda y_1 \\ -y_1 + 2y_2 - y_3 = \lambda y_2 \\ -2y_2 + 2y_3 = \lambda y_3 \end{array}\right\} \tag{2.9a}$$

In matrix form, these equations may be written

$$\mathbf{Ay} = \lambda\mathbf{y} \tag{2.9b}$$

where $\mathbf{y} = \begin{bmatrix} y_1 \\ y_2 \\ y_3 \end{bmatrix}$ and $\mathbf{A}$ is the matrix $\begin{bmatrix} 2 & -1 & 0 \\ -1 & 2 & -1 \\ 0 & -2 & 2 \end{bmatrix}$.

We have reduced the differential equation to a set of linear simultaneous equations. The original problem always was an eigenvalue problem even if cast in a different light.

Obviously, one solution of (2.9b) is $\mathbf{y} = \mathbf{0}$ which corresponds to no deflection of the strut; this is a possible equilibrium profile. The other possible equilibrium profiles must correspond to the eigenvectors of matrix $\mathbf{A}$, since these are the only non-zero solutions of (2.9).

The eigenvalues of $\mathbf{A}$ are $(2 - \sqrt{3})$, 2 and $(2 + \sqrt{3})$. Note the useful checks

(i) sum of eigenvalues is the trace of $\mathbf{A} = 2 + 2 + 2 = 6$
(ii) product of eigenvalues is $|\mathbf{A}| = 2$.
Both checks are satisfied.

To 2 d.p. the eigenvalues are 0.27, 2 and 3.73 (these values satisfy check (i) exactly but not exactly check (ii)).

We therefore have three critical values of P: $\dfrac{9.72EI}{l^2}$, $\dfrac{72EI}{l^2}$, and $\dfrac{134.28EI}{l^2}$.

The interpretation of these results is that the strut can assume a symmetrical curved profile under an axial load P and remain in equilibrium only if P has one of the three values quoted.

What do these deflected profiles look like? We need to find the eigenvectors of **A**. An eigenvector corresponding to $\lambda = 2 - \sqrt{3}$ is $(1, \sqrt{3}, 2)^T$; an eigenvector for $\lambda = 2$ is $(1, 0, -1)^T$; an eigenvector corresponding to $\lambda = 2 + \sqrt{3}$ is $(1, -\sqrt{3}, 2)^T$. Sketches of the three profiles are given in Figure 2.7.

Figure 2.7

However, we might have chosen other eigenvectors from each class; certainly we could have reversed all the signs to obtain profiles which were mirror images of the ones shown. A more serious problem is that we may only determine eigenvectors within a scalar multiple. The actual values of y_1, y_2 and y_3 will not result from this analysis.

You should by now be asking yourself what would have happened if the original finite difference approximation were improved, e.g. if we had divided the strut into 12 equal parts. We should then have had 11 internal points and, employing symmetry, 6 equations to solve in 6 unknowns. This would give rise to a 6×6 matrix with 6 eigenvalues. In theory there would be 6 critical loads P.

The analytical solution to the problem is $y = B_n \sin \dfrac{n\pi x}{l}$ where $\sqrt{\dfrac{P}{EI}} = \dfrac{n\pi}{l}$

for $n = 1, 2, 3, \ldots$ (The case $n = 0$ gives the undeflected profile $y \equiv 0$). There

is an infinite number of theoretical critical loads given by $P = \dfrac{n^2 \pi^2 EI}{l^2}$. The one

of most practical interest is the smallest one where $n = 1$. Then $P = \pi^2 EI/l^2$. Compare this with the approximation we produced of $9.72EI/l^2$. Would we expect the approximation to improve as the number of internal points increases? For the case of 11 internal points after a lot of tedious algebra the approximation had improved only to $9.79EI/l^2$. It was hardly worthwhile; in such a situation we might well ask if a numerical technique for finding eigenvalues would not have been better. We take up this idea in the next sections.

2.6 Finding the Eigenvalue of Largest Modulus

We have so far chosen very simple physical systems. The number of degrees of freedom has been small giving rise to at most a 3×3 matrix and we have been careful to make masses and stiffnesses equal. We shall continue to illustrate our ideas with 3×3 matrices because these are small enough to allow easy computation and presentation of results, yet are large enough to prevent trivial cases dominating our conclusions.

Consider the system of Figure 2.8. It is effectively that of Figure 2.1, but the masses of the disks are no longer equal and the stiffness of different parts of the shaft differ; further, the left-hand end is fixed so that rigid body rotation is no longer possible.

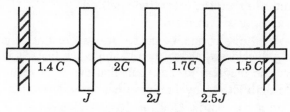

Figure 2.8

For consistency with previous notation, we let x_1, x_2, x_3 be the angles turned through by the disks (reading from left to right). The equations of motion are

$$
\left.
\begin{aligned}
J\ddot{x}_1 &= -1.4Cx_1 - 2C(x_1 - x_2) \\
2J\ddot{x}_2 &= 2C(x_1 - x_2) - 1.7C(x_2 - x_3) \\
2.5J\ddot{x}_3 &= 1.7C(x_2 - x_3)
\end{aligned}
\right\} \tag{2.10}
$$

Applying the ideas of Section 2.2, we put $\ddot{x}_i = -\omega^2 x_i$ and write $\lambda = J\omega^2/C$ to obtain the equations in matrix form as

$$
\left.
\begin{bmatrix} 3.4 & -2 & 0 \\ -2 & 3.7 & -1.7 \\ 0 & -1.7 & 1.7 \end{bmatrix}
\begin{bmatrix} x_1 \\ x_2 \\ x_3 \end{bmatrix}
= \lambda
\begin{bmatrix} 1 & 0 & 0 \\ 0 & 2 & 0 \\ 0 & 0 & 2.5 \end{bmatrix}
\begin{bmatrix} x_1 \\ x_2 \\ x_3 \end{bmatrix}
\right\} \tag{2.11}
$$

or
$$
\mathbf{A}\mathbf{x} = \lambda \mathbf{B}\mathbf{x} \tag{2.12}
$$

Make sure that you can derive this result yourself.

Although there are certain standard programs available which will handle the eigenvalue problem in the form (2.12) it is easy in this case to form the matrix

$$\mathbf{B}^{-1} = \begin{bmatrix} 1 & 0 & 0 \\ 0 & 0.5 & 0 \\ 0 & 0 & 0.4 \end{bmatrix}$$ and then multiply (2.12) by $\mathbf{B}^{-1}$ to produce the equation

$\mathbf{Hx} = \lambda \mathbf{x}$ where $\mathbf{H} = \mathbf{B}^{-1} \mathbf{A}$. We have now recovered the more usual form of the

eigenvalue problem. In our case we find $\mathbf{H} = \begin{bmatrix} 3.4 & -2 & 0 \\ -1 & 1.85 & -0.85 \\ 0 & -0.68 & 0.68 \end{bmatrix}$. If we were

to form the characteristic equation for $\mathbf{H}$ we should have to resort to numerical methods of root-finding. We seek a different approach.

Algebraic foundation of the iterative method

We first state a very useful theorem.

If $\mathbf{A}$ is a real $n \times n$ matrix and λ_1 and λ_2 are two distinct eigenvalues with corresponding eigenvectors $\mathbf{x}_1$ and $\mathbf{x}_2$ then $\mathbf{x}_1$ and $\mathbf{x}_2$ are linearly independent.

We generalise the theorem to the result that the eigenvectors of a matrix corresponding to **distinct** eigenvalues are **linearly independent**.

If, therefore, an $n \times n$ matrix $\mathbf{A}$ has n distinct eigenvalues $\lambda_1, ..., \lambda_n$ with corresponding eigenvectors $\mathbf{x}_1, ..., \mathbf{x}_n$ then these eigenvectors form a basis for the vector space R^n.

Therefore, any vector $\mathbf{y}_0 \in R^n$ can be expressed as a linear combination of these:

$$\mathbf{y}_0 = d_1 \mathbf{x}_1 + d_2 \mathbf{x}_2 + ... + d_n \mathbf{x}_n \qquad (2.13)$$

Now consider, $\mathbf{y}_1 = \mathbf{Ay}_0 = d_1 \mathbf{Ax}_1 + d_2 \mathbf{Ax}_2 + ... + d_n \mathbf{Ax}_n$

$$= \lambda_1 d_1 \mathbf{x}_1 + \lambda_2 d_2 \mathbf{x}_2 + ... + \lambda_n d_n \mathbf{x}_n$$

Similarly $\mathbf{y}_2 = \mathbf{Ay}_1 = \lambda_1^2 d_1 \mathbf{x}_1 + \lambda_2^2 d_2 \mathbf{x}_2 + ... + \lambda_n^2 d_n \mathbf{x}_n$

and, in general, $\mathbf{y}_r = \mathbf{Ay}_{r-1} = \lambda_1^r d_1 \mathbf{x}_1 + \lambda_2^r d_2 \mathbf{x}_2 + ... + \lambda_n^r d_n \mathbf{x}_n$

If the eigenvalue of largest modulus is λ_1, then the ratios $\left| \dfrac{\lambda_2}{\lambda_1} \right|, \left| \dfrac{\lambda_3}{\lambda_1} \right|, ..., \left| \dfrac{\lambda_n}{\lambda_1} \right|$

are all less than 1. We write

$$\mathbf{y}_r = \lambda_1^r \left[d_1 \mathbf{x}_1 + \left[\frac{\lambda_2}{\lambda_1} \right]^r d_2 \mathbf{x}_2 + ... + d_n \left[\frac{\lambda_n}{\lambda_1} \right]^r \mathbf{x}_n \right] \qquad (2.14)$$

For sufficiently large r we might expect the ratios $\left[\dfrac{\lambda_2}{\lambda_1} \right]^r, ..., \left[\dfrac{\lambda_n}{\lambda_1} \right]^r$ to be small

enough to allow $\mathbf{y}_r \cong \lambda_1^r d_1 \mathbf{x}_1$ to be a reasonable approximation to the truth.

The ratio $\left|\dfrac{y_{r+1}}{y_r}\right|$ becomes progressively closer to λ_1 as r becomes larger.

All we have to do is to form the sequence $\{y_0, Ay_0, A^2y_0, A^3y_0, ...\}$. We then form a sequence of ratios of successive terms. The limit of this sequence is the eigenvalue of largest modulus.

Before we apply the method we shall add one modification which will allow us the luxury of also finding the eigenvector that corresponds to this eigenvalue.

It helps for the moment to assume we are dealing with a real symmetric matrix with distinct eigenvalues. The corresponding eigenvectors are orthogonal and, in effect, represent a rotated set of Cartesian axes. As we proceed down the sequence $\{y_0, y_1, y_2, y_3, ...\}$ we effectively produce a sequence of vectors whose directions become ever more closely aligned to the axis represented by x_1.

The choice of y_0 is to some extent arbitrary. Let us arrange that one component (which for the purpose of this discussion we may assume is the first) is equal to 1. Then $y_1 = Ay_0$ can be written as $b_1 y_1'$ where the first component is also equal to 1.

For example, if $y_0 = \begin{bmatrix} 1 \\ 2 \\ 3 \end{bmatrix}$ and $y_1 = \begin{bmatrix} 4 \\ 10 \\ 12 \end{bmatrix}$ then $b_1 = 4$ and $y_1' = \begin{bmatrix} 1 \\ 2.5 \\ 3 \end{bmatrix}$. Thus we can see how closely y_0 and y_1' agree, component for component. Later on in the sequence, when $y_r \cong \lambda_1^r d_1 x_1$ and $y_{r+1} \cong \lambda_1^{r+1} d_1 x_1$ we should expect that the modified vectors would agree quite closely. If we have modified the y_r' so that the sequence we really obtain is $y_0, y_1', y_2', ...$ then we shall have the relationship $Ay_r' = y_{r+1} = b_r y'_{r+1}$; b_r is an approximation to λ_1.

If you study the following two examples, the technique should become clearer.

Example 1

The matrix $A = \begin{bmatrix} 1 & -1 & 0 \\ -1 & 2 & -1 \\ 0 & -1 & 1 \end{bmatrix}$ has eigenvalues 0, 1, 3. Let us guess that

$y_0 = \begin{bmatrix} 1 \\ -1 \\ 2 \end{bmatrix}$. Why? We know that an eigenvector corresponding to $\lambda = 3$ is $\begin{bmatrix} 1 \\ -2 \\ 1 \end{bmatrix}$

and for purposes of demonstration we want to be neither too close to the end result nor too far away. In general, the further away our starting vector the more iterations we require to reach a reasonably accurate approximation.

Then $y_1 = \begin{bmatrix} 1 & -1 & 0 \\ -1 & 2 & -1 \\ 0 & -1 & 1 \end{bmatrix}\begin{bmatrix} 1 \\ -1 \\ 2 \end{bmatrix} = \begin{bmatrix} 2 \\ -5 \\ 3 \end{bmatrix} = 2\begin{bmatrix} 1 \\ -2.5 \\ 1.5 \end{bmatrix} = b_1 y_1'$

We performed the last step so that the first components of the input y_0 and output y_1' are equal; we can compare the second and third components. Repeating the step, we obtain

$$y_2 = \begin{bmatrix} 1 & -1 & 0 \\ -1 & 2 & -1 \\ 0 & -1 & 1 \end{bmatrix} \begin{bmatrix} 1 \\ -2.5 \\ 1.5 \end{bmatrix} = \begin{bmatrix} 3.5 \\ -7.5 \\ 4 \end{bmatrix} = 3.5 \begin{bmatrix} 1 \\ -2.143 \\ 1.143 \end{bmatrix} = b_2\, y_2'$$

and　　$$y_3 = \begin{bmatrix} 1 & -1 & 0 \\ -1 & 2 & -1 \\ 0 & -1 & 1 \end{bmatrix} \begin{bmatrix} 1 \\ -2.143 \\ 1.143 \end{bmatrix} = \begin{bmatrix} 3.143 \\ -6.429 \\ 3.286 \end{bmatrix} = 3.143 \begin{bmatrix} 1 \\ -2.045 \\ 1.045 \end{bmatrix} = b_3\, y_3'$$

We continue the calculations in Table 2.1

The method has converged fairly quickly to the eigenvalue 3 to 3 d.p. But, before we pass the technique with flying colours, consider a second example.

Table 2.1

r	y_r	y_{r+1}	b_{r+1}	y'_{r+1}
0	(1, –1, 2)	(2, –5, 3)	2	(1, –2.5, 1.5)
1	(1, –2.5, 1.5)	(3.5, –7.5, 4)	3.5	(1, –2.143, 1.143)
2	(1, –2.143, 1.143)	(3.143, –6.429, 3.286)	3.143	(1, –2.045, 1.045)
3	(1, –2.045, 1.045)	(3.045, –6.135, 3.090)	3.045	(1, –2.015, 1.015)
4	(1, –2.015, 1.015)	(3.015, –6.045, 3.030)	3.015	(1, –2.005, 1.005)
5	(1, –2.005, 1.005)	(3.005, –6.015, 3.010)	3.005	(1, –2.002, 1.002)
6	(1, –2.002, 1.002)	(3.002, –6.006, 3.004)	3.002	(1, –2.001, 1.001)
7	(1, –2.001, 1.001)	(3.001, –6.003, 3.002)	3.001	(1, –2.000, 1.000)
8	(1, –2.000, 1.000)	(3.000, –6.000, 3.000)	3.000	(1, –2.000, 1.000)

Example 2

The matrix $D = \begin{bmatrix} 4 & 2 & -2 \\ 1 & 3 & 1 \\ -1 & -1 & 5 \end{bmatrix}$ has eigenvalues 2, 4, 6. Let us make an initial

guess of $y_0 = (1,1,1)^T$ for the eigenvector associated with the eigenvalue of largest modulus.

Then　　$$y_1 = \begin{bmatrix} 4 & 2 & -2 \\ 1 & 3 & 1 \\ -1 & -1 & 5 \end{bmatrix} \begin{bmatrix} 1 \\ 1 \\ 1 \end{bmatrix} = \begin{bmatrix} 4 \\ 5 \\ 3 \end{bmatrix} = 4 \begin{bmatrix} 1 \\ 1.25 \\ 0.75 \end{bmatrix} = b_1\, y_1'$$

Further　　$$y_2 = \begin{bmatrix} 4 & 2 & -2 \\ 1 & 3 & 1 \\ -1 & -1 & 5 \end{bmatrix} \begin{bmatrix} 1 \\ 1.25 \\ 0.75 \end{bmatrix} = \begin{bmatrix} 5 \\ 5.5 \\ 1.5 \end{bmatrix} = 5 \begin{bmatrix} 1 \\ 1.1 \\ 0.3 \end{bmatrix} = b_2\, y_2'$$

After 10 iterations

$y_r = (1, 0.068, -0.943)^T$, $b_{r+1} = 6.022$, $y'_{r+1} = (1, 0.043, -0.960)^T$.

You can see that this time convergence is much slower. Is this because of a poor initial approximation or some other factor? Looking back at formula (2.14) we might expect the rate of convergence to be influenced by the ratios $|\lambda_2/\lambda_1|$ and $|\lambda_3/\lambda_1|$. In the previous example, these ratios were 1/3 and 0, whereas in this example the ratios are 2/3 and 1/3.

The worst initial approximation we could take is one at 'right angles' to the eigenvector sought, since the sequence of approximating vectors has 'further to go'. In Example 1, the initial approximation was by no means perpendicular to the required eigenvector, but in Example 2 the initial approximation $(1,1,1)^T$ was perpendicular to the eigenvector $(1,0,-1)^T$.

The iteration is really on the eigenvector and *not* on the eigenvalue. In Example 2 we actually obtained an estimate of the largest eigenvalue correct to 3 d.p. as early as the 8th iteration but the eigenvector was nowhere in sight.

Breakdown of the iterative technique

The process may break down in certain circumstances.

By a fluke, we may choose as our starting approximation an eigenvector corresponding to an eigenvalue of smaller modulus.

For example, if in Example 1 we start with $y_0 = (1,0,-1)^T$ we should find $b_1 = 1$, $y_1 = (1,0,-1)^T$.

Trouble arises if $|\lambda_1| = |\lambda_2|$ which will be the case if $\lambda_2 = -\lambda_1$ or if λ_1 and λ_2 are complex conjugates.

Consider the matrix $C = \begin{bmatrix} 1 & 0 & 0 \\ -2 & 2 & -1 \\ 0 & 1 & 2 \end{bmatrix}$. If we start with $y_0 = (0,1,1)$ we find that

the iterations do not tend toward any limit.

In fact the eigenvalues of C are 1, 2 − i, 2 + i and the eigenvectors which correspond to them are $(1,1,-1)^T$, $(0,1,i)^T$ and $(0,1,-i)^T$ respectively.

2.7 Determination of Other Eigenvalues

In practical situations there is often more interest in the lowest eigenvalue; sometimes it is important to know *all* the eigenvalues. In this section we examine some ways by which the other eigenvalues may be found.

Lowest Eigenvalue

Let A be a non-singular matrix, λ one of its eigenvalues and x a corresponding eigenvector, so that $Ax = \lambda x$. Premultiplying by A^{-1} produces $A^{-1}Ax = \lambda A^{-1}x$, i.e. $x = \lambda A^{-1}x$. Dividing both sides by λ (which we may do since A is non-singular so that $\lambda \neq 0$) we obtain $(1/\lambda)x = A^{-1}x$. This last equation may be written as $A^{-1}x = (1/\lambda)x$ which demonstrates that $1/\lambda$ is an eigenvalue of A^{-1} and x is a corresponding eigenvector.

Example

The matrix $D = \begin{bmatrix} 4 & 2 & -2 \\ 1 & 3 & 1 \\ -1 & -1 & 5 \end{bmatrix}$ had eigenvalues 2, 4 and 6 (page 50).

Corresponding eigenvectors were $(1,-1,0)^T$, $(0,1,1)^T$ and $(1,0,-1)^T$ respectively.

Therefore the matrix $D^{-1} = \dfrac{1}{24}\begin{bmatrix} 8 & -4 & 4 \\ -3 & 9 & -3 \\ 1 & 1 & 5 \end{bmatrix}$ will have eigenvalues $\dfrac{1}{2}, \dfrac{1}{4}, \dfrac{1}{6}$ with

corresponding eigenvectors $(1,-1,0)^T$, $(0,1,1)^T$ and $(1,0,-1)^T$ respectively. A procedure to find the lowest eigenvalue of a matrix then is as follows. Invert the matrix and find the eigenvalue of largest modulus of the inverse by the iterative technique developed in the last section. The reciprocal of the result is the eigenvalue we seek and the eigenvector obtained by the iterations will serve as an eigenvector for our purposes.

Apply the iterative technique to D^{-1}.

The rate of convergence is governed by the rules described in the last section.

However, the explicit determination of the inverse matrix is a lengthy process. Instead of finding y_r from the equation $y_{r+1} = A^{-1}y_r$, we could solve the set of equations $Ay_{r+1} = y_r$ by Gauss Elimination.

Example

We wish to find the lowest eigenvalue of matrix D as above. We shall find y_r
via $$Dy_r = y'_{r-1} \tag{2.15}$$

If we start with an initial guess $y_0 = \begin{bmatrix} 1 \\ 1 \\ 1 \end{bmatrix}$ then $\begin{bmatrix} 4 & 2 & -2 \\ 1 & 3 & 1 \\ -1 & -1 & 5 \end{bmatrix}\begin{bmatrix} y_1 \\ y_2 \\ y_3 \end{bmatrix} = \begin{bmatrix} 1 \\ 1 \\ 1 \end{bmatrix}$

Gaussian elimination with partial pivoting produces the equations

$$\begin{bmatrix} 4 & 2 & -2 \\ 0 & 2.5 & 1.5 \\ 0 & 0 & 4.8 \end{bmatrix}\begin{bmatrix} y_1 \\ y_2 \\ y_3 \end{bmatrix} = \begin{bmatrix} 1 \\ 3/4 \\ 1.4 \end{bmatrix}$$

Hence $y_3 = 7/24$, $y_2 = 3/24$, $y_1 = 8/24$
so that $y_1 = (8/24, 3/24, 7/24) = 8/24(1, 3/8, 7/8) = b_1 y_1'$.

Now we need to solve the equations $\begin{bmatrix} 4 & 2 & -2 \\ 1 & 3 & 1 \\ -1 & -1 & 5 \end{bmatrix}\begin{bmatrix} y_1 \\ y_2 \\ y_3 \end{bmatrix} = \begin{bmatrix} 1 \\ 3/8 \\ 7/8 \end{bmatrix}$

Since we shall be carrying out the same sequence of operations each time that we use the iterative formula, we can establish the row operations in the first place and then simply apply these rules to the sequence of vectors $y_1', y_2', ...$ You continue the computation of the lowest eigenvalue of **D**. When producing y_r' from y_r we divide by the component of largest modulus, which is denoted b_r.

Eigenvalue nearest a prescribed value

Suppose we know that there is an eigenvalue approximately equal to some number p. Let the matrix **A** have eigenvalues $\lambda_1, \lambda_2, ..., \lambda_n$ with corresponding eigenvectors $x_1, x_2, ..., x_n$. Then if $Ax_1 = \lambda_1 x_1$ it follows that $(A - pI)x_1 = (\lambda_1 - p)x_1$. $Ax_2 = \lambda_2 x_2$ implies that $(A - pI)x_2 = (\lambda_2 - p)x_2$ and so on. The eigenvalues of the matrix $(A - pI)$ are $(\lambda_1 - p), (\lambda_2 - p), ..., (\lambda_n - p)$; the eigenvector corresponding to $(\lambda_i - p)$ is the same as that which corresponds to λ_i. For example if a matrix **D** has eigenvalues 2, 4, 6 and we seek the middle of these, then the matrix $(D - 3.5I)$ has eigenvalues -1.5, 0.5 and 2.5. The iteration procedure to find the eigenvalue of least modulus will produce the value $\lambda_2 = 0.5$ with associated eigenvector x_2. Then we add back 3.5 to λ_2 to obtain the required value 4.0; x_2 is the required eigenvector. This method does require knowledge of the approximate location of the eigenvalues.

Acceleration of convergence

We may use the above technique to accelerate the convergence of the iterative method. For example if the eigenvalues of a matrix **A** are 5, 4, 3, 2, 1 then the iterative method applied to **A** will converge at a rate which depends upon the speed at which $(4/5)^r \to 0$. If we form $(A - 2.5I)$ then the appropriate eigenvalues are 2.5, 1.5, 0.5, -0.5 and -1.5 and the rate of convergence is determined by the speed at which $[(1.5)/(2.5)]^r \to 0$. Note that if we form $(A - 3.5I)$ so that the eigenvalues are now 1.5, 0.5, -0.5, -1.5 and -2.5 the rate of convergence depends upon $(0.5/1.5)^r$.

Problems

Section 2.2

1 Find the eigenvalues and eigenvectors of the following matrices.

(i) $\begin{bmatrix} 4 & 2 \\ -1 & 1 \end{bmatrix}$ (ii) $\begin{bmatrix} 1 & 0 & 0 \\ 1 & 2 & 0 \\ 2 & -2 & 3 \end{bmatrix}$ (iii) $\begin{bmatrix} 2 & 2 & 1 \\ 1 & 3 & 1 \\ 1 & 2 & 2 \end{bmatrix}$ (iv) $\begin{bmatrix} 1 & 0 & -1 \\ 1 & 2 & 1 \\ 2 & 2 & 3 \end{bmatrix}$

(v) $\begin{bmatrix} 2 & 2 & 0 \\ 2 & 2 & 0 \\ 0 & 0 & 1 \end{bmatrix}$ (vi) $\begin{bmatrix} 2 & 1 & 1 \\ 1 & 2 & 1 \\ 0 & 0 & 1 \end{bmatrix}$ (vii) $\begin{bmatrix} 1 & -1 & -1 \\ 1 & -1 & 0 \\ 1 & 0 & -1 \end{bmatrix}$ (viii) $\begin{bmatrix} -2 & -8 & -12 \\ 1 & 4 & 4 \\ 0 & 0 & 1 \end{bmatrix}$

(ix) $\begin{bmatrix} 1 & 1 & -2 \\ -1 & 2 & 1 \\ 0 & 1 & -1 \end{bmatrix}$ (x) $\begin{bmatrix} 3 & 2 & 2 & -4 \\ 2 & 3 & 2 & -1 \\ 1 & 1 & 2 & -1 \\ 2 & 2 & 2 & -1 \end{bmatrix}$

2   (i)   Indicate briefly some physical examples of eigenvalue problems.

     (ii)   Construct the matrix A whose eigenvalues are 0,1,3, with corresponding eigenvectors

$$\begin{bmatrix} 1 \\ 0 \\ 1 \end{bmatrix}, \begin{bmatrix} 0 \\ 3 \\ 0 \end{bmatrix}, \begin{bmatrix} 1 \\ 1 \\ 2 \end{bmatrix}.$$ Verify your method by determining the eigenvalues and eigenvectors of

the matrix A.   (EC)

3   Find the eigenvalues and eigenvectors of the system

$$2x_1 - 2x_2 + 3x_3 - \lambda x_1 = 0; \quad x_1 + x_2 + x_3 - \lambda x_2 = 0; \quad x_1 + 3x_2 - x_3 - \lambda x_3 = 0$$

Section 2.3

4   Find the eigenvalues and linearly independent eigenvectors of the following matrices.

(i) $\begin{bmatrix} 2 & 5 \\ 4 & 3 \end{bmatrix}$ (ii) $\begin{bmatrix} 3 & -2 \\ 2 & 3 \end{bmatrix}$ (iii) $\begin{bmatrix} 4 & 0 & 0 \\ 9 & -2 & 0 \\ 0 & 8 & 7 \end{bmatrix}$ (iv) $\begin{bmatrix} 7 & 4 & -4 \\ 4 & -8 & -1 \\ 4 & -1 & -8 \end{bmatrix}$

5   Find the eigenvalues and corresponding eigenvectors of $\begin{bmatrix} 2 & 2 & 0 \\ 2 & 2 & 0 \\ 0 & 0 & 1 \end{bmatrix}$. Verify that the eigen-

vectors are mutually orthogonal and comment on the reason for this.

6   If $C = \begin{bmatrix} 1 & 0 & 1 \\ 0 & 2 & 2 \\ 1 & 2 & 3 \end{bmatrix}$ and x is a column vector, show that the equation $Cx = \lambda x$ has solutions

in which $x \neq 0$, for each of three different values of λ. Find the corresponding solutions and, if these are denoted by x_1, x_2, x_3, show that $x_1^T x_2 = x_2^T x_3 = x_3^T x_1 = 0$.   (LU)

7   Find the characteristic roots of the matrix $\begin{bmatrix} 2 & 1 & 0 \\ 1 & 2 & 1 \\ 0 & 1 & 2 \end{bmatrix}$. Find also a characteristic vector for

each root and verify that these vectors are mutually perpendicular.

8   Find the eigenvalues of the matrix $\begin{bmatrix} 3 & 2 & 3 \\ -1 & 0 & -3 \\ 1 & -2 & 1 \end{bmatrix}$ and the eigenvector corresponding to the

smaller of the two positive eigenvalues.

9   Show that the latent vectors of a real symmetric matrix, corresponding to distinct latent roots, are orthogonal.

     The 3×3 symmetric matrix A has latent roots 3, 6, –9. The vector $(-2,2,1)$ corresponds to the root 3 and the vector $(1,2,-2)$ corresponds to the root –9. Find a latent vector corresponding to the root 6 and calculate A from the relation $P = MAM^{-1}$ where M is a matrix having the latent vectors as its columns and P is the diagonal matrix of latent roots. (LU)

10 Given the matrices $A_1 = \begin{bmatrix} 1 & -3 & 3 \\ 3 & -5 & 3 \\ 6 & -6 & 4 \end{bmatrix}$ and $A_2 = \begin{bmatrix} -3 & 1 & -1 \\ -7 & 5 & -1 \\ -6 & 6 & -2 \end{bmatrix}$ show that A_1 and A_2 have

the same characteristic polynomial but different eigenvectors. Show which of these matrices can be diagonalised by a transformation of the form $P^{-1} AP = D$. (EC)

11 The matrix $B = \begin{bmatrix} 1 & 0 & 0 \\ 0 & 1 & 0 \\ 1 & 1 & 2 \end{bmatrix}$ has characteristic equation $(\lambda - 1)^2(\lambda - 2) = 0$; find three

linearly independent eigenvectors for B.

12 The Cayley-Hamilton Theorem states that any square matrix satisfies its own characteristic

polynomial. Verify for $\begin{bmatrix} 2 & 3 & -5 \\ -3 & 1 & 2 \\ 1 & -3 & 4 \end{bmatrix}$.

13 Repeat Problem 12 for the matrices of Problem 4.

14 Use the Cayley-Hamilton Theorem to find A^3, A^4 and A^{-1} for the matrix A given by

(i) $\begin{bmatrix} 1 & 2 \\ 1 & 1 \end{bmatrix}$ (ii) $\begin{bmatrix} 1 & 1 & 2 \\ 3 & 1 & 1 \\ 2 & 3 & 1 \end{bmatrix}$ (iii) $\begin{bmatrix} 3 & 1 & -2 \\ 2 & 4 & -4 \\ 2 & 1 & -1 \end{bmatrix}$

15 A system of vibrating particles satisfies the differential equations:

$$\frac{d^2 x}{dt^2} = -5x - y - z; \quad \frac{d^2 y}{dt^2} = -x - 3y - z; \quad \frac{d^2 z}{dt^2} = -x - y - 3z$$

Using a trial solution of the form

$$x = x_0 \sin \omega t; \quad y = y_0 \sin \omega t; \quad z = z_0 \sin \omega t$$

show that ω^2 is an eigenvalue and (x_0, y_0, z_0) an eigenvector of a matrix A. Hence find three independent solutions to the equations.

16 Let the distinct eigenvalues of an $n \times n$ matrix A be $\lambda_1, \lambda_2, ..., \lambda_n$. If there exists a non-singular matrix P such that $C = P^{-1} AP$ is a matrix whose only non-zero elements are its diagonal ones which are the eigenvalues of A, then we may define matrices B_i whose only non-zero element is a 1, corresponding to λ_i in C. Further, let $E_i = PB_iP^{-1}$; then we may write $A = \lambda_1 E_1 + \lambda_2 E_2 + ... + \lambda_n E_n$; this is called the *spectral decomposition of* A. (The definition can be modified if the eigenvectors are not all distinct.) Obtain this decomposition for matrices (i), (iv) and (v) of Problem 1, Section 2.2.

Section 2.4

17 Write out in full the quadratic form in x_1, x_2, x_3 whose matrix is

(i) $\begin{bmatrix} 2 & -3 & 1 \\ -3 & 2 & 4 \\ 1 & 4 & -5 \end{bmatrix}$ (ii) $\begin{bmatrix} 3 & 0 & 2 \\ 0 & 4 & -5 \\ 2 & -5 & 6 \end{bmatrix}$ (iii) $\begin{bmatrix} 1 & -2 & 4 \\ -2 & 0 & 3 \\ 4 & 3 & 2 \end{bmatrix}$

18 Write the following quadratic forms in matrix notation:

(i) $2x_1^2 - 6x_1 x_2 + x_3^2$ (ii) $3x_1^2 - 8x_1 x_2 + 4x_2^2 + 5x_1 x_3$

(iii) $x_1^2 - 2x_2^2 - 3x_3^2 + 4x_1x_2 + 6x_1 x_3 - 8x_2 x_3$

19 Reduce the following quadratic forms to sums of squares

(i) $\begin{bmatrix} 1 & -1 & 0 \\ -1 & 2 & -1 \\ 0 & -1 & 2 \end{bmatrix}$ (ii) $\begin{bmatrix} 1 & 0 & -2 \\ 0 & 0 & 1 \\ -2 & 1 & 3 \end{bmatrix}$ (iii) $\begin{bmatrix} 4 & -4 & 2 \\ -4 & 3 & -3 \\ 2 & -3 & 1 \end{bmatrix}$

(iv) $\begin{bmatrix} 1 & -2 & -1 \\ 2 & 4 & 2 \\ -1 & 2 & 3 \end{bmatrix}$ (v) $\begin{bmatrix} 0 & 2 & 0 & 1 \\ 2 & 4 & 3 & 1 \\ 0 & 3 & 1 & 1 \\ 1 & 1 & 1 & 1 \end{bmatrix}$ (vi) $\begin{bmatrix} 1 & 2 & 3 & 1 \\ 2 & -4 & 6 & 2 \\ 3 & 6 & 9 & 3 \\ 1 & 2 & 3 & 1 \end{bmatrix}$

20 Given that one characteristic root of the matrix $A = \begin{bmatrix} 5 & -1 & -1 \\ 1 & 3 & 1 \\ -2 & 2 & 4 \end{bmatrix}$ is 2, find the other two, and characteristic vectors corresponding to each of the three roots. Hence find a matrix H such that $H^{-1} AH$ is diagonal, verifying that your choice is correct.

21 Find the eigenvalues and eigenvectors of the matrix $B = \begin{bmatrix} 4 & -3 \\ -3 & 4 \end{bmatrix}$. Find a matrix P such that $P^{-1} BP$ is diagonal and hence show that $u(x, y) \equiv 4x^2 - 6xy + 4y^2$ is positive definite. Show that $B^2 - 8B + 7I = 0$, where 0 is the appropriate zero matrix, and hence find B^{-1}.

22 Find the eigenvalues and three linearly independent eigenvectors of the matrix A, where

$A = \begin{bmatrix} -1 & 2 & 2 \\ 2 & 2 & -1 \\ 2 & -1 & 2 \end{bmatrix}$. Find the matrix H which diagonalises A. Is A positive definite, negative definite, or indefinite?

23 Verify that the quadratic form $5x_1^2 + 5x_2^2 + 2x_3^2 + 8x_1 x_2 + 4x_1 x_3 + 4x_2 x_3$ is positive definite.

24 Find the eigenvalues and eigenvectors for the matrix $A = \begin{bmatrix} 5 & 1 & 1 \\ 1 & 3 & 1 \\ 1 & 1 & 3 \end{bmatrix}$. Hence construct a matrix H such that $H^{-1} AH$ is a diagonal matrix and verify your result.

25 Find three linearly independent characteristic vectors of the matrix $A = \begin{bmatrix} -3 & 0 & 6 \\ 0 & 3 & 6 \\ 6 & 6 & 0 \end{bmatrix}$ and

hence determine a non-singular matrix **P** such that the matrix P^{-1} **AP** is diagonal, verifying the result.

26 Express the quadratic form $4x_1{}^2 + 3x_2{}^2 + 4x_3{}^2 - 4x_1 x_2 - 6x_1 x_3 + 2x_2 x_3$ in matrix form **X'AX** where **A** is a symmetric matrix and **X'** is the transpose of **X**. Test whether the given quadratic form is positive definite.

Section 2.5

27 Using the finite-difference approach, we can treat the buckling of a column as an eigenvalue-eigenvector problem.

$$\begin{bmatrix} 7 & -4 & 1 & 0 \\ -4 & 6 & -4 & 1 \\ 1 & -4 & 5 & -2 \\ 0 & 1 & -2 & 1 \end{bmatrix}\begin{bmatrix} w_b \\ w_c \\ w_d \\ w_e \end{bmatrix} = \begin{bmatrix} 2k & -k & 0 & 0 \\ -k & 2k & -k & 0 \\ 0 & -k & 2k & -k \\ 0 & 0 & -k & k \end{bmatrix}\begin{bmatrix} w_b \\ w_c \\ w_d \\ w_e \end{bmatrix}$$

Find the buckling load $P = 16kEI/l^2$, where the k's are the eigenvalues, and relative values of w_b, w_c, w_d and w_e are represented by the eigenvectors.

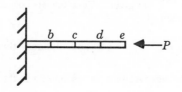   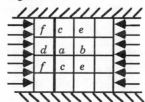

28 The buckling of a plate can be treated as an eigenvalue problem:

$$\begin{bmatrix} 20 & -8 & -16 & -8 & 4 & 4 \\ -8 & 19 & 4 & 1 & -16 & 0 \\ -16 & 4 & 44 & 4 & -16 & -16 \\ -8 & 1 & 4 & 19 & 0 & -16 \\ 4 & -16 & -16 & 0 & 42 & 2 \\ 4 & 0 & -16 & -16 & 2 & 42 \end{bmatrix}\begin{bmatrix} w_a \\ w_b \\ w_c \\ w_d \\ w_e \\ w_f \end{bmatrix} = k\begin{bmatrix} -2 & 1 & 0 & 1 & 0 & 0 \\ 1 & -2 & 0 & 0 & 0 & 0 \\ 0 & 0 & -4 & 0 & 2 & 2 \\ 1 & 0 & 0 & -2 & 0 & 0 \\ 0 & 0 & 2 & 0 & -2 & 0 \\ 0 & 0 & 2 & 0 & 0 & -2 \end{bmatrix}\begin{bmatrix} w_a \\ w_b \\ w_c \\ w_d \\ w_e \\ w_f \end{bmatrix}$$

Find the eigenvalues (in terms of k) and the eigenvectors.

29 Determine the frequency ω and the modes of the system for which the equation of motion is

$$\begin{bmatrix} y_1 \\ y_2 \\ y_3 \end{bmatrix} = \frac{\omega^2 \delta_{11}}{27}\begin{bmatrix} 27 & 14 & 4 \\ 14 & 8 & 2.5 \\ 4 & 2.5 & 1 \end{bmatrix}\begin{bmatrix} M_1 & 0 & 0 \\ 0 & M_2 & 0 \\ 0 & 0 & M_3 \end{bmatrix}\begin{bmatrix} y_1 \\ y_2 \\ y_3 \end{bmatrix}$$

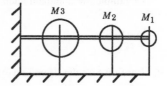

where $M_1 = M_2 = 2$, $M_3 = 3$ and δ_{11} is a constant.

Section 2.6

30 Find by the iterative method the largest eigenvalue and the associated eigenvector for the matrices of Problem 1, (i), (iv), (v), (vi), (viii), (x) in Section 2.2.

31 Given that the largest eigenvalue of a matrix A is real and simple (i.e. is not a multiple root) describe an iterative method of finding its value.

Write a program to carry out the determination of the largest eigenvalue of the following matrix, using the iterative method.
$$\begin{bmatrix} 1 & 1 & 1 \\ 1 & 2 & 3 \\ 1 & 3 & 6 \end{bmatrix}$$
(LU)

32 If $C = \begin{bmatrix} 5 & 2 & 0 \\ 2 & 10 & 2 \\ 0 & 2 & 10 \end{bmatrix}$ apply an iterative method to find the dominant eigenvalue and eigenvector (work in fractions). Use an initial approximation of $(1,2,3)^T$ for the eigenvector.

What factors determine the rate of convergence of this method? Explain briefly how you would find the two other eigenvalues of C.

33 The matrix $A = \begin{bmatrix} b & c & 0 & 0 \\ a & b & c & 0 \\ 0 & a & b & c \\ 0 & 0 & a & b \end{bmatrix}$ where $\begin{matrix} a = -0.539 \\ b = 1.173 \\ c = -0.634 \end{matrix}$ are given. Show that its characteristic equation is $(\lambda - b)^4 - 3ca(\lambda - b)^2 + c^2 a^2 = 0$ and deduce the roots

$\lambda = b \pm \sqrt{\frac{1}{2}ca(3 \pm \sqrt{5})}$. Determine also the eigenvector x corresponding to the largest eigenvalue and verify that the equation $Ax = \lambda x$ is satisfied.

34 If $B = \begin{bmatrix} 10.0 & 2.0 & 0.0 \\ 2.0 & 10.4 & 2.0 \\ 0.0 & 2.0 & 6.4 \end{bmatrix}$ find its dominant eigenvalue and eigenvector working to 2 d.p.

35 Show that if the latent roots of a matrix A are real and distinct the iteration process $z_{i+1} = Ay_i;\ y_{i+1} = z_{i+1}/k_{i+1}$ where k_{i+1} is the numerically largest element of z_{i+1} and y_0 is an arbitrary vector, can be used to obtain the dominant latent root and latent vector of A.

Determine the dominant latent root of the matrix $A = \begin{bmatrix} 9 & 10 & 8 \\ 10 & 5 & -1 \\ 8 & -1 & 3 \end{bmatrix}$ to the nearest whole number.

36 Verify the Cayley-Hamilton theorem for the matrix $A = \begin{bmatrix} -1 & 2 & 4 \\ 2 & 2 & -2 \\ 4 & -2 & -1 \end{bmatrix}$. Show also that in

this case A satisfies an equation of lower degree than its characteristic equation. Hence, or otherwise, determine the inverse of A.

Further, apply direct iteration $x^{(k+1)} = \dfrac{1}{\lambda_{k+1}} A x^{(k)}$, $x^{(0)} = (-1, 0, 1)$, to obtain

approximate values for the dominant eigenvalue (-6) and eigenvector $(-1, \frac{1}{2}, 1)$ of A. (EC)

37 Specify clearly as an algorithm *or* in the form of a brief flow diagram the iterative method

suggested by $x_{k+1} = \dfrac{1}{\lambda_{k+1}} A x_k$ for determining the dominant eigenvalue and eigenvector

of the system $Ax = \lambda x$.

In a three-dimensional stress problem we have the matrix $\begin{bmatrix} \sigma_x & \tau_{xy} & \tau_{zx} \\ \tau_{xy} & \sigma_y & \tau_{yz} \\ \tau_{zx} & \tau_{yz} & \sigma_z \end{bmatrix}$. Using the

above method, or otherwise, write a program in a specified code to determine the largest principal stress (eigenvalue) and its associated direction ratios (eigenvector). (EC)

Section 2.7

38 Find by inverse iteration the smallest eigenvalue and associated eigenvector of the matrices of Section 2.2, Problem 1 (iv), (x), Problem 3; Section 2.3, Problem 4 (iii).

39 If we have an approximate eigenvector x and corresponding eigenvalue $\lambda^{(n)}$ we may obtain an improved estimate of the eigenvalue by means of the *Rayleigh Quotient*

$\lambda^{(n+1)} = \dfrac{x^T A x}{x \cdot x}$. Given the approximate eigenvector $x = (0.731, 0.233, 1)^T$ for matrix

$A = \begin{bmatrix} 2 & 1 & 3 \\ 1 & -1 & 1 \\ 3 & 1 & 4 \end{bmatrix}$ obtain an improved estimate of λ.

40 Find all the eigenvalues and eigenvectors of the matrices:

(i) $\begin{bmatrix} 2 & -1 & 0 \\ -1 & 2 & -1 \\ 0 & -1 & 2 \end{bmatrix}$ (ii) $\begin{bmatrix} 1 & 1 & 1 & 1 \\ 1 & 2 & 3 & 4 \\ 1 & 3 & 6 & 10 \\ 1 & 4 & 10 & 20 \end{bmatrix}$

41 The moment of inertia matrix of a 3-dimensional rigid body is

$mr^2 \begin{bmatrix} 10.000 & 0.134 & -0.866 \\ 0.134 & 6.500 & -1.000 \\ -0.866 & -1.000 & 7.500 \end{bmatrix}$, the eigenvalues being the principal moments of inertia

and the eigenvectors the directions of principal axes.

Determine by iteration (from the characteristic equation) correct to 3 d.p. the numerically smallest eigenvalue and hence find its associated eigenvector. How can the remaining solutions be obtained?

42 Find by direct iteration correct to 2 d.p. the dominant eigenvalue and eigenvector of the matrix

$$A = \begin{bmatrix} 9 & 10 & 8 \\ 10 & 5 & -1 \\ 8 & -1 & 3 \end{bmatrix}, \text{ taking } x^{(0)} = \begin{bmatrix} 5 \\ 3 \\ 2 \end{bmatrix}$$

Then using the fact that if λ is an eigenvalue of A, $\lambda - a$ is an eigenvalue of $A - aI$; similarly obtain a second eigenvalue and associated eigenvector.

43 The equations of motion of a particular 3-mass system lead to the eigenvalue formulation

$$Ax = \lambda Bx, \text{ where } A = \begin{bmatrix} 2 & -1 & 0 \\ -1 & 2 & -1 \\ 0 & -1 & 2 \end{bmatrix} \text{ and } B = \begin{bmatrix} 1 & 0 & 0 \\ 0 & 2 & 0 \\ 0 & 0 & 3 \end{bmatrix}.$$

Form the characteristic equation and hence determine the eigenvalues and eigenvectors (fundamental frequencies and modes of vibration).

Expressing $B = G^2$ where G is a diagonal matrix, form the matrix $Q = G^{-1} A G^{-1}$ and verify that the eigenvalues of Q are exactly those obtained above while each eigenvector of Q is G times the corresponding eigenvector of $Ax = \lambda Bx$.

44 Find by direct iteration the dominant eigenvalue and eigenvector of the matrix

$$A = \begin{bmatrix} 9 & 10 & 8 \\ 10 & 5 & -1 \\ 8 & -1 & 3 \end{bmatrix} \text{ taking } x_0 = (1,1,0)^T \text{ as a starting vector and working to one decimal}$$

place with 3 iterations.

Using the fact that if λ is an eigenvalue of A, then $\lambda - p$ is an eigenvalue of $A - pI$ where I is the unit matrix, obtain a second eigenvalue and associated eigenvector of A, again working to one decimal place and starting with $(-1,1,1)^T$. (Only two iterations are required.)

45 The largest eigenvalue, λ_1, of the matrix $C = \begin{bmatrix} 5 & -4 & 1 \\ -4 & 6 & -4 \\ 1 & -4 & 7 \end{bmatrix}$ is sought.

Starting with $x^{(0)} = (1, 1, 1)$ verify that successive iterations yield:

$\lambda_1^{(1)} = 2$; $x^{(1)} = (1, -1, 2)$;

$\lambda_1^{(2)} = 11$; $x^{(2)} = (1, -1.63, 1.72)$;

$\lambda_1^{(3)} = 13.24$; $x^{(3)} = (1, -1.56, 1.48)$;

$\lambda_1^{(8)} = 12.265$; $x^{(8)} = (1, -1.485, 1.325)$.

Use the Rayleigh Quotient to produce a better estimate of λ_1 from $\lambda_1^{(1)}$.

3

OPTIMIZATION

3.1 Linear Programming – Graphical Solution

Many engineering problems are concerned with optimizing the performance of a system. In this section we look at a special class of such problems, namely those in which we optimize a **linear objective function** subject to **linear constraints**. These problems are named **linear programming** problems. Although the restriction of linearity may seem severe, there are many problems in which cost is to be minimised or profit maximised where this technique is applicable.

We shall first look at two linear programming problems which can be solved graphically, in order to study the main features of this class of optimization problems. In the next section we shall find an algebraic solution and develop the **simplex algorithm**. Finally, we shall mention some of the modifications to our basic algorithm which will enable a wider range of linear problems to be tackled. We stress that we are merely providing you with an *introduction* to linear programming.

Example 1

A company markets two mixes of concrete, X and Y. The company is capable of producing up to 10 truckloads per hour of X or up to 5 truckloads per hour of Y. The trucks available can carry up to 6 loads per hour of X and up to 10 loads per hour of Y, because of the different distances that have to be covered. However, no more than 7 loads per hour can be handled at the loading phase of the operation. The profit on a truckload of mix X is £4 and on a truckload of mix Y is £6.

Let the number of truckloads of mixes X and Y that are made per hour be x and y respectively. Clearly, the aim is to maximise the profit $4x + 6y$; however there are certain constraints on the values that x and y can take. First of all, the limit on production means that $\frac{x}{10} + \frac{y}{5} \le 1$; note that $\frac{x}{10}$ is the fraction of an hour occupied in producing mix X and $\frac{y}{5}$ is the fraction of an hour occupied in producing mix Y. Similarly, the limitation on delivery implies that $\frac{x}{6} + \frac{y}{10} \le 1$;

finally, the limitation on loading implies that $x + y \leq 7$. Perhaps 'finally' was a little hasty: for completeness x and y are to be non-negative, i.e. $x \geq 0$, $y \geq 0$. To summarise, we wish to maximise the linear **objective function**:

$$P = 4x + 6y \qquad (3.1)$$

subject to the constraints

$$x + 2y \leq 10; \quad 5x + 3y \leq 30; \quad x + y \leq 7 \qquad (3.2)$$

and $x \geq 0 \quad y \geq 0$ $\qquad (3.3)$

The combination x and y can be represented by a point (x,y) in the Cartesian plane. The constraints (3.3) restrict the possible points to the first quadrant and it is relatively straightforward to see that the **region of feasible solutions**, i.e. those points which also satisfy the constraints (3.2) is the shaded region in Figure 3.1(a). Each constraint specifies a set of points and the feasible region is the intersection of these sets.

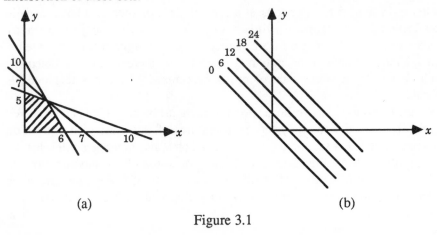

(a)   (b)

Figure 3.1

Figure 3.1(b) shows five lines of the form $4x + 6y = P$ where the appropriate value of P is attached to each line. Along each of these **iso-profit** lines any combination of x and y gives rise to the same profit, P. The further away from the origin we draw one of these lines, the greater the profit it represents.

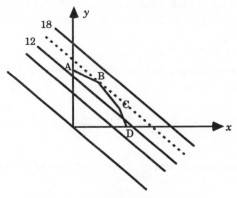

Figure 3.2

We require a point (or points) which lies in the feasible region and which lies on the furthest possible iso-profit line from the origin. It is easiest to see where this optimum point (or **optimal solution**) occurs by superimposing the basic features of Figure 3.1; this is done in Figure 3.2.

Consider the broken line which passes through vertex B of the feasible region. Any iso-profit lines further from the origin will have no contact with the feasible region, whereas one nearer the origin will have contact but will provide a lower profit. The point B represents the optimal solution. To find the corresponding values of x and y we may construct the features of Figure 3.2 on squared paper or solve the equations of sides AB and BC simultaneously. These equations are $x + 2y = 10$ and $x + y = 7$, respectively. It is straightforward to obtain the solution $x = 4$, $y = 3$ and therefore the company's policy should be to produce 4 truckloads of mix X and 3 truckloads of mix Y thereby providing an hourly profit of £34. Notice that the loading time is fully used and production is fully maintained; however, the delivery time is not totally used.

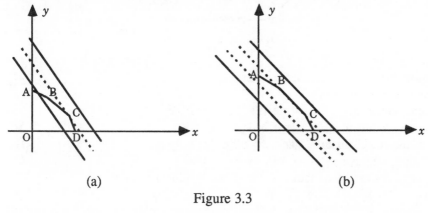

(a)   (b)

Figure 3.3

Having obtained the optimum solution, we may carry out a **sensitivity analysis** and determine the effect on the optimum of altering constraints or the coefficients in the profit expression. For example consider the effect of two further profit expressions. In Figure 3.3(a) we use the expression $P = 4x + 3y$ and in Figure 3.3(b) we use the expression $P = 4x + 4y$. In Figure 3.3(a) the optimum point is now C which has coordinates (4.5, 2.5). In Figure 3.3(b), the iso-profit lines are parallel to the side BC: any point on that side will give the same profit; in order to determine a unique optimum point, we must bring other considerations into account. We may well question the appearance of non-integer coordinates for point C: as long as the working day is an even number of hours, we shall be able to produce complete truckloads of each mix. You should examine our model, detailed in equations (3.1) to (3.3) from a practical standpoint in this and other respects. It is interesting to note before passing to a second example, that in all three cases considered the maximum profit point was at a vertex of the feasible region.

Example 2

Figure 3.4 shows a three-membered rigid frame; it is loaded as shown. We require to find the minimum frame weight; the plastic-moment capacities of the beam and the columns are called x and y respectively. There are three possible collapse mechanisms: beam mechanism, sway mechanism and combined mechanism. The mechanism constraints can be obtained using the principles of virtual work. We may state the problem algebraically as follows:

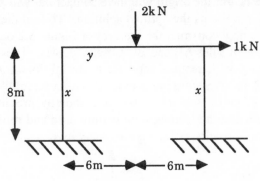

Figure 3.4

Minimise	$16x + 12y$	(3.4)

subject to the constraints

beam mechanism	$4y \geq 12$; $2x + 2y \geq 12$	(3.5a)
sway mechanism	$2x + 2y \geq 8$; $4x \geq 8$	(3.5b)
combined mechanism	$4x + 2y \geq 20$; $2x + 4y \geq 20$	(3.5c)
In addition, we have	$x \geq 0$, $y \geq 0$	(3.6)

Figure 3.5 shows the main features of the graphical solution.

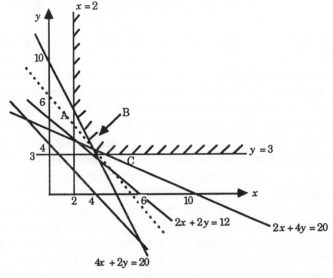

Figure 3.5

The broken line indicates that the optimum point is B (note again this occurs at a vertex); the shaded feasible region is not bounded. This time we require the **iso-cost** line nearest the origin. The point B has the coordinates (10/3, 10/3) and yields a minimum frame weight of 280/3.

Of course, we have chosen particularly simple examples where we can express the problem in terms of two variables x and y and hence obtain a graphical solution. In more complicated problems a different approach is needed. We shall head in the general direction in an intuitive fashion and then put the method on a reasonably rigorous footing.

3.2 The Simplex Algorithm

We shall examine Example 1 of the last section with a view to developing an algorithm capable of handling more than two variables. The clues are that the maximum points in the three cases considered occurred at vertices of the feasible region and that at each side of this polygonal region either one of the $\leq$ constraints is an equality or one of the original problem variables is zero: see Figure 3.6.

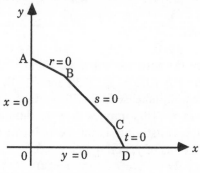

Figure 3.6

Consider the inequality $x + 2y \leq 10$ which is represented by AB. If we introduce a variable r which represents unused production time then the constraint can be written $x + 2y + r = 10$ where we impose the restriction $r \geq 0$. We can similarly introduce variables s and t, satisfying $x + y + s = 7$, $5x + 3y + t = 30$, $s \geq 0$, $t \geq 0$. The variables r, s and t are called **slack variables** and we can see from Figure 3.6 that the feasible region has as many sides as there are variables and that each side represents the restriction that one of the variables is zero. Hence, a vertex, where an optimum is to be found, represents the restriction that two of the variables are zero. It might seem that we have made the problem more complicated by increasing the number of variables from two to five, but we are now able to express the three constraints as equalities as long as we remember that we must impose the restriction that all variables are positive.

The second feature to note is that, since the optimum appears to be found at a vertex, all we need do is to evaluate the objective function at each vertex and see where it takes its least value. In practice it will be quicker to start at O and proceed round the vertices, either clockwise or anti-clockwise until the maximum is reached; we detect this by finding a smaller value of the objective function at the next vertex. (If two adjacent vertices provide the same maximum value, then all points on the side joining them will provide the same value.) If the maximum value occurs at B, it will be more efficient to proceed clockwise from O, whereas, if the maximum occurs at C, an anti-clockwise journey will be the shorter. We need, therefore, a means of checking which direction will be the quicker; this will be especially important in problems with several variables.

One final point: the feasible region can be defined as that region where all the five variables are non-negative simultaneously.

Development of the algorithm

Let us now summarise the equations we have obtained

$$4x + 6y \quad = P \tag{3.1}$$
$$x + 2y + r = 10 \tag{3.7a}$$
$$x + y + s = 7 \tag{3.7b}$$
$$5x + 3y + t = 30 \tag{3.7c}$$

We also have the restrictions

$$x \geq 0, \quad y \geq 0, \quad r \geq 0, \quad s \geq 0, \quad t \geq 0 \tag{3.8}$$

We start from O where the zero variables are x and y. The values of other variables at this vertex are $r = 10$, $s = 7$, $t = 30$ and $P = 0$. Increasing either x or y will increase P, but a unit increase in y will produce a more rapid increase in P than a unit increase in x. This implies that we should hold x at zero and increase y until one of the constraints is on the point of violation. First we express (3.7a) to (3.7c) in terms of y, noting that $x = 0$:

$$r = 10 - 2y \qquad s = 7 - y \qquad t = 30 - 3y$$

As y increases from zero, so r, s and t decrease. The danger signals are when they become zero. When $y = 5$, r becomes zero, when $y = 7$, s becomes zero and when $y = 10$, t becomes zero.

Note that, for example, when $y = 7$, r is negative. Therefore we must take the least of these values for y, viz 5, and our new vertex, A, is where $x = r = 0$. The next step is to write the profit P in terms of x and r. We need an equation connecting x, y and r, viz (3.7a), and we obtain $y = 5 - \frac{1}{2}x - \frac{1}{2}r$ which we substitute into (3.1) to get

$$30 + x - 3r = P \tag{3.9}$$

Now an increase in x *will* produce an increase in P, whereas an increase in r *will not*. There is no choice to be made this time. We express (3.7a) to (3.7c) in terms of x, noting that r is zero at the vertex. We have

$$y = 5 - \frac{1}{2}x \qquad s = 7 - y - x \qquad t = 30 - 3y - 5x$$

We have expressed y in terms of x from the first constraint and we must substitute this into the other constraints to obtain

$$y = 5 - \frac{1}{2}x \qquad s = 2 - \frac{1}{2}x \qquad t = 15 - \frac{7}{2}x$$

As x increases from zero we see that the first variable to become zero is s, when $x = 4$. The new vertex, B, is where $r = s = 0$. The profit P is now expressed in terms of r and s via (3.9) and (3.7a) and (3.7b). We have first, from these last two equations, $x + 2s - r = 4$ and then $30 + (4 + r - 2s) - 3r = P$ i.e. $34 - 2r - 2s = P$. This time, increasing either r or s *reduces* the value of P and hence we have arrived at the optimum point. Here $r = s = 0$ and from (3.7a) and (3.7b) we find that $x = 4$, $y = 3$, as before. We started from a vertex O where $x = y = 0$ and the **basic variables** in our solution were r, s, t. We then moved to vertex A where $x = r = 0$ and our basic variables were y, s and t; finally, we moved to vertex B where $r = s = 0$ and our basic variables were x, y and t. Try this approach for the two other profit lines considered.

In this simple example we were able to trace the progress of the method graphically; however, for problems involving more variables, this may not be possible. We need therefore to obtain the basic algebraic steps on a general level so that we do not need to appeal to geometry.

Matrix formulation

If we rewrite equation (3.1) as $P - 4x - 6y = 0$ then this form together with equations (3.7) may be written in matrix form as

$$
\begin{array}{ccccccc}
P & x & y & r & s & t & \\
\end{array}
$$
$$
\begin{array}{c}
\\ r \\ s \\ t
\end{array}
\begin{bmatrix}
1 & -4 & -6 & 0 & 0 & 0 & 0 \\
0 & 1 & 2 & 1 & 0 & 0 & 10 \\
0 & 1 & 1 & 0 & 1 & 0 & 7 \\
0 & 5 & 3 & 0 & 0 & 1 & 30
\end{bmatrix}
$$

where we have appended the variables to which the first six columns and last three rows relate. We may **partition** this matrix as follows

$$
\begin{bmatrix}
1 & \vdots & -4 & -6 & \vdots & 0 & 0 & 0 & \vdots & 0 \\
0 & \vdots & 1 & 2 & \vdots & 1 & 0 & 0 & \vdots & 10 \\
0 & \vdots & 1 & 1 & \vdots & 0 & 1 & 0 & \vdots & 7 \\
0 & \vdots & 5 & 3 & \vdots & 0 & 0 & 1 & \vdots & 30
\end{bmatrix}
$$

In practice, we tend to omit the first column since this gives us no useful information as regards the solution of the problem. The reduced matrix is known as the **Simplex Tableau**.

$$
\begin{bmatrix}
-4 & -6 & \vdots & 0 & 0 & 0 & \vdots & 0 \\
1 & 2 & \vdots & 1 & 0 & 0 & \vdots & 10 \\
1 & 1 & \vdots & 0 & 1 & 0 & \vdots & 7 \\
5 & 3 & \vdots & 0 & 0 & 1 & \vdots & 30
\end{bmatrix}
$$

The first step is to decide which of the non-basic variables to convert to a basic variable. Scanning the top row, we notice that both non-zero elements are negative and we choose the more negative of these. This concentrates our attention on the second column. To select the variable which will be replaced as a basic variable, we look down the second column and divide the number in that position in each of the rows into the element in the last column of that row. We select the least resulting number. In this example the second row gives $10/2 = 5$, the third gives $7/1 = 7$ and the fourth yields $30/3 = 10$. The least of these numbers came from the second row and we circle the element in row 2 column 2 (i.e. the rth row and and the yth column); this is to indicate that y is to take over from r as a basic variable. The circled element is the **pivot**.

$$
\begin{bmatrix}
-4 & -6 & 0 & 0 & 0 & \vdots & 0 \\
1 & ② & 1 & 0 & 0 & \vdots & 10 \\
1 & 1 & 0 & 1 & 0 & \vdots & 7 \\
5 & 3 & 0 & 0 & 1 & \vdots & 30
\end{bmatrix}
$$

We divide the row which contains the pivot by the value of the pivot.

$$
\begin{bmatrix}
-4 & -6 & 0 & 0 & 0 & \vdots & 0 \\
1/2 & 1 & 1/2 & 0 & 0 & \vdots & 5 \\
1 & 1 & 0 & 1 & 0 & \vdots & 7 \\
5 & 3 & 0 & 0 & 1 & \vdots & 30
\end{bmatrix}
\begin{matrix}
\text{Row 1 + 6 Row 2} \\
\\
\text{Row 3 − Row 2} \\
\text{Row 4 − 3 Row 2}
\end{matrix}
$$

Then we subtract multiples of this row from the other rows so that the entries in their y column become zero.

$$
\begin{bmatrix}
-1 & 0 & 3 & 0 & 0 & \vdots & 30 \\
1/2 & 1 & 1/2 & 0 & 0 & \vdots & 5 \\
1/2 & 0 & -1/2 & 1 & 0 & \vdots & 2 \\
7/2 & 0 & -3/2 & 0 & 1 & \vdots & 15
\end{bmatrix}
$$

From this tableau we find that, at the next vertex, the basic variables are y, s and t (1 in the appropriate column with zero entries) and the profit, P, is 30 (top right element). Furthermore, the values of y, s and t at the vertex are 5, 2, 15 respectively (last column) and, of course x and r are both zero. Now we repeat the procedure.

We scan the top row and find only one negative entry and we concentrate on the first column. As we move down, the division process gives $5/\frac{1}{2} = 10$ from

row 2, $2/\frac{1}{2} = 4$ from row 3 and $15/\frac{7}{2} = \frac{30}{7}$ from row 4. The least of these being

4, we select row 3 and circle the **pivot** $\frac{1}{2}$ in that row.

$$\begin{bmatrix} -1 & 0 & 3 & 0 & 0 & \vdots & 30 \\ 1/2 & 1 & 1/2 & 0 & 0 & \vdots & 5 \\ \textcircled{1/2} & 0 & -1/2 & 1 & 0 & \vdots & 2 \\ 7/2 & 0 & -3/2 & 0 & 1 & \vdots & 15 \end{bmatrix}$$

First we divide the third row by the value of the pivot.

$$\begin{bmatrix} -1 & 0 & 3 & 0 & 0 & \vdots & 30 \\ 1/2 & 1 & 1/2 & 0 & 0 & \vdots & 5 \\ 1 & 0 & -1 & 2 & 0 & \vdots & 4 \\ 7/2 & 0 & -3/2 & 0 & 1 & \vdots & 15 \end{bmatrix} \begin{array}{l} \text{Row } 1 + \text{Row } 3 \\ \text{Row } 2 - 1/2 \text{ Row } 3 \\ \\ \text{Row } 4 - 7/2 \text{ Row } 3 \end{array}$$

We subtract suitable multiples of row 3 from the other rows to reduce the entries in their first column to zero.

$$\begin{bmatrix} 0 & 0 & 2 & 2 & 0 & \vdots & 34 \\ 0 & 1 & 1 & -1 & 0 & \vdots & 3 \\ 1 & 0 & -1 & 2 & 0 & \vdots & 4 \\ 0 & 0 & 4/2 & -7 & 1 & \vdots & 1 \end{bmatrix}$$

From this tableau we see that the values of the basic variables at this new vertex are $y = 3$, $x = 4$, $t = 1$ and the value of P is 34. Scanning the top row, we see that both entries are positive and this is the clue that no further improvement in the solution can be made.

Example in three variables

So far we have considered only a maximisation problem in two variables, which we were able to accomplish graphically. We now consider an example in three variables.

A manufacturer of plastic products makes three kinds of article X, Y and Z, each of which requires three stages in its manufacture: moulding, painting and assembling. These operations for one article of X take 4, 2 and 3 hours respectively. The corresponding times for one article of Y are 3, 2 and 2 hours, respectively and for one article of Z are 5, 2 and 3 hours, respectively. There are available 400 hours of moulding time, 200 hours of painting time and 300 hours of assembling time. The profit per article of X is £4, per article of Y is £4 and per article of Z is £5. How many articles of each type should be manufactured so as to maximise the total profit?

We let the number of articles of each type manufactured be x_1, x_2 and x_3 respectively. Then we formulate the problem as follows.

Maximise $z = 4x_1 + 4x_2 + 5x_3$ subject to the constraints $4x_1 + 3x_2 + 5x_3 \le 400$, $2x_1 + 2x_2 + 2x_3 \le 200$, $3x_1 + 2x_2 + 3x_3 \le 300$ with $x_1, x_2, x_3 \ge 0$.

To employ the simplex algorithm we introduce slack variables r, s and t, all positive, so that

$$4x_1 + 3x_2 + 5x_3 + r = 400$$
$$2x_1 + 2x_2 + 2x_3 + s = 200$$
$$3x_1 + 2x_2 + 3x_3 + t = 300$$

The simplex tableau becomes as follows. We indicate the main tableaux in the development of the solution. Pivots are circled. The solution starts at vertex O.

$$
\begin{bmatrix}
-4 & -4 & -5 & 0 & 0 & 0 & : & 0 \\
4 & 3 & ⑤ & 1 & 0 & 0 & : & 400 \\
2 & 2 & 2 & 0 & 1 & 0 & : & 200 \\
3 & 2 & 3 & 0 & 0 & 1 & : & 300
\end{bmatrix}
$$

$$
\begin{bmatrix}
-4 & -4 & -5 & 0 & 0 & 0 & : & 0 \\
4/5 & 3/5 & 1 & 1/5 & 0 & 0 & : & 80 \\
2 & 2 & 2 & 0 & 1 & 0 & : & 200 \\
3 & 2 & 3 & 0 & 0 & 1 & : & 300
\end{bmatrix}
\begin{array}{l}
\\ \text{Row } 1 + 5 \text{ Row } 2 \\
\\ \text{Row } 3 - 2 \text{ Row } 2 \\
\text{Row } 4 - 3 \text{ Row } 2
\end{array}
$$

$$
\begin{bmatrix}
0 & -1 & 0 & 1 & 0 & 0 & : & 400 \\
4/5 & 3/5 & 1 & 1/5 & 0 & 0 & : & 80 \\
2/5 & ④/⑤ & 0 & -2/5 & 1 & 0 & : & 40 \\
3/5 & 1/5 & 0 & -3/5 & 0 & 1 & : & 60
\end{bmatrix}
$$

$$
\begin{bmatrix}
0 & -1 & 0 & 1 & 0 & 0 & : & 400 \\
4/5 & 3/5 & 1 & 1/5 & 0 & 0 & : & 80 \\
1/2 & 1 & 0 & -1/2 & 5/4 & 0 & : & 50 \\
3/5 & 1/5 & 0 & -3/5 & 0 & 1 & : & 60
\end{bmatrix}
\begin{array}{l}
\text{Row } 1 + \text{Row } 3 \\
\text{Row } 2 - 3/5 \text{ Row } 3 \\
\\ \text{Row } 4 - 1/5 \text{ Row } 3
\end{array}
$$

$$
\begin{bmatrix}
1/2 & 0 & 0 & 1/2 & 5/4 & 0 & : & 450 \\
1/2 & 0 & 1 & 1/2 & -3/4 & 0 & : & 50 \\
1/2 & 1 & 0 & -1/2 & 5/4 & 0 & : & 50 \\
1/2 & 0 & 0 & -1/2 & -1/4 & 1 & : & 50
\end{bmatrix}
$$

Let us now outline the steps taken. In the original profit formula, increasing x_3 gave the most rapid return and this governed the choice of the third column of the matrix. The pivot was found to be 5 and row 2 was divided by this value; other entries in column 3 were reduced to zero by row operations and the third tableau represents the situation at the new vertex, which is E. At this vertex the profit is £400 and the values of the basic variables x_3, s and t are 80, 40 and 60 respectively. There being only one negative entry in row 1 of this tableau, we are restricted to column 2 where we find the pivot as 4/5. We divide row 3 by this pivot and then reduce other entries in column 2 to zero by row operations. We are left with the fifth tableau which represents the state of affairs at the next vertex, D. At this vertex the profit is £450 and the values of the basic variables x_2, x_3 and t are 50, 50 and 50 respectively (x_1, r and s are zero). It is perhaps worth remarking that in proceeding from E to D we have maintained zero wastage r, reduced wastage s to zero and decreased wastage t.

In the final solution then, the manufacturer should produce 50 articles each of Y and Z, none of X and make a profit of £450. He will fully use the time available for moulding and painting but will have 50 hours of assembly time idle.

3.3 Non-linear Optimization

We first collect some basic results concerning functions of one variable. There is a **local minimum** of $f(x)$ at $x = a$ if $f(a + h) > f(a)$ for small values of h (small enough not to stray into regions where other features of $f(x)$ occur) and a similar definition applies to a **local maximum**. At a **point of inflection** the quantity $[f(a + h_1) - f(a)].[f(a - h_2) - f(a)] < 0$, where h_1 and h_2 are small positive numbers. A **stationary point** is one at which $f'(x) = 0$. Let one of these stationary points be a. If the first non-vanishing derivative at $x = a$ is of *even* order, then we have a local minimum if the value of this derivative is positive and a local maximum if the value is negative. However, if the first non-vanishing derivative at $x = a$ is of *odd* order, we have a point of inflection.

In optimization problems we are often concerned with **global maxima** or **global minima**. Sometimes we restrict the values of x to an interval, and sometimes we place no restriction on the values of x.

We have sketched in Figure 3.7 a function which, in the range $a \leq x \leq b$ exhibits a global maximum (b) and global minimum (x_1), local maxima (x_6 and a), a local minimum (x_4) and two points of inflection (x_2 and x_5). In addition, we have a **valley** (x_7) and a **ridge** (x_3): these are points where $f''(x) = 0$, but unlike a point of inflection the second derivative does not change sign through the point.

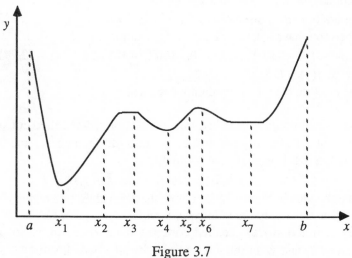

Figure 3.7

The function $f(x)$ has a global maximum in the interval $a \leq x \leq b$ at one of the following three places
(i)   where $f'(x) = 0$
(ii) at a boundary
(iii) where $f'(x)$ has a discontinuity.

Now consider Figure 3.8; it depicts the function $f(x) = (x^3 - 2x^2)^{1/5}$ in the range $-1 \leq x \leq 5$.

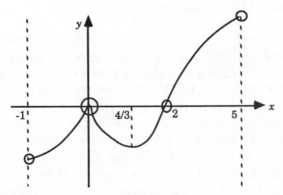

Figure 3.8

We can isolate the global minimum (*least value*) of $f(x)$ by examining the value of the function at each of the three classes of place listed above.

(i) $f'(x) = \dfrac{1}{5}\dfrac{(3x^2 - 4x)}{(x^3 - 2x^2)^{4/5}} = \dfrac{1}{5}\dfrac{x(3x - 4)}{x^{8/5}(x - 2)^{4/5}} = \dfrac{1}{5}\dfrac{(3x - 4)}{x^{3/5}(x - 2)^{4/5}}$

Therefore $f'(x) = 0$ where $x = 4/3$

(ii) The boundaries are $x = -1$ and $x = 5$

(iii) $f'(x)$ possibly has a discontinuity at $x^3 - 2x^2 = 0$ i.e. $x = 0$ or $x = 2$.
We therefore have a set of values of x to examine, viz $\{-1, 0, 4/3, 2, 5\}$
$f(-1) = (-3)^{1/5} = -(3)^{1/5}$, $f(0) = 0$, $f(4/3) = (-32/27)^{1/5} = -(32/27)^{1/5}$,
$f(2) = 0$, $f(5) = (75)^{1/5}$
In this instance the global minimum occurs at $x = -1$.

A further pair of definitions is helpful. A function $f(x)$ is called **convex** over some interval of x if, for any two values x_1 and x_2 in the interval and for all λ which satisfy $0 \leq \lambda \leq 1$, we have the inequality

$$f[\lambda x_2 + [1 - \lambda]x_1] \leq \lambda f(x_2) + (1 - \lambda)f(x_1) \qquad (3.10)$$

We illustrate the idea in Figure 3.9(a). Note that the values of a convex function are always overestimated by *linear interpolation*.

Similarly, a function $f(x)$ is called **concave** over some interval of x if, for any two values x_1 and x_2 in the interval and for all λ which satisfy $0 \leq \lambda \leq 1$, we have the inequality

$$f[\lambda x_2 + [1 - \lambda]x_1] \geq \lambda f(x_2) + (1 - \lambda)f(x_1) \qquad (3.11)$$

This concept is depicted in Figure 3.9(b), and we see that linear interpolation provides an underestimate.

Note that, in both cases, equality holds at x_1 and x_2.

One word of caution. We have blithely stated that we must find the values of x for which $f'(x) = 0$; however, it is not always that easy. Consider the example $f(x) = 5x^6 - 4x^4 + 3x^2 - x + 7$. The derivative vanishes where

$f'(x) = 30x^5 - 16x^3 + 6x - 1 = 0$. To solve this equation requires techniques such as that of Newton-Raphson.

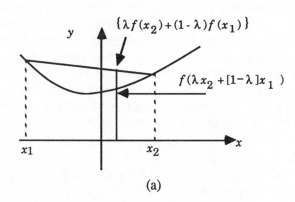

(a)

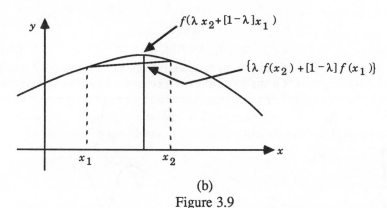

(b)

Figure 3.9

If we know that a function is convex in a closed interval we can straight away use the result that a local minimum of the function in that interval is also the global minimum there. Similarly if a function is concave in a closed interval, a local maximum of the function in that interval is also the global maximum there. We complement these results with the following: (i) the global maximum of a convex function $f(x)$ in a closed interval is taken at one or both boundaries of the interval; (ii) the global minimum of a concave function $f(x)$ in a closed interval is taken at one or both boundaries of the interval.

Example

Find the least value of $f(x) = x^2 + 3x + 2$ in the interval $[-2,4]$.

It should first be shown that the function is convex. (Verify this for yourself.)

Since $f(x)$ is a convex function, it is convex in the closed interval $[-2,4]$. Now $f'(x) = 2x + 3$ and hence $f'(x) = 0$ when $x = -(3/2)$. This local minimum is therefore the global minimum and the value of the function at the

point in question is $\dfrac{9}{4} - \dfrac{9}{2} + 2$ i.e. $-\dfrac{1}{4}$.

3.4 Search Techniques in One Variable

Sometimes a mathematical relationship between variables cannot be found and in order to obtain values of $f(x)$ we have to carry out an experiment or a series of calculations. This means that we want to keep the number of evaluations to a minimum. The strategy will be to start from a *base point* and select where next to make an evaluation of $f(x)$; depending on whether we get a value of $f(x)$ nearer to the optimum (e.g. a lower value if we seek a minimum) we make a decision as to where next to choose x for a further evaluation. The values of x that we select will form a sequence $\{x_i\}$. The problem then arises as to how we detect convergence; this can be a complex problem and there is a danger in using too simple a criterion for convergence.

In addition to the sequence $\{x_i\}$, we have a companion sequence $\{f(x_i)\}$. Therefore, we have two possible criteria: for seeking a least value of $f(x)$ these are

$$|f(x_i) - f(x_{i+1})| < \varepsilon \qquad (3.12a)$$

and
$$|x_i - x_{i+1}| < \varepsilon^* \qquad (3.12b)$$

where ε and ε^* are prescribed positive values. Consider Figure 3.10; the first diagram 3.10(a) represents a shallow valley where large changes in x give rise to only small changes in $f(x)$ and the criterion (3.12b) would be the more useful. Figure 3.10(b) shows a steep-sided valley and small changes in x now cause large changes in $f(x)$; here the criterion (3.12a) would be the better choice.

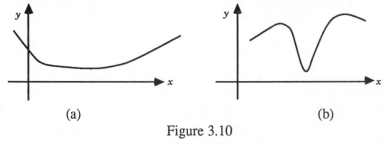

(a)   (b)

Figure 3.10

A possible ploy would be to require both criteria to hold for each of a given number of iterations. Whilst not completely safe, it should nevertheless be reasonably safe; indeed it may, with some problems, be too safe.

Search techniques fall into two main categories: those which specify an interval in which the optimum value lies and which seek to reduce successively the interval and those which specify the position of the optimum point by a point which approximates to it. We shall consider an example of each class; the **Fibonacci Search** for the former and **Powell's algorithm** for the latter.

We first assume that the function in question is **unimodal**, i.e. there is only one stationary value in the interval of interest. We further assume that the stationary value x^* is a minimum.

[Note: if we wish to find the maximum of a function we can consider the equivalent problem of minimising $-f(x)$.]

Search strategy

If a function $f(x)$ is unimodal in an interval (a,b) then it is necessary to evaluate the function at two internal points before the location of the stationary value can be confined to a subinterval of (a,b). Try and prove this result. Suppose we let the current interval be (a,b) then we evaluate the function at the internal points x_1 and x_2 where $a < x_1 < x_2 < b$.

Consider Figure 3.11(a). We have evaluated the function $f(x)$ at two internal points x_1 and x_2. Note that $b - x_2 = x_1 - a$. The result $f(x_1) > f(x_2)$ leads us to reject the interval $a \leq x \leq x_1$.

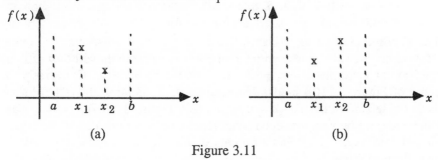

(a) (b)

Figure 3.11

Similarly, if we obtain a situation as depicted in Figure 3.11(b), we shall reject the interval $x_2 < x \leq b$. This suggests the following rule:

$$\left. \begin{array}{ll} \text{If } x_1 < x_2 \quad \text{then} & f(x_1) > f(x_2) \Rightarrow x^* > x_1 \\ \qquad \text{and} & f(x_1) < f(x_2) \Rightarrow x^* < x_2 \end{array} \right\} \tag{3.13}$$

What do you deduce if $f(x_1) = f(x_2)$?

If we reject the subinterval (a, x_1) we relabel x_1 as a; if we reject the subinterval (x_2, b) we relabel x_2 as b. In either event we have a new interval (a,b) and it would be helpful if we were to keep the internal point $(x_2$ or $x_1)$ so that we needed just one further function evaluation at the next stage.

Fibonacci search

We seek an algorithm which gives the largest ratio of initial to final interval length for a fixed number of function evaluations. For simplicity take the final interval as being of length 1 and find L_n, the length of the largest interval which can be reduced to 1 after n function evaluations.

First find an upper bound for L_n, the length of the interval (a,b). As before we evaluate the function at the internal points x_1 and x_2. If the minimum lies in (a, x_1) then we have only $(n - 2)$ evaluations left to refine our interval of uncertainty and hence $x_1 - a \leq L_{n-2}$. On the other hand, if the minimum lies in (x_1, b) we already have one point, x_2, in this subinterval and therefore the interval (x_1, b) can be refined by $(n - 1)$ evaluations provided we use x_2 in our strategy; hence $b - x_1 \leq L_{n-1}$. Adding these inequalities, we obtain $(b - a) \equiv L_n \leq L_{n-2} + L_{n-1}$. Further, for consistency we must define L_0

and L_1 to be equal to 1, since we need *two* function evaluations to reduce the interval of uncertainty.

Suppose we could find a sequence of numbers F_n such that

$$F_n = F_{n-1} + F_{n-2} \qquad (3.14)$$

with
$$F_1 = F_2 = 1$$

Then we would have the ideal optimum for the interval reduction (L_n would have achieved its upper bound).

The sequence of numbers governed by (3.14) are the **Fibonacci numbers**; the first few are

$$1, 1, 2, 3, 5, 8, 13, 21, 34, 55, ...$$

In Problem 14(a) we ask you to derive a formula for F_n.

The **Fibonacci strategy** is as follows. If the case illustrated in Figure 3.11(a) occurs then we choose an internal point x_3 as far to the left of b as x_2 is to the right of x_1 so that $b - x_3 = x_2 - x_1$; see Figure 3.12(a). Conversely, if the case of Figure 3.11(b) holds, then we choose x_3 as far to the right of a as x_1 is to the left of x_2 i.e. $x_3 - a = x_2 - x_1$; see Figure 3.12(b). Notice that, in either case, we only require one further function evaluation at each stage after the initial one, since one of the four points at any stage remains an intermediate point at the next stage.

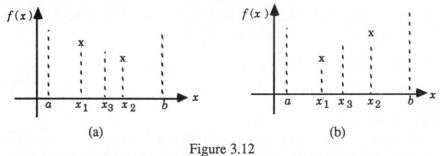

                (a)   (b)

Figure 3.12

It remains to show how this strategy fits in with the requirement of fewest function evaluations. Since the initial interval is an integer multiple of the final interval we may consider the initial interval (a,b) divided by r internal points equally spaced.

Let the points x_1 and x_2 be two of these. We have to decide where these points will be.

Once this decision is made then the above strategy will select the remaining points. The set $\{x_1, x_2, x_3, ..., x_s\}$ will be a subset of the internal points.

In Figure 3.13 we depict successive stages in the reduction of the interval of uncertainty. An interval of length 13 ($= F_7$) is divided into two intervals of length 5 and one of length 3 (Figure 3.13(a)). Note that $x_1 - a = 5 = b - x_2$ and that $x_1 - a = 5 = F_5$ whilst $b - x_1 = 8 = F_6$. Suppose we consistently discard the right-hand sub-interval†. Then a point is chosen $F_4 = 3$ units from the left-hand end of the remaining interval to be placed symmetrically with the

---

† The results derived hold quite generally.

remaining internal point; Figure 3.13(b).

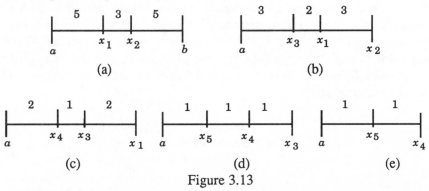

Figure 3.13

At the stage of diagram (e) we need one further function evaluation to locate the minimum. We place this, not at the same place as x_5 as theory would suggest, but slightly to the left as shown in Example 2.

Example 1

Suppose $f(x)$ has a minimum in the interval $(a,b) = (0,10)$. Further suppose that we wish to determine an interval containing the minimum which is of length $d = 0.2$ at most, i.e. which is $\leq 2\%$ of the given interval. The pertinent question is how many stages are necessary to reduce the length of the interval of uncertainty from 10 to ≤ 0.2?

It would be fortuitous if there *was* a Fibonacci number F_N such that $F_N d$ was *exactly* equal to $(b - a)$. We want the first Fibonacci number F_N for which $F_N d > (b - a)$ *because*, working back to the end product, we would then have an interval of uncertainty $< d$. In our current example, we require the first Fibonacci number > 50; this is $F_8 = 55$.

Example 2

We programmed a computer to find the minimum of $f(x) = x^2 - 7x + 4$ given that $A = 0, B = 10$ and the percentage accuracy, $P = 5\%$. We printed out the value of N, such that $F_N > 100/P$ for the first time. The values of A, L, R, B are printed above the corresponding values of $f(A), f(L), f(R)$ and $f(B)$ for each of the $(N - 2)$ stages. We present the results in Table 3.1 and work through the first 2 stages. (Via calculus we find that the minimum value of $f(x)$ occurs at $x = 3.5$.)

At stage 1 we notice that $f(L) < f(R)$; therefore we reject the interval $[R, B]$. We then relabel B at 6.19048, R at 3.80952 and calculate L as 2.38095; we note that $6.19048 - 3.80952 = 2.38096$, but this discrepancy occurs primarily because of round-off in printing the numbers. At stage 2 we see that $f(R) < f(L)$ and we discard the interval $[A, L]$, relabel A at 2.38095, L at 3.80952 and calculate R as 4.7619 which is as far from 6.19048 as 3.80952 is from 2.38095. At stage 6, both numbers L and R should be 3.33333, which is equidistant from 2.85714 and 3.80952; in fact we took the other internal point as

$3.80952 - 0.495 \times (3.80952 - 2.85714)$. Notice that *all* the other values of L and R are original internal points (as closely as round-off allows). At stage 6 we conclude that the interval of uncertainty is $(3.33333, 3.80952)$, which does indeed include the true value 3.5. Further the length of this interval is 0.47619; although this is slightly larger than half the previous interval because of the displaced value L, it is less than the required length 0.5, since we used $F_7 = 21$ instead of 19 internal points.

Table 3.1

The Value of N is 7

Stage	A	L	R	B	
1	0	3.80952	6.19048	10	X-values
	4	−8.1542	−1.01134	34	Function
2	0	2.38095	3.80952	6.19048	X-values
	4	−6.99773	−8.1542	−1.01134	Function
3	2.38095	3.80952	4.7619	6.19048	X-values
	−6.99773	−8.1542	−6.6576	−1.01134	Function
4	2.38095	3.33333	3.80952	4.7619	X-values
	−6.99773	−8.22222	−8.1542	−6.6576	Function
5	2.38095	2.85714	3.33333	3.80952	X-values
	−6.99773	−7.83673	−8.22222	−8.1542	Function
6	2.85714	3.33333	3.35714	3.80952	X-values
	−7.83673	-8.22222	−8.22959	−8.1542	Function

Powell's algorithm

The function is evaluated at an initial point x_1 and at $x_2 = x_1 + d$; let the corresponding function values be f_1 and f_2. We choose $x_3 = x_2 + 2d$ if $f_1 > f_2$ and $x_3 = x_1 - d$ if $f_1 < f_2$. The optimum of the quadratic fitted through the three points is given by

$$x_m = \frac{1}{2} \frac{(x_2^2 - x_3^2)f_1 + (x_3^2 - x_1^2)f_2 + (x_1^2 - x_2^2)f_3}{(x_2 - x_3)f_1 + (x_3 - x_1)f_2 + (x_1 - x_2)f_3}$$

If the smallest of the values $|x_1 - x_m|$, $|x_2 - x_m|$, $|x_3 - x_m|$ is less than the required distance, we have approximated the optimum by x_m. If this is not so then we evaluate the function at x_m and discard that point of $\{x_1, x_2, x_3\}$ which corresponds to the largest function value. The cycle is repeated until the desired accuracy is achieved. The method can lead to a point far distant from the minimum and we shall then return to the minimum only slowly. (See Problem 15.)

3.5 Functions of Several Variables: Direct Search Methods

As soon as we move into more than one dimension, the problem of finding greatest or least values of a function becomes much more complicated. We shall try and give you an insight into the problem by adopting a parallel approach: we shall give examples of two variables, thus allowing geometrical interpretation, side by side with formulae relating to the general case of n dimensions. In the general case, the point $(x_1, x_2, x_3, ...x_n)$ is conveniently referenced as $\mathbf{x}$ and we shall employ vector notation extensively. It will be helpful to plot *contours* of a function of two variables, a contour being a curve connecting points where the function values are equal. In Figure 3.14 we depict a set of contours which illustrate some of the main features to be encountered.

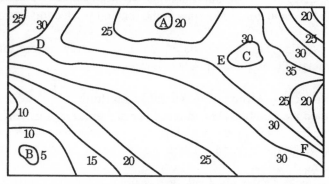

Figure 3.14

Points A and B represent local minima, point C represents a local maximum and points D, E and F represent possible **saddle points**. Can you detect any other features?

Suppose we consider the problem of cutting down uncertainty. In Figure 3.15(a) we see an **interval of uncertainty** of 10% of the original interval. If we now consider a function of two variables, each of which is restricted to a 10% interval of uncertainty, we see that the **area of uncertainty** in evaluating $f(x,y)$ is 1%. However, the situation is really the reverse: we specify the area of uncertainty as a percentage of the original area and then determine the uncertainty in each variable. Since we aim for 10% uncertainty in this discussion we are led to a percentage uncertainty of $(100/\sqrt{10}) \cong 31.62\%$. This is depicted in Figure 3.15(b); here we see how much less precisely the values of x_1 and x_2 are located. If we consider a function of three variables, the same 10% 'volume' of accuracy would mean a percentage uncertainty in each variable of 46.5%. If we take the case of 50 variables, the percentage uncertainty in each variable is 91%. We therefore are forced to seek regions of uncertainty which are very small fractions of the original region in which the problem was defined.

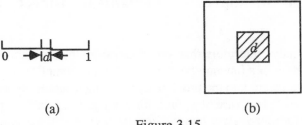

(a)   (b)

Figure 3.15

Methods of simultaneous search

We could set up a uniform grid over the problem region and evaluate at each node. We could, but we would be ill-advised to do so; with 10 variables and each variable divided into 10 sub-intervals we would need 10^{10} evaluations. Alternatively, we could generate points in the region by selecting sets of random numbers.

However, we shall study methods of sequential search, where information already accumulated is used to help select new points.

Sequential search – Alternating variable method

In this method the variables are examined one at a time and a *univariate search* made along the line of the chosen variable. Any univariate search technique may be employed. When the univariate search produces a minimum (to the prescribed accuracy) the next variable in order is chosen and a search performed for it. If there are only two variables involved, then searches take place in each direction alternately; if there are three variables, the sequence of direction is $\{x_1, x_2, x_3, x_1, x_2, x_3, x_1, ...\}$. In vector notation, we start by labelling the unit vectors parallel to the axes $\mathbf{e}_i$, $i = 1, ..., n$ and denote d_i, $i = 1, ..., n$ as scaling factors. The starting point is labelled $\mathbf{x}_0$; searches are made in the directions $\mathbf{e}_1, \mathbf{e}_2, ..., \mathbf{e}_n$ in turn. At each stage we move to a new point given by $\mathbf{x} = \mathbf{x}_s + d_i \mathbf{e}_i$, where $\mathbf{x}_s$ is the starting point.

The method is very simple but, naturally, has its drawbacks. It is at its best when the function under consideration is the analogue of a sphere in n-dimensions, called a **hypersphere**.

Consider Figure 3.16(a). Here we show contours of $f(x,y) = x^2 + y^2$.

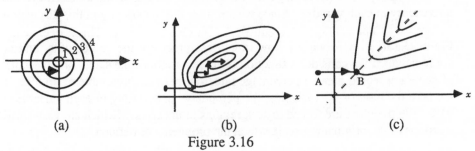

(a)   (b)   (c)

Figure 3.16

We start our search from $x_0 = (-3.5, -1.6)$ and move along the line of e_1, keeping y fixed. A search procedure will reveal the minimum in this direction to be at $x = (0, -1.6)$. Keeping x fixed, we move along the line of e_2 and locate the minimum at $(0,0)$. We have achieved our aim in a set of two searches. (We could have been lucky and started on either axis, or very lucky and started on the origin.) There is a general result that in n dimensions, the hypersphere will require at *most n* searches to locate the optimum point. Notice that if we use a practical search technique, we shall not locate minimum values exactly and we may need an extra set of searches. Figure 3.16(b) depicts a function where the variables x and y are strongly inter-related; here, the minimum point requires many searches. There are cases where the method fails: in Figure 3.16(c) we have a sharp ridge which rises to the north-east. Once at the point B we shall remain there.

Produce a flow chart for this method.

Rosenbrock's Method

This method modifies the alternating variable method. A step of a given length is taken in each of the coordinate directions in turn. If a success is encountered in a direction insofar as a lower function value is achieved, then the step length in that direction is increased for the next search in that direction. However, if a failure is encountered that step is decreased and the next search in that direction takes place in the opposite sense. When a complete cycle is accomplished, a new cycle takes place with the revised step lengths. The next direction is then searched. When in each direction, a failure has been recorded after a success, the time has come to realign the axes to take account of the local geometry of the function surface.

The idea behind the realignment is to retain the direction of most progress and select directions orthogonal to it. Suppose that we have an objective function of three variables and that from the previous base point we have advanced d_1, d_2 and d_3 in the directions x_1, x_2, x_3. We choose $q_1 = d_1 x_1 + d_2 x_2 + d_3 x_3$, $q_2 = d_1 x_1 + d_2 x_2$, $q_3 = d_1 x_1$ and we must then select a set of three orthogonal directions which we do by means of the **Gram-Schmidt ortho-gonalisation process**. (Refer back to Section 1.2 for details.) We now advance from the latest point along these directions. This method will also handle constrained functions. In Figure 3.17 we show the algorithm schematically in two dimensions. Try and produce a flow chart for the method.

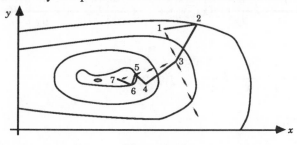

Figure 3.17

3.6 Calculus Approach to Functions of Several Variables

The behaviour of functions of one variable near stationary points can be determined by considering the Taylor's series for the function, centred at the stationary point currently under study. We now examine the analogous result for a function of two variables.

The **tangent plane approximation** is

$$f(x_0 + h, y_0 + k) \cong f(x_0, y_0) + hf_x(x_0, y_0) + kf_y(x_0, y_0) \qquad (3.15)$$

A stationary point occurs at (x_0, y_0) when the tangent plane is horizontal there, i.e.
$$f_x(x_0, y_0) = 0 = f_y(x_0, y_0) \qquad (3.16)$$

We need a more refined approximation than (3.15); this ensured that the plane given by the right-hand side passed through the point $(x_0, y_0, f(x_0, y_0))$ with the correct slopes in the x and y directions. In order to get the curvature correctly matched we must take a *quadratic approximation* whose second derivatives agree with those of $f(x,y)$ at the point (x_0, y_0). We take the general quadratic function in two variables:

$$q(x,y) = \alpha + \beta_1 x + \beta_2 y + \gamma_1 x^2 + \gamma_2 xy + \gamma_3 y^2$$

Now we obtain the coefficients by matching the 1st and 2nd derivatives at (x_0, y_0).

We also require that $q(x_0, y_0) = f(x_0, y_0)$.

It can be shown that we have, on setting $x = x_0 + h, y = y_0 + k$, the approximation

$$f(x_0 + h, y_0 + k) \cong f(x_0, y_0) + hf_x(x_0, y_0) + kf_y(x_0, y_0)$$

$$+ \tfrac{1}{2}[h^2 f_{xx}(x_0, y_0) + 2hk f_{xy}(x_0, y_0) + k^2 f_{yy}(x_0, y_0)] \qquad (3.17)$$

As we were able to extend the quadratic approximation for a function of one variable, so we can extend equation (3.17). Taylor's Theorem for a function of two variables states that if the function is continuous in some closed domain D of the x-y plane and possesses partial derivatives up to order $(n + 1)$ in D, then

$$f(x_0 + h, y_0 + k) = f(x_0, y_0) + hf_x(x_0, y_0) + kf_y(x_0, y_0)$$

$$+ \frac{1}{2!}[h^2 f_{xx}(x_0, y_0) + 2hk f_{xy}(x_0, y_0) + k^2 f_{yy}(x_0, y_0)]$$

$$+ \frac{1}{3!}\left[h\frac{\partial}{\partial x} + k\frac{\partial}{\partial y}\right]_0^3 f(x_0, y_0) + \ldots + \frac{1}{n!}\left[h\frac{\partial}{\partial x} + k\frac{\partial}{\partial y}\right]_0^n f(x,y)$$

$$+ \frac{1}{(n+1)!}\left[h\frac{\partial}{\partial x} + k\frac{\partial}{\partial y}\right]_0^{n+1} f(\xi, \eta) \qquad (3.18)$$

where for example,

$$\left[h\frac{\partial}{\partial x}+k\frac{\partial}{\partial y}\right]_0^3 f(x,y) \equiv h^3\frac{\partial^3 f(x,y)}{\partial x^3}+3h^2k\frac{\partial^3 f(x,y)}{(\partial x^2)\partial y}$$

$$+3hk^2\frac{\partial^3 f(x,y)}{\partial x\,\partial y^2}+k^3\frac{\partial^3 f(x,y)}{\partial y^3}$$

evaluated at (x_0, y_0) and where (ξ, η) is a point in the interior of D.

(It should be clear how to extend this result for a function of three or more variables.) If we add the condition that

$$\lim_{n\to\infty}\left[\frac{1}{(n+1)!}\left[h\frac{\partial}{\partial x}+k\frac{\partial}{\partial y}\right]_0^{n+1} f(\xi,\eta)\right]=0 \text{ for all } (x,y) \text{ in } D$$

then we can obtain the infinite series expansion. Note that it is convenient to write $f(x_1, x_2, x_3, ..., x_n)$ as $f(\mathbf{x})$ and therefore we may generalise the quadratic approximation to

$$f(\mathbf{x}) \cong f(\mathbf{x}_0) + (\mathbf{x} - \mathbf{x}_0)\cdot\nabla_0 f + \tfrac{1}{2}(\mathbf{x} - \mathbf{x}_0)^\mathrm{T} H_0(\mathbf{x} - \mathbf{x}_0) \qquad (3.19)$$

We shall quote the meanings of the new symbols for two variables

$\mathbf{x}_0 \equiv (x_0, y_0)$ and $\mathbf{x} \equiv (x,y) = (x_0 + h, y_0 + k)$

$$\nabla f \equiv \left[\frac{\partial f}{\partial x}, \frac{\partial f}{\partial y}\right] \text{ and is called the \textbf{gradient vector} of } f, \text{ pronounced 'grad } f\text{'}$$

$$H \equiv \begin{bmatrix} \dfrac{\partial^2 f}{\partial x^2} & \dfrac{\partial^2 f}{\partial x\,\partial y} \\[2mm] \dfrac{\partial^2 f}{\partial y\,\partial x} & \dfrac{\partial^2 f}{\partial y^2} \end{bmatrix}$$ is called the **Hessian matrix** of f (it is the matrix of second partial derivatives and is symmetrical);

the suffix $_0$ means that the derivatives are evaluated at (x_0, y_0).

Maximum and Minimum values

We quote a theorem due to Weierstrass.

If a function $f(\mathbf{x})$ is continuous in a closed domain D then it has a greatest and a least value either in D or on the boundary of D.

We know that a necessary condition for $f(x,y)$ to have a local maximum or a local minimum at (x_0, y_0) is that the tangent plane should be horizontal there, i.e. $\nabla f(x_0, y_0) = \mathbf{0}$. Therefore we may re-write (3.19) as

$$\delta f \equiv f(\mathbf{x}) - f(\mathbf{x}_0) = \tfrac{1}{2}(\mathbf{x} - \mathbf{x}_0)^T H_0(\mathbf{x} - \mathbf{x}_0)$$

We shall develop the appropriate conditions for two variables and quote the generalisation. The resulting quadratic form $(h, k)^T H_0 (h, k)$ can be symbolised as $Ah^2 + 2Bhk + Ck^2$ where A, B and C are constants. If the point (x_0, y_0) is a local maximum then δf will be negative for all values of h and k [small] i.e. the quadratic form is **negative definite**. Correspondingly, if the point (x_0, y_0) is a local minimum then δf will be positive for all small values of h and k i.e. the quadratic form is **positive definite**. Some indefinite forms will be geometrically described as **saddle points**, so that passing over the point (x_0, y_0) in some directions gives the impression of it being a local maximum and in other directions, the impression created is that of a local minimum. See Figure 3.18.

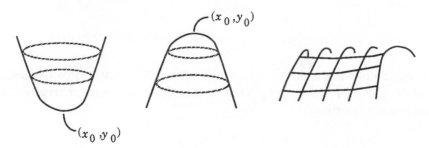

Figure 3.18

Other more esoteric features exist. If the quadratic form is identically zero, then we should consider third-order terms.

The conditions for classifying stationary points are:

for a local maximum, $f_{xx} < 0$ and $f_{xx}f_{yy} - (f_{xy})^2 > 0$ (3.20a)

for a local minimum, $f_{xx} > 0$ and $f_{xx}f_{yy} - (f_{xy})^2 > 0$ (3.20b)

If $f_{xx}f_{yy} - (f_{xy})^2 < 0$, we designate this a saddle point.

We note that for two variables $D = f_{xx}f_{yy} - (f_{xy})^2$ is the determinant of the Hessian matrix. [Why could we substitute $f_{yy} > 0$ for $f_{xx} > 0$ in the conditions for a local minimum and similarly for a local maximum?]

Example 1

Consider $f(x,y) = x^3 + y^3 - 12x^2 - 9y^2 + 256$; $f_x = 3x^2 - 24x$; $f_y = 3y^2 - 18y$; $f_{xx} = 6x - 24$; $f_{xy} = 0$; $f_{yy} = 6y - 18$.

We have stationary points where $f_x = 0$ and $f_y = 0$, i.e. where $3x(x - 8) = 0$ *and* $3y(y - 6) = 0$. This gives us four possibilities: $(0,0)$, $(8,0)$, $(0,6)$, $(8,6)$. To examine the nature of these stationary points we are well advised to draw up a table as shown on the following page.

Point	(0,0)	(8,0)	(0,6)	(8,6)
f_{xx}	−24	24	−24	24
f_{yy}	−18	−18	18	18
f_{xy}	0	0	0	0
D	>0	<0	<0	>0
	Local Maximum	Saddle Point	Saddle Point	Local Minimum

Example 2

$f(x,y) = x^2y - 4x^2 - 2y^2 + 16y$; $f_x = 2xy - 8x$; $f_y = x^2 - 4y + 16$; $f_{xx} = 2y - 8$; $f_{xy} = 2x$; $f_{yy} = -4$

Stationary points occur when $2x(y - 4) = 0$ and $x^2 = 4y - 16$. Then $x = 0$ and hence $y = 4$ is one possibility; if $y = 4$, $x^2 = 0$. This means we can find only one stationary point at (0,4); there $f_{xx} = 0$, $f_{xy} = 0$ and $f_{yy} = -4$. This gives $D = 0$. We need further investigation. The quadratic approximation is

$$f(h, 4 + k) \cong f(0,4) + hf_x(0,4) + kf_y(0,4) + \frac{1}{2!}[h^2 f_{xx}(0,4)$$

$$+ 2hk f_{xy}(0,4) + k^2 f_{yy}(0,4)]$$

i.e. $\delta f \equiv f(h, 4 + k) - f(0,4) \cong \frac{1}{2}k^2(-4) = -2k^2$

This suggests the value of δf is always negative, which implies $f(x,y)$ has a local maximum at (0,4). However you should find the next term in the series and show that we have, in fact, a saddle point.

Example 3

A rectangular tank is to be constructed to contain $\frac{1}{2}$ cu.m. of hot liquid; in order to minimise heat loss we must so construct the tank that the surface area is least. In addition, we require the base to be of double thickness. What dimensions do we choose?

Let the lengths of three mutually perpendicular edges be x, y, z metres.

Then the volume restriction implies $\qquad xyz = \frac{1}{2}$ (3.21)

The surface area is given by $\qquad S = 3xy + 2xz + 2yz$ (3.22)

It is symmetrical with respect to x and y but not z.

If we substitute for z from (3.21) into (3.22) we obtain

$$S = 3xy + (x + y)/xy = 3xy + (1/y) + (1/x)$$

To locate the stationary points, we find S_x and S_y and equate to zero.

$$S_x = 3y - 1/x^2 = 0 = 3x - 1/y^2 = S_y$$

Hence we obtain $3x^2y - 1 = 0$ and $3xy^2 - 1 = 0$ which produce, on subtraction, $3x^2y - 3xy^2 = 0$, i.e. $3xy(x - y) = 0$. We reject the possibilities $x = 0$ and $y = 0$ since they do not represent a physical solution to our problem; these, in theory, give a maximum surface area. The remaining possibility is

$y = x$ which leads to $x = y = \sqrt[3]{(1/3)}$, $z = \frac{1}{2}\sqrt[3]{9}$ or to 3 s.f. $x = y = 0.693$,

$z = 1.05$ and a minimum surface area of $S_{min} = \sqrt[3]{3} + 2\sqrt[3]{3} = 4.33\text{m}^2$.

Constrained Problems

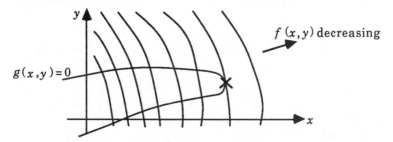

Figure 3.19

Example 3 contained a constraint, viz the volume of the tank was a fixed quantity; we overcame the constraint by making a substitution to remove one of the variables. In this subsection we now develop a technique – the method of **Lagrange multipliers** which effectively delays the substitution. We approach the technique from a geometrical viewpoint for a function of two variables. Figure 3.19 shows a set of contours of the function $f(x,y)$ and the curve representing the constraint $g(x,y) = 0$. The point marked with a cross is the minimum value of $f(x,y)$ subject to the constraint $g(x,y) = 0$. This point lies on a particular contour $f(x,y) = \text{constant}$ and it should be clear that this contour and the constraint curve have a common tangent at this point.

We can represent the constraint curve in parametric form, with the parameter chosen as s, the length along the curve from a given point. Hence $x = x(s)$ and $y = y(s)$ so that the objective function can be written in terms of s, as follows

$$f(x,y) \equiv f[x(s), y(s)] = F(s)\dagger \qquad (3.23)$$

By the chain rule $\qquad F'(s) = f_x \dfrac{dx}{ds} + f_y \dfrac{dy}{ds} \qquad (3.24)$

and the maximum and minimum points of $f(x,y)$ on the constraint curve can be found by solving $F'(s) = 0$. Now (3.24) can be written as $F'(s) = (\nabla f).\mathbf{t}$

---

† Note that since the expression in s will be different from that in x and y we use a different symbol for it.

where $\mathbf{t} = \dfrac{dx}{ds}\mathbf{i} + \dfrac{dy}{ds}\mathbf{j}$. The condition $F'(s) = 0 \Rightarrow \nabla f . \mathbf{t} = 0$ and therefore ∇f is normal to $\mathbf{t}$. However, $\mathbf{t}$ is a tangent to the constraint curve. (Why?) Therefore ∇f must be normal to this curve at a maximum or minimum point. But the vector ∇g is also normal to the constraint curve. Because we have two vectors normal to a given curve at a given point, it follows that one is a scalar multiple of the other i.e.

$$\nabla f + \lambda \nabla g = 0 \qquad (3.25)$$

where λ is a scalar. Hence, to find the maxima and minima of $f(x,y)$ subject to $g(x,y) = 0$ we must solve the equations

$$f_x + \lambda g_x = 0; \quad f_y + \lambda g_y = 0; \quad g(x,y) = 0 \qquad (3.26)$$

If there are three original variables, then the procedure is as below.

Example 1

Consider the rectangular tank problem of page 85 where we seek to minimise

$$f(x,y,z) = 3xy + 2yz + 2xz \text{ subject to } g(x,y,z) = xyz - \frac{1}{2} = 0.$$

It helps to form $\quad F(x,y,z) \equiv 3xy + 2yz + 2xz + \lambda(xyz - \frac{1}{2})$

$$\equiv f(x,y,z) + \lambda g(x,y,z)$$

Then $\qquad F_x = f_x + \lambda g_x = 3y + 2z + \lambda yz = 0 \qquad (3.27a)$

$$F_y = f_y + \lambda g_y = 3x + 2z + \lambda xz = 0 \qquad (3.27b)$$

$$F_z = f_z + \lambda g_z = 2y + 2x + \lambda xy = 0 \qquad (3.27c)$$

and $\qquad g(x,y,z) = xyz - \frac{1}{2} = 0 \qquad (3.27d)$

We have apparently introduced a fourth variable λ, but we have four equations to solve for the three variables x, y, z and the extra variable λ. We use the first three equations to express x, y and z in terms of λ and hence the ratio $x : y : z$ and we use the fourth equation to find the actual values of these variables at any stationary point. We have already remarked on the symmetry between x and y and we can make use of this (we ought to add that the quick solution of such equations depends on the relative skill and experienced insight of the solver). We subtract (3.27b) from (3.27a) to obtain

$$3(y - x) + \lambda z(y - x) = 0 \quad \text{i.e.} \quad (y - x)[3 + \lambda z] = 0$$

We have two alternatives to consider and we substitute the possibility $y = x$ into (3.27c) to obtain $4x + \lambda x^2 = 0$. This yields $x = 0$ (rejected – why?) or $\lambda x = -4$. The case $\lambda x = -4$ gives $x = y = -4/\lambda$ and substitution into either (3.27a) or (3.27b) yields $-(12/\lambda) + 2z - 4z = 0$ i.e. $z = -6/\lambda$. We must now return to the condition $3 + \lambda z = 0$. If we substitute into (3.27a) we obtain the equation $3y + 2z - 3y = 0$ i.e. $z = 0$ which we reject.

We have as our only possibility $x : y : z = 2 : 2 : 3$. Substituting in (3.27d) we obtain $x = y = \sqrt[3]{(1/3)}$, $z = \frac{1}{2}\sqrt[3]{9}$. We argue for a minimum on physical grounds.

Example 2

Find the least value of the function $f(x,y,z) = x^2 + y^2 + z^2$ along the line of intersection of the planes $2x + y - z = 2$ and $2x - y + z = -2$. We seek to minimise $f(x,y,z) = x^2 + y^2 + z^2$ subject to $g(x,y,z) = 2x + y - z - 2 = 0$ and $h(x,y,z) = 2x - y + z + 2 = 0$. We form

$F(x,y,z) \equiv f(x,y,z) + \lambda g(x,y,z) + \mu h(x,y,z)$ where λ and μ are scalars.

Then
$$F_x = 2x + 2\lambda + 2\mu = 0 \tag{3.28a}$$
$$F_y = 2y + \lambda - \mu = 0 \tag{3.28b}$$
$$F_z = 2z - \lambda + \mu = 0 \tag{3.28c}$$
$$g(x,y,z) = 2x + y - z - 2 = 0 \tag{3.28d}$$
$$h(x,y,z) = 2x - y + z + 2 = 0 \tag{3.28e}$$

We solve the first three equations to obtain x, y, z in terms of λ, μ. Hence $x = -\lambda - \mu$ and $y = -(\lambda/2) + (\mu/2)$ and $z = (\lambda/2) - (\mu/2)$. Substitution in the last two equations produces $\lambda = -1$, $\mu = 1$ and hence $x = 0$, $y = 1$, $z = -1$ giving a minimum value for $f(x,y,z)$ of 2. Can you see geometrically that this means that the nearest distance from the origin of any point on the line of intersection is 2? Note why we can claim the stationary point is a minimum and note further that if we seek such least distances, it is better to work with the square of the distance.

Although it is not discussed here, it should be noted that the method of Lagrange multipliers can logically be extended to cover the case of an objective function having three variables, subject to one constraint.

3.7 Methods Using the Gradient of a Function

The previous search techniques that we have mentioned used only the properties of continuous functions. We now examine methods which call upon further information from a function, namely the derivatives of the function. The simplest method works on the principle of choosing the quickest way down.

Method of steepest descent

In this technique we effectively fit a tangent plane to the function at specified points. In each tangent plane we find the line of steepest slope, and proceed down this line either by a fixed step size or until we find the lowest point (which can be accomplished by a one-variable search). At this lowest point we find the new direction of steepest descent and repeat the procedure. Let us first see how to choose this direction analytically.

We consider a small change $(\delta x_1, \delta x_2, ..., \delta x_n)$ from the present point. The first approximation to the change in the value of the function is given by

$$df \cong \sum_{j=1}^{n} f_{x_j} \delta x_j$$

where the derivatives are evaluated at the present point. Of all possible changes of a given magnitude

$$D = \sqrt{\sum_{j=1}^{n} [\delta x_j]^2}$$

we seek that one which causes greatest change in $f(\mathbf{x})$. We work the idea through for two independent variables; x_1 and x_2. We want to maximise

$$df = \frac{\partial f}{\partial x_1} \cdot \delta x_1 + \frac{\partial f}{\partial x_2} \cdot \delta x_2 \quad \text{subject to} \quad D^2 - (\delta x_1)^2 - (\delta x_2)^2 = 0$$

This time, our variables are δx_1 and δx_2; if this confuses you, call them l and m, for example. We use the method of Lagrange multipliers and form

$$F(\delta x_1, \delta x_2, \lambda) = \frac{\partial f}{\partial x_1} \delta x_1 + \frac{\partial f}{\partial x_2} \delta x_2 + \lambda[D^2 - (\delta x_1)^2 - (\delta x_2)^2]$$

The equations to be solved are

$$\frac{\partial f}{\partial x_1} - 2\lambda \, \delta x_1 = 0 \qquad \frac{\partial f}{\partial x_2} - 2\lambda \, \delta x_2 = 0 \qquad \text{and} \qquad (\delta x_1)^2 + (\delta x_2)^2 = D^2$$

Therefore $\dfrac{1}{2\lambda} = \dfrac{\delta x_1}{\partial f / \partial x_1} = \dfrac{\delta x_2}{\partial f / \partial x_2}$ and this tells us to choose the direction of

change so that the components of the change are proportional to the partial derivatives of $f(\mathbf{x})$ i.e. we proceed along the gradient vector ∇f. The constant of proportionality, λ, will be chosen to be negative so that we make a *descent*.

Note that the gradient vector is orthogonal to the contours. Suppose we move from our present point along the contour which passes through it by a change $\mathbf{d} = (d_1, d_2)$.

Then $df = 0 = \dfrac{\partial f}{\partial x_1} d_1 + \dfrac{\partial f}{\partial x_2} d_2$; but $\nabla f \cdot \mathbf{d} = \dfrac{\partial f}{\partial x_1} \cdot d_1 + \dfrac{\partial f}{\partial x_2} \cdot d_2 = 0$,

hence orthogonality is proved.

In general, the method of Lagrange multipliers shows that the direction of local steepest descent is given by ∇f and the orthogonality property still holds.

A variant on the method is to form the **normalised gradient vector** $\dfrac{\nabla f}{|\nabla f|}$

and step a fixed distance, repeating this until no reduction in function value is obtained. Then the step size is reduced and the method restarted from the previous best point.

At each minimum point, the direction of steepest descent is orthogonal to the most recent search direction. For functions of two variables, only two directions are used and this procedure is akin to an alternating variable search. Methods of speeding up the process have been suggested but the trend is towards employing different techniques.

Example

$f(x,y) = x^2 + 4y^2 + 2xy$

The contours of the function are shown in Figure 3.20.

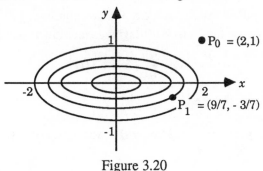

Figure 3.20

We start at the point $P_0 = (2,1)$. Now $f_x = 2x + 2y$ and $f_y = 2x + 8y$, and at $(2,1)$ the values are 6 and 12 respectively. Hence we find that the local

direction of steepest descent is given by the unit vector $\left(\dfrac{1}{\sqrt{5}}, \dfrac{2}{\sqrt{5}}\right)$ [Why?]. We

have to proceed along the line through $(2,1)$ with gradient 2, i.e. along the line $y = 2x - 3$, until we come to the lowest point. In practice we would start with a given step length and reduce the step length when necessary until it obtained a critical size; the point then reached would be an approximation to the point we shall obtain by calculus.

We seek the minimum of $x^2 + 4y^2 + 2xy$ subject to $y = 2x - 3$. Using the method of Lagrange multipliers, we minimise $F(x,y,\lambda) = x^2 + 4y^2 + 2xy + \lambda(y - 2x + 3)$. We solve $2x + 2y - 2\lambda = 0$ and $2x + 8y + \lambda = 0$.

Eliminating λ, we obtain $6x + 18y = 0$, i.e. $y = -x/3$ and hence we obtain the point $P_1 = (9/7, -3/7)$. At this point $f_x = 12/7$ and $f_y = -6/7$.

The local direction of steepest descent is given by the unit vector $\left(\dfrac{2}{\sqrt{5}}, \dfrac{-1}{\sqrt{5}}\right)$ and the cycle repeated. Further results are shown in Table 3.2. You can see the approach to the minimum at (0,0).

Table 3.2

Stage	Start Point	f_x	f_y	Unit Vector	End Point
0	$(2, 1)$	6	12	$\left(\dfrac{1}{\sqrt{5}}, \dfrac{2}{\sqrt{5}}\right)$	$\left(\dfrac{9}{7}, \dfrac{-3}{7}\right)$
1	$\left(\dfrac{9}{7}, \dfrac{-3}{7}\right)$	$\dfrac{12}{7}$	$\dfrac{-6}{7}$	$\left(\dfrac{2}{\sqrt{5}}, \dfrac{-1}{\sqrt{5}}\right)$	$\left(\dfrac{6}{28}, \dfrac{3}{28}\right)$
2	$\left(\dfrac{6}{28}, \dfrac{3}{28}\right)$	$\dfrac{9}{14}$	$\dfrac{18}{14}$	$\left(\dfrac{1}{\sqrt{5}}, \dfrac{2}{\sqrt{5}}\right)$	$\left(\dfrac{27}{196}, \dfrac{-9}{196}\right)$
3	$\left(\dfrac{27}{196}, \dfrac{-9}{196}\right)$	$\dfrac{18}{98}$	$\dfrac{-9}{98}$	$\left(\dfrac{2}{\sqrt{5}}, \dfrac{-1}{\sqrt{5}}\right)$	$\left(\dfrac{18}{784}, \dfrac{9}{784}\right)$
4	$\left(\dfrac{18}{784}, \dfrac{9}{784}\right)$	$\dfrac{27}{392}$	$\dfrac{54}{392}$	$\left(\dfrac{1}{\sqrt{5}}, \dfrac{2}{\sqrt{5}}\right)$	$\left(\dfrac{81}{5488}, \dfrac{-27}{5488}\right)$

If a direct search method is used at each stage, progress may be slower.

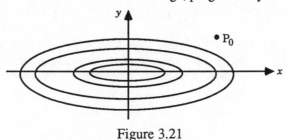

Figure 3.21

The efficiency of this method is not high and is related to the scales that we use. It works best when the contours are circular (or the analogue in higher dimensions) and the closer the contours are to being circular, the more rapid the convergence. Consider the function $z = x^2 + 36y^2$, some contours of which are shown in Figure 3.21. Starting from the point P_0, the direction of steepest descent is decidedly *not* towards the lowest point, which is the origin. However, if we make the transformation $v = 6y$ the function becomes $z = x^2 + v^2$ and the direction of steepest descent is *directly* towards the origin. This choice of scale is not simply graphical; if the two variables represent different physical quantities such as temperature and pressure it may not be possible to achieve a suitable set of contours. Attempts have been made to accelerate the convergence, but since the method of steepest descent is not a practical contender, we shall not discuss them.

Method of Newton-Raphson

Near a local minimum, the contours are nearly elliptical (in two dimensions). Many methods make use of this quadratic property. If we return to the quadratic approximation

$$f(\mathbf{x}) \cong f(\mathbf{x}_0) + (\mathbf{x} - \mathbf{x}_0) \cdot \nabla f_0 + \tfrac{1}{2}(\mathbf{x} - \mathbf{x}_0)^T H_0(\mathbf{x} - \mathbf{x}_0) \qquad (3.29)$$

we know that at the minimum, which we assume is $\mathbf{x}_0$, $\nabla f = 0$.

For two variables, we have

$$f(x,y) \cong f(x_0, y_0) + (x - x_0)f_x(x_0, y_0) + (y - y_0)f_y(x_0, y_0)$$

$$+ \tfrac{1}{2}[(x - x_0)^2 f_{xx}(x_0, y_0) + 2(x - x_0)(y - y_0)f_{xy}(x_0, y_0) + (y - y_0)^2 f_{yy}(x_0, y_0)]$$

Differentiating this with respect to x and with respect to y we obtain

$$f_x(x,y) \cong 0 + f_x(x_0, y_0) + (x - x_0)f_{xx}(x_0, y_0) + (y - y_0)f_{xy}(x_0, y_0)$$

and $f_y(x,y) \cong 0 + f_y(x_0, y_0) + (x - x_0)f_{xy}(x_0, y_0) + (y - y_0)f_{yy}(x_0, y_0)$

But if (x,y) is the local minimum then $f_x(x,y) = 0 = f_y(x,y)$.

Suppose we are at the point (x_0, y_0) seeking the minimum. Then let the minimum point be expressed at $(x_0 + h, y_0 + k)$ where h and k are found approximately by solving the equations

$$\left. \begin{array}{l} 0 = f_x(x_0, y_0) + h f_{xx}(x_0, y_0) + k f_{xy}(x_0, y_0) \\[2mm] 0 = f_y(x_0, y_0) + h f_{xy}(x_0, y_0) + k f_{yy}(x_0, y_0) \end{array} \right\} \qquad (3.30)$$

We generalise to the case of a function of n variables; let $\mathbf{l}_0$ be the vector of step sizes $(h_1, h_2, ..., h_n)$. The equations to be solved are $\nabla f_0 = -H_0 \mathbf{l}_0$ where H is the Hessian matrix. The solution is obtained by matrix inversion as $\mathbf{l}_0 = -H_0^{-1} \nabla f_0$ and we move to the point $\mathbf{x}_0 + \mathbf{l}_0$. For a quadratic function this will give an exact minimum in one move. For a function near the minimum, we set up an iterative scheme.

$$\mathbf{x}_{i+1} = \mathbf{x}_i - H_i^{-1} \nabla f_i$$

where H_i and ∇f_i are evaluated at $\mathbf{x}_i$.

Note the analogy to the Newton-Raphson method for locating roots of non-linear equations in one variable.

Example

Consider $f(x,y) = x^3 + y^3 - 12x^2 - 9y^2 + 256$

$f_x = 3x^2 - 24x$; $f_y = 3y^2 - 18y$; $f_{xx} = 6x - 24$; $f_{xy} = 0$; $f_{yy} = 6y - 18$

Hence $\nabla f = (3x^2 - 24x, 3y^2 - 18y)$ and $\mathbf{H} = \begin{bmatrix} 6x - 24 & 0 \\ 0 & 6y - 18 \end{bmatrix}$

At the point $\mathbf{x}_0 = (1,1)$ which we can select as our initial point,

$$\nabla f = (-21, -15) \text{ and } \mathbf{H} = \begin{bmatrix} -18 & 0 \\ 0 & -12 \end{bmatrix}$$

Hence $\mathbf{H}^{-1} = \begin{bmatrix} \dfrac{-1}{18} & 0 \\ 0 & \dfrac{-1}{12} \end{bmatrix}$ and $\mathbf{l} = \begin{bmatrix} \dfrac{-1}{18} & 0 \\ 0 & \dfrac{-1}{12} \end{bmatrix} \begin{bmatrix} -21 \\ -15 \end{bmatrix} = \begin{bmatrix} \dfrac{7}{6} \\ \dfrac{5}{4} \end{bmatrix}$

We therefore advance to the point $\mathbf{x}_1 = (1,1) - \left(\dfrac{7}{6}, \dfrac{5}{4} \right) = \left(-\dfrac{1}{6}, -\dfrac{1}{4} \right)$

At $\mathbf{x}_1$, $\nabla f = \left(\dfrac{147}{36}, \dfrac{75}{16} \right)$ and $\mathbf{H} = \begin{bmatrix} -25 & 0 \\ 0 & \dfrac{-39}{2} \end{bmatrix}$.

Hence $\mathbf{H}^{-1} = \begin{bmatrix} \dfrac{-1}{25} & 0 \\ 0 & \dfrac{-2}{39} \end{bmatrix}$ and $\mathbf{l} = \begin{bmatrix} -0.163 \\ -0.240 \end{bmatrix}$.

We advance to the point $\mathbf{x}_2 = (-0.167, -0.25) - (-0.163, -0.240) = (-0.004, -0.010)$. Further iterations on a computer terminal yielded

$\mathbf{x}_3 = (-1.39 \times 10^{-6}, -1.54 \times 10^{-5})$, $\mathbf{x}_4 = (0, 4.00 \times 10^{-11})$, $\mathbf{x}_5 = (0,0)$.

There are certain drawbacks to this method, apart from the fact that it relies upon the function being approximately quadratic. The matrix $\mathbf{H}$ must be positive definite which will only for certain be true for a convex function. As the method stands, we require $\mathbf{H}^{-1}$ at every step and this may require a numerical inversion which demands $n(n + 3)/2$ evaluations; then the solution of $\mathbf{l} = \mathbf{H}^{-1} \nabla f$ may demand a further number of operations of $0(n^3)$. Other difficulties also exist.

Again, modifications have been suggested, including a method due to Davidon. In effect, he approximates $\mathbf{H}^{-1}$ in such a way that the successive approximations are positive definite.

Methods using conjugate directions

These methods use the properties of ellipses (or their n-dimensional analogues). They aim to collect information on the curvature of the objective function, but do not calculate the second derivatives of the function explicitly. In this way, they are as rapid as Newton's method (when it works), but they avoid its drawbacks. Provided the initial point is near the minimum, the contours will be closed and almost elliptical. The origin of the conjugate direction method lies in a simple result concerning the ellipse.

Theorem

The line joining the points of contact of two parallel tangents to an ellipse will pass through the centre of the ellipse.

The direction of the tangents is said to be **conjugate** to the direction of the line through the centre of the ellipse.

In n-dimensions the result may be generalised: if P_1 is the point of contact to a quadric (generalisation of an ellipse) of an $(n - 1)$ dimensional tangent subspace and P_2 is the point of contact of a parallel tangent subspace, then the centre of the quadric lies on the line P_1P_2. It is no use trying to appeal to geometry in n-dimensions so we need an algebraic definition of conjugacy. The general quadric is of the form

$$F(\mathbf{x}) = f_0 + \mathbf{x}^T \mathbf{a} + \tfrac{1}{2}\mathbf{x}^T \mathbf{A}\mathbf{x}.$$

For example $f(x,y) = 4 + 3x + 2y + x^2 + 2xy + 4y^2$ can be written as above

with $\mathbf{x} = \begin{bmatrix} x \\ y \end{bmatrix}$, $\mathbf{a} = \begin{bmatrix} 3 \\ 2 \end{bmatrix}$, $\mathbf{A} = \begin{bmatrix} 1 & 1 \\ 1 & 4 \end{bmatrix}$.

If two vectors $\mathbf{x}$ and $\mathbf{y}$ are such that $\mathbf{x}^T \mathbf{A}\mathbf{y} = 0$ then $\mathbf{x}$ and $\mathbf{y}$ are defined to be *conjugate* to $\mathbf{A}$.

The algebraic definition is a generalisation of the geometric result.

A set of directions $\mathbf{t}_k$, $k = 1, 2, 3, ..., n$ is said to be mutually conjugate with respect to a matrix $\mathbf{A}$ if $\mathbf{t}_i^T \mathbf{A}\mathbf{t}_j = 0$ for all $i, j \in \{0, 1, ..., n\}$ where $i \neq j$.

The great advantage of a method employing conjugate directions is embodied in the following theorem.

If linear searches are performed along a set of n directions which are mutually conjugate with respect to the Hessian matrix of a quadratic function then the minimum will be located in at most n searches.

Example

$$f(x,y) = x^2 + 2xy + 4y^2 - 8x - 6y + 16$$

The Hessian matrix is $\begin{bmatrix} 2 & 2 \\ 2 & 8 \end{bmatrix}$. Suppose we start from the point $(0,0)$ along the x-axis, i.e. in the direction given by $(1,0)$. We seek the conjugate direction to $(1,0)$; let this be given by (u, v). Then

$$(1,0)\begin{bmatrix} 2 & 2 \\ 2 & 8 \end{bmatrix}\begin{bmatrix} u \\ v \end{bmatrix} = 0 \quad \text{i.e.} \quad (2,2)\begin{bmatrix} u \\ v \end{bmatrix} = 0 \quad \text{i.e.} \quad 2u + 2v = 0;$$

hence a vector representing the conjugate direction is $(1, -1)$.

In the direction $(1,0)$ the function becomes $x^2 - 8x + 16$ and this has a minimum at $(4,0)$. We now start from $(4,0)$ along the direction given by $(1, -1)$ i.e. along the line $y = 4 - x$; the function along this line can be written

$$x^2 + 2x(4 - x) + 4(4 - x)^2 - 8x - 6(4 - x) + 16 \quad \text{i.e.} \quad 3x^2 - 26x + 56.$$

This has a minimum at $x = \dfrac{13}{3}$ (and $y = \dfrac{-1}{3}$). We have located the point $(\dfrac{13}{3}, \dfrac{-1}{3})$ after two linear searches. You can check that this is the minimum by partially differentiating $f(x,y)$.

Alternatively, suppose we find the gradient vector at (0,0). This is (–8, –6). If we choose the tangent at (0,0) as one direction, i.e. (6, –8) or (3, –4) then the conjugate direction (u,v) is given by

$$(3, -4)\begin{bmatrix} 2 & 2 \\ 2 & 8 \end{bmatrix}\begin{bmatrix} u \\ v \end{bmatrix} = 0 \quad \text{i.e.} \quad (-2, -26)\begin{bmatrix} u \\ v \end{bmatrix} = 0 \quad \text{hence} \quad \begin{bmatrix} u \\ v \end{bmatrix} = \begin{bmatrix} 13 \\ -1 \end{bmatrix}$$

In the direction (6, –8) or $y = \dfrac{-4}{3}x$ the function becomes $\dfrac{49}{9}x^2 + 16$ which has a minimum at (0,0). Starting from this point, we move along the line $y = \dfrac{-1}{13}x$ and find the minimum at $\left(\dfrac{13}{3}, \dfrac{-1}{3}\right)$ as before.

Practical use of algorithms

A measure of the efficiency of an algorithm is the number of function evaluations needed to locate the minimum of the objective function to a specified accuracy. Some algorithms require the estimation of derivatives and allowances must be made for this additional calculation.

When setting up a subroutine to specify the objective function, it is important to avoid cancellation errors, especially when we use function values to estimate derivatives.

The scaling of variables is also an important factor. If there is a disparity in the values of the variables of x then x^2 can be swamped by just one of the variables. Care should be taken to ensure that the magnitudes of the variables are roughly on a par. The gradient vector is quite dependent on scaling and therefore steep-sided valleys may cause severe problems if the variables are not well scaled. Of course, if the variables are similar physical quantities for example, lengths, there should be no problem of scale; if, however, the variables are temperature and pressure, this may not be so easy. An advantage of scaling is that the Hessian matrix may be quite close to the unit matrix of the same size.

It is important not to demand too much accuracy for the final solution or else rounding errors may become of comparable magnitude.

We can also carry out a sensitivity analysis on the final solution to see how the optimum is related to small changes in the objective function.

There is no 'best' algorithm for all optimization problems and, if a library of computer routines is available, there is usually a user's guide at hand.

Problems

Section 3.1

1   For each of the following problems, sketch the feasible region, and a few profit lines and hence find the optimal solution: specify the optimal value of the objective function and the corresponding values of the basic variables. If the solution does not possess integer coordinates choose that point with integer coordinates which is optimal. In all cases you may assume $x \geq 0, y \geq 0$, unless otherwise stated.

(a)   Maximise $x + y$, subject to $x + 3y \leq 9$, $10x + 7y \leq 44$

(b)   Repeat (a) for the objective functions $x + 2y$, $x + 3y$, $2x + y$

(c)   Maximise $2x + 3y$ subject to $x + 2y \leq 82$, $8x + 5y \leq 370$

(d)   Maximise and minimise $2x + y$ subject to $6x + y \leq 6$, $40x + 11y \leq 99$, $4x + 3y \leq 18$, $x \leq 3/2$, $y \geq 2$

(e)   Maximise $3x + y$ subject to $15x + 4y \geq 120$, $20x + 3y \leq 120$, $y \leq 14$

(f)   Maximise $2x + y$ and minimise $x + 2y$ subject to $3 \leq y \leq 15$, $5x + 2y \leq 50$, $6x + y \geq 15$, $x + 2y \geq 10$

2   Sketch the feasible regions for the following problems. Again, $x, y \geq 0$ unless stated otherwise.

(a)   Constraints are $4x_1 + 3x_2 \leq 6$, $3x_1 + 4x_2 \leq 6$, $3x_1 + 5x_2 \leq 7.5$. Note the redundancy of one of the constraints. The problem is said to be **degenerate**.

(b)   Constraints are $5x_1 + 2x_2 \leq 10$, $x_1 + 3x_2 \leq 6$, $2x_1 + 5x_2 \leq 10$
Comment on the region. What happens if we wish to maximise $4x_1 + 3x_2$?

(c)   Constraints are $3x_1 + 2x_2 \leq 8$, $3x_1 + 2x_2 \geq 10$

(d)   Constraints are $3x_1 - x_2 \leq -4$, $x_1 - x_2 \geq 1$

(e)   Constraints are $5x_1 + 4x_2 \leq 10$, $x_1 + x_2 \geq 1.5$, $4x_1 + x_2 \geq 2$, $7x_1 + 10x_2 \leq 17.5$, $x_1 + 4x_2 \geq 2$

(f)   Constraints are $3x_1 - 2x_2 \geq -2$, $x_1 - 2x_2 \geq -6$. What happens if we try to maximise $5x_1 + 4x_2$?

Section 3.2

3   Solve Problem 1(a), (b) and (c) of Section 3.1 by the Simplex method.

4   Solve the following problem by the simplex method.
Maximise $5x_1 + 2x_2$ subject to $2x_1 + x_2 \leq 9$, $x_1 - 2x_2 \leq 2$, $-3x_1 + 2x_2 \leq 3$; $x_i \geq 0$. Interpret your result graphically.

5   (a)   Maximise $3x_1 + x_2 + 2x_3$ subject to
$x_1 - 2x_2 - x_3 \leq 10$, $2x_1 + x_2 + 2x_3 \leq 12$, $x_1 - x_2 + x_3 \leq 5$; $x_i \geq 0$

(b)   Maximise $3x_1 + 2x_2 + x_3$ subject to
$4x_1 + x_2 + x_3 \leq 8$, $3x_1 + 3x_2 + 2x_3 \leq 9$; $x_i \geq 0$

(c)   Maximise $x_1 + 3x_2 + 2x_3$ subject to
$2x_1 + x_3 \leq 10$, $2x_1 + x_2 \leq 12$, $x_1 + 2x_2 + 3x_3 \leq 30$; $x_i \geq 0$

(d)   Maximise $5x_1 + 3x_2 + 2x_3$ subject to
$x_1 + 2x_2 + 3x_3 \leq 25$, $2x_1 + 3x_2 + x_3 \leq 6$, $3x_1 + x_2 + 2x_3 \leq 20$; $x_i \geq 0$

(e) Maximise $4x_1 + 5x_2 + x_3$ subject to

$$2x_1 + x_2 + x_3 \le 100, \quad 2x_1 + 3x_2 \le 120, \quad x_1 + 2x_3 \le 80$$

6   A manufacturer has three products which each require the same three stages in their manufacture. Product X requires 4 hours in stage I, 3 hours in stage II and 1 hour in stage III per article. The corresponding times for an article of Y are 1, 2 and 3 hours and for an article of Z are 3, 2 and 4 hours. However, in a given week, there is a limited amount of time available for each stage and these are 39 hours for stage I, 34 hours for stage II and 28 hours for stage III. The profit on an article of X is £200, of Y is £300 and of Z is £400. Find how many articles of each product should be made.

7   Using the simplex method, solve the following problems.
(a) Maximise $2x_1 + x_2 + 6x_3 + 2x_4$, subject to $2x_1 + 3x_2 + x_3 + 7x_4 \le 13$, $5x_1 + x_2 + 3x_3 + 2x_4 \le 5$, $3x_1 + 5x_2 + 2x_3 + x_4 \le 4$; $x_i \ge 0$
(b) Maximise $x_1 + x_2 + 2x_3 + x_4$, subject to $x_2 + x_3 + x_4 \le 2$, $x_1 + 2x_2 + x_4 \le 4$, $2x_1 - x_2 - x_3 + x_4 \le 3$, $-x_1 + x_2 + x_3 - 2x_4 \le 3$; $x_i \ge 0$

8   A factory is required to produce 100 kg per week of an alloy containing 40% lead, 30% zinc, 30% tin. Raw materials available are unlimited supplies of pure lead at £90 per kg, zinc at £120 per kg, tin at £150 per kg, and limited quantities of two types of scrap alloy. Scrap alloy A contains 20% lead, 70% zinc, 10% tin; scrap alloy B contains 10% lead, 50% zinc, 40% tin. One week there is available 90 kg of A at £30 per kg and 110 kg of B at £33 per kg. (Note that prices are fictitious.)
How much of each of the raw materials should the factory use in the week to minimise the cost, based on the available data?

9   A rolling mill produces three products, A, B and C. Before going into the rolling mill the products must be heated in a soaking pit which can heat product A at the rate of 1200 kg/shift, product B at 600 kg/shift and C at 400 kg/shift. The rolling mill proper can roll A at the rate of 600 kg/shift, B at 750 kg/shift and C at 1000 kg/shift. The output of A and B together must exceed 500 kg/shift. The profits per kg on A, B and C are in the ratio 1:3:4. Use the simplex method to derive the production plan for one shift which gives the maximum profit within the operating constraints.

Section 3.3
10 Sketch the curve $x^3 - 6x^2 + 9x + 4$ in $-1 \le x \le 5$. Indicate local and global maxima and minima. Repeat for $x^3 - 18x^2 + 96x$ in the interval $(0,9)$.

11 Why will calculus not find the minimum of
(i) $f(x) = |x|$ (ii) $f(x) = x^2 - 2x + 2; \ x \ge 2$

12 Show that $(ax + b)/(cx + d)$ has no local extrema for any values of a, b, c, d.

Section 3.4
13 (a) Uniform search evaluates $f(x)$ at points equally spaced in the original interval of uncertainty. Show that the functions of the original interval within which the optimum lies after N evaluations is given by $F = 2/(N + 1)$.

(b) Uniform dichotomous search divides the original interval into $(N/2) + 1$ intervals of width $L/[(N/2) + 1]$. At each of the $(N/2)$ points, two function evaluations are performed, one on either side of the points, separated by a small distance 2ϵ. Find the fraction F for this search method.

(c) Sequential dichotomous search commences by evaluating the function at two points distance ϵ on either side of the mid-point of the interval of uncertainty. Almost half this interval can now be rejected. The next two evaluations are performed close to the mid-point on the new interval of uncertainty, and so on. Show that $F = (1/2)^{N/2}$.

(d) Show that for the Fibonacci search, $F = 1/F_N$.

14 (a) Assuming that the recurrence relation $F_n = F_{n-1} + F_{n-2}$ has a solution of the form

$$F_n = k^n, \text{ derive the result } F_n = \frac{1}{\sqrt{5}}\left\{\left[\frac{1+\sqrt{5}}{2}\right]^{n+1} - \left[\frac{1-\sqrt{5}}{2}\right]^{n+1}\right\}.$$

Show that for large values of n, the successive reductions in interval are asymptotically approximately $\frac{1}{2}(1 + \sqrt{5}) = \tau$. **Golden Section** search has a constant interval reduction of τ. Hence $F = 1/\tau^N$. Show that the ratio of interval reduction by Fibonacci search to that by Golden Section search $F_N/\tau^{N-1} \cong 1.17$.

(b) Since the ratio of the whole interval at any stage to the larger sub-interval is equal to the ratio of the larger sub-interval to the smaller one, show that this result leads to the equation $\tau^2 = \tau + 1$. Solve this equation.

(c) Repeat Example 2 on page 77 by Golden Section search.

(d) Write a computer program for (c).

(e) Write a program to evaluate the first 100 Fibonacci numbers.

15 A second disadvantage of Powell's method is that x_m will correspond to a maximum if

$$\frac{(x_2 - x_3)f_1 + (x_3 - x_1)f_2 + (x_1 - x_2)f_3}{(x_1 - x_2)(x_2 - x_3)(x_3 - x_1)} \geq 0. \quad \text{Prove this statement.}$$

[Note: if either disadvantage arises, we take a maximum permissible step in the direction of decreasing function values and the new point replaces one of x_1, x_2 or x_3 as in the main statement of the algorithm.]

16 Minimise the following functions by Fibonacci search.
(a) $|\cos x|$ over $[-1,1]$ (b) $\cos x$ over $[0,2]$
(c) $e^{3x} - 3x + 1 - (x + 1)\ln(x + 1)$ in $[0,4]$
(d) $x^4 - 2x^2 - 4x + 3$ in $[1,2]$ (e) $(x - 1)^3 \ln x$

17 Repeat Problem 16 using Powell's algorithm.

18 Draw a flow diagram to represent in more detail the algorithm outlined below for locating by direction search the absolute maximum of a function $f(x)$ in the range $a \le x \le b$.
(a) Select an interval of search h
(b) Compute points $x_i = a + ih$, $f_i = f(x_i)$, $i = 1, 2, ...$
(c) At each stage in (b) examine whether f_i is greater than the previous maximum.
Write a program to determine by the above method the absolute maximum of
$f(x) = e^x \sin 6x + \ln(1 + \cos x)$ in the range $0 \le x \le \pi/3$.   (EC)

19 A function $f(x)$ can be evaluated for any given value of x in the range $a \le x \le b$ but its derivatives are not available. Derive an algorithm for finding a maximum from the following principles.
Take steps of length h from the point a. When a maximum turning point is located, take smaller steps from a suitable point and repeat. Stop when the step length is reduced below a given value, or if no turning point is detected in the first search.   (EC)

Section 3.5

20 Find the minimum of $z = 3x^2 + 2y^2 - 4xy - 3x - 7y + 10$, starting from the origin. Check your answer by calculus.   (LU)

21 Use a computer program to find the minimum of the function $f(x,y) = x^3 + y^3 - 12x^2 - 9y^2 + 256$. Try plotting the contours of $f(x,y)$ and mark on these the successive points reached by the method.

22 Minimise the following functions by alternating variables:
(a) $8x^2 - 4xy + 5y^2$; start from $(5,2)$

(b) $(x - 1)^2 + (y - 1)^2 + \left[\dfrac{x}{4} + \dfrac{y}{4} - 1 \right]^2$ from $(0,0)$

(c) $100(y - x^2)^2 + (1 - x)^2$ from $(0,0)$ {Rosenbrock's function}
(d) $100(y - x^3)^2 + (1 - x)^2$ from $(0,0)$
(e) $(x^2 + xy^3 - 9)^2 + (3x^2y - y^3 - 4)^2$ from $(0, 0.2)$
(f) $(x^2 + y^2 - 1)^2 + x^2y^2$ from $(0,1)$, $(0.8, 0.2)$, $(0.8, -0.2)$, $(1.2, 0.4)$

23 Repeat Problem 22 using the Rosenbrock method.

Section 3.6
24 How do the following functions behave at the origin
(i) $x^2 + xy + y^2 + x^3 + x^2y + y^3$ (ii) $x^2 + 2xy + y^2 + x^4 + x^2y^2 + y^4$?

25 Show that $z = x^3 + y^3 - 3\lambda(x + y) + 6xy$ has four stationary values where $\lambda > 3$. Show that when $\lambda = 4$, z has one minimum value which is $28 - 20\sqrt{5}$.   (LU)

26 Find the absolute maximum and minimum values of $f(x,y) = 5x + 2y$ in the region $x^2 + 4y^2 \le 1$.

27 Show that $z = x^2 + y^2 - 4xy^2$ has one stationary point and find its nature.   (LU)

28   Find the stationary values of $f(x,y) = x^3 + ay^2 - 6axy$, where $a > 0$ and determine whether they are maxima, minima or saddle points.   (LU)

29   Show that the sum of the squares of the distances of the point (x_0, y_0) from the vertices of a triangle in the x-y plane is least when (x_0, y_0) is the centroid of the triangle.   (LU)

30   Find the local maximum of the function $x^3 + y^3 + z^3 + 47.3$ subject to $x^2 - y^2 + z^2 = 1$.   (LU)

31   Find that point on the sphere $x^2 + y^2 + z^2 = 1$ furthest from the point $(1,3,2)$.   (LU)

32   A rectangular block with edges x, y, z is cut from a sphere of radius b so that the volume xyz has a maximum value. Find x, y, z and show that the maximum value is $8b^3/3\sqrt{3}$.   (LU)

33   State conditions sufficient to ensure that a function $f(x,y)$ of the two independent variables x, y should have a minimum value at the point (a,b).

A closed capsule is to be constructed of material of negligible thickness to have a constant volume V. It comprises a hollow right circular cylinder of fixed radius R but variable length z, surmounted at the ends by two right circular cones of base radius R and variable heights x and y. Show that the surface area S of the capsule may be written in the form

$$S = \frac{2V}{R} - \frac{2}{3}\pi R(x + y) + \pi R(x^2 + R^2)^{1/2} + \pi R(y^2 + R^2)^{1/2}.$$

If $V = 2\pi R^3 \sqrt{5}/3$ show that S assumes a minimum value when $x = y = z = 2R/\sqrt{5}$.

   (LU)

34   (i)   Show that the function $\dfrac{x^2 + y^2 + 1}{(x + y + 1)^2}$ is a minimum at the point $(1,1)$.

     (ii)   Use Lagrange's method of undetermined multipliers to prove that a closed tank, in the form of a circular cylinder with plane ends, has maximum volume for given surface area when the length is equal to the diameter of the ends.   (LU)

35   Using Lagrange multipliers find a point (x,y,z) on the unit sphere $x^2 + y^2 + z^2 = 1$ which minimises the function $x + y^2 + yz + 2z^2$.   (LU)

36   (i)   The lengths a, b, c vary so that $a + b + c = 3k$, where k is constant.

        Find the maximum volume enclosed by the ellipsoid $\dfrac{x^2}{a^2} + \dfrac{y^2}{b^2} + \dfrac{z^2}{c^2} = 1$.

     (ii)   Find the greatest value of the function $(x^2 - y^2)\exp[-x^2 - 2y^2]$.   (LU)

37   (i)   If $z = x^4 + y^4 - 2(x - y)^2$ show that z has minimum values at $(\sqrt{2}, -\sqrt{2})$ and $(-\sqrt{2}, \sqrt{2})$.

     (ii)   Positive numbers x, y and z are subject to the condition $x + y + z =$ constant. Prove that their product is greatest when they are equal.   (LU)

38 (i) Prove that the function $xy(3 - x - y)$ has a maximum when $x = y = 1$.

(ii) Use Lagrange's method of undetermined multipliers to find the shortest distance from the origin to the plane $2x - y + z = 12$. Check your result by the methods of coordinate geometry. (LU)

39 Describe briefly how the method of Lagrange's multipliers can be used in the determination of extrema of a function $f(x,y,z)$ subject to the constraint $g(x,y,z) = 0$.
A solid has the form of a right circular cylinder of radius R and height H surmounted by a right circular cone of height h whose base coincides with the upper end of the cylinder. If the solid is to have a given constant volume, use the above method to show that when its surface area is a minimum, $H : R : h = (\sqrt{5} + 1) : \sqrt{5} : 2$. (EC)

40 A closed right circular cylindrical storage tank is designed to hold 1000 m^3 of hot liquid. It is proposed to minimise the heat loss by making the surface area as small as possible. Use the method of Lagrange multipliers to obtain the optimum dimensions of the tank. (EC)

41 Use the method of Lagrange's multipliers to find the minimum of the function $f(x,y,z,t) = x^2 + y^2 + z^2 + t^2$ subject to the conditions $x + y - z + 2t = 2$, $2x - y + z + 3t = 3$. (EC)

42 A garbage bin is designed to fit into a corner and its shape is that of a quarter of a circular cylinder. It has a base and three sides but no lid and is made from thin uniform metal plate. It is required to hold a given volume. If the minimum amount of metal plate is to be used, show (using the Lagrange multiplier technique) that
$$h(4 + \pi) = \pi r$$
Show also that in this case the base of the bin accounts for one-third of the total metal plate. (EC)

43 A storage tank is in the form of a closed right cylinder with a regular hexagonal cross-section. The tank is to hold 4500 m^3 of liquid. Find, using Lagrange multipliers, the dimensions of the tank if the total surface area is to be minimised. (EC)

Section 3.7
44 Find the unit vectors which represent the directions of steepest descent for the function $z = 3x^2 + y^3 - 2x^2y + 10$ at the points $(1,1)$, $(2,1)$, $(1,2)$ (LU)

45 Using the steepest descent find the least value of $z = x^2 + xy + y^2 + x^3 + xy^2 + y^3 + 10$, starting from $(1,1)$. Why can we not reach $(0,0)$ in one step? (LU)

46 Repeat Problem 22, Section 3.5, using any of the methods described in this section.

47 Distinguish between the classical steepest descent method and a typical gradient method for minimisation of a function of several variables.
Starting from the origin locate, correct to 1 decimal place, by one of the above methods the minimum value of $f(x_1, x_2) = 2x_1^2 - 2x_1x_2 + 2x_2^2 - 6x_1 + 6$ in the region $x_1 \geq 0$, $x_2 \geq 0$, $x_1 + x_2 \leq 2$. (EC)

48 From the starting point $(1,1,1)$, take one step of the steepest descent method towards finding

the minimum of the function $f(x,y,z) = 2x^2 + 2xy + y^2 + yz + z^2$.   (EC)

49 Describe in detail a gradient method for minimizing the unconstrained function $f(x_1, x_2, ..., x_n)$. Starting from the point $(-1, 2)$, find the value, correct to 2 decimal places, of the function $F = (x^2 + y^2 - 1)^2 + (x + y - 1)^2$ after taking one step using the Newton-Raphson method.   (EC)

50 Two stages of the Newton-Raphson algorithm for the minimization of a function of several variables are as follows:
(i)   Form $h = Hg$ where h and g are vectors and H is a matrix
(ii) If all elements of h are not greater in magnitude than a given quantity ϵ, stop; otherwise replace x by $x - h$, where x is also a vector.
Write a section of program in a stated code to perform these two stages, assuming the vectors have m elements.
Also write a section of program to compute the gradient vector g in the particular case when the function to be minimized is $(\sin(x_1 x_2) - x_2)^2 + (e^{2x} - 1)\sin x_2$.   (EC)

51 Derive the equations for the components of the vector ξ where $x + \xi$ is a single step from the point x in the Newton-Raphson direction for maximising (or minimising) the function $f(x)$.
Starting at the point $(0, -1)$ find the new value of the function

$$f(x,y) = (x - y)^2 + \frac{1}{9}(x + y - 10)^2$$

after taking
(a) one step using the above method   (b) two steps using the gradient method.
Comment briefly on these results.   (EC)

52 The iteration $x^{(k+1)} = x^{(k)} + Af(x^{(k)})$ where x is a vector, f is a vector of functions, and A is a constant matrix, converges in certain circumstances to a root of $f(x) = 0$.

If $x^{(0)} = \begin{bmatrix} 2 \\ 2 \end{bmatrix}$, $f(x) = \begin{bmatrix} x^2 + y^2 - 9 \\ x - y^2 + 1 \end{bmatrix}$ and $A = -\frac{1}{10}\begin{bmatrix} 2 & 2 \\ 1 & -2 \end{bmatrix}$

use the above algorithm to find a solution of $f(x) = 0$ near $x^{(0)}$ correct to two decimal places. State concisely how a suitable matrix A could be found when f is (a) differentiable and (b) non-differentiable, and state how the convergence rate could be improved.   (EC)

4

ORDINARY DIFFERENTIAL EQUATIONS

4.1 Introduction

This chapter is concerned with numerical methods of solving initial-value ordinary differential equations and, via difference equations, the stability of these methods. We introduce examples of **predictor-corrector** methods which comprise two parts: a predictor formula which generates an estimate of a new value for the dependent variable and a corrector formula which aims to improve upon the predicted value. Then we examine features of systems of differential equations related to problems arising in practice. Finally, we turn our attention to boundary-value problems.

The numerical methods are presented for solving equations in the form $y' = f(x,y)$. Equations of higher order can be solved in the equivalent form of a system of first order equations. For example, the equation

$$y'' + P(x)y' + Q(x)y = R(x)$$

can be formulated as the system

$$y' = z \qquad z' = R(x) - Q(x).y - P(x).z$$

This is an example of the general formulation $y' = f(x,y,z)$, $z' = g(x,y,z)$.

4.2 One-Step and Multistep Methods

The objective in solving numerically an initial-value differential equation $y' = f(x,y)$, with initial condition $y(0) = y_0$, is to generate a sequence of points on the solution curve, viz, (x_1, y_1), (x_2, y_2), (x_3, y_3), ..., (x_m, y_m). **One step methods** use only the previously-obtained solution point (x_n, y_n) to generate the next point (x_{n+1}, y_{n+1}).

The most popular one-step methods are the **Runge-Kutta methods.** They are designed to give the same values as a Taylor series truncated to the term in h^p, where h is the (constant) step size $(x_{n+1} - x_n)$, and p is the **order** of the method. The advantage they possess is that they require only the evaluation of $f(x,y)$ for certain pairs of values (x,y), and then take a weighted average of these values to approximate the change $y_{n+1} - y_n$; they do not require derivatives.

The simplest such method is **Euler's method** which effectively approximates the solution via Taylor's series up to the term in h. It is therefore a linear approximation. The working formula is

$$y_{n+1} = y_n + hf(x_n, y_n) \tag{4.1}$$

Second-order Runge-Kutta Methods

For second-order methods, the assumption made is that the value y_{n+1} can be computed from a formula of the general kind

$$y_{n+1} = y_n + ak_1 + bk_2 \tag{4.2}$$

where $k_1 = hf(x_n, y_n)$ and $k_2 = hf(x_n + \alpha h, y_n + \beta k_1)$

The four unknowns a, b, α and β have to be chosen so that the formula (4.2) is equivalent to a quadratic approximation

$$y_{n+1} \cong y_n + h\, y'(x_n) + \tfrac{1}{2}h^2\, y''(x_n) \tag{4.3}$$

We can expand k_2 by a Taylor's series and when we match formulae (4.2) and (4.3) the following relationships are obtained

$$a + b = 1 \qquad a\alpha = b\beta = \tfrac{1}{2} \tag{4.4}$$

There are only three equations connecting four unknowns and, therefore, one degree of freedom. Choosing $a = \tfrac{1}{2}$ produces $b = \tfrac{1}{2}$, $\alpha = \beta = 1$. Formula (4.2) then becomes

$$y_{n+1} = y_n + \frac{h}{2}\left\{ f(x_n, y_n) + f(x_n + h, y_n + hf[x_n, y_n]) \right\} \tag{4.5}$$

which is the **Improved Euler Method**.

It can be shown that the local truncation error is given by

$$y_{n+1} - y_n = \frac{h^3}{12}(f_{xx} + 2f \cdot f_{xy} + f^2 \cdot f_{yy} - 2f_x f_y - 2f \cdot f_y^2) + O(h^4)$$

This is so complicated as to render error estimation very tedious. However, we see that the local truncation error is of order h^3 as opposed to Euler's method where this error is of order h^2. Hence the extra evaluation of $f(x,y)$ is compensated by the larger step size that we can use to maintain a required accuracy.

Fourth-order methods

By far the most popular Runge-Kutta methods are those of the fourth order, which implies a local truncation error of order h^5. One algorithm is as follows.

$$\left.\begin{aligned}
k_1 &= hf(x_n, y_n) & k_2 &= hf(x_n + \tfrac{1}{2}h,\ y_n + \tfrac{1}{2}k_1) \\[2mm]
k_3 &= hf(x_n + \tfrac{1}{2}h,\ y_n + \tfrac{1}{2}k_2) & k_4 &= hf(x_n + h,\ y_n + k_3) \\[2mm]
y_{n+1} &= y_n + \tfrac{1}{6}(k_1 + 2k_2 + 2k_3 + k_4)
\end{aligned}\right\} \tag{4.6}$$

The difficulty is that without a knowledge of the truncation error it is not easy to choose a suitable step-size h. Collatz suggested that if the quantity $|(k_2 - k_3)/(k_1 - k_2)|$ becomes greater than a few hundredths it is wise to decrease h.

Merson derived a modification of this fourth-order method which kept an error of order h^5, but which allowed an error estimate to be made relatively easily. The snag is that $f(x,y)$ has to be evaluated five times. The equations are:

$$\left.\begin{array}{ll} k_1 = hf(x_n, y_n) & k_2 = hf(x_n + \tfrac{1}{3}h, \ y_n + \tfrac{1}{3}k_1) \\[2mm] k_3 = hf(x_n + \tfrac{1}{3}h, \ y_n + \tfrac{1}{6}k_1 + \tfrac{1}{6}k_2) \quad k_4 = hf(x_n + \tfrac{1}{2}h, \ y_n + \tfrac{1}{8}k_1 + \tfrac{1}{8}3k_3) \\[2mm] k_5 = hf(x_n + h, \ y_n + \tfrac{1}{2}k_1 - \tfrac{1}{2}3k_3 + 2k_4) \\[2mm] y_{n+1} = y_n + \tfrac{1}{6}(k_1 + 4k_4 + k_5) \end{array}\right\} \quad (4.7)$$

We may estimate the local truncation error ε_τ by

$$\varepsilon_\tau \cong \frac{1}{30}(2k_1 - 9k_3 + 8k_4 - k_5) \qquad (4.8)$$

If $f(x,y)$ is of the form $ax + by + c$, where a, b and c are constants, then formula (4.8) is exact.

Suppose the evaluation of $f(x,y)$ is quite involved. Then the extra calculation in (4.7) may become quite expensive in computation time. In addition, the Runge-Kutta methods, being one-step methods, are quite wasteful since, having proceeded from (x_n, y_n) to (x_{n+1}, y_{n+1}), we have to start from scratch to proceed to (x_{n+2}, y_{n+2}); all previous information (except for y_{n+1}) is abandoned.

There is another disadvantage to Runge-Kutta methods. For algebraic convenience we illustrate the point via a second-order method. Consider the equation $dy/dx = -8y$ with the condition $y(0) = 1$; this has the analytical solution $y = e^{-8x}$. The second-order formula (4.5) becomes

$$y_{n+1} = y_n + \frac{h}{2}[-8y_n - 8(y_n - h \cdot 8y_n)] = y_n(1 - 8h + 32h^2)$$

which is equivalent to the first three terms of the expansion $y_n(e^{-8h})$. It follows from the initial condition that $y_{n+1} = (1 - 8h + 32h^2)^{n+1}$.

This result can be obtained from any second-order Runge-Kutta method. Its significance lies in the fact that $(1 - 8h + 32h^2) > 1$ if $h > 0.25$ and y_{n+1} grows indefinitely larger as n increases. However, the true solution, $y = e^{-8x}$, *decreases* as x, and hence n, increases. To match such behaviour we would have to choose a step size smaller than 0.25. This feature is known as **partial instability**. The phenomenon is *not* restricted to cases where the true solution shows an exponential decay.

Multistep methods

We attempt to use information previously calculated in making our next step.

An example of such a method is the Adams-Bashforth formula

$$y_{n+1} = y_n + h(1 + \frac{1}{2}\nabla + \frac{5}{12}\nabla^2 + \frac{3}{8}\nabla^3 + \frac{251}{720}\nabla^4 + ...)f_n \tag{4.9}$$

or its equivalent *ordinate* form (as far as the term in $\nabla^3 f_n$)

$$y_{n+1} = y_n + \frac{h}{24}(55f_n - 59f_{n-1} + 37f_{n-2} - 9f_{n-3}) \tag{4.10}$$

You will notice that this formula requires three values prior to f_n to get the method started. A useful ploy is to use a Runge-Kutta method three times with the initial value to obtain y_0, y_1, y_2 and y_3, and hence f_0, f_1, f_2 and f_3. The Adams-Bashforth formula (4.10) can then be used for further values y_n.

The local truncation error in formula (4.10) is given by $\varepsilon_\tau = \frac{251}{720}y^{(v)}(\xi)$ where $x_n < \xi < x_{n+1}$.

Example

The equation $dy/dx = x - y^2$ is given, together with the initial condition $y(0) = 0$. We require to estimate y for $x = 0.2, 0.4, 0.6, ..., 1.0$.

We can obtain the Maclaurin series for y:

$$y = \frac{1}{2}x^2 - \frac{1}{20}x^5 + \frac{1}{160}x^8 - ... \tag{4.11}$$

We could then use this series to calculate y for $x = 0.2, 0.4$ and 0.6 and then embark with formula (4.10). However, we shall assume that we are using a digital computer throughout and estimate these values via the fourth-order Runge-Kutta method (4.6). Notice that if we use (4.10) we are predetermining the local truncation error; we could alternatively use formula (4.9) with as many terms as a prescribed accuracy required.

Let us work the first step with the Runge-Kutta method.

Here $h = 0.2$ and $f(x,y) = x - y^2$.

Then $k_1 = 0.2(0 - 0^2) = 0$

$k_2 = 0.2(0.1 - 0^2) = 0.02$

$k_3 = 0.2(0.1 - (0 + 0.01)^2) = 0.2(0.0999) = 0.01998$

$k_4 = 0.2(0.2 - (0 + 0.01998)^2) = 0.2(0.19961) = 0.03992$

$y_1 \cong 0 + \frac{1}{6}(0 + 0.04 + 0.03996 + 0.03992) = \frac{1}{6}(0.11988) = 0.01998$

We can compute two further values y_2 and y_3 and thence we obtain $f_0 = 0$, $f_1 = 0.199600$, $f_2 = 0.393682$ and $f_3 = 0.568952$. Using (4.10) we have $y_4 = 0.304957$.

We continued the application of (4.10) to obtain the remaining value of y_n. Results are tabulated in Table 4.1, to 6 decimal places. How many of these decimal places are justified?

Table 4.1

x_n	y_n	$f_n = x_n - y_n^2$
0.2	0.01998	0.199600
0.4	0.079484	0.393682
0.6	0.176204	0.568952
0.8	0.304957	0.707001
1.0	0.455680	

Comparison of approaches

Runge-Kutta methods do not require information from previously calculated data points and are therefore self-starting. This allows an easy change in the step size. However, they require several evaluations of $f(x,y)$ at each step and demand much computing time. Furthermore, they do not easily provide information about the local truncation error.

Multistep methods, on the other hand, *do* require information about previously calculated data points. They can also provide a reasonable estimate of the local truncation error. A drawback is that to change the step size requires reversion to a Runge-Kutta method to provide sufficient starting values.

A point to bear in mind is that when a Runge-Kutta method is used to provide starting values for a multistep method, the former should have a local truncation error of the same order as the multistep method. However, if the truncation error of the multistep method is of order higher than h^5, a fourth-order Runge-Kutta method *can* be used with a step size which is an integer divisor of the step size of the multistep method.

4.3 Predictor-Corrector Methods

In this section we study two methods from a general class of predictor-corrector methods. First, however, we mention some general points.

Most of the formulae can be derived by integrating the differential equation between two values of x at which tabulated values of y are required.

A **predictor** formula can be based on the integral

$$y(x_{n+1}) - y(x_{n-s}) = \int_{x_{n-s}}^{x_{n+1}} f(x,y)\mathrm{d}x \qquad (4.12a)$$

If we approximate $f(x,y)$ by an interpolating polynomial $P(x)$ (the interpolating points being $x_n, x_{n-1}, x_{n-2}, ..., x_{n-s+1}, ...$) we obtain the formula

$$y^P_{n+1} = y_{n-s} + \int_{x_{n-s}}^{x_{n+1}} P(x)\mathrm{d}x \qquad (4.12b)$$

where y_{n-s} is an approximation to $y(x_{n-s})$ and y^P_{n+1} is the obtained approximation to $y(x_{n+1})$. This is an **explicit** formula since x_{n+1} is not used in defining $P(x)$. The integration extends past x_n and therefore we are extrapolating $P(x)$ to x_{n+1}; hence (4.12b) is a predicting formula.

Having obtained a predicted value y^P_{n+1} we specify a new interpolating polynomial $Q(x)$ which is defined also at x_{n+1}. Then we obtain a corrector formula from

$$y_{n+1} - y_{n-r} = \int_{x_{n-r}}^{x_{n+1}} f(x,y)\mathrm{d}x$$

where r may not be the same as s in (4.12).

The formula is $\qquad y^c_{n+1} = y_{n-r} + \int_{x_{n-r}}^{x_{n+1}} Q(x)\mathrm{d}x \qquad (4.13)$

This is an **implicit** or **closed** formula. (y^c_{n+1} is the corrected value.)

We can use the predictor formula once to estimate y_{n+1} and then apply (4.13) as an iterative formula to generate successively better approximations to y_{n+1}. It is usually preferred, however, to use the corrector formula once only and to choose a small enough step size to allow for reasonably accurate representation of y_{n+1}.

Milne-Simpson method

Starting from the equation $y' = f(x,y)$ we integrate between x_{n+1} and x_{n-3} as follows.

$$\int_{x_{n-3}}^{x_{n+1}} \frac{\mathrm{d}y}{\mathrm{d}x}\,\mathrm{d}x = \int_{x_{n-3}}^{x_{n+1}} f(x,y)\mathrm{d}x \cong \int_{x_{n-3}}^{x_{n+1}} P(x)\mathrm{d}x \qquad (4.14)$$

where $P(x)$ is a quadratic interpolating polynomial agreeing with $f(x,y)$ at x_n, x_{n-1} and x_{n-2}. Notice that we extrapolate to x_{n+1} and to x_{n-3} which could give rise to a sizeable error. The polynomial $P(x)$ is given by the formula

$$P(x) \equiv P(x_n + rh) = f_n + r\nabla f_n + \frac{r(r+1)}{2}\nabla^2 f_n$$

$$= \tfrac{1}{2}(r+2)(r+1)f_n - r(r+2)f_{n-1} + \tfrac{1}{2}r(r+1)f_{n-2}$$

Hence $\displaystyle\int_{x_{n-3}}^{x_{n+1}} P(x)\mathrm{d}x \;=\; \int_{-3}^{1} P(x_n + rh).h\ \mathrm{d}r$

$$= h \int_{-3}^{1} (f_n + r\nabla f_n + \tfrac{1}{2}r(r+1)\ \nabla^2 f_n)\mathrm{d}r$$

$$= h(4f_n - 4\nabla f_n + \frac{8}{3}\nabla^2 f_n) = h(\frac{8}{3}f_n - \frac{4}{3}f_{n-1} + \frac{8}{3}f_{n-2})$$

Since $\displaystyle\int_{x_{n-3}}^{x_{n+1}} \frac{\mathrm{d}y}{\mathrm{d}x}\ \mathrm{d}x = y_{n+1} - y_{n-3}$ we establish the Milne predictor formula

$$y_{n+1} = y_{n-3} + \frac{4h}{3}(2f_n - f_{n-1} + 2f_{n-2}) \tag{4.15}$$

By recourse to the error in the interpolation polynomial it can be shown that the truncation error is given by

$$\varepsilon_\tau \equiv (28/90)h^5\ y^{(v)}\ (\xi_1), \quad \text{where } x_{n-3} < \xi_1 < x_{n+1} \tag{4.16}$$

We may then estimate $f(x_{n+1}, y_{n+1})$.

The corrector formula which completes the Milne-Simpson method is found by integrating the differential equation over the range $[x_{n-1}, x_{n+1}]$. The interpolating polynomial is not now extrapolated; it agrees with $f(x,y)$ at x_{n-1}, x_n and x_{n+1}. The **Simpson corrector formula**, so called because of its resemblance to the Simpson integration rule, is

$$y_{n+1} = y_{n-1} + \frac{h}{3}(f_{n+1} + 4f_n + f_{n-1}) \tag{4.17}$$

We quote the truncation error

$$\xi_\tau \equiv \frac{1}{90}h^5\ y^{(v)}\ (\xi_2), \quad \text{where } x_{n-1} < \xi_2 < x_{n+1} \tag{4.18}$$

Example

We apply the method to the differential equation $\mathrm{d}y/\mathrm{d}x = x - y^2$ with the starting values given on page 106. The results are shown in Table 4.2.

Table 4.2

x_n	y_n	$f_n = x_n - y_n^2$	y^p_{n+1}	f^p_{n+1}	y^c_{n+1}
0	0	0			
0.2	0.01998	0.199600			
0.4	0.079484	0.393682			
0.6	0.176204	0.568952	0.304913	0.707028	0.304586
0.8	0.304586	0.707228	0.455411	0.792601	0.455568
1.0	0.455568				

Adams-Moulton method

The derivation of the predictor and corrector formulae for this method follows a similar line to that used for the Milne-Simpson method. For the predictor formula we approximate $f(x,y)$ by a cubic agreeing with it at $x_{n-3}, x_{n-2}, x_{n-1}$ and x_n, and we perform the integration over $[x_n, x_{n+1}]$. The formula we obtain is

$$y_{n+1} = y_n + (h/24)(55f_n - 59f_{n-1} + 37f_{n-2} - 9f_{n-3}) \qquad (4.10)$$

which is just the Adams-Bashforth predictor formula, obtained earlier.

As we have stated, the truncation error is given by

$$\varepsilon_\tau = \frac{251}{720} y^{(v)} (\xi_1), \quad \text{where } x_n < \xi_1 < x_{n+1} \qquad (4.19)$$

Having predicted y_{n+1}, we may estimate $f_{n+1} \equiv f(x_{n+1}, y_{n+1})$ and then apply the **Adams-Moulton correction formula**. In this case we approximate $f(x,y)$ by a cubic agreeing at x_{n-2}, x_{n-1}, x_n and x_{n+1} and again integrate over $[x_n, x_{n+1}]$. The formula obtained is

$$y_{n+1} = y_n + (h/24)(9f_{n+1} + 19f_n - 5f_{n-1} + f_{n-2}) \qquad (4.20)$$

with truncation error given by

$$\varepsilon_\tau = \frac{19}{720} y^{(v)} (\xi_2), \quad \text{where } x_{n-2} < \xi_2 < x_{n+1} \qquad (4.21)$$

We saw that (4.10) could be expressed in finite difference form as

$$y_{n+1} = y_n + h(f_n + \frac{1}{2}\nabla f_n + \frac{5}{12}\nabla^2 f_n + \frac{3}{8}\nabla^3 f_n)$$

In the same way, (4.20) can be written as

$$y_{n+1} = y_n + h(f_{n+1} - \frac{1}{2}\nabla f_{n+1} - \frac{1}{12}\nabla^2 f_{n+1} - \frac{1}{24}\nabla^3 f_{n+1})$$

Example

Consider again the differential equation $dy/dx = x - y^2$; $y(0) = 0$. This time we need four starting values. On page 106 we really did half the stages. Let us assume that we have already calculated y_1, y_2 and y_3 by the Runge-Kutta method. We then estimate y_4 by (4.10) as before to obtain the value 0.304957.

Now we are able to estimate $f_4 = x_4 - y_4^2 = 0.707001$.

The corrector formula (4.20) is used to obtain a revised estimate for y_4. It produces $y_4 = 0.304573$.

In Table 4.3 we have carried forward the solution and we show the predicted and corrected values of y_{n+1}.

Table 4.3

x_n	y_n	f_n	$y^p{}_{n+1}$	$f^p{}_{n+1}$	$y^c{}_{n+1}$
0	0	0			
0.2	0.01998	0.199600			
0.4	0.079484	0.393682			
0.6	0.176204	0.568952	0.304957	0.707001	0.304573
0.8	0.304573	0.707235	0.455403	0.792608	0.455572
1.0	0.455572				

Accuracy and Convergence

It has been stated that we could, if we wished, use the corrector formula repeatedly to obtain successively better approximations of y_{n+1}. Does the corrector formula necessarily converge, and if it does, will it be to the true value of $y(x_{n+1})$?

In the Milne-Simpson and Adams-Moulton methods the practice is not to use the corrector formula iteratively. Instead we try to use a step size h which is small enough to render re-corrections unnecessary. For the Milne-Simpson method the condition that successive re-corrections converge is:

$$h < 3/|f_y(x_n, y_n)| \qquad (4.22)$$

The condition that the first corrected value is adequate to N decimal places and will not be altered in the Nth decimal place by further corrections is:

$$|y^c{}_{n+1} - y^p{}_{n+1}| < \frac{3}{10^N \, h \, |f_y(x_n, y_n)|} \qquad (4.23a)$$

where $y^c{}_{n+1}$ and $y^p{}_{n+1}$ are the corrected and predicted values of y_{n+1} via (4.17) and (4.15). Finally, a consideration of the predictor and corrector formulae would suggest that when the predicted and corrected values of y_{n+1} do not agree, the true value is likely to lie between them and nearer to the corrector value. A working criterion for accuracy to N decimal places is

$$|y^c{}_{n+1} - y^p{}_{n+1}| < 29/10^N \qquad (4.23b)$$

For the **Adams-Moulton method** the criteria are, respectively,

$$h < \frac{24}{9 \, |f_y(x_n, y_n)|} \qquad (4.24a)$$

$$|y^c_{n+1} - y^p_{n+1}| < \frac{24}{9 \cdot 10^N \, h \, |f_y(x_n, y_n)|} \qquad (4.24b)$$

and
$$|y^c_{n+1} - y^p_{n+1}| < \frac{14}{10^N} \qquad (4.24c)$$

Example

The criteria for the Adams-Moulton method are checked on the results obtained on page 111 and extended to $x_n = 1.8$.

x_n	y_n	x_{n+1}	y^p_{n+1}	y^c_{n+1}
1.8	1.071014	2.0	1.194004	1.192770

Since $y' = x - y^2$, $f_y(x,y) = -2y$. Hence $f_y(x_n, y_n) = -2.142028$
Criterion (4.24a) requires $h < 1.2449$, which it is.
Criterion (4.24b) requires $10^N < 24/[9|y^c_{n+1} - y^p_{n+1}|(0.2)(2.142028)]$
     i.e. $N < 3.7$.
Criterion (4.24c) similarly requires $N < 4.05$, so that we can rely on 3 d.p. only.

It should be noted that the criteria do **not** apply in a straight-forward extension to systems of first-order equations. The criteria for such systems are very complicated and beyond our scope.

If we are working with numbers markedly different from 1, a criterion to be used in place of (4.24c) might be

$$|y^c_{n+1} - y^p_{n+1}| / |y^c_{n+1}| < \frac{14}{10^N}$$

For a general program we might adopt the following check at each step. If the relative error exceeds a prescribed upper bound, halve the step, compute four starting values and return to the predictor-corrector formulae. If the relative error is less than a second prescribed value, indicating that the step size is so small that we are obtaining too much accuracy for our purposes, double the step size, compute four starting values and return to the predictor-corrector formulae.

4.4 Linear Difference Equations

In studying the stability of numerical methods of solving ordinary differential equations we encounter linear difference equations with constant coefficients. These are equations of the form

$$y_{k+r} + a_{r-1} y_{k+r-1} + \dots + a_1 y_{k+1} + a_0 y_k = f(k) \qquad (4.25)$$

where k and r are integers.

The theory of linear difference equations is very closely aligned to the corresponding theory of linear differential equations.

Assuming that $a_0 \neq 0$ then equation (4.25) is said to be of **order** r.

(a) **First-order homogeneous equations**

Consider a typical equation $y_{k+1} = Ay_k$ for A constant.

Suppose that we are given the initial condition $y_0 = \alpha$, a constant.

Then $y_1 = Ay_0 = A\alpha$, $y_2 = Ay_1 = A^2\alpha$ and, in general,

$$y_k = A^k \alpha \qquad (4.26)$$

This solution satisfies the difference equation and the initial condition; it is therefore *the* solution (we have not *proved* this statement).

Example $y_{n+1} = 3y_n$; $y_0 = 1$. The solution is $y_n = 3^n$.

(b) **First-order non-homogeneous equations**

Consider $y_{k+1} = Ay_k + B$, where A and B are constants.

Then $y_1 = Ay_0 + B$, $y_2 = Ay_1 + B = A^2y_0 + AB + B$, $y_3 = Ay_2 + B = A^3y_0 + A^2B + AB + B$ and so $y_k = A^ky_0 + B(A^{k-1} + A^{k-2} + ... + A + 1)$

Hence $\qquad y_k = \begin{cases} A^ky_0 + B(A^k - 1)/(A - 1), & A \neq 1 \\ y_0 + Bk, & A = 1 \end{cases}$ $\qquad (4.27)$

Example $\qquad 2y_{k+1} = y_k + 4$, $y_0 = 3$.

First we write the equation in the form $y_{k+1} = \frac{1}{2}y_k + 2$ and then apply

(4.27) to obtain $y_k = 4 - (\frac{1}{2})^k$.

(c) **Second-order homogeneous equations**

The equations are of the form $y_{k+2} + \alpha y_{k+1} + \beta y_k = 0$ where α and β are constant. Two initial conditions y_0 and y_1 are required to determine the solution uniquely. We look for two linearly independent solutions. The technique is to substitute $y_k = m^k$ to obtain the **auxiliary equation** $m^2 + \alpha m + \beta = 0$.

Examples

(i) $\qquad y_{k+2} - 3y_{k+1} + 2y_k = 0$; $y_0 = 3/2, y_1 = 2$ $\qquad (4.28)$

Substitution of $y_k = m^k$ produces the auxiliary equation $m^2 - 3m + 2 = 0$ which has roots $m = 1$ and $m = 2$.

The general solution is $y_k = A.2^k + B.1^k \equiv A.2^k + B$ where A and B are constants. Since $y_0 = 3/2$, it follows that $3/2 = 1.A + B$. Similarly, from $y_1 = 2$ we obtain $2 = 2A + B$. Hence $A = 1/2$, $B = 1$ and $y_k = 2^{k-1} + 1$.

(ii) $y_{k+2} + 6y_{k+1} + 9y_k = 0$; $y_0 = 1, y_1 = 2$ (4.29)
The auxiliary equation is $m^2 + 6m + 9 = 0$ which has a repeated root
$m = -3$. By analogy with differential equations we may quote the general
solution as $y_k = (Ak + B)(-3)^k$.
From $y_0 = 1$ and $y_1 = 2$ we find that $1 = (B)1$ and $2 = (A + B)(-3)$ so
that $B = 1, A = -5/3$ to give the solution $y_k = (1 - 5k/3)(-3)^k$.

(iii) $y_{k+2} - 2y_{k+1} + 2y_k = 0$; $y_0 = 0, y_1 = 1$ (4.30)
The auxiliary equation is $m^2 - 2m + 2 = 0$ which has roots
         $m = 1 + i = \sqrt{2}(\cos \pi/4 + i \sin \pi/4) = \sqrt{2}\ e^{i\pi/4}$
and $m = 1 - i = \sqrt{2}(\cos \pi/4 - i \sin \pi/4) = \sqrt{2}\ e^{-i\pi/4}$
The general solution is $y_k = (\sqrt{2})^k (A \cos k\pi/4 + B \sin k\pi/4)$ and, from
the initial conditions, $A = 0$ and $B = 1$, so that $y_k = (\sqrt{2})^k \sin k\pi/4$.

(d)   **Second-order non-homogeneous equations**
       We look for *the* solution in the form: *a* particular solution + *the* comple-
       mentary solution. To find a particular solution we employ the method of
       trial solutions. For the equation $y_{k+2} + \alpha y_{k+1} + \beta y_k = f(k)$ we present
       in Table 4.4 suitable trial solutions for several forms of $f(k)$.

Table 4.4

$f(k)$	Trial solution
a^k, a constant	Aa^k, A constant
$\left.\begin{matrix}\sin bk \\ \cos bk\end{matrix}\right\}$	$A \sin bk + B \cos bk$, A, B constant
k^h	$A_0 + A_1 k + ... + A_h\ k^h$, A_i constant
$k^h \cdot a^k$	$a^k(A_0 + A_1 k + ... + A_h\ k^h)$
$\left.\begin{matrix}a^k \sin bk \\ a^k \cos bk\end{matrix}\right\}$	$a^k(A \sin bk + B \cos bk)$
$f_1(k) + f_2(k)$	Trial solution for $f_1(k)$ + trial solution for $f_2(k)$

Example $y_{k+2} - 6y_{k+1} + 9y_k = 2k + 3k^2$
The auxiliary equation is $m^2 - 6m + 9 = 0$ which leads to the
complementary solution $y = (Ak + B)3^k$.
We try a particular solution of the form $y_k = (\alpha_0 + \alpha_1 k + \alpha_2 k^2)$
Then $y_{k+1} = \alpha_0 + \alpha_1(k + 1) + \alpha_2(k + 1)^2$
and $y_{k+2} = \alpha_0 + \alpha_1(k + 2) + \alpha_2(k + 2)^2$
Substitution into the difference equation produces

$$\alpha_0 + \alpha_1(k+2) + \alpha_2(k+2)^2 - 6\alpha_0 - 6\alpha_1(k+1) - 6\alpha_2(k+1)^2$$
$$+ 9\alpha_0 + 9\alpha_1 k + 9\alpha_2 k^2 = 2k + 3k^2.$$

Equating the coefficients of like powers of k we obtain the conditions

$$\alpha_2 - 6\alpha_2 + 9\alpha_2 = 3 \Rightarrow \alpha_2 = 3/4$$
$$\alpha_1 + 4\alpha_2 - 6\alpha_1 - 12\alpha_2 + 9\alpha_1 = 2 \Rightarrow \alpha_1 = 2$$
$$\alpha_0 + 2\alpha_1 + 4\alpha_2 - 6\alpha_0 - 6\alpha_1 - 6\alpha_2 + 9\alpha_0 = 0 \Rightarrow \alpha_0 = 19/8$$

A particular solution is $y_k = (19/8) + 2k + (3/4)k^2$

Therefore the general solution is $y_k = (Ak + B)3^k + (19/8) + 2k + (3/4)k^2$.

4.5 Stability of Numerical Procedures

Section 4.2 introduced the phenomenon of partial instability. Roughly speaking, stability is the phenomenon of the computed solution behaving like the true solution of the differential equation. Stability analyses are concerned with the way errors propagate through the solution. Much of the study of stability has been confined to the equation $dy/dx = \lambda y$, for a constant λ; this is because the general problem is so very complicated. The basic results are summarised below.

Stability may depend upon the differential equation under study, the method used for solution, and the step size, h.

There is the possibility of **inherent instability** where small variations in the initial conditions attached to the problem give rise to large variations in the true solution. Such ill-conditioning must be reflected in the numerical scheme and will occur whatever method is used.

Consider the differential equation $dy/dx = y - 2x$. This has the analytical solution $y = Ae^x + 2x + 2$. Suppose that the initial conditions are $y = 2$ when $x = 0$; then $A = 0$ and the particular solution becomes $y = 2x + 2$. The term Ae^x, which would dominate the solution for large values of x, is absent. However, round-off error and truncation error will ensure that e^x will appear in the solution, albeit with a small coefficient. This will mean that for large values of x, this spurious term will swamp the true solution. Since the phenomenon is a feature of the differential equation, the only hope is to reformulate the problem.

Partial instability occurs when the numerical method concerned represents the true solution badly for step sizes above a certain value.

Certain finite difference schemes are unsuitable for use with some differential equations, no matter how small the step size may be. For other differential equations, the schemes may work quite well. This phenomenon is called **weak instability**.

There are situations in which the method, despite a low truncation error, introduces spurious solutions which increase, in a way which cannot be controlled by the step size, for any differential equation. This is the disastrous property of **strong instability**. This often results because a high-accuracy corrector formula has been employed which results in several spurious solutions, at least one of which grows uncontrollably.

General test for stability

A general test for stability is the following. Find the appropriate difference equation when the numerical method is applied to a particular equation. Let the roots of the difference equation be $m_1, m_2, ..., m_k$ where k is the order of the difference equation. Then the general solution of the difference equation may be written $y_r = a_1 m_1{}^r + a_2 m_2{}^r + ... + a_k m_k{}^r$ where a_i are constants. One of the roots, say m_1, will lead to the exact solution as $h \to 0$. If the roots satisfy $|m_i| < 1$ then the method is said to be **strongly stable**. If any of the extraneous roots satisfy $|m_i| > 1$ then errors will grow exponentially and the method is **unstable**.

Schemes which are weakly stable should not be integrated over long intervals.

In the next section we apply a predictor-corrector method of Hamming to a practical problem. Hamming's method is to compromise between stability and accuracy. The price that is paid for good stability is a fairly high truncation error. The problem involves a *system* of first-order equations arising from a second-order differential equation.

4.6 Case Study – Surge Tank

In a hydro-electric scheme, water may be conveyed to the turbines by a very long pipe. The very large mass of water will require considerable forces to retard or accelerate it when the demands of the turbines change. There might be a change in demand of output or there might be a sudden failure in the system which cannot be coped with, except by the incorporation of a surge tank near the power station. In Figure 4.1(a) we represent schematically a typical set-up. If the flow to the turbine increases, the level of water in the surge tank falls and hence an acceleration head is produced in the pipe; conversely, a reduction in flow to the turbines causes an increase in water level in the surge tank which in turn produces a retarding head in the pipe. Another beneficial effect is the reduction of water-hammer in the pipe. When the setting of the valve controlling flow to the turbine is altered there is an oscillation of the water level in the surge tank until a steady state is reached. It is important often to know the maximum and minimum water levels in the tank and also the frequency of oscillation; there is the possibility of prolonged oscillations leading to resonance. A further problem is whether the system is in stable equilibrium.

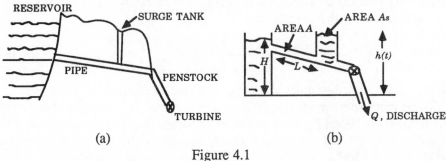

Figure 4.1

The pipe is assumed to be rigid and the water to be incompressible. This latter assumption is acceptable if the duration of an oscillation in the tank is much greater than the time taken for a pressure wave to travel along the pipe. Refer to Figure 4.1(b) during the derivation of the governing model equations.

The elevation of the water surface in the surge tank will be given by $h(t) = h_0 + y(t)$ where h_0 is the equilibrium elevation and we let the velocity of water in the pipe be given by $v(t) = v_0 + x(t)$ where v_0 is the velocity corresponding to steady discharge.

Two new variables $y(t)$ and $x(t)$ are introduced which represent respectively departures from equilibrium values of head in the surge tank and velocity of water in the pipe. The momentum equation is found by applying Newton's second law to the water in the pipe. Assuming that there is an upward velocity $u(t)$ of water in the surge tank, it can be written

$$\frac{L}{g}\frac{dv}{dt} + (h - h_0) + F_p\,v|v| + F_\tau\,u|u| = 0 \qquad (4.31)$$

where F_p is the friction coefficient for the pipe and F_τ is the so-called 'throttle' friction coefficient. Note that

$$u = \frac{dh}{dt} \qquad (4.32)$$

Finally, we have the equation of continuity applied at the base of the surge tank;

viz. $$vA = \frac{dh}{dt}A_s + Q, \text{ where } Q \text{ is the discharge.} \qquad (4.33)$$

We shall only consider the case where $F_\tau = 0$. Introducing the variables $y(t)$ and $x(t)$,

$$\frac{L}{g}\frac{dx}{dt} = -y - F_p\,v|v| \qquad (4.34)$$

$$A_s\frac{dy}{dt} = Ax + Av_0 - Q \qquad (4.35)$$

If we specify the discharge Q and the initial values h_0, v_0 we can determine $x(t)$ and $y(t)$ from these last two equations. For the particular problem of

demanding constant efficiency for a variable power output it can be shown that
for a simple tank, with F_τ negligible

$$Q = \frac{h_0 + y_f}{h_0 + y} Q_f \tag{4.36}$$

where y_f and Q_f are the values of y and Q when the oscillations have decayed.

To be more specific let $L = 2000$m, $A = 4.909$m^2, h_0 (the static head over
the turbines) $= 100$m, $F_p = 0.5$ and $A_s = 78.54$m^2. Suppose that the flow was
initially 30m^3 s^{-1} and that the power requirement is altered so that there is a
steady flow of 15m^3 s^{-1} ($= Q_f$). We find the maximum surge height.

In the final steady state

$$\frac{dx}{dt} = 0 \text{ so that } y_f = -F_p \, v_f |v_f| \text{ and } \frac{dy}{dt} = 0 \text{ so that } A(x_f + v_0) = Av_f = Q_f$$

Therefore

$v_f = Q_f/A = 15/4.909 = 3.056$ms^{-1}, $y_f = -0.5(3.056)|3.056| = -4.67$m

Initially, $Av_0 = Q_0$ so that $v_0 = 30/4.909 = 6.11$ms^{-1};

also $y_0 = -F_p \, v_0 |v_0| = -18.67$m.

During the oscillations $Q = \dfrac{100 - 4.67}{100 + y} \times 15 = \dfrac{15 \times 95.33}{100 + y}$

We now particularise the equations (4.34) and (4.35) to the forms

$$\left. \begin{array}{l} \dfrac{dx}{dt} = \dfrac{9.81}{2000} \left[-y - 0.5(x + v_0) \, |x + v_0| \right] \\[4mm] \dfrac{dy}{dt} = \dfrac{1}{78.54} \left[4.909(x + v_0) - \dfrac{15 \times 95.33}{100 + y} \right] \end{array} \right\} \tag{4.37}$$

These are special cases of the equations

$$\frac{dx}{dt} = f_1(x,y,t) \qquad \frac{dy}{dt} = f_2(x,y,t) \tag{4.38}$$

When f_1 and f_2 are independent of time t, the equations describe an
autonomous system.

Hamming's method

Hamming's method has been designed to choose the corrector formula (from a
general class of formulae) which would produce accurate values for polynomials
up to degree four and has a much stronger stability than the Milne-Simpson
method.

Step 1 We have starting values y_0, y_1, y_2, y_3 (hence we can calculate
corresponding values f_0, f_1, f_2, f_3). Usually y_0 will be the given
initial condition and y_1, y_2, y_3 can be found by a fourth-order Runge-
Kutta method.

Step 2 y_{n+1} is predicted by the Milne formula

$$y^p_{n+1} = y_{n-3} + \frac{4h}{3}(2f_n - f_{n-1} + 2f_{n-2}) \tag{4.15}$$

Step 3 The predicted value is modified by assuming that the local truncation error does not change appreciably on successive intervals. The modification is

$$y^m_{n+1} = y^p_{n+1} + \frac{112}{121}(y_n^c - y_n^p) \tag{4.39}$$

For $n = 3$ there is no corrected value y_n^c; this step is then by-passed.

Step 4 The corrected value for y_{n+1} is obtained from the Hamming formula

$$y^c_{n+1} = \frac{9}{8}y_n - \frac{1}{8}y_{n-2} + \frac{3h}{8}(f_{n+1} + 2f_n - f_{n-1}) \tag{4.40}$$

It is usual to choose the step size such that one application will ensure that the required accuracy is obtained.

Step 5 The truncation error for the corrector formula is calculated from

$$\varepsilon_c \cong \frac{9}{121}(y^c_{n+1} - y^p_{n+1}) = e_t \tag{4.41}$$

(This formula is quoted without derivation.)

Step 6 The final value for y_{n+1} is obtained from
$$y_{n+1} = y^c_{n+1} - e_t \tag{4.42}$$
The value of f_{n+1} is also calculated.

Step 7 If the local truncation error is acceptably small, then we return to step 2, provided more values of y are required. If the estimate (4.41) lies outside the acceptable limits then we reduce the step size and return to step 2.

Application to a system of differential equations

Suppose we wish to apply Hamming's method to the system of first-order ordinary differential equations

$$\left.\begin{array}{l} \dfrac{dy_1}{dx} = f_1(x, y_1, y_2, ..., y_n) \\[2mm] \dfrac{dy_2}{dx} = f_2(x, y_1, y_2, ..., y_n) \\[2mm] \vdots \\[2mm] \dfrac{dy_n}{dx} = f_n(x, y_1, y_2, ..., y_n) \end{array}\right\} \tag{4.43}$$

In essence, we simply perform a step on each equation before proceeding to the next step. The only problem comes in storing the required values efficiently.

Application to the surge problem

We applied Hamming's method to equations (4.37) using a step size of 10 seconds. Some results are shown in Table 4.5 to 3 d.p. They look reasonable; but how reliable do you think they are?

Table 4.5

t	10	20	30	100	200
x	−0.035	−0.129	−0.267	−1.354	−2.902
y	−7.085	−15.488	−13.924	−3.162	12.219

4.7 Phase-Plane Diagrams

Engineering systems are often governed by second-order differential equations that do not involve the independent variable *explicitly*; the governing equation can be reduced to a first order equation. For example, the system governed by

the equation $\ddot{x} + x\dot{x} + x^2 = 0$ can be studied more easily if we put $y = \dot{x} = \dfrac{dx}{dt}$

so that $\ddot{x} = \dot{y} \equiv \dfrac{dy}{dt} = \dfrac{dy}{dx} \cdot \dfrac{dx}{dt} = y'y$. We then have as governing equation

$yy' + xy + x^2 = 0$ which is a first order equation in y. The initial conditions

to the original problem could have specified $x(0)$ and $\dot{x}(0)$. For the reduced equation we have a corresponding pair of values (x,y) at the initial time which allows a solution to be found. The graph of $y = y(x)$, is called a **phase-plane diagram**. As t varies, both x and y will change and the point (x,y) will move along the diagram.

Example 1

Consider a system experiencing simple harmonic motion of frequency ω,

where the governing equation is $\ddot{x} + \omega^2 x = 0$. If $y = \dot{x}$, then $yy' + \omega^2 x = 0$.

Integrating, $\frac{1}{2}y^2 + \frac{1}{2}\omega^2 x^2 = $ constant. Suppose that initially $x = a$ and $y = \dot{x} = 0$. Then the solution is

$$y^2 + \omega^2 x^2 = \omega^2 a^2 \tag{4.44}$$

In Figure 4.2 the phase-plane diagram is sketched.

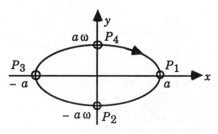

Figure 4.2

Notice that above the x-axis $y = \dot{x}$ is positive and therefore x increases with time; this suggests that the imposed arrow marks the progress of the motion.

The initial point is P_1. From the diagram we see that x oscillates between $-a$ and $+a$ and $y = \dot{x}$ between $-a\omega$ and $+a\omega$. The closed contour implies a periodic motion.

Example 2

Consider a simple pendulum of length l moving in a vacuum. The equation of motion is $\ddot{x} + \omega^2 \sin x = 0$, where $\omega^2 = g/l$ and x is the angle made with the vertical; the equation can be converted to $yy' + \omega^2 \sin x = 0$. Integration produces the equation $\frac{1}{2}y^2 - \omega^2 \cos x = $ constant. Now let $x = 0$ and $y = \dot{x} = u$ at $t = 0$. Then $\frac{1}{2}u^2 - \omega^2 = $ constant so that

$$y^2 = 2\omega^2 \cos x + (u^2 - 2\omega^2) \tag{4.45}$$

There are three cases to consider

(i) $u^2 > 4\omega^2$ so that $y^2 > 2\omega^2(\cos x + 1)$ and $y^2 > 0$ for all values of x.

(ii) $u^2 = 4\omega^2$ so that $y^2 = 2\omega^2(\cos x + 1) = 4\omega^2 \cos^2 \frac{1}{2}x$.

     Therefore $y = \pm 2\omega \cos \frac{1}{2}x$, except that $y = +2\omega \cos \frac{1}{2}x$ if $x > 0$ and $y = -2\omega \cos \frac{1}{2}x$ if $x < 0$.

(iii) $u^2 < 4\omega^2$ so that y^2 can be negative for some values of x and therefore the solution is not defined for all x.

Check that the phase-plane diagram is as shown in Figure 4.3. Notice that for small initial velocities the motion is periodic whereas, for larger velocities the motion is that of a progressively increasing angle x or a progressively decreasing one.

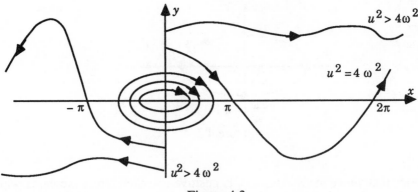

Figure 4.3

Try to interpret these cases physically. Try and extend this diagram for the general case $x = a$ and $y = \dot{x} = u$ at $t = 0$.

Sometimes the phase-plane diagram is not easy to draw and the use of isoclines is often helpful.

Example 3

Consider the system governed by the equation of motion $\ddot{x} + x\dot{x} + x^2 = 0$ which indicates a non-linear restoring force and a non-linear damping term. The usual substitution $y = \dot{x}$ gives rise to the equation

$$yy' = -xy - x^2 \quad \text{or} \quad y' = -x - x^2/y \quad (4.46)$$

The **isoclines** are curves where $y' = $ constant, in this case $-x - x^2/y = C$, C a constant. The equation for isoclines may be written $y = -x^2/(x + C)$. On each isocline, the solution curves pass through with the slope equal to C. In Figure 4.4(a) the isoclines are drawn and the local slopes of the solution curves are indicated; in Figure 4.4(b) some solution curves are shown.

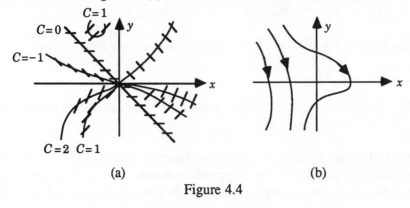

(a)   (b)

Figure 4.4

Check the isoclines and subsequent development. Note that $x = 0$ corresponds to $y' = 0$ from equation (4.46).

Finally we show an example of a **limit cycle** which is a closed contour to which all solution curves tend as $t \to \infty$, no matter what the initial conditions.

Example 4 – Van der Pol's equation

Figure 4.5(a) shows a circuit diagram for a simple electronic oscillator. We state, without proof, that a suitable governing equation is

$$\ddot{x} - \varepsilon(1 - x^2)\dot{x} + x = 0 \qquad (4.47)$$

where $\varepsilon > 0$ is a constant of the circuit and x is a function of the current i. The equation represents a sustained periodic oscillation. The phase-plane diagram shown in Figure 4.5(b) is for the case $\varepsilon = 1$.

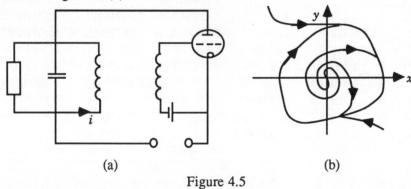

(a)   (b)

Figure 4.5

4.8 Boundary Value Problems

In this section we examine briefly two methods of solving ordinary differential equations where the attached conditions refer to different values of the independent variable.

(a) Shooting Methods

This class of methods effectively converts the boundary-value problem into an initial-value problem. We illustrate its strategy via a simple example. Consider the boundary-value problem

$$y'' + y = 0 ; \quad y(0) = 0, \quad y(1) = 1 \qquad (4.48)$$

It is easy to show that the analytical solution is $y = A \sin x$ where $A = 1.1884$ (4 d.p.). First we split the given second order equation into two simultaneous first order equations. If we write $y_1 = y$ and $y_2 = y'$ then it is straightforward to obtain the coupled equations

$$y_1{}' = y_2, \quad y_2{}' = -y_1$$

The boundary conditions can now be written

$$y_1(0) = 0, \quad y_1(1) = 1$$

We now make a guess for $y_2(0)$ and compute the values of y_1 and y_2 at suitable intervals of x until we reach $x = 1$. We can then compare the estimated value of $y_1(1)$ with the known value of 1. If we are too much in error we must choose a different value of $y_2(0)$ and repeat the procedure. The name 'shooting methods' derives from the analogy with firing a projectile to hit a ground target; if the initial angle of projection fails to produce a hit then we can adjust the angle and try again.

Table 4.6

x	y_1	y_2	y_1	y_2	y_1	y_2	Analytical
0	0	0	0	2	0	1.133134191	0
0.1	0	0	0.2	2	0.113313419	1.133134191	0.118641543
0.2	0	0	0.4	1.98	0.226626838	1.12180285	0.23609766
0.3	0	0	0.598	1.94	0.338807123	1.09914017	0.351194767
0.4	0	0	0.792	1.8802	0.448721140	1.06525945	0.462782852
0.5	0	0	0.98002	1.80100	0.555247085	1.02038734	0.569746963
0.6	0	0	1.16012	1.702998	0.657285819	0.96486263	0.671018351
0.7	0	0	1.3304198	1.5869860	0.753772082	0.899134048	0.765585146
0.8	0	0	1.4891184	1.45394402	0.843685486	0.82375684	0.852502467
0.9	0	0	1.6345128	1.30503218	0.926061171	0.739388292	0.930901865
1.0	0	0	1.76501602	1.14158090	1	0.646782175	1

Table 4.6 shows in its first five columns the results of using two initial values for y_2, viz. 0 and 2 respectively; these are chosen fairly arbitrarily, the second being a response to the undershoot achieved by the first. Euler's method has been used to obtain the estimated values, viz.

$$y_1^{\text{new}} = y_1^{\text{old}} + hy_2^{\text{old}} = y_1^{\text{old}} + 0.1y_2^{\text{old}}$$
$$y_2^{\text{new}} = y_2^{\text{old}} - hy_1^{\text{old}} = y_2^{\text{old}} - 0.1y_1^{\text{old}}$$

Since we have found that by using $y_2(0) = 0$, $y_1(1) = 0$ and with $y_2(0) = 2$, $y_1(1) = 1.76501602$, we employ interpolation to make a more successful estimate of $y_2(0)$. Given the linearity of the boundary value problem we can rely upon theorems which are well-established to make use of linear interpolation to obtain an accurate estimate. (In the case of non-linear equations methods of root-finding need to be used.) In this example we refer to Figure 4.6 to help visualise the process.

Employing linear interpolation we obtain $y_2(0) = 1.331419$. Taking this value for $y_2(0)$ we produce columns six and seven of Table 4.6. It can be seen that $y_1(1) = 1$ to the accuracy of the computer. It would be a useful exercise for you to compute the analytical values of y at each of the x values in the table.

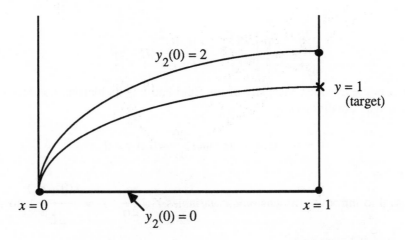

Figure 4.6

(b)   Finite Difference Methods

The technique, of substituting finite differences for derivatives, is dealt with more fully in the chapter on partial differential equations. We indicate briefly its working here.

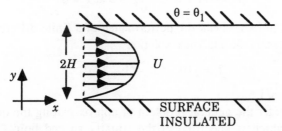

Figure 4.7

In Figure 4.7 is depicted the steady flow of a highly viscous fluid between two stationary, flat, parallel plates. There is a pressure gradient acting in the x-direction and this gives rise to a parabolic velocity distribution as shown. The flow is everywhere in the x-direction and the maximum velocity U occurs at $y = H$. The general velocity is denoted by u. It can be readily shown that

$$u = \frac{U}{H^2} (2Hy - y^2) \qquad (4.49)$$

Now let the upper plate be maintained at a fixed temperature and let the lower plate be thermally insulated. It can be shown that the temperature distribution in the fluid, $\theta(y)$, is given by

$$\frac{d^2\theta}{dy^2} = -\frac{\mu}{k}\left[\frac{du}{dy}\right]^2 = -\frac{\mu}{k}\frac{4U^2}{H^4}(H-y)^2 \qquad (4.50)$$

where μ is the dynamic viscosity of the fluid and k is its thermal conductivity. The boundary conditions are

$$\theta = \theta_1 \text{ at } y = 2H, \text{ and } \frac{d\theta}{dy} = 0 \text{ at } y = 0$$

It is usual to introduce dimensionless variables $Y = \dfrac{y}{2H}$, $T = \dfrac{k(\theta - \theta_1)}{\mu U^2}$ so that equation (4.50) becomes

$$\frac{d^2T}{dY^2} = -16(1 - 2Y)^2 \qquad (4.50a)$$

and the boundary conditions are

$$T = 0 \text{ at } Y = 1, \quad \frac{dT}{dY} = 0 \text{ at } Y = 0 \qquad (4.50b)$$

Now we can, in fact, check the performance of a finite difference technique because we can obtain the analytical solution

$$T = \tfrac{1}{3}[9 - 8Y - (1 - 2Y)^4] \qquad (4.51)$$

Note that $T(0) = 8/3$.

First we consider the general problem of approximating the derivatives of a function $y(x)$ which is specified at the equally spaced points $x_0, x_1, x_2, ...,$ x_n as shown in Figure 4.8.

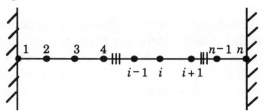

Figure 4.8

Let $y(x_i)$ be denoted y_i, $y'(x_i)$ be denoted y_i', etc. Then via Taylor's series we obtain

$$y_{i+1} = y_i + hy_i' + (h^2/2!)y_i'' + (h^3/3!)y_i''' + ...$$
$$y_{i-1} = y_i - hy_i' + (h^2/2!)y_i'' - (h^3/3!)y_i''' + ...$$

Adding these equations we obtain

$$y_{i+1} + y_{i-1} = 2y_i + h^2 y_i'' + 0(h^4)$$

and this may be rearranged to give

$$y_i'' = [y_{i+1} - 2y_i + y_{i-1}]/h^2 + 0(h^2) \qquad (4.52)$$

It may, in fact, be shown that the error term $0(h^2)$ is actually $-\dfrac{h^2}{12} y^{(iv)}(x_i)$.

Subtracting, we obtain

$$y_i = (y_{i+1} - y_{i-1})/2h + 0(h^2) \qquad (4.53)$$

Equations (4.52) and (4.53) can be applied at the points $i = 1, 2, 3, ..., (n-1)$. For example the equation

$$y'' + 4y' + 3y = e^x$$

becomes

$$(y_{i+1} - 2y_i + y_{i-1})/h^2 + 4(y_{i+1} - y_{i-1})/h + 3y_i = e^{x_i} \qquad (4.54)$$

Applying the boundary conditions
(a) y specified on the boundary
If, for example, $y_0 = A$ then we may substitute this value into the appropriate equation. With the current example and $i = 1$ we obtain

$$(y_2 - 2y_1 + A)/h^2 + 4(y_2 - A)/h + 3y_i = e^{x_i}$$

(b) y' specified on a boundary
If we consider $i = 0$ the equation becomes

$$(y_1 - 2y_0 + y_{-1})/h^2 + 4(y_1 - y_{-1})/h + 3y_0 = e^{x_0} \qquad (4.55)$$

Now suppose that $y_0' = B$ is given. Then we approximate this condition by

$$(y_1 - y_{-1})/h = B$$

and we eliminate the fictitious value y_{-1} from equation (4.55). Obviously, the same technique could be applied at the other boundary.

(c) Linear combination of y and y' specified at a boundary
This is treated in a manner similar to (b). When the fictitious values have been eliminated the resulting system of equations becomes tri-diagonal and can be solved by a standard procedure.

Example $y'' = -y; \quad y(0) = 0, \quad y(1) = 1$
We divide the interval $(0,1)$ into 5 equal sub-intervals, i.e. $h = 0.2$. Then the equation is approximated at $x = x_i = ih = 0.2i$ by

$$(y_{i+1} - 2y_i + y_{i-1}).25 = -y_i$$

i.e.

$$y_{i-1} - 1.96y_i + y_{i+1} = 0 \qquad (4.56)$$

When $i = 1$, (4.56) becomes

$$y_0 - 1.96y_1 + y_2 = 0$$

i.e.

$$-1.96y_1 + y_2 = 0$$

When $i = 4$, (4.56) becomes

$$y_3 - 1.96y_4 + y_5 = 0$$

i.e. $y_3 - 1.96y_4 = -1$

The system of equations is written in matrix form as

$$\begin{bmatrix} -1.96 & 1 & 0 & 0 \\ 1 & -1.96 & 1 & 0 \\ 0 & 1 & -1.96 & 1 \\ 0 & 0 & 1 & -1.96 \end{bmatrix} \begin{bmatrix} y_1 \\ y_2 \\ y_3 \\ y_4 \end{bmatrix} = \begin{bmatrix} 0 \\ 0 \\ 0 \\ -1 \end{bmatrix} \qquad (4.57)$$

The solution is, recorded to 4 d.p.,

$$(y_1, y_2, y_3, y_4) = (0.2362, 0.4630, 0.6753, 0.8527)$$

which compares favourably with the results in Table 4.6, bearing in mind that there $h = 0.1$.

Table 4.7

Number of steps n	$T(0)$ (4 d.p.)	Analytical value (4 d.p.)	IErrorI (4 d.p.)
8	2.3125	2.6667	0.3542
16	2.5703	2.6667	0.0964
32	2.6416	2.6667	0.0351
64	2.6603	2.6667	0.0064
128	2.6651	2.6667	0.0016

Returning to the example provided by equations (4.50), Table 4.7 shows the values of $T(0)$ obtained by using the finite difference approach with n subdivisions of the interval [0,1] for different values of n.

The finite difference approximation to (4.50a) is

$$T_{i-1} - 2T_i + T_{i+1} = h^2\{-16(1 - 2Y_i)^2\}$$

At $Y = 1$ this becomes

$$T_{n-2} - 2T_{n-1} = h^2\{-16(1 - 2Y_{n-1})^2\}$$

The condition at $Y = 0$ is approximated as

$$T_1 - T_{-1} = 0$$

which together with the equation

$$T_{-1} - 2T_0 + T_1 = h^2\{-16(1 - Y_0)^2\}$$

gives, remembering that $Y_0 = 0$,

$$-2T_0 + 2T_1 = -16h^2$$

Table 4.8 shows some values of T for the case $n = 128$; these are compared with the analytical values obtained from (4.51). Both results are recorded to 4 d.p.

Table 4.8

Y	0	0.25	0.5	0.75	1.0
$T(Y)$ computed	2.6667	2.3203	1.6862	1.0004	0.0412
$T(Y)$ analytical	2.6667	2.3125	1.6667	0.9792	0

Shooting methods do present problems of stability; on the other hand, the finite difference methods possess better stability behaviour but usually require more computational effort to achieve a specified accuracy. The analytical results for Example 1 bear this out.

4.9 Further Remarks

In this section we mention briefly two concepts which are of increasing importance in the study and application of ordinary differential equations.

(a) Stiff differential equations

A stiff differential equation is one in which the exact solution contains two or more widely differing time constants. In the same way we may define a stiff system of differential equations. Consider the equation

$$y'' + 101y' + 100y = 100 \tag{4.58}$$

which has a general solution of

$$y = Ae^{-100x} + Be^{-x} + 1$$

To be particular, we add the initial conditions

$$y(0) = 2.01, \quad y'(0) = -2$$

so that the particular solution is

$$y = 0.01e^{-100x} + e^{-x} + 1$$

In order to obtain an accurate solution of the early stages when the more rapidly decaying transient is present we need to use very small integration steps. It would be tempting to think that after, say, $x = 0.1$ when

$$y \cong 0.01(4.54 \times 10^{-5}) + 0.905 + 1$$

that we could increase the integration step to take account of the more slowly changing nature of the solution. However, the problem is that the differential equation is the same and it is this that the numerical method will be attempting to solve. Consequently the same problems of stability remain throughout the integration process; the term in e^{-x} needs to be approximated accurately even though it contributes relatively very little to the solution in the later stages. It should be noted that the second order equation (4.58) may well be converted to the stiff system on the following page.

$$\left. \begin{array}{l} y_1' = y_2 \\ y_2' = -101y_2 - 100y_1 + 100 \end{array} \right\} \qquad (4.59)$$

One class of methods which is designed to avoid the need for the expensive short integration steps is that of **backward integration**. One very simple method in this class is the backward Euler method for $y' = f(x,y)$, viz.

$$y_{n+1} = y_n + hf(x_{n+1}, y_{n+1}).$$

It has to be pointed out that in many cases the solution process involves the solution of a system of equations (linear or non-linear) at each step. Most reputable computer packages for the solution of differential equations incorporate a suitable method for handling stiff systems.

(b)   Extrapolation methods

Extrapolation methods are used in general to derive more accurate approximations from low-order formulae. An example will suffice to show the general idea behind such methods. Consider the initial-value problem

$$y' = -y \; ; \quad y(0) = 1$$

so that the solution is $y = e^{-x}$. The Euler method with step size h is

$$y_{n+1} = y_n(1 - h)$$

In Table 4.9 we show the results of integrating over the inteval $[0,1]$ with step sizes $h = 0.1$ and $h = 0.05$; in the latter case only those values at multiples of 0.1 are shown.

Table 4.9

x	$\exp(-x)$	$h = 0.1$	$h = 0.05$	w
0.1	0.904837418	0.9	0.9025	0.905
0.2	0.818730753	0.81	0.81450625	0.8190125
0.3	0.740818221	0.729	0.735091891	0.741183781
0.4	0.670320046	0.656	0.663420431	0.670740862
0.5	0.60653066	0.59049	0.598736939	0.606983878
0.6	0.548811636	0.531441	0.540360087	0.549279175
0.7	0.496585304	0.4782969	0.487674979	0.497053058
0.8	0.449328964	0.43046721	0.440126668	0.449786127
0.9	0.40656966	0.387420489	0.397214318	0.407008147
1.0	0.367879441	0.34867844	0.358485922	0.368193404

The extrapolation technique is applied at the points shown in the table. It is effectively

$$w = \frac{0.1y_{0.05} - 0.05y_{0.1}}{0.1 - 0.05} = 2y_{0.05} - y_{0.1}$$

where $y_{0.1}$ refers to the approximate value of y using $h = 0.1$ etc. It can readily be seen how much more accurate the w values are. It can be checked that this process produces estimates of order h^2, compared to the usual Euler accuracy of $O(h)$.

Problems

Section 4.2

1   We are given the equation $y' = -2y$ with the condition $y(0) = 1$. We require the value of $y(0.2)$. With a step size $h = 0.1$ apply in turn the formulae (4.1), (4.5), (4.6), (4.7), (4.9). Compare the results with the analytical solution. With a step size $h = 1$ estimate $y(1.0)$ via (4.7). Comment.

2   The response x of a given hydraulic valve subject to sinusoidal input variation is given by

$$\frac{dx}{dt} = \sqrt{2\left\{1 - \frac{x^2}{\sin^2 t}\right\}} \quad \text{with } x = 0 \text{ at } t = 0.$$

Show whether a Taylor series solution of this equation is possible.

Show that $\left.\dfrac{dx}{dt}\right|_{x=0} = \sqrt{\dfrac{2}{3}}$ and hence use the Runge-Kutta fourth-order method to obtain

a solution at $t = 0.2$.   (EC)

3   Use the method of isoclines to sketch the solution curves of $dy/dx = x^2 + y^2$ in the region $-2 < x < 2$, $-2 < y < 2$.
A second-order Runge-Kutta formula in the usual notation is

$$k_1 = hf(x,y), \quad k_2 = hf(x + h, y + k_1), \quad y(x + h) = y(x) + \frac{1}{2}(k_1 + k_2)$$

Illustrate how this is used by taking two steps with $h = 0.2$ starting from $(0,1)$ for the above differential equation. Compare this with a Taylor series solution and explain any discrepancy. Would you have found a discrepancy if a fourth-order method had been used?
                                                                                                        (EC)

4   Solve the following differential equations via the Runge-Kutta second- and fourth-order methods and the Adams-Bashforth methods.
   (i) $y' = x + y + xy + 1$ $y(0) = 2$
   (ii) $y' = x/y$ $y(0) = 1$
   (iii) $y' = 1/(x + y)$ $y(0) = 2$
   (iv) $y' = x^3 + y^2$ $y(0) = 0$
In each case estimate $y(1)$ using various step sizes h. Draw conclusions about the most suitable method for each equation.

5   Write a computer program for the Merson method. Use it on any of the equations in
     Problem 4.

Section 4.3

6   Use a Taylor series method to obtain values of y to three decimal places at $x = -0.1, 0.1$
     and 0.2, if $y'' + (x + 1)y^2 = 0$ with $y(0) = 1$ and $y'(0) = 1$. Then use the following
     predictor-corrector pair of formulae to obtain the value of y at $x = 0.3$.

Predictor: $y_{n+1} = 2y_{n-1} - y_{n-3} + \dfrac{4h^2}{3}(y''_n + y''_{n-1} + y''_{n-2})$

Corrector: $y_{n+1} = 2y_n - y_{n-1} + \dfrac{h^2}{12}(y''_{n+1} + 10y''_n + y''_{n-1})$

Describe *briefly* how such forward integration formulae can be used to solve a boundary
value problem.   (EC)

7   By means of the first 3 non-zero terms of a Taylor's series obtain the solution of the
     differential equation $dz/dt = e^{2t} - z^2$ for the values $t = 0.10$, $t = 0.20$, $t = 0.30$, given
     that $z = 1$ when $t = 0$. Using the Adams-Bashforth method for continuing the solution by

     means of the backward difference equation $z_{i+1} = z_i + h[1 + \frac{1}{2}\nabla + \frac{5}{12}\nabla^2 + \frac{3}{8}\nabla^3]\left(\dfrac{dz}{dt}\right)_i$

     obtain the value of z when $t = 0.4$.   (LU)

8   The table below gives the starting values for the solution of $dy/dx = 1 + y^2$. Use the

     formula (from Simpson's rule) $y_1 = y_{-1} + 2h\left\{1 + \frac{1}{6}\delta^2\right\}f_0$ as a 'corrector' formula

     to obtain the value of y for $x = 0.6$. Give your reasons for your choice of the 'predicted'
     value from which you commence. Give your answer correct to 3 decimal places.

x	y	$2hf$
0.0	0.000	0.4000
0.2	0.203	0.4164
0.4	0.423	0.4715

                                                                                      (LU)

9   A first-order differential equation may be solved numerically by using one of the pairs (a) or
     (b) of predictor and corrector formulae.

(a) $\begin{cases} y_1 = y_0 + h(1 + \dfrac{1}{2}\nabla + \dfrac{5}{12}\nabla^2 + \dfrac{3}{8}\nabla^3 + \dfrac{251}{720}\nabla^4 + ...)y'_0 \\[4mm] y_1 = y_0 + h(1 - \dfrac{1}{2}\nabla - \dfrac{1}{12}\nabla^2 - \dfrac{1}{24}\nabla^3 - \dfrac{19}{720}\nabla^4 + ...)y'_1 \end{cases}$

(b)
$$
\begin{cases}
y_4 = y_0 + \dfrac{4}{3} h(2y'_1 - y'_2 + 2y'_3) \\[2ex]
y_4 = y_2 + \dfrac{1}{3} h(y'_2 + 4y'_3 + y'_4)
\end{cases}
$$

Prove *one only* of the predictor formulae.

The differential equation $y' = x - \frac{1}{10}y^2$ is to be solved with the initial value $y = 1$ when $x = 0$. Assuming that the following starting values have been obtained, find the value of y at $x = 0.3$.

x	-0.2	-0.1	0.1	0.2	
y	1.04068	1.01513	0.99507	1.00013	(LU)

10 Obtain, by the Adams-Moulton formulae or otherwise, the next line, beginning $x = 0.5$, of the table giving the start of the numerical solution of $dy/dx = (x + y)^{1/2}$ given the values below. First form a difference table as far as $\nabla^2 f$.

x	0.1	0.2	0.3	0.4	
y	0.3071	0.3771	0.4587	0.5511	(LU)

Use the Adams-Moulton formulae in finite difference form to extend the table below by one step, in the step-by-step solution of the differential equation $dy/dx = f(x,y) = \sqrt{(x + y)}$.

x	0	0.1	0.2	0.3	
y	0.2500	0.3071	0.3771	0.4587	(LU)

11 The table below gives the starting values for the numerical solution of the differential equation $dy/dx = 1 + y^2$, with $y = 0$ when $x = 0$. Form a difference table as far as $\nabla^3 y'$. Explain how the two formulae

(a) $\quad y_{n+1} = y_n + h[1 + \frac{1}{2}\nabla + \frac{5}{12}\nabla^2 + \frac{3}{8}\nabla^3]y'_n$

(b) $\quad y_{n+1} = y_n + h[1 - \frac{1}{2}\nabla - \frac{1}{12}\nabla^2 - \frac{1}{24}\nabla^3]y'_{n+1}$

are used to continue the solution, and obtain the value of y at $x = 0.3$

x	-0.2	-0.1	0.0	0.1	0.2	
y	-0.20271	-0.10033	0.00000	0.10033	0.20271	(LU)

12 Describe a predictor-corrector method for solving the differential equation $dy/dx = f(x,y)$ with initial condition $y = y_0$ at $x = x_0$.
Solve numerically the differential equation $dy/dx = x - y^2$, $y = 1$ at $x = 0$ for the values of $x = 0.2$ and $x = 0.4$, specifying your results to 3 decimal places.   (EC)

Section 4.4

13 Obtain a solution of the difference equation $y_{n+2} - 2r \cos \theta \cdot y_{n+1} + r^2 y_n = 0$ in the form $y = r^n[A \cos n\theta + B \sin n\theta]$.
Solve the equation $y_{n+2} - 3y_{n+1} + 9y_n = 13n + 3^n$.   (LU)

14 (a) Solve the difference equation $y_{n+2} + 6y_{n+1} - 7y_n = 3^n$.

    (b) Show that the equation $y_{n+1} \cdot y_n + 2y_{n+1} + 8y_n + 9 = 0$ can be reduced by the substitution $y_n = (V_{n+1}/V_n) - 2$ to an equation with constant coefficients. Hence solve the given equation when $y_0 = -5$. (LU)

15 (a) Solve the difference equation $y_{n+2} - 4y_{n+1} + 4y_n = 3^n$.

    (b) Obtain the solution of the simultaneous difference equations $x_{n+1} = 2x_n + y_n + n$, $y_{n+1} = 2x_n + 3y_n - n$, for which $x_0 = 3$ and $y_0 = 0$. (LU)

16 (a) Solve the difference equation $y_{n+2} - 6y_{n+1} + 9y_n = 2^n + n$.

    (b) Show that $y_n = n$ is a solution of the difference equation

$$n(n + 1)y_{n+2} - 5n(n + 2)y_{n+1} + 4(n + 1)(n + 2)y_n = 0.$$

By substituting $y_n = nu_n$, obtain the general solution of this equation and show that

the solution for which $y_1 = 12$ and $y_2 = 60$ is $y_n = \frac{3}{2}n.4^n + 6n$. (LU)

17 Solve the difference equations

    (a) $y_{n+2} - 6y_{n+1} + 13y_n = 2^n$

    (b) $\Delta^2 y_n - 4\Delta y_n + 4y_n = n \cdot 2^n + 3^n$ (LU)

18 A long row of trucks, each of mass M, are standing in a line, touching one another, joined by couplings which become tight when the distance between trucks is a. A steady force $P = Mf_0$ is applied to the end truck in a direction moving it away from its neighbour. In the subsequent motion, v_n denotes the common velocity of the moving truck immediately after the nth truck starts to move. Assuming that frictional resistance can be neglected, show that

$$(n + 1)^2 v^2_{n+1} = n^2 v^2_n + 2naf_0 \text{ and deduce that } v^2_{n+1} = \frac{n}{n + 1}af_0$$ (LU)

19 A mechanical model of a band-pass filter comprises a series of identical disks each connected to a fixed base by an identical spring. If θ_x is the angle turned through by the xth disk, the equation of its motion is

$$c(\theta_{x-1} - \theta_x) + c(\theta_{x+1} - \theta_x) - k\theta_x = I\frac{d^2\theta_x}{dt^2} \quad \text{where } c, k \text{ and } I \text{ are constants.}$$

Substitute $\theta_x = \Theta_x \sin \omega t$, put $\beta = 1 + \frac{k}{2c} - \frac{I\omega^2}{2c}$ and discuss the solutions of the resulting difference equation for the cases $-1 \le \beta \le 1$, $\beta < -1$ and $\beta > 1$.

Section 4.5

20 Consider the equation $y' = y - x$ with initial condition $y(0) = 1$. Write down the analytical solution. Let the value of $y(0)$ vary by $\pm 1\%$ and obtain the analytical solutions. Comment on the variation in $y(x)$ for $x = 5$ and deduce what will happen for larger values of x. This is an example of *inherent instability*.

21 Consider the equation $y' = -10y$ with initial condition $y(0) = 1$. The analytical solution is $y = e^{-10x}$. Show that a Runge-Kutta second-order method leads to $y_{m+1} = (1 - 10h + 50h^2)^m$ and show that the expression in parentheses exceeds 1 if $h > 0.2$. What effect will this have on computed values of x? Repeat the study for a Runge-Kutta fourth-order method. This is an example of *partial instability*.

22 Repeat the principles of Problem 21 on the equation $y' = -5y$ with $y(0) = 1$.

23 Use a second-order Runge-Kutta method on the equation $y' = -10y$ with $y(0) = 1$ and a step size of 0.2. Comment.

24 Apply the Milne-Simpson corrector formula to the equation $y' + 40y = \frac{1}{2} + 20x$ with $y(1) = 0.5$ and a step size of 0.1. Tabulate y for $x = 1.0, 1.1, ..., 1.5, 2.0, 3.0, 4.0$ and 5.0 and compare with the analytical solution. Apply the formula to the equation $y' + \lambda y = 0$, $\lambda > 0$. You should obtain a second order difference equation. This will have *two* roots, yet it should be an approximation to a *first-order* differential equation. We have introduced an unwanted solution which, in fact, will grow even if the step size $h \to 0$. If, however, $\lambda < 0$ the spurious solution decays exponentially. This shows that the Milne-Simpson method is *weakly unstable* for this equation.

Section 4.7

25 There are various ways of investigating the behaviour of systems described by the Van der Pol equation $\dfrac{d^2x}{dt^2} - \varepsilon(1 - x^2)\dfrac{dx}{dt} + x = 0$.

Using the substitution $y = dx/dt$, express the equation in the (phase-plane) form $dy/dx = f(x,y)$, sketch the isoclines $f(x,y) = \lambda$ for $\varepsilon = 1$ and $\lambda = 0, \pm 1, \pm 2, \infty$, and hence graphically deduce the nature of the solution.   (EC)

26 Sketch the phase-plane diagram for each of the following

(i) $\ddot{x} - 9x = 0$ (ii) $\ddot{x} + 2x - 1.5x^2 = 0$

(iii) $\ddot{x} - 2x + 3x^3 = 0$ (iv) $\ddot{x} + \dot{x}\,|\dot{x}| + x = 0$

27 Use isoclines to sketch the phase-plane diagram in the cases

(i) $\ddot{x} + 2x\dot{x} + x^2 = 0$ (ii) $\ddot{x} + x^2 - 2x = 0$ (iii) $\ddot{x} + 2\dot{x} + 8x = 0$

Section 4.8

28 Describe how a second order boundary-value problem (second order differential equation) may be solved over a given interval using the Euler method with trapezoidal correction.

For the boundary-value problem $yy'' + y'^2 + 1 = 0$, $y(0) = 1$, $y(1) = 2$, solve for y consistent to two decimal places at $x = 0.5$ and $x = 1.0$ using the above method and a trial value $y'(0) = 1$. From the result estimate a new value for $y'(0)$.

Note: The Euler-trapezoidal formulae applied to the *first order* differential equation $y' = f(x,y)$ are

$$y_{r+1}^{(0)} = y_r + hf(x_r, y_r), \quad y_{r+1}^{(n)} = y_r + \frac{1}{2}h[f(x_r, y_r) + f(x_{r+1}, y_{r+1}^{(n-1)})] \qquad \text{(EC)}$$

29 The Blasius problem relating to boundary layer fluid flow past a flat plate can be reduced to the dimensionless form $f(\eta)f''(\eta) + 2f'''(\eta) = 0$ with boundary conditions $f(0) = f'(0) = 0, f'(\infty) = 1$. Express this as a set of first-order differential equations. Assuming $f''(0) = 0.4$, take one step $\Delta\eta = 1$ in the solution of the resulting initial-value problem, using Runge-Kutta or any equivalent method.

Indicate briefly how the original problem could be solved using this approach. (EC)

30 Solve $y'' + xy' - 3y = 4.2x - 3$ with $y(0) = 1$, $y(1) = 2.9$. Use $h = 0.25$ and compare the values obtained with the analytical solution.

Verify that this latter is $y = x^3 + 0.9x + 1$.

31 Solve the equation $y'' - yy' = e^x$ with $y(0) = 1$, $y(1) = -1$.

32 The lateral deflection of a pin-ended strut is governed by the equation

$$\frac{d^2y}{dx^2} + \frac{P}{EI}\left[1 + \left[\frac{dy}{dx}\right]^2\right]^{3/2} y = 0, \quad \text{where } P \text{ is the axial load and } E \text{ and } I \text{ are constants.}$$

Given that $y(0) = y(L) = 0$ find the deflection at the mid-point, using suitable test data for P, E, I and L.

33 Show that the boundary-value problem $\dfrac{d^2y}{dx^2} = \dfrac{0.06912x(x-1)}{I(x)}$, $y(0) = y(1) = 0$, can be

approximated at the points $x_1, x_2, ..., x_r = x_0 + rh, ...$ by the algebraic equations

$$y_{r-1} - 2y_r + y_{r+1} = \frac{0.06912\, h^2\, x_r(x_r - 1)}{I(x_r)}, \text{ for } r = 1, 2, ... \left[\frac{1}{h} - 1\right]$$

Write down these equations given

x	1/6	1/3	1/2	2/3	5/6
$I(x)$	0.01281	0.01327	0.01342	0.01327	0.01281

and taking $h = 1/6$.

Given the initial approximations $(y_1, y_2, y_3, y_4, y_5) = (0.07, 0.12, 0.14, 0.12, 0.07)$, obtain more accurate results by applying the Gauss-Seidel iteration. (The actual number of equations to be solved may be reduced by assuming symmetry of y about $x = 1/2$.) (EC)

34 Given $y'' + (x^2 - 1)y = 0$, $y(0) = 0$, $y'(0) = 1$, find $y(0.1)$, $y(0.2)$ from the Taylor series for y, and $y(0.3)$, from the central difference formula

$$\delta^2 y = (1 + \frac{1}{12}\delta^2 + ...)(h^2y''), \text{ retaining four decimal places.} \qquad \text{(LU)}$$

35 Solve the equation $y'' + y = 0$ with $y(0) = 0$, $y(\pi/2) = 1$. Use a step size $h = 0.25$ after normalising the interval.

36 Solve the boundary-value problem $y'' + xy' - xy = 2x$, $y(0) = 1$, $y(1) = 0$ using $h = 0.2$.

37 Solve the boundary-value problem $y'' + xy' + y = 2x$, $y(0) = 1$, $y(1) = 0$.

38 Solve the boundary-value problem $y'' + 2y' + y = x$, $y(0) = y(1) = 0$.

5

SPECIAL FUNCTIONS

5.1 Introduction: A Problem in Heat Transfer

The reason that we study the functions *sine, cosine, logarithm, exponential,* etc. is that the behaviours they describe occur very frequently in model solutions. Certain ordinary differential equations occur often in various branches of engineering – the solutions of which cannot be expressed in compact formulae involving standard functions of the calculus. Their importance has led to the creation and study of special functions by means of which we may express their solutions. In this chapter we shall briefly examine some of these functions, after discussing the power series solutions of ordinary differential equations.

Cooling Fins

In order to speed up the rate of convective heat transfer from a surface, such as a domestic radiator, thin metal strips or fins are attached to the primary surface and conduct heat away from the main construction unit, thereby increasing the area available for cooling by the surrounding fluid. A first analysis of the effect of a fin can often be made by assuming a one-dimensional flow of heat. In Figure 5.1 we depict a one-dimensional fin with variable cross-section. We shall assume that steady-state conditions prevail.

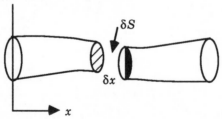

Figure 5.1

$A(x)$ is the cross-section area at a general point and δS represents the surface area of the fin between the planes x and $x + \delta x$. Note that δS is a function of x. Consider the heat flow in this section of the fin.

Let the rate of flow of heat, dQ/dt, be denoted by $R(x)$. Then the equation of heat balance for the section of fin is

$$R(x) = R(x + \delta x) + dQ_c/dt \tag{5.1}$$

where Q_c is the heat convected through the surface element δS. Expanding

$R(x + \delta x)$ as a Taylor series we obtain

$$R(x + \delta x) = R(x) + \delta x\, R'(x) + \frac{(\delta x)^2}{2!} R''(x) + \ldots \qquad (5.2)$$

Let the temperature of the surrounding fluid be zero and let $\theta(x)$ be the temperature at any point inside the fin. Then

$$\frac{dQ_c}{dt} = h\, \delta S\, \theta \qquad (5.3)$$

$$R(x) = -kA(x)[d\theta/dx] \qquad (5.4)$$

where h is the coefficient of heat transfer and k is the thermal conductivity of the fin material.

Substituting (5.2), (5,3) and (5.4) into (5.1) we find that as $\delta x \to 0$ the heat balance equation becomes

$$\frac{1}{A(x)} \frac{d}{dx}\left[-kA(x) \frac{d\theta}{dx} \right] + \frac{h}{A(x)} \frac{dS}{dx}\theta = 0 \qquad (5.5)$$

or, expanding the first term and dividing by $(-k)$,

$$\frac{d^2\theta}{dx^2} + \frac{1}{A(x)} \frac{dA}{dx} \cdot \frac{d\theta}{dx} - \frac{h}{k} \frac{1}{A(x)} \frac{dS}{dx}\theta = 0 \qquad (5.6)$$

Consider a truncated triangular fin as shown in Figure 5.2.

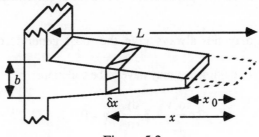

Figure 5.2

Let the fin be of unit width so that $A(x) = bx/L$. Furthermore,

$$S(x) = 2\left\{ \left[x^2 + \left[\frac{bx}{2L}\right]^2 \right]^{1/2} - \left[x_0^2 + \left[\frac{bx_0}{2L}\right]^2 \right]^{1/2} \right\}$$

$$= 2(x - x_0)\left[1 + \left[\frac{b}{2L}\right]^2 \right]^{1/2}$$

Substituting these expressions for $A(x)$ and $S(x)$ into (5.6) we obtain

$$\frac{d^2\theta}{dx^2} + \frac{1}{x}\frac{d\theta}{dx} - \frac{r^2}{x}\theta = 0 \tag{5.7}$$

where $r^2 = \frac{2hL}{kb}\left[1 + \left[\frac{b}{2L}\right]^2\right]^{1/2}$. Further, we impose the boundary conditions

that the temperature at the base of the fin is given i.e.

$$\theta = \theta_0 \quad \text{at} \quad x = L \tag{5.8a}$$

and convection occurs at the tip of the fin, i.e.

$$-k\frac{d\theta}{dx} = h_0\,\theta \quad \text{at} \quad x = x_0 \tag{5.8b}$$

5.2 Series Solution of Ordinary Differential Equations

We shall attempt to solve equation (5.7) by a series method.

It is a particular case of $\dfrac{d^2y}{dx^2} + P(x)\dfrac{dy}{dx} + Q(x)\,y = 0$. We have to be

careful in our choice of expansion. The straightforward substitution

$y = \displaystyle\sum_{m=0}^{\infty} a_m x^m$ may not always give the most general solution and the

method of variation of parameters may have to be called in.

Example $\qquad\qquad 2x\dfrac{d^2y}{dx^2} + 3\dfrac{dy}{dx} + y = 0 \tag{5.9}$

If we cast this in the form $\dfrac{d^2y}{dx^2} + \dfrac{3}{2x}\dfrac{dy}{dx} + \dfrac{1}{2x}y = 0$ we see that we cannot

expand $P(x)$ and $Q(x)$ as power series involving only positive integral powers

of x. Nevertheless we try the substitution $y = \displaystyle\sum_{m=0}^{\infty} a_m x^m$.

Assuming that this series converges and can be differentiated twice term by term, from (5.9), we obtain

$$2x(2a_2 + 6a_3x + 12a_4x^2 + ...) + 3(a_1 + 2a_2x + 3a_3x^2 + ...)$$
$$+ (a_0 + a_1x + a_2x^2 + ...) = 0$$

Equating coefficients of powers of x

x^0: $3a_1 + a_0 = 0$ hence $a_1 = -a_0/3$

x^1: $4a_2 + 6a_2 + a_1 = 0$ hence $a_2 = -a_1/10$

x^2: $12a_3 + 9a_3 + a_2 = 0$ hence $a_3 = -a_2/21$ and so on.

You can show that equating the coefficient of x^m to zero gives

$$2m(m + 1)a_{m+1} + 3(m + 1)a_{m+1} + a_m = 0$$

i.e. $\dfrac{a_{m+1}}{a_m} = \dfrac{-1}{(2m + 3)(m + 1)}$ for all $m \geq 0$.

Hence we may deduce that

$$\frac{a_{m+1}}{a_0} = \left[\frac{-1}{(2m + 3)(m + 1)}\right]\left[\frac{-1}{(2m + 1) \cdot m}\right] \cdots \left[\frac{-1}{3.1}\right]$$

$$= (-1)^{m+1} \frac{2^{m+1}}{(2m + 3)(2m + 2)(2m + 1)(2m) ... 3.2} = (-1)^{m+1}.2^{m+1}/(2m + 3)!$$

It is customary to work in terms of a_m and therefore $a_m = a_0(-1)^m 2^m/(2m + 1)!$

We have a solution of (5.9) in the form

$$y = a_0 \sum_{m=0}^{\infty} (-1)^m 2^m x^m /(2m + 1)! \qquad (5.10)$$

Now (5.10) contains only one arbitrary constant, a_0. To obtain the general solution of (5.9) we shall have to resort to variation of parameters. First we try and cast (5.10) in terms of elementary functions: *this procedure may not often be possible.*

$$y = a_0\left[1 - \frac{2x}{3!} + \frac{(2x)^2}{5!} - \frac{(2x)^3}{7!} + ...\right]$$

$$= \frac{a_0}{\sqrt{(2x)}}\left[(2x)^{1/2} - \frac{(2x)^{3/2}}{3!} + \frac{(2x)^{5/2}}{5!} - \frac{(2x)^{7/2}}{7!} + ...\right]$$

$$= \frac{a_0}{\sqrt{(2x)}} \sin\sqrt{(2x)} \qquad (5.11)$$

(We had to use some insight to obtain the result.)

We now put $y = \dfrac{v(x)}{\sqrt{(2x)}}\sin\sqrt{(2x)}$ and differentiate twice to find expressions

for dy/dx and d^2y/dx^2. Substituting these three expressions into (5.9) and simplifying,

$$2x \sin \sqrt{(2x)} \frac{d^2 v}{dx^2} + [2 \sqrt{2} \sqrt{x} \cos \sqrt{(2x)} + \sin \sqrt{(2x)}] \frac{dv}{dx} = 0 \qquad (5.12)$$

This equation can be solved by the integrating factor technique, since it is essentially of first order in dv/dx. The final solution is

$$y = A \frac{\sin \sqrt{(2x)}}{\sqrt{(2x)}} + B \frac{\cos \sqrt{(2x)}}{\sqrt{(2x)}} \qquad (5.13)$$

where A and B are constants. We have therefore found a second solution

$$\cos \sqrt{(2x)} / (\sqrt{2x}) = \frac{x^{-1/2}}{\sqrt{2}} \left[1 - \frac{(2x)}{2!} + \frac{(2x)^2}{4!} + \ldots \right]$$

$$= \frac{1}{\sqrt{2}} \left[x^{-1/2} - x^{1/2} + \frac{x^{3/2}}{3!} - \ldots \right]$$

and we see that the second solution involves fractional powers of x. The following method of solution allows this possibility from the outset.

The Frobenius method

A solution is sought in the form

$$y = x^c \sum_{n=0}^{\infty} a_n x^n = \sum_{n=0}^{\infty} a_n x^{n+c} \qquad (5.14)$$

where c is not necessarily a positive integer. It is implicit that $a_0 \neq 0$. This allows the flexibility we require.

Singular points of a differential equation

In the general form

$$\frac{d^2 y}{dx^2} + P(x) \frac{dy}{dx} + Q(x) . y = 0 \qquad (5.15)$$

if both $P(x)$ and $Q(x)$ can be expanded as a Taylor series in the neighbourhood of $x = x_0$ then (5.15) is said to have an **ordinary point** at $x = x_0$. If the expansions are valid over an interval $(x_0 - \alpha, x_0 + \alpha)$ then it can be shown that the series solution, obtained as above, is valid for that interval. If either $P(x)$ or $Q(x)$ cannot be expanded as a Taylor series about a point $x = x_1$ then x_1 is said to be a **singular point** for the differential equation.

If x_1 is a singular point and if both $(x - x_1)P(x)$ and $(x - x_1)^2 Q(x)$ can be expanded as Taylor series about $x = x_1$ then x_1 is said to be a **regular singular point**. In such a case (5.15) can be written as

$$(x - x_1)^2 \frac{d^2y}{dx^2} + (x - x_1) P_1(x) \frac{dy}{dx} + Q_1(x)y = 0 \tag{5.16}$$

where $P_1(x)$ and $Q_1(x)$ can be expanded as Taylor series about $x = x_1$.

Many practical problems have governing differential equations of this form. Two that we shall study are:

Bessel's equation:

$$x^2 \frac{d^2y}{dx^2} + x \frac{dy}{dx} + (x^2 - v^2)y = 0 \tag{5.17}$$

Legendre's equation:

$$(1 - x^2) \frac{d^2y}{dx^2} - 2x \frac{dy}{dx} + v(v + 1)y = 0 \tag{5.18}$$

Bessel's equation has a regular singular point at $x = 0$, while Legendre's equation has one at $x = 1$ and a second at $x = -1$. The latter case is not so obvious, but we can rewrite (5.18) as

$$(1 - x)^2 \frac{d^2y}{dx^2} - \frac{2x(1 - x)}{(1 + x)} \frac{dy}{dx} + \frac{v(v + 1)(1 - x)}{(1 + x)} y = 0$$

so that $P_1(x) = 2x/(1 + x)$ and $Q_1(x) = v(v + 1)(1 - x)/(1 + x)$

Both $P_1(x)$ and $Q_1(x)$ can be expanded as Taylor series about $x = 1$ and therefore $x = 1$ is a regular singular point. Similarly, we may show that $x = -1$ is a regular singular point.

In effect, the series expansion of y about a regular singular point is valid for those values of x for which the Taylor series expansions of $P_1(x)$ and $Q_1(x)$ are valid.

Example

Find the singular points in the finite plane of

(a) $\quad x^2 \frac{d^2y}{dx^2} - 4 \frac{dy}{dx} + 2y = 0$ (b) $\quad x(x - 2) \frac{d^2y}{dx^2} + (x - 2) \frac{dy}{dx} + 6xy = 0$

(c) $\quad 5 \frac{d^2y}{dx^2} + 4x \frac{dy}{dx} + 3y = 0$

We are concerned with the finite plane since we do not wish to deal with points at infinity.

(a) The singular point is at $x = 0$, $P(x) = -4/x^2$ and $Q(x) = 2/x^2$ and therefore it is not regular.

(b) The singular points are at $x = 0$ and $x = 2$; both are regular.

(c)   There are no singular points.

The indicial equation

We expect that the Frobenius method will yield two linearly independent solutions, both in the form of infinite series. Apart from the two coefficients a_n, the unknown is c. As with the auxiliary equation employed in the method of complementary function where we may have two real roots, one real root or no real roots, so we must expect different possibilities for the values of c. To see how Frobenius' method works, we shall attempt to solve equation (5.7); first we multiply it by x and change the dependent variable to y to obtain

$$x\frac{d^2y}{dx^2} + \frac{dy}{dx} - r^2 y = 0 \tag{5.19}$$

We now seek a solution in the form

$$y = x^c \sum_{n=0}^{\infty} a_n x^n = \sum_{n=0}^{\infty} a_n x^{n+c} \tag{5.14}$$

Therefore

$$\left. \begin{array}{c} \dfrac{dy}{dx} = \displaystyle\sum_{n=0}^{\infty} a_n \cdot (n+c)x^{n+c-1} \\[3em] \dfrac{d^2y}{dx^2} = \displaystyle\sum_{n=0}^{\infty} a_n(n+c)(n+c-1)x^{n+c-2} \end{array} \right\} \tag{5.20}$$

Substituting into (5.19) gives

$$\sum_{n=0}^{\infty} a_n(n+c)(n+c-1)x^{n+c-1} + \sum_{n=0}^{\infty} a_n(n+c)x^{n+c-1} - r^2 \sum_{n=0}^{\infty} a_n x^{n+c} = 0$$

or
$$\sum_{n=0}^{\infty} a_n(n+c)^2 \, x^{n+c-1} - r^2 \sum_{n=0}^{\infty} a_n x^{n+c} = 0 \tag{5.21}$$

We can expand the series in (5.21). Since the results hold for all x, we equate the coefficient of each power of x on the left-hand side to zero. The smallest power of x is x^{c-1} and the coefficient of this is

$$a_0 c^2 = 0 \tag{5.22}$$

(There is no contribution from the second summation.)

Equation (5.22) is called the **indicial equation**, and is always found by equating the coefficient of the lowest power of x to zero. Since $a_0 \neq 0$, we have the repeated root $c = 0$. This will lead us only to one solution and we have hit trouble. We shall now look at various possibilities for the indicial equation by selecting four arbitrary examples. Then we shall return to this problem.

Roots of indicial equation differ by a non-integer

$$\text{Consider the equation} \quad 3x\frac{d^2y}{dx^2} + 2\frac{dy}{dx} + y = 0 \tag{5.23}$$

If we substitute (5.14) and (5.20) we obtain

$$\sum_{n=0}^{\infty} a_n(n + c)(3[n + c - 1] + 2)x^{n+c-1} + \sum_{n=0}^{\infty} a_n x^{n+c} = 0$$

Equating the coefficient of the lowest power of x, viz. x^{c-1}, to zero gives the indicial equation $a_0 c(3c - 1) = 0$. This has two roots $c = 0$ and $c = 1/3$ which differ by a non-integer.

We equate the coefficient of x^{n+c-1}, $n > 1$ to zero to obtain

$$a_n(n + c)(3n + 3c - 1) + a_{n-1} = 0 \tag{5.24}$$

Now we take the two values of c in turn.

$\underline{c = 0}$ The general equation (5.24) becomes $a_n n(3n - 1) + a_{n-1} = 0$. Therefore

$$a_n = \frac{-a_{n-1}}{n(3n - 1)} = \frac{(-1)^2}{n(3n - 1)}\frac{a_{n-2}}{(n - 1)(3n - 4)} = \cdots = \frac{(-1)^n a_0}{n!(3n - 1)(3n - 4) \ldots 5.2}$$

and from (5.14)

$$y = a_0\left\{1 - \frac{x}{2.1!} + \frac{x^2}{2.5.2!} - \frac{x^3}{2.5.8.3!} + \ldots\right\} \tag{5.25}$$

$\underline{c = 1/3}$ Equation (5.24) is $a_n(n + 1/3)(3n) + a_{n-1} = 0$ so that

$$a_n = \frac{-a_{n-1}}{n(3n + 1)} = \frac{(-1)^2 a_{n-2}}{n(3n + 1)(n - 1)(3n - 2)} = \frac{(-1)^n a_0}{n!(3n + 1)(3n - 2) \ldots 7.4}$$

and we have a second solution

$$y = a_0 x^{1/3}\left\{1 - \frac{x}{4.1!} + \frac{x^2}{4.7.2!} - \frac{x^3}{4.7.10.3!} + \ldots\right\} \tag{5.26}$$

The general solution is a linear combination of the two, thus we obtain the result

$$y = A \left\{ 1 - \frac{x}{2.1!} + \frac{x^2}{2.5.2!} - \frac{x^3}{2.5.8.3!} + ... \right\}$$

$$+ Bx^{1/3} \left\{ 1 - \frac{x}{4.1!} + \frac{x^2}{4.7.2!} - \frac{x^3}{4.7.10.3!} + ... \right\} \qquad (5.27)$$

Roots of indicial equation equal

The general solution for equations of the last category may be written
$$y = Au(x, c_1) + Bu(x, c_2)$$

$$= (A + B) u(x, c_1) + B(c_2 - c_1) \left\{ \frac{u(x, c_2) - u(x, c_1)}{c_2 - c_1} \right\}$$

$$= \alpha \, u(x, c_1) + \beta \left\{ \frac{u(x, c_2) - u(x, c_1)}{c_2 - c_1} \right\}$$

Now take the limit as $c_2 \rightarrow c_1$ to obtain

$$y = \alpha \, u(x, c_1) + \beta \left. \frac{\partial u}{\partial c} \right|_{c = c_1} \qquad (5.28)$$

Example 1

Consider the equation $\qquad x \frac{d^2 y}{dx^2} + \frac{dy}{dx} - xy = 0 \qquad (5.29)$

Substituting for y and its derivatives we obtain

$$\sum_{n=0}^{\infty} a_n (n + c)(n + c - 1 + 1)x^{n+c-1} - \sum_{n=0}^{\infty} a_n x^{n+c+1} = 0$$

The indicial equation is $a_0 c^2 = 0$ which has a repeated root $c = 0$.

The coefficient of x^c equated to zero produces $a_1 (c + 1)^2 = 0$ i.e. $a_1 = 0$.

The coefficient of x^{n+c-1} ($n > 1$) equated to zero gives $a_n (n + c)^2 - a_{n-2} = 0$

i.e. $\qquad\qquad\qquad\qquad a_n = \frac{a_{n-2}}{(n + c)^2}$

Hence $a_3 = a_5 = a_7 = ... = 0$ and $a_2 = a_0/(2 + c)^2$, $a_4 = a_2/(4 + c)^2 = a_0/[(2 + c)^2 . (4 + c)^2]$ etc. so that the only solution we can obtain is found by taking $c = 0$ in

$$y = a_0 x^c \left\{ 1 - \frac{x^2}{(2+c)^2} + \frac{x^4}{(2+c)^2 (4+c)^2} - \ldots \right\} = a_0 u(x, c) \quad (5.30)$$

The second solution is found from $y = \dfrac{\partial u}{\partial c}(x, c)$ i.e.

$$y = a_0 x^c \ln x \left\{ 1 - \frac{x^2}{(2+c)^2} + \frac{x^4}{(2+c)^2 (4+c)^2} - \ldots \right\}$$

$$+ a_0 x^c \left\{ \frac{2x^2}{(2+c)^3} - \frac{2x^4}{(2+c)^3 (4+c)^2} - \frac{2x^4}{(2+c)^2 (4+c)^3} + \ldots \right\} \quad (5.31)$$

Let $c \to 0$ in (5.30) and (5.31) and put $a_0 = 1$ to obtain

$$y_1 = 1 - \frac{x^2}{2^2} + \frac{x^4}{2^2 . 4^3} - \ldots \quad (5.32a)$$

and $y_2 = \ln x \left\{ 1 - \dfrac{x^2}{2^2} + \dfrac{x^4}{2^2 . 4^2} - \ldots \right\} + \dfrac{2x^2}{2^3} - \dfrac{2x^4}{2^3 . 4^2} - \dfrac{2x^4}{2^2 . 4^3} + \ldots$

i.e. $y_2 = \ln x \left\{ 1 - \dfrac{x^2}{2^2} + \dfrac{x^4}{2^2 . 4^2} - \ldots \right\} + \dfrac{x^2}{2^2} - \dfrac{x^4}{2^2 . 4^2} (1 - \dfrac{2}{4}) + \ldots \quad (5.32b)$

The general solution of (5.29) is $y = Ay_1 + By_2$.

Example 2

We saw that the equation for the fin

$$x \frac{d^2 y}{dx^2} + \frac{dy}{dx} - r^2 y = 0 \quad (5.19)$$

had an indicial equation with repeated root $c = 0$.

Equating the coefficient of x^c to zero produces $a_1 (1 + c)^2 - r^2 a_0 = 0$ and the general recurrence relationship is $a_n (n + c)^2 - r^2 a_{n-1} = 0$.

Hence one solution is

$$y_1 = 1 + \frac{r^2 x}{1^2} + \frac{r^4 x^2}{1^2 . 2^2} + \ldots$$

The second solution is

$$y_2 = \ln x \left\{ 1 + \frac{r^2 x}{1^2} + \frac{r^4 x^2}{1^2 \cdot 2^2} + \ldots \right\} - 2 \left\{ \frac{r^2 x}{1^3} + \frac{r^4 x^2}{1^3 \cdot 2^3}(2 + 1) + \ldots \right\}$$

Roots of indicial equation differing by an integer

Trouble arises in cases where the roots of the indicial equation differ by an integer. The trouble can be of two kinds.

(i) One of the coefficients a_n becomes infinite

(ii) One of the coefficients a_n becomes indeterminate.

We merely outline the way each of these troubles can be overcome.

Example 1

Consider the equation

$$x^2 \frac{d^2 y}{dx^2} + x \frac{dy}{dx} - (x^2 - 9)y = 0 \tag{5.33}$$

The indicial equation is $a_0(c - 3)(c + 3) = 0$ of which the roots are $c_1 = -3$, $c_2 = 3$. The equation for the coefficient of x^{c+1} is $a_1(c - 2)(c + 4) = 0$ and hence $a_1 = 0$. The recurrence relation for the coefficients a_n, $n \geq 2$ is $(c + n - 3)(c + n + 3)a_n + a_{n-2} = 0$. Therefore $a_3 = a_5 = a_7 = \ldots = 0$.

Further $a_2 = -a_0/(c - 1)(c + 5)$, $a_4 = -a_2(c + 1)(c + 7)$, $a_6 = -a_4/(c + 3)(c + 9)$ and so on. The trouble is that a_6, and all subsequent 'even' coefficients will have $(c + 3)$ in their denominator. This renders the case $c = -3$ invalid. Note that it is the smaller root which causes trouble.

We can, of course, carry on with $c_2 = 3$ to obtain the first solution

$$y = a_0 u(x, c_2) = a_0 x^3 \left\{ 1 - \frac{x^2}{2.8} + \frac{x^4}{2.4.8.10} - \frac{x^6}{2.4.6.8.10.12} + \ldots \right\} \tag{5.34}$$

We quote, without proof, that the way to find a second solution of (5.33) is to find $\dfrac{\partial}{\partial c}\{(c - c_1)u(x, c)\}$ and then evaluate the expression at $c = c_1 = -3$.

Now $\quad u(x, c) = x^c \left\{ 1 - \dfrac{x^2}{(c - 1)(c + 5)} + \dfrac{x^4}{(c - 1)(c + 1)(c + 5)(c + 7)} \right.$

$$\left. - \frac{x^6}{(c - 1)(c + 1)(c + 3)(c + 5)(c + 7)(c + 9)} + \ldots \right\}$$

and
$$(c+3)u(x, c) = x^c \left\{ 1 - \frac{(c+3)x^2}{(c-1)(c+5)} + \frac{(c+3)x^4}{(c-1)(c+1)(c+5)(c+7)} \right.$$

$$\left. - \frac{x^6}{(c-1)(c+1)(c+5)(c+7)(c+9)} + \ldots \right\}$$

Notice we have effectively removed $(c+3)$ from the denominators. Then

$$\frac{\partial}{\partial c} \{(c+3)u(x, c)\} = x^c \ln x \left\{ 1 - \frac{(c+3)x^2}{(c-1)(c+5)} + \frac{(c+3)x^4}{(c-1)(c+1)(c+5)(c+7)} \right.$$

$$\left. - \frac{x^6}{(c-1)(c+1)(c+5)(c+7)(c+9)} + \ldots \right\}$$

$$+ x^c \left[\frac{-(c+3)x^2}{(c-1)(c+5)} \left\{ \frac{1}{(c+3)} - \frac{1}{(c+5)} - \frac{1}{(c-1)} \right\} \right.$$

$$+ \frac{(c+3)x^4}{(c-1)(c+1)(c+5)(c+7)} \left\{ \frac{1}{(c+3)} - \frac{1}{(c-1)} - \frac{1}{(c+1)} - \frac{1}{(c+5)} - \frac{1}{(c+7)} \right\}$$

$$\left. - \frac{x^6}{(c-1)(c+1)(c+5)(c+7)(c+9)} \left\{ \frac{-1}{(c-1)} - \frac{1}{(c+1)} - \frac{1}{(c+5)} - \frac{1}{(c+7)} - \frac{1}{(c+9)} \right\} + \ldots \right]$$

If we now put $c = -3$ we obtain

$$v(x, c_1) = x^{-3} \ln x \left\{ 1 - \frac{x^6}{(-4)(-2)2.4.6} + \ldots \right\} + x^{-3} \left\{ \frac{-x^2}{(-4).2} + \right.$$

$$\left. \frac{x^4}{(-4)(-2).2.4} - \frac{x^6}{(-4).(-2)(2.4.6)} \left[-\frac{1}{(-4)} - \frac{1}{(-2)} - \frac{1}{2} - \frac{1}{4} - \frac{1}{6} \right] + \ldots \right\} \quad (5.35)$$

The general solution of (5.33) is a linear combination of expressions (5.34) and (5.35).

Example 2

Consider the equation

$$(1 - x^2) \frac{d^2 y}{dx^2} - 4x \frac{dy}{dx} + 4y = 0 \quad (5.36)$$

The indicial equation is $a_0 c(c-1)$. The coefficient of x^{c-1} is
$$a_1 (c+1) = 0 \quad (5.37)$$
and the general recurrence relation for $n > 2$ reduces to

$$a_n = \frac{a_{n-2}(c + n + 2)(c + n - 3)}{(c + n)(c + n - 1)} \qquad (5.38)$$

The roots of the indicial equation are $c = 1$ or $c = 0$. However, if we take $c = 0$ then (5.37) does not give us a value for a_1; a_1 is indeterminate. Let us assume it is finite. Then consider $c = 0$ so that (5.38) becomes

$$a_n = a_{n-2}(n + 2)(n - 3)/n(n - 1).$$

Hence $a_2 = \dfrac{a_0 \, 4.(-1)}{2.1}$, $a_4 = \dfrac{a_2 \, .6.1}{4.3} = \dfrac{a_0 \, .6.4.1.(-1)}{4.3.2.1}$, $a_6 = \dfrac{a_0 \, .8.6.4.3.1.(-1)}{6.5.4.3.2.1}$,

and $a_3 = a_1.0$, making $a_5 = a_7 = ... = 0$.

We use a_1 and a_0 as the two arbitrary constants. The general solution is

$$y = a_0 \left\{ 1 - \frac{2x^2}{1} - \frac{3x^4}{3} - \frac{4x^6}{5} - \frac{5x^8}{7} - ..... \right\} + a_1 x \qquad (5.39)$$

We could replace a_0 and a_1 by A and B respectively, to be consistent.

Note that the second part of the solution is a finite series and that the power of x which appears is the value of c not yet considered, viz. 1.

To justify the statement that (5.39) is the general solution of (5.36) we could now consider the case $c = 1$. In fact, we obtain merely $y = a_1 x$ which is covered by (5.39).

Convergence of power series

We have so far tacitly assumed that the series we derive will converge. Strictly speaking we should test them by, for example, the ratio test to decide over what range of values of x they will be valid expansions. More importantly, we should like to know over what range of x the series are useful for practical computation.

Expansion about other points

So far we have considered examples where $x = 0$ was a regular singular point. If, for example, $x = 1$ is a regular singular point then we could attempt a Frobenius solution about $x = 1$. The easiest way perhaps to accomplish this is to change the independent variable by the transformation $x = 1 - z$.

Furthermore, there may be cases where the Frobenius method fails and then it is sometimes possible to expand effectively about $x = \infty$ by transforming the variables via $x = 1/z$ and hoping that $x = 0$ is a regular singular point of the transformed differential equation. In such a case we would end up with an expansion in descending powers of x.

5.3 The Gamma Function

In this section we examine a function which is useful in describing the Bessel and Legendre functions to which the rest of this chapter is directed.

In essence, the **gamma function** $\Gamma(x)$ extends the idea of factorial from the range of non-negative integers. For positive values of x we may define

$$\Gamma(x) = \int_0^\infty t^{x-1}\, e^{-t}\, dt \tag{5.40}$$

and extend the definition by the additional rule (which can be shown to be true for $x > 0$ from (5.40)),

$$\Gamma(x + 1) = x\, \Gamma(x) \tag{5.41}$$

We emphasise that $\Gamma(x)$ is not defined for $x = 0, -1, -2, -3, \ldots$

Since we can show from (5.40) that

$$\Gamma(1) = 1 \tag{5.42}$$

it follows by induction that

$$\Gamma(n + 1) = n! \tag{5.43}$$

if n is a positive integer, so that we see the relationship with factorial function.

Example

Tables of $\Gamma(x)$ are usually given for the range $1 < x < 2$ and other values can be found from applying (5.41) sufficiently often.

For example, if we know that to 4 d.p. then $\Gamma(1.27) = 0.9126$, and hence

$$\Gamma(2.27) = 1.27\Gamma(1.27) = 1.1590$$

and

$$\Gamma(-0.73) = \frac{\Gamma(-0.73 + 1)}{-0.73} = -\frac{\Gamma(0.27)}{0.73} = \frac{-\Gamma(1.27)}{(0.73)(0.27)} = -4.6301 \quad (4 \text{ d.p.})$$

Further properties

We can show directly that $\qquad \Gamma(\tfrac{1}{2}) = \sqrt{\pi}$ $\qquad\qquad\qquad$ (5.44)

Also it can be proved that

$$\Gamma(n + \tfrac{1}{2}) = \frac{1.3.5. \ldots\ldots (2n - 1)\sqrt{\pi}}{2^n} \quad \text{for } n = 1, 2, 3, \ldots \tag{5.45}$$

and that

$$\Gamma(x)\, \Gamma(1 - x) = \pi \cosec \pi x \quad \text{for } x \neq 0, \pm 1, \pm 2, \ldots\ldots \tag{5.46}$$

In Figure 5.3 on the following page we show the graph of $\Gamma(x)$.
Note the discontinuities at $x = 0, -1, -2$, etc.

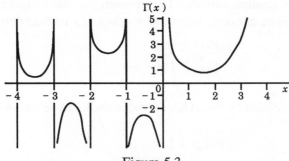

Figure 5.3

5.4   Bessel Functions of the First and Second Kind

The subject of Bessel Functions can be approached from many directions. One major application lies in the solution of certain partial differential equations where circular symmetry exists.

Bessel's equation of order v is

$$x^2 \frac{d^2y}{dx^2} + x \frac{dy}{dx} + (x^2 - v^2)y = 0 \qquad (5.47)$$

Note that this is a second-order differential equation; the parameter v refers to the order of the function which is a solution of (5.47). The equation has a regular singular point at the origin and we may use the Frobenius method to effect a solution. We shall not go into details here, but merely present results.

We first consider the case when v is a non-integer. The solution of (5.47) can then be written as

$$y = AJ_v(x) + BJ_{-v}(x) \qquad (5.48)$$

where A and B are constants,

$$J_v(x) = \sum_{s=0}^{\infty} (-1)^s \frac{(x/2)^{v+2s}}{\Gamma(v+s+1).s!} \qquad (5.49a)$$

which is called a **Bessel function of the first kind of order v**, and

$$J_{-v}(x) = \sum_{s=0}^{\infty} (-1)^s \frac{(x/2)^{2s-v}}{\Gamma(s+1-v).s!} \qquad (5.49b)$$

When $v = 0, J_{-0}(x) \equiv J_0(x)$ and we seem to have only one solution

$$J_0(x) = \sum_{s=0}^{\infty} (-1)^s \frac{(x/2)^{2s}}{(s!)^2} \qquad (5.50)$$

which is the Bessel function of the first kind of order zero. A second solution will emerge via the Frobenius method; it is given by

$$Y_0(x) = \frac{2}{\pi}\left[\{\ln(\tfrac{1}{2}x) + \gamma\} J_0(x) - \sum_{s=0}^{\infty}\left\{\frac{(-1)^s (x/2)^{2s}}{(s!)^2}(1 + \frac{1}{2} + \dots + \frac{1}{s})\right\}\right] \quad (5.51)$$

γ is the **Euler constant**; $\gamma = \lim_{n \to \infty} (1 + \frac{1}{2} + \dots + \frac{1}{n} - \ln n) = 0.5772 \quad$ (4 d.p.)

This complicated expression is called the **Bessel function of the second kind of order zero**. The general solution of (5.47) is then

$$y = AJ_0(x) + BY_0(x) \quad (5.52)$$

These functions $J_v(x)$, $J_0(x)$, $Y_0(x)$ are solutions of equations which crop up frequently in models of engineering situations. Consequently, just as we give the series $x - x^3/3! + x^5/5! - \dots$ the name $\sin x$ we give the series (5.49), (5.50) and (5.51) simple notations. If we can find the main properties of these series for different values of x we shall be able to understand the qualitative nature of the solution of equations like (5.47) in addition to calculating approximate values from the series expansions. Since tables of Bessel functions exist we merely need to be able to express solutions in terms of these just as we could do for $\sin x$ and $\cos x$.

One case of (5.47) remains; when v is a non-zero integer, n. The solution is

$$y = AJ_n(x) + BY_n(x) \quad (5.53)$$

where

$$J_n(x) = \sum_{s=0}^{\infty} \frac{(-1)^s (x/2)^{2s+n}}{(s+n)! \, s!} \quad (5.54)$$

and

$$Y_n(x) = \frac{2}{\pi}\left[\{\ln(\tfrac{1}{2}x) + \gamma\}J_n(x) - \frac{1}{2}\sum_{s=0}^{n-1}\frac{(n-s-1)!}{s!}\left[\frac{x}{2}\right]^{2s-n}\right.$$

$$\left. - \frac{1}{2}\sum_{s=0}^{\infty}\frac{(-1)^s (x/2)^{2s+n}}{s!(n+s)!}\left\{1 + \frac{1}{2} + \frac{1}{3} + \dots + \frac{1}{s} + 1 + \frac{1}{2} + \dots + \frac{1}{n+s}\right\}\right]$$

$$(5.55)$$

We note that the term with $s = 0$ in the last summation is $\dfrac{(x/2)^n}{n!}\{1 + \frac{1}{2} + \dots + \frac{1}{n}\}$.

All the series converge for all x.

You must not be put off by these formulae. We have written them down simply to show the advantages of developing properties of the Bessel functions and merely dealing with expressions like $J_{1/2}(x)$, $Y_3(x)$ etc. Now we shall produce graphs of some Bessel functions of positive integer order and look at some main features.

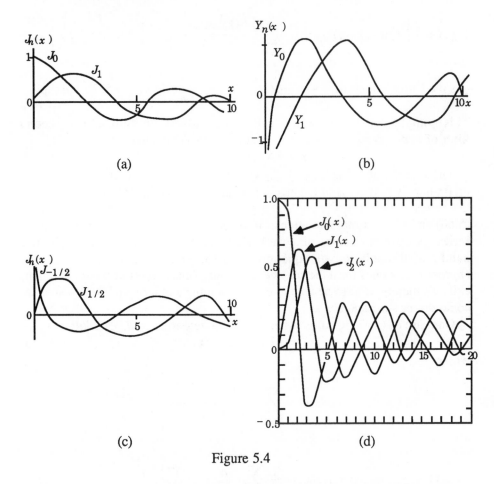

Figure 5.4

In Figure 5.4 we see the graphs of $J_0(x)$ and $Y_0(x)$; note that both are oscillatory with oscillations of decaying amplitude, which is a characteristic of all these Bessel functions. Whereas $J_0(x)$ is finite at $x = 0$, $Y_0(x)$ is not, because of the presence of a logarithm in its definition. From graph (d) we see that between any two consecutive values of x for which $J_n(x) = 0$ there will be one zero of $J_m(x)$ where $m \neq n$. This is called *interlacing of zeros*. The larger n the larger are the values of the corresponding zeros. Note that apart from $J_0(x)$ all $J_n(x)$ are zero at $x = 0$. Further, each $J_n(x)$ oscillates with smaller amplitude than its predecessor. In graph (b) we see a similar set-up for the $Y_n(x)$. In graph (c) we show $J_{1/2}(x)$ and $J_{-1/2}(x)$ which are involved in the solution of the wave equation in spherical polar coordinates. They also have the advantage that they can be expressed as

$$J_{1/2}(x) = \sqrt{2/(\pi x)} \cdot \sin x \qquad (5.56)$$

and $$J_{-1/2}(x) = \sqrt{2/(\pi x)} \cdot \cos x \qquad (5.57)$$

Properties of Bessel functions

We now quote some useful formulae which are helpful when equations have to be reduced to a form of Bessel's equation. We shall assume that n is an integer and v is any real number.

(a) $\quad J_{-n}(x) = (-1)^n J_n(x)$ $\qquad (5.58)$

(b) **Generating function**

$$e^{(1/2)x(t-t^{-1})} = \sum_{n=-\infty}^{\infty} J_n(x) \cdot t^n \qquad (5.59)$$

(c) **Recurrence relations**

$$J_v(x) = \frac{x}{2v} \{J_{v-1}(x) + J_{v+1}(x)\} \qquad (5.60)$$

$$J'_v(x) = \frac{1}{2} \{J_{v-1}(x) - J_{v+1}(x)\} \qquad (5.61)$$

$$\frac{d}{dx} \{x^v J_v(x)\} = x^v J_{v-1}(x) \qquad (5.62)$$

(d) **Integral representation**

$$J_0(x) = \frac{2}{\pi} \int_0^{\pi/2} \cos(x \sin \theta) d\theta \qquad (5.63)$$

(e) **Asymptotic behaviour**

For large positive x, the series expansions of $J_v(x)$ and $Y_v(x)$ converge very slowly. For large enough x,

$$J_v(x) \sim \left[\frac{2}{\pi x}\right]^{1/2} \cos\left[x - \frac{v\pi}{2} - \frac{\pi}{4}\right] \qquad (5.64a)$$

$$Y_v(x) \sim \left[\frac{2}{\pi x}\right]^{1/2} \sin\left[x - \frac{v\pi}{2} - \frac{\pi}{4}\right] \qquad (5.64b)$$

which shows the oscillatory behaviour with decreasing amplitude. For small values of x, it may be shown that

$$J_v(x) \sim (x/2)^v / \Gamma(v + 1) \qquad (5.65)$$

The symbol $\sim$ means asymptotically equivalent to.

(f)   **Zeros**

Using exact equivalents of (5.64) we locate the zeros of $J_n(x)$ by iteration. In Table 5.1 we show the first zeros of $J_0(x)$, $J_1(x)$, $J_2(x)$ to 2 d.p.

Table 5.1

n	1st root	2nd root	3rd root
0	2.40	5.52	8.65
1	3.83	7.02	10.17
2	5.14	8.42	11.62

(g)   **Orthogonal property**

If α_1 and α_2 are two unequal positive roots of $J_n(\alpha a) = 0$ then

$$\int_0^a x\, J_n(\alpha_1 x)\, J_n(\alpha_2 x)\, \mathrm{d}x = 0 \qquad (5.66)$$

The factor x is said to be a **weighting factor**. Under the condition that α_r are the roots of $J_n(\alpha a) = 0$ it follows that a function $f(x)$ can be expanded as

$$f(x) = \sum_{r=0}^{\infty} A_r\, J_n(\alpha_r x); \qquad \int_0^a x\, J_n^{\ 2}(\alpha x)\, \mathrm{d}x = \tfrac{1}{2} a^2 [J'_n(\alpha a)]^2 A_r \qquad (5.67)$$

where the A_r are coefficients. This property will prove useful in the solution of certain partial differential equations. We note in passing that there are other conditions under which orthogonal expansion is possible.

5.5 Modified Bessel Functions

Sometimes, when solving Laplace's equation in cylindrical coordinates we obtain

$$x^2 \frac{\mathrm{d}^2 y}{\mathrm{d}x^2} + x \frac{\mathrm{d}y}{\mathrm{d}x} - (x^2 + v^2)y = 0 \qquad (5.68)$$

and this is called the **modified Bessel equation** and for v non-integral it has a general solution in the form

$$y = A I_v(x) + B I_{-v}(x) \qquad (5.69)$$

where A and B are constants and $I_v(x)$ is called the modified Bessel function of the first kind, of order v. It can be shown that

$$I_v(x) = (i)^{-v} J_v(ix) \tag{5.70}$$

For the case v integral we have a **modified Bessel function of the second kind,** $K_n(x)$ which has a role relative to $I_n(x)$ which is similar to the role of $Y_n(x)$ relative to $J_n(x)$. The general solution of (5.68) with v integral is

$$y = AI_v(x) + BK_v(x) \tag{5.71}$$

5.6 Transformations of Bessel's Equation

Many problems can be modelled by a differential equation which can be transformed to Bessel's equation.

We quote the result that the differential equation

$$x^2 \frac{d^2 y}{dx^2} + (1 - 2\alpha)x \frac{dy}{dx} + [\beta^2 \gamma^2 x^{2\gamma} + (\alpha^2 - v^2 \gamma^2)]y = 0 \tag{5.72}$$

has general solution

$$y = x^\alpha [AJ_v(\beta x^\gamma) + BJ_{-v}(\beta x^\gamma)], \; v \text{ a non-integer} \tag{5.73a}$$

or

$$y = x^\alpha [AJ_n(\beta x^\gamma) + BY_n(\beta x^\gamma)], \; v = n, \text{ an integer} \tag{5.73b}$$

Example 1

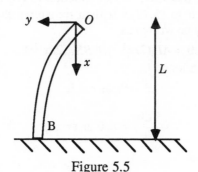

Figure 5.5

Consider the thin strut shown in Figure 5.5. We wish to find what length it can be without buckling under its own weight. It can be shown that the governing equation is

$$\frac{d^2 p}{dx^2} + k^2 xp = 0 \tag{5.74}$$

where $p = dy/dx$ and $k^2 = w/EI$, w being the weight per unit length. The boundary conditions are

$$p'(0) = 0, p'(L) = 0 \tag{5.75}$$

We can either use the results of (5.72) directly or carry out three transformations of variables. We shall take the easy way out. Multiplying (5.74) by x^2 we obtain the following equation

$x^2 \dfrac{d^2 p}{dx^2} + k^2 x^3 p = 0$ which compares with (5.72) if $1 - 2\alpha = 0$, $\beta^2 \gamma^2 x^{2\gamma} =$

$k^2 x^3$ and $\alpha^2 - v^2 \gamma^2 = 0$. Therefore $\alpha = \dfrac{1}{2}$, $\gamma = \dfrac{3}{2}$, $\beta = 2k/3$, $v = \dfrac{1}{3}$.

Therefore, from (5.73a) we have as general solution

$$p = x^{1/2} \left[A J_{1/3}(\tfrac{2}{3} kx^{3/2}) + B J_{-1/3}(\tfrac{2}{3} kx^{3/2}) \right] \tag{5.76}$$

It can be shown after much algebra that the condition $p' = 0$ at $x = 0$ implies that $A = 0$ and the condition $p'(L) = 0$ means that if the column has buckled (in the sense that its vertical equilibrium position is unstable) then $B \neq 0$ and hence

$J_{-1/3}(\tfrac{2}{3} kL^{3/2}) = 0$. From tables we find that the smallest positive root of this

last equation is approximately $2/3kL^{3/2} = 1.866$ and hence $L \approx 1.986 k^{-2/3} = 1.986 \, [EI/w]^{1/3}$. For greater lengths, the column will buckle under its own weight.

Example 2

In the same way, the modified Bessel equation (5.68) may be generalised to give an equation similar to (5.72). It then transpires that the solution of the cooling fin problem may be written as

$$\theta = A I_0(2rx^{1/2}) + B K_0(2rx^{1/2}) \tag{5.77}$$

From the boundary conditions

$$\theta = \theta_0 \text{ at } x = L \tag{5.78a}$$

and
$$-\frac{K d\theta}{dx} = h_0 \theta \text{ at } x = x_0 \tag{5.78b}$$

we may evaluate A and B.

5.7 An introduction to Legendre Polynomials

Any solution of Laplace's equation $\nabla^2 u = 0$ is called a **harmonic function**. In spherical polar coordinates (r, θ, ϕ), Laplace's equation becomes

$$\frac{\partial}{\partial r} \left\{ r^2 \frac{\partial u}{\partial r} \right\} + \frac{1}{\sin \theta} \frac{\partial}{\partial \theta} \left(\sin \theta \frac{\partial u}{\partial \theta} \right) + \frac{1}{\sin^2 \theta} \frac{\partial^2 u}{\partial \phi^2} = 0 \tag{5.79}$$

For problems which have axial symmetry, i.e. are independent of ϕ, we obtain

$$\frac{\partial}{\partial r}\left\{r^2\frac{\partial u}{\partial r}\right\}+\frac{1}{\sin\theta}\frac{\partial}{\partial\theta}\left(\sin\theta\frac{\partial u}{\partial\theta}\right)=0 \qquad (5.80)$$

This equation has solutions $r^n P_n(\mu)$ and $r^{-n-1}P_n(\mu)$ where n is a positive integer $\mu=\cos\theta$ and $P_n(\mu)$ satisfies **Legendre's equation** viz.

$$\frac{d}{d\mu}\left\{(1-\mu^2)\frac{dP_n}{d\mu}\right\}+n(n+1)P_n=0 \qquad (5.81)$$

The functions $P_n(\mu)$ are called Legendre polynomials and their counterparts $r^n P_n(\mu)$ and $r^{-n-1}P_n(\mu)$ are called spherical harmonics and are especially useful in solving problems with spherical boundaries. Note that $-1\le\mu\le 1$. It can be shown that the Legendre polynomials are linearly independent.

Generating Function

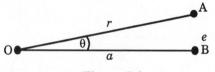

Figure 5.6

In some potential problems it is useful to expand the potential as a series in r or r^{-1}. Figure 5.6 shows a point electric charge e at B which is a distance a from 0. The electrostatic potential V at A due to the charge B is given by

$$V=e/AB=e[r^2-2ra\cos\theta+a^2]^{-(1/2)}=e[r^2-2ra\mu+a^2]^{-(1/2)}$$

Hence
$$V=\begin{cases}(e/a)\,[(r/a)^2-2(r/a)\mu+1]^{-(1/2)} & 0\le r\le a\\[2mm](e/r)\,[1-2(a/r)\mu+(a/r)^2]^{-(1/2)} & a\le r\end{cases}$$

It can be shown that

$$V=\begin{cases}(e/a)\displaystyle\sum_{n=0}^{\infty}P_n(\mu)\left[\dfrac{r}{a}\right]^n & 0\le r\le a\\[6mm](e/r)\displaystyle\sum_{n=0}^{\infty}P_n(\mu)\left[\dfrac{a}{r}\right]^n & a\le r\end{cases} \qquad (5.82)$$

This means that $P_n(\mu)$ is the coefficient of h^n in the expansion of $\{1-2h\mu+h^2\}^{-(1/2)}$. (In the case $0\le r\le a$ we used $h=r/a$ and for $r\ge a$ we used $h=a/r$.)

It needs to be shown that this definition of $P_n(\mu)$ allows it to satisfy equation (5.81). In other words the definition

$$\{1 - 2h\mu + h^2\}^{-(1/2)} = \sum_{n=0}^{\infty} h^n P_n(\mu) \qquad (5.83)$$

where $-1 \leq \mu \leq 1$ and $h \leq 1$ and the equation (5.81) are equivalent.

Examples of $P_n(\mu)$

We can obtain expressions for the $P_n(\mu)$ from (5.83). The general formula for $P_n(\mu)$ is

$$P_n(\mu) = \frac{(2n)! \, \mu^n}{2^n (n!)^2} \left\{ 1 - \frac{n(n-1)}{2(2n-1)} \frac{1}{\mu^2} + \frac{n(n-1)(n-2)(n-3)}{2.4(2n-1)(2n-3)} \frac{1}{\mu^4} + \cdots \right\} \qquad (5.84)$$

which could have been derived by applying the Frobenius technique to Legendre's equation (5.81). We can obtain the first few polynomials directly from (5.83).

$$(1 - 2h\mu + h^2)^{-(1/2)} = 1 + \frac{\left(-\frac{1}{2}\right)}{1}(-2\mu h + h^2) + \frac{\left(-\frac{1}{2}\right)\left(-\frac{3}{2}\right)}{1.2}(-2\mu h + h^2)^2 + \cdots$$

$$= 1 + \mu h + \left(-\frac{1}{2} + \frac{3}{2}\mu^2\right)h^2 + \left(-\frac{3}{2}\mu + \frac{5}{2}\mu^3\right)h^3 + \cdots$$

Hence

$$P_0(\mu) = 1, \; P_1(\mu) = \mu, \; P_2(\mu) = \frac{1}{2}(3\mu^2 - 1), \; P_3(\mu) = \frac{1}{2}(5\mu^3 - 3\mu) ,\ldots$$

Note that $P_{2n+1}(\mu)$ involves only odd powers of μ, whilst $P_{2n}(\mu)$ involves only even powers. Also $P_n(\mu)$ is a polynomial in μ of degree n.

Example

We can, in fact, find $P_n(0)$, $P_n(1)$ and $P_n(-1)$ easily. From (5.83) we obtain

$$\sum_{n=0}^{\infty} P_n(0)h^n = (1 + h^2)^{-(1/2)} = 1 + \frac{\left(-\frac{1}{2}\right)}{1}h^2 + \frac{\left(-\frac{1}{2}\right)\left(-\frac{3}{2}\right)}{1.2}h^4 + \cdots$$

so that $P_n(0) = 0$ for n odd and $P_0(0) = 1$, $P_2(0) = -\frac{1}{2}$, $P_4(0) = \frac{3}{8}$, and

$$P_{2n}(0) = (-1)^n \frac{1.3 \ldots (2n-1)}{2.4 \ldots \ldots 2n}$$

In the same way we can show that $P_n(1) = 1$ for all n and $P_n(-1) = (-1)^n$. We could have deduced this last result as a special case of $P_n(-\mu) = (-1)^n P_n(\mu)$ which follows directly from (5.84).

In Figure 5.7(a) we graph P_0, P_1, P_2 and P_3 for $0 < \mu < 1$; in Figure 5.7(b) we show the graphs with θ the horizontal coordinate.

The graph of $P_{2n}(\mu)$ is symmetrical about the vertical axis since it is an even function; $P_{2n+1}(\mu)$ is an odd function.

The expressions for higher order P_n are best found by means of recurrence relations which we are now going to study.

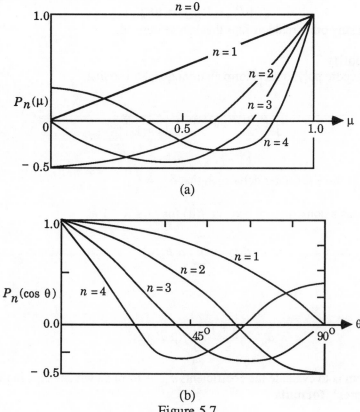

(a)

(b)

Figure 5.7

Recurrence relations

From (5.83) we may derive two main recurrence relations for $P_n(\mu)$.
Differentiating with respect to h we obtain

$$(\mu - h)\left(1 - 2h\mu + h^2\right)^{-(3/2)} = \sum_{n=0}^{\infty} nh^{n-1}P_n(\mu)$$

i.e. $\quad (\mu - h)\left(1 - 2h\mu + h^2\right)^{-(1/2)} = \left(1 - 2h\mu + h^2\right) \sum_{n=0}^{\infty} nh^{n-1}P_n(\mu)$

or $\quad (\mu - h) \sum_{n=0}^{\infty} h^n P_n(\mu) = \left(1 - 2h\mu + h^2\right) \sum_{n=0}^{\infty} nh^{n-1}P_n(\mu)$

Equating coefficients of h^n we find that

$$\mu P_n - P_{n-1} = (n+1)P_{n+1} - 2n\mu P_n + (n-1)P_{n-1}$$

i.e. $(2n+1)\mu P_n = (n+1)P_{n+1} + nP_{n-1}$ (5.85)

Similarly, by differentiating (5.83) with respect to μ we can derive

$$\mu P'_n - P'_{n-1} = nP_n \qquad (5.86)$$

There are many other relationships that can be derived.

Orthogonality

The Legendre polynomials form an orthogonal set so that

$$\int_{-1}^{1} P_n(\mu)P_m(\mu)d\mu = \frac{2}{2n+1}\,\delta_{mn} \qquad (5.87)$$

where δ_{mn} is the Kronecker delta such that $\delta_{mn} = \begin{cases} 0 & n \neq m \\ 1 & n = m \end{cases}$

We can, as a consequence, expand $f(\mu)$ for $-1 \leq \mu \leq 1$ as

$$f(\mu) = \sum_{n=0}^{\infty} a_n P_n(\mu) \qquad (5.88)$$

where

$$a_n = \frac{2n+1}{2} \int_{-1}^{1} f(\mu)P_n(\mu)d\mu \qquad (5.89)$$

The problem is to evaluate the coefficients a_n. A formula which helps in this task is **Rodrigues' formula**

$$P_n(\mu) = \frac{1}{2^n\,n!}\,\frac{d^n}{d\mu^n}\left\{(\mu^2-1)^n\right\}$$

Example of application to Potential Theory

A point charge Q is situated at a point A which is distance h from the origin of a sphere $r = a$ which is constructed from material of dielectric constant K. Find the potentials ϕ_1 outside the sphere and ϕ_2 inside the sphere; see Figure 5.8.

Now the potentials ϕ_1 and ϕ_2 must satisfy Laplace's equation and the

conditions that at $r = a$, $\phi_1 = \phi_2$ and $\dfrac{\partial\phi_2}{\partial r} = K\,\dfrac{\partial\phi_1}{\partial r}$; the derivation of these

results can be found in a textbook on electrostatics.

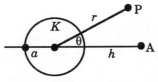

Figure 5.8

Let P be a point outside the sphere. The potential at P due to the charge at A

is $\dfrac{Q}{\left\{h^2 + r^2 - 2hr\cos\theta\right\}^{1/2}}$ which equals $\displaystyle\sum_{n=0}^{\infty} \frac{Qr^n}{h^{n+1}} P_n(\mu), \quad r < h.$

The contribution to the potential outside the sphere due to the dielectric $\to 0$ as

$r \to \infty$ and therefore has to be of the form $\displaystyle\sum_{n=0}^{\infty} \frac{A_n P_n(\mu)}{r^{n+1}}$ which leads to

$$\phi_2 = \sum_{n=0}^{\infty} \left[\frac{Qr_n}{h^{n+1}} + \frac{A_n}{r^{n+1}} \right] P_n(\mu) \tag{5.90}$$

Inside the sphere the potential must be finite at $r = 0$ and therefore it must be

$$\phi_2 = \sum_{n=0}^{\infty} B_n r^n P_n(\mu) \tag{5.91}$$

If we apply the boundary conditions at $r = a$, we find (as you can verify) that

$$A_n = \frac{Qn(1-K)a^{2n-1}}{[n(K+1)+1]h^{n+1}} \qquad B_n = \frac{Q(2n+1)}{[n(K+1)+1]h^{n+1}} \tag{5.92}$$

5.8 Solution of Partial Differential Equations

(This section should be read after consulting Chapter 7.)

We conclude this chapter on special functions with a brief glance at their role in the solution of partial differential equations. This section is included in this chapter for completeness.

Example 1

A solid cylinder $0 \le r \le a$, $0 \le \theta \le 2\pi$, $0 \le z \le L$ has its bottom face insulated, its top edge held at constant temperature T_0 and its vertical sides maintained at zero temperature. In steady-state, the temperature $T(r, z)$ is governed by the equation

$$\frac{\partial^2 T}{\partial r^2} + \frac{1}{r} \frac{\partial T}{\partial r} + \frac{\partial^2 T}{\partial z^2} = 0 \qquad (5.93)$$

(We have assumed that the temperature is independent of the angle θ.) The boundary conditions can be stated

$$\frac{\partial T}{\partial z} (r, 0) = 0, \qquad 0 \le r \le a$$

$$T(r, L) = T_0, \qquad 0 \le r \le a$$
$$T(a, z) = 0, \qquad 0 \le z \le L$$

Applying the method of separation of variables as described in Chapter 7, we look for solutions in the form $T(r, z) = R(r). Z(z)$ and obtain the equations

$$R'' + \frac{1}{r} R' - \lambda R = 0$$

$$Z'' + \lambda Z = 0$$

where λ is a constant. If we write $\lambda = -\mu^2$, then the equations become

$$r^2 R'' + r R' + \mu^2 r^2 R = 0 \qquad (5.94a)$$
$$Z'' - \mu^2 Z = 0 \qquad (5.94b)$$

Equation (5.94a) is Bessel's equation with $\nu = 0$ and has solution

$$R = A J_0(\mu r) + B Y_0(\mu r)$$

Further, (5.94b) has the solution

$$Z = C \cosh \mu z + D \sinh \mu z$$

where A, B, C and D are constants. Hence a solution of (5.93) has the form

$$T = (A J_0(\mu r) + B Y_0(\mu r))(C \cosh \mu z + D \sinh \mu z) \qquad (5.95)$$

It is implicit in the problem as stated that T stays finite as $r \to 0$. This implies that we must take $B = 0$. We can also take $A = 1$ without loss of generality. Then (5.95) becomes

$$T = J_0(\mu r) [C \cosh \mu z + D \sinh \mu z]$$

If $T = 0$ when $r = a$ it follows that

$$J_0(\mu a) = 0$$

Therefore μ is one of a set of values μ_i such that $\mu_i a$ is a zero of J_0. Applying the insulation boundary condition gives

$$J_0(\mu r) (\mu) [C \sinh \mu z + D \cosh \mu z] \big|_{z=0} = 0$$

from which we find that $D = 0$ and hence

$$T = C J_0(\mu r) \cosh \mu z.$$

Because the p.d.e. is linear, the general solution to date is

$$T = \sum_{i=1}^{\infty} C_i J_0(\mu_i r) \cosh(\mu_i z)$$

at $z = L$
$$\sum_{i=1}^{\infty} C_i J_0 (\mu_i r) \cosh (\mu_i L) = T_0$$

If we write $C_i \cosh (\mu_i L) = E_i$ then we may use the properties of Bessel functions to determine the coefficients of E_i.

A function $f(r)$ can be expanded as a series of Bessel functions viz.

$f(r) = E_1 J_0(\mu_1 r) + E_2 J_0(\mu_2 r) + ...$

where the coefficients are given by

$$E_i = \int_0^a f(r) \, r J_0(\mu_i r) dr \Big/ \int_0^a r J_0^2(\mu_i r) dr \tag{5.96}$$

Now the denominator is equal to

$$J_1^2(\mu_i a) \int_0^a r dr = \frac{1}{2} a^2 J_1^2(\mu_i a)$$

Further, the numerator is
$$\int_0^a T_0 r J_0(\mu_i r) dr = \left[\frac{r J_1 (\mu_i r)}{\mu_i} \right]_0^a T_0$$

Hence $E_i = \dfrac{2T_0}{\mu_i a J_1 (\mu_i a)}$

so that
$$T = \frac{2T_0}{a} \sum_{i=1}^{\infty} \frac{J_0 (\mu_i r) \cosh (\mu_i z)}{\mu_i J_1 (\mu_i a) \cosh (\mu_i L)} \tag{5.97}$$

Example 2

A circular drum of radius a is depressed at its centre and released. The transverse displacement $z(r, t)$ of the skin distance r from the centre of the drum at time t is governed by

$$\frac{\partial^2 z}{\partial t^2} = \frac{T}{\rho} \left\{ \frac{\partial^2 z}{\partial r^2} + \frac{1}{r} \frac{\partial z}{\partial r} \right\}$$

where T is skin tension and ρ is the mass per unit area. The boundary conditions are as follows

$$z(a, t) = 0, t > 0$$

$$\frac{\partial z}{\partial t} (r, 0) = 0, 0 \le r \le a$$

$$z(r, 0) = 0.1(a^2 - r^2), 0 \le r \le a$$

Proceeding in a manner similar to that of Example 1 we obtain the general solution as

$$z(r, t) = \sum_{i=1}^{\infty} E_i \cos cp_i t \, J_0(p_i r)$$

where $c^2 = T/\rho$ and p_i are values such that $p_i a$ are zeros of J_0. When $t = 0$

$$\sum_{i=1}^{\infty} E_i J_0 (p_i r) = 0.1(a^2 - r^2)$$

Then the coefficients are given by

$$E_i = \frac{\displaystyle\int_0^a 0.1 \, (a^2 - r^2) \, r J_0 \, (p_i \, r) \, dr}{\displaystyle\int_0^a r J_0^{\,2} (p_i \, r) \, dr}$$

The denominator is $\frac{1}{2} a^2 J_1^{\,2}(p_i a)$ and the numerator is

$$\left[(a^2 - r^2) r \, \frac{J_1 (p_i r)}{p_i} \right]_0^a + \int_0^a \frac{2 r^2 J_1 (p_i r) \, dr}{p_i} = 0 - 0 + \frac{2}{p_i} \int_0^a r^2 J_1(p_i r) dr$$

$$= \frac{2}{p_i} \left[\frac{r^2 J_2 (p_i r)}{p_i} \right]_0^a = \frac{2a^2}{p_i^{\,2}} J_2(p_i a)$$

Hence $E_i = 0.4 \, J_2(p_i a)/p_i^2 J_1^{\,2}(p_i a)$

Finally, $z(r, t) = 0.4 \displaystyle\sum_{i=1}^{\infty} \frac{J_2 (p_i \, a) \, J_0 \, (p_i \, r)}{p_i^{\,2} J_1^2(p_i a)} \cos cp_i t$, where $J_0(p_i a) = 0$.

Example 3

The electric potential V inside a hemisphere $0 \leq r \leq a, 0 \leq \theta \leq \frac{\pi}{2}, 0 \leq \phi \leq 2\pi$

is given by $\dfrac{\partial}{\partial r} \left(r^2 \dfrac{\partial V}{\partial r} \right) + \dfrac{1}{\sin \theta} \dfrac{\partial}{\partial \theta} \left(\sin \theta \dfrac{\partial V}{\partial \theta} \right) = 0$

where it has been assumed that V is independent of ϕ.
Solutions are sought in the form

$$V = R(r) \, \Theta(\theta)$$

The boundary conditions are

$$\frac{\partial V}{\partial \theta} = 0 \text{ at } \theta = \pi/2 \,, 0 \leq r \leq a$$

$$V(a, \theta) = 100 \sin^2\theta \,, 0 \leq \theta \leq \frac{\pi}{2}$$

Separation of variables leads to the equations

$$r^2 R'' + 2rR' - n(n + 1)R = 0 \qquad (5.98)$$

$$\Theta'' + \cot \theta \, \Theta' + n(n + 1)\Theta = 0 \qquad (5.99)$$

where the constant of separation has been chosen as $\lambda = n(n + 1)$.

Equation (5.98) is Euler's equation. Substituting $R = Cr^m$ leads to the solution $\qquad\qquad R = Ar^n + Br^{-n-1} \qquad (5.100)$

Also (5.99) is a form of Legendre's equation. Substituting $\omega = \cos \theta$

then $\dfrac{d\Theta}{d\theta} = \dfrac{d\omega}{d\theta} \dfrac{d\Theta}{d\omega} = -\sin \theta \dfrac{d\Theta}{d\omega} = -(1 - \omega^2)^{1/2} \dfrac{d\Theta}{d\omega}$

and $\qquad \dfrac{d^2\Theta}{d\theta^2} = -(1 - \omega^2)^{1/2} \dfrac{d}{d\omega} \left\{ -(1 - \omega^2)^{1/2} \dfrac{d\Theta}{d\omega} \right\}$

$$= (1 - \omega^2) \frac{d^2\Theta}{d\omega^2} - \omega \frac{d\Theta}{d\omega}$$

Hence (5.99) becomes

$$(1 - \omega^2) \frac{d^2\Theta}{d\omega^2} - \omega \frac{d\Theta}{d\omega} - \omega \frac{d\Theta}{d\omega} + n(n + 1) \, \Theta = 0$$

This is Legendre's equation with solution

$$\Theta = C_n P_n(\omega), \quad n = 0, 1, 2, \ldots$$

$$= C_n P_n(\cos \theta), \quad n = 0, 1, 2, \ldots$$

Hence $V(r, \theta) = (Ar^n + \dfrac{B}{r^{n+1}})P_n (\cos \theta)$ where D was taken as 0 since we require a finite solution at $\theta = 0$ and C was taken as 1 without loss of generality. Using the linearity of the p.d.e. we obtain the solution

$$V(r, \theta) = \sum_{n=0}^{\infty} (A_n r^n + \frac{B_n}{r^{n+1}}) P_n (\cos \theta)$$

Applying the boundary conditions leads to $B = 0$. Then

$$\frac{\partial V}{\partial \theta} (r, \theta) = \sum_{n=0}^{\infty} A_n r^n P_n{}'(\cos \theta) \sin \theta; \quad \text{at } \theta = \frac{\pi}{2} \sum_{n=0}^{\infty} A_n r^n P_n{}'(0) = 0$$

However, $P_n{}'(0) \begin{cases} = 0 &, n \text{ even} \\ \neq 0 &, n \text{ odd} \end{cases}$

so that $A_1 = A_3 = .. = 0$. Then $V(r, \theta) = A_0 P_0(\cos \theta) + A_2 r^2 P_2(\cos \theta)$. The final boundary condition gives

$$100 \sin^2\theta = A_0 P_0(\cos \theta) + A_2 a^2 P_2(\cos \theta)$$

or $\qquad\qquad 100 (1 - \omega^2) = A_0 P_0(\omega) + A_2 a^2 P_2(\omega)$

Since $P_0(\omega) = 1$, $P_2(\omega) = (3\omega^2 - 1)/2$ it follows that $A_0 = 200/3$, $A_2 = -200/3a^2$, and

$$V(r, \theta) = \frac{200}{3} \left\{ P_0 (\cos \theta) - \left(\frac{r}{a}\right)^2 P_2 (\cos \theta) \right\}$$

(We could use the Fourier-Legendre series approach but this would be longer.)

Problems

Section 5.2

1   Use any method of Frobenius to solve the following differential equations.

(a) $4xy'' + (1 - 4x)y' - 4y = 0$ (b) $x^2 y'' - \frac{1}{2}xy' + \frac{1}{2}(1 - x^2)y = 0$

(c) $x(1 - x)y'' - 3y' + 2y = 0$ (d) $4xy'' + (4x + 3)y' + y = 0$ (LU)

(e) $2x^2 y'' + x(2x - 1)y' + y = 0$ (LU)   (f) $xy'' + 2y' + xy = 0$ (LU)

(g) $3x(1 - 3x)y'' - 4y' + 4y = 0$ (LU)   (h) $xy'' + (1 + x)y' + 2y = 0$ (LU)

2   Find the general series solution of the differential equation $4xy'' + 6y' + y = 0$ and show that it can be expressed in the form $(A \cos \sqrt{x} + B \sin \sqrt{x})/\sqrt{x}$.   (EC)

3   Use the method of Frobenius to obtain the general solution of $2xy'' + y' + y = 0$ up to the term in x^3, and deduce the solution which satisfies the initial conditions $y = -y' = 1$

at $x = 0$. (EC)

4　What is meant by a regular singular point of a differential equation? Obtain the solution of the differential equation $xy'' - (3 + x)y' + 2y = 0$ in the form of a series. Give the general term of the infinite series. (EC)

5　Show that the differential equation $xy'' + (1 + x)y' + 2y = 0$ has a regular singular point

at $x = 0$. Hence show that $y = 1 - 2x + \frac{3}{2}x^2 - \frac{2}{3}x^3 + \dots$ is a solution. (EC)

6　(i)　Classify the singular points in the finite plane for the equation
$$x^4(x^2 + 1)(x - 1)^2\, y'' + 4x^3(x - 1)y' + (x + 1)y = 0.$$

(ii)　Show that the solution of the differential equation $xy'' + (1 - x)y' - y = 0$ can be expressed in the form

$$y = Ae^x + B\left[e^x \ln x - \sum_{n=0}^{\infty} \frac{x^n}{n!}\left(1 + \frac{1}{2} + \frac{1}{3} + \dots + \frac{1}{n}\right)\right]$$

where A and B are arbitrary constants. (EC)

7　Find a series solution $u(x)$ of the differential equation $xy'' + y' - y = 0$ which is finite for $x = 0$. A second, independent solution of this equation has the form

$$y = u(x) \ln x + \sum_{r=1}^{\infty} b_r x^r.$$

Show that $b_1 + 2a_1 = 0$ and $b_{r+1}(r + 1)^2 - b_r + 2a_{r+1}(r + 1) = 0$ $(r = 1, 2, \dots)$.
By substituting $b_r = k_r a_r$ show that $k_{r+1} - k_r = -2/(r + 1)$ and obtain an expression for the second solution. (LU)

8　Show that the indicial equation of the differential equation $xy'' + (\lambda - x)y' + 3y = 0$ is $c(c + \lambda - 1) = 0$ and show further that $(c + r)(c + r + \lambda - 1)a_r = (c + r - 4)a_{r-1}$.

Show that one solution is $1 - \frac{3}{\lambda}x + \frac{3}{\lambda(\lambda + 1)}x^2 - \frac{1}{\lambda(\lambda + 1)(\lambda + 2)}x^3$, and obtain the

first three terms of the other solution (a) when λ is not an integer, (b) when $\lambda = 2$. (LU)

9　By putting $y = u \exp(-\frac{1}{4}x^2)$, transform the equation $y'' + (n + \frac{1}{2} - \frac{1}{4}x^2)y = 0$, into

$$\frac{d^2u}{dx^2} - x\frac{du}{dx} + nu = 0,$$ and obtain two solutions of the second equation in the form of a

series of ascending powers of x.
Show that if n is a positive integer, one of these series terminates. Hence obtain one

solution of the equation $y'' + \left[\frac{9}{2} - \frac{x^2}{4}\right]y = 0$, such that $y = 1$ when $x = 0$ and $y \to 0$

as $x \to \infty$. (LU)

10   A tubular gas preheater works by drawing cool air through a cylindrical heated tube. For a particular tube the governing equation for the temperature of the air is

$$\frac{d^2 T}{dx^2} - 7500 \frac{dT}{dx} - 3500 \frac{T}{x^{1/2}} = 0$$

where T is the difference between the temperature of the wall of the tube and the air at a distance x from the inlet. Find the series solution for T *as far as the term in* $x^{5/2}$. (It helps to substitute $x = z^2$.) The second derivative term is due to the thermal conductivity of the air. Can this be neglected to leave a reasonably good approximate model equation?

Section 5.3

11   Show that $\displaystyle\int_0^{\pi/2} \sin^n x \; dx = \frac{1}{2} \sqrt{\pi} \; \Gamma\left[\frac{n+1}{2}\right] \Big/ \Gamma\left[\frac{n}{2} + 1\right]$

12   Evaluate in terms of Γ (1.6) the following: Γ (0.6), Γ (2.6), Γ (12.6), Γ (−18.6)

13   Find the length of the lemniscate $r^2 = \cos 2\theta$.

14   Find the area bounded by the curve $y^2 = 1 - x^4$.

15   Find an expression for $\displaystyle\int_0^a \sqrt{a^n - x^n} \; dx$ and verify the answer directly in the case $a = 10$,

    $n = 2$.

16   By means of a suitable substitution, express $\displaystyle\int_0^{(1/2)\pi} \left[\frac{1}{\sin^2 \theta} - \frac{1}{\sin \theta}\right]^{1/3} \cos \theta \; d\theta$ in terms of

    the Γ function.

Section 5.4

17   Defining the Bessel function $J_n(x)$ by means of the generating function

$$\exp\{\tfrac{1}{2}x(t - t^{-1})\} = \sum_{n=-\infty}^{\infty} t^n J_n(x) \text{ show that, if } n \text{ is an integer,}$$

(a) $J_n(x) = (\tfrac{1}{2}x)^n \displaystyle\sum_{r=0}^{\infty} \frac{(-x^2/4)^r}{r! \, (n+r)!}$ (b) $J_{-n}(x) = (-1)^n J_n(x)$

(c) $J_{n-1}(x) + J_{n+1}(x) = \dfrac{2n}{x} J_n(x)$ (d) $J_{n-1}(x) - J_{n+1}(x) = 2J'_n(x)$

    Deduce that $J_n(x)$ satisfies the differential equation $x^2 y'' + xy' + (x^2 - n^2)y = 0$.

                                                                         (LU)

18   Show that a solution of Bessel's equation $x^2 y'' + xy' + (x^2 - v^2)y = 0$ is

$$J_v(x) = \sum_{r=0}^{\infty} \frac{(-1)^r}{r!\,\Gamma(v+r+1)} \left\{\frac{x}{2}\right\}^{v+2r}$$ where v is a constant. Hence obtain

$$J_{1/2}(x) = \left[\frac{2}{\pi x}\right]^{1/2} \sin x.$$ Prove the recurrence relation $\dfrac{v}{x}J_v(x) - J'_v(x) = J_{v+1}(x)$

and deduce that $J'_0(x) = -J_1(x)$. (EC)

19 Show that the following relationships hold.

(i) $J_{1/2}(x) = \sin x (2/\pi x)^{1/2}$ (ii) $J_{-1/2}(x) = \cos x (2/\pi x)^{1/2}$

(iii) $J_{3/2}(x) = \dfrac{\sin x}{x} - \cos x (2/\pi x)^{1/2}$

20 In the formula $\exp\{\frac{1}{2}x(t - 1/t)\}$ put $t = e^{i\theta}$ and deduce that

$\cos(x \sin \theta) = J_0(x) + 2J_2(x) \cos 2\theta + 2J_4(x) \cos 4\theta + \dots$
$\sin(x \sin \theta) = 2[J_1(x) \sin \theta + J_3(x) \sin 3\theta + J_5(x) \sin 5\theta + \dots]$.

21 (i)   Express $J_5(x)$ in terms of $J_0(x)$ and $J_1(x)$.

   (ii) Express $J_{-3/2}(x)$ in terms of $\sin x$ and $\cos x$.

   (iii) Find the first derivative with respect to x of the following:
      $x^2 J_3(2x),\ xJ_0(x^2),\ J_2(x)$.

22 Find the indefinite integrals of the following: $J_0(x) \cos x,\ J_0(x) \sin x,\ J_1(x) \cos x,$
$J_1(x) \sin x,\ xJ_0(x) \cos x,\ xJ_0(x) \sin x,\ xJ_1(x) \cos x,\ xJ_1(x) \sin x,\ xJ_0(x),$
$x^2 J_0(x),\ x^3 J_0(x),\ J_1(x),\ xJ_1(x)$.

23 (i)   Prove that $\dfrac{d}{dx}[x^n J_n(ax)] = ax^n J_{n-1}(ax)$

   (ii) Show that $\left[\dfrac{1}{x}\dfrac{d}{dx}\right]^r [x^v J_v(x)] = x^{v-r} J_{v-r}(x)$

      $\left[\dfrac{1}{x}\dfrac{d}{dx}\right]^r [x^{-v} J_v(x)] = (-1)^r x^{-v-r} J_{v+r}(x)$ where r is a positive integer.

24 The current in a rectifying valve is $a \exp(b \cos \omega t)$ where a and b are constants. Prove that the mean value of the current over one period is $aJ_0(ib)$. (LU)
(Hint: Use a similar approach to Problem 20 with $t = ie^{i\theta}$.)

25 Show that the following results hold.

(i) $J_n(x) Y_{n+2}(x) - Y_n(x) J_{n+2}(x) = \dfrac{-4(n+1)}{\pi x^2}$

(ii) $J_\nu(x) Y'_\nu(x) - Y_\nu(x) J'_\nu(x) = \dfrac{2}{\pi x}$

(iii) $Y_{1/2}(x) = -\left[\dfrac{2}{\pi x}\right]^{1/2} \cos x$ (iv) $Y_{n+(1/2)}(x) = (-1)^{n+1} J_{-n-(1/2)}(x)$

26 Compute the values of $J_0(\tfrac{1}{2}), J_1(\tfrac{1}{2}), J_2(\tfrac{1}{2})$ directly from the series. Comment on the practical convergence of the series.

27 If the positive roots of the equation $J_0(x) = 0$ are $\alpha_1, \alpha_2, \ldots$ show that for $0 < x < 1$,

$$\sum_{r=1}^{\infty} \frac{J_0(\alpha_r x)}{\alpha_r J_1(\alpha_r)} = \frac{1}{2}.$$

Section 5.5

28 Show that $\exp[(1/2)x(t + 1/t)] = I_0(x) + I_1(x)t + I_2(x)t^2 + \ldots + I_{-1}(x)\left\{\dfrac{1}{t}\right\} +$

$I_{-2}(x)\left\{\dfrac{1}{t}\right\}^2 + \ldots$. Hence deduce the properties

(i) $I_{-n}(x) = I_n(x)$ (ii) $2I'_n(x) = I_{n-1}(x) + I_{n+1}(x)$
(iii) $2nI_n(x) = x[I_{n-1}(x) - I_{n+1}(x)]$ (iv) $xI'_n(x) = xI_{n-1}(x) - nI_n(x)$
(v) $xI'_n(x) = nI_n(x) + xI_{n+1}(x)$ (vi) $I'_0(x) = I_1(x)$
(vii) $I_{1/2}(x) = \sinh x[2/(\pi x)]^{1/2}$ (viii) $I_{-1/2}(x) = \cosh x[2/(\pi x)]^{1/2}$

(ix) $I_{-1/2}(x) - I_{1/2}(x) = \dfrac{2}{\pi} K_{1/2}(x)$ (x) $K_{1/2}(x) = \left[\dfrac{\pi}{2x}\right]^{1/2} e^{-x}$

Section 5.6
29 Transform the following seven equations into Bessel-type equation and solve.
(i) $xy'' + y' + xy = 0$ (ii) $x^2 y'' + xy' + (x^2 - 9)y = 0$
(iii) $y'' - (1/x)y' + y = 0$ (iv) $x^2 y'' + xy' + 9(x^2 - 1)y = 0$
(v) $y'' + (3/x)y' + y = 0$ (vi) $9xy'' + 9y' + y = 0$
(vii) $y'' + (k^2 e^x - n^2)y = 0$

30 By means of the substitutions $y = x^{1/2}u$, $z = kx$ reduce the equation

$$y'' + \left[k^2 + \frac{1}{4x^2}\right]y = 0 \text{ to the form } \frac{d^2u}{dz^2} + \frac{1}{z}\frac{du}{dz} + u = 0. \text{ If } x^{-1/2}y \text{ is finite when}$$

$x = 0$ and if $y' = a^{1/2}$ when $x = a$ show that the solution is $y = \dfrac{2ax^{\frac{1}{2}} J_0(kx)}{J_0(ka) - 2kaJ_1(ka)}$

(LU)

31 In a problem on the stability of a tapered strut the displacement y satisfies the equation

$$y'' + \frac{K^2}{4x}y = 0 \text{ where } K \text{ has to be determined. By writing } y = x^{1/2}u, z = Kx^{1/2},$$

the equation can be reduced to the form $z^2\dfrac{d^2u}{dz^2} + z\dfrac{du}{dz} + (z^2 - 1)u = 0$.

If $dy/dx = 0$ at $x = a$ and at $x = l$, show that the equation for K is
$J_0(Ka^{1/2})Y_0(Kl^{1/2}) = J_0(Kl^{1/2})Y_0(Ka^{1/2})$. (LU)

32 Show that $J_n(x)/x^n$ is the solution of $y'' + ([1 + 2n]/x)y' + y = 0$ and that

$\sqrt{x}\, J_n(kx)$ is a solution of $y'' + (k^2 - (4n^2 - 1)/4x^2)y = 0$. In both cases, n is a positive integer.

33 Two thin-walled metal pipes, joined by circular flanges, carry steam. The exposed surfaces of the flanges lose heat to the surrounding air which is at T_1 °C. The conductivity of the flange metal is k and the heat transfer coefficients is h. Let the inner radius of the flanges be a and the outer radius b. Further, let the steam be at T_0 °C. The heat balance

equation can be written $r\dfrac{d^2T}{dr^2} + \dfrac{T}{r} - \dfrac{2h}{k}r(T - T_1) = 0$ where T is the temperature of

the flange metal at radius r from the centre of the flanges. Write down the boundary conditions and find the rate of heat loss from a pipe through the flanges and the proportion of this which escapes from the rim.

Section 5.7

34 Find the expressions for $P_4(x)$, $P_5(x)$ and $P_6(x)$ directly and deduce expressions for $P_7(x)$, $P_8(x)$ and $P_9(x)$.

35 Prove that $P'_5(x) = 9P_4(x) + 5P_2(x) + P_0(x)$.

36 Using Rodrigues' formula obtain expressions for the polynomials $P_3(x)$, $P_4(x)$.

37 In terms of Legendre polynomials, find expressions for $1, x, x^2, x^3, x^4$. Comment on your results.

38 Show that

(i) $\displaystyle\int_{-1}^{1} x P_n(x) P_{n+1}(x)\, dx = 2(n+1)/(2n+1)(2n+3)$

(ii) $\displaystyle\int_{-1}^{1} \left\{ P_n(x) \right\}^2 dx = \frac{2}{2n+1}$

39 Charges $e, -e$ are placed at points A, B respectively with $AB = 2a$. Polar coordinates are established with pole at O, the mid-point of AB and the angle θ is measured from OA. Show that the electrostatic potential at the point $P = (r, \theta)$ is given by

$$\phi(r, \theta) = \begin{cases} \dfrac{2e}{r} \displaystyle\sum_{n=0}^{\infty} \left\{ \dfrac{a}{r} \right\}^{2n+1} P_{2n+1}(\cos\theta), & (r > a) \\[4mm] \dfrac{2e}{a} \displaystyle\sum_{n=0}^{\infty} \left\{ \dfrac{r}{a} \right\}^{2n+1} P_{2n+1}(\cos\theta), & (r < a) \end{cases}$$

40 Show that
(i) $P'_n(x) = (2n-1)P_{n-1}(x) + (2n-5)P_{n-3}(x) + (2n-9)P_{n-5}(x) + \ldots$
(ii) $xP'_n(x) = nP_n(x) + (2n-3)P_{n-2}(x) + (2n-7)P_{n-4}(x) + \ldots$

41 Given that the steady state temperature, T, in a solid sphere of radius a is given by

$$\sin\theta = \frac{\partial}{\partial r} \left\{ r^2 \frac{\partial T}{\partial r} \right\} + \frac{\partial}{\partial \theta} \left\{ \sin\theta \frac{\partial T}{\partial \theta} \right\} = 0$$ find the solution which satisfies the

boundary conditions that the top hemispherical surface is maintained at a constant temperature T_a whilst the lower hemispherical surface is kept at a constant temperature T_b.

42 A circular wire of radius a carries a charge of constant line density e and surrounds an earthed spherical conductor of radius c, the centre of the circle coinciding with the centre O of the sphere. Taking O as origin and choosing the polar axis to be perpendicular to the plane of the wire, show that the surface density of charge at the general point of the surface of the conductor is given by

$$\sigma = -\frac{e}{2c} \left\{ 1 - 5.\frac{1}{2}\left\{\frac{c}{a}\right\}^2 P_2(\cos\theta) + 9.\frac{1.3}{2.4}\left\{\frac{c}{a}\right\}^4 P_4(\cos\theta) + 13.\frac{1.3.5}{2.4.6}\left\{\frac{c}{a}\right\}^6 P_6(\cos\theta) \right\}$$

43 Two similar rings of radius a lie opposite each other in parallel planes, so spaced that the radius of one ring subtends an angle α at the other. Show that when the rings carry charges e and e' the repulsion between them is

$$\frac{ee'}{a^2} \sum_{n=0}^{\infty} \frac{(-1)^n (2n+1)!}{2^{2n}(n!)^2} (\sin \alpha)^{2n+2} P_{2n+1}(\cos \alpha). \tag{LU}$$

Section 5.8

44 A function V of x and t is to satisfy the equation

$$\frac{\partial V}{\partial t} = k\left(\frac{\partial^2 V}{\partial x^2} + \frac{1}{x}\frac{\partial V}{\partial x}\right)$$

subject to the conditions
(i) $V \to 0$ as $t \to +\infty$
(ii) V is finite as $x \to 0$
(iii) $\partial V/\partial x = -\lambda V$ when $x = a$, λ being a constant
(iv) $V = f(x)$ when $t = 0$ for $0 < x < a$
Using the method of separation of the variables show that

$$V = \Sigma A_n \exp(-k\alpha_n^2 t/a^2) J_0\left(\frac{x\alpha_n}{a}\right)$$

where
$$\frac{1}{2}A_n a^2[J_0^2(\alpha_n) + J_1^2(\alpha_n)] = \int_0^a xf(x) J_0\left(\frac{x\alpha_n}{a}\right) dx$$

and where α_n is a root of the equation

$$\lambda a J_0(\alpha_n) = \alpha_n J_1(\alpha_n)$$

whilst the summation is taken over all the values of α. (LU)

45 A finite cylinder of radius a and length l has one end and the curved surface maintained at a constant temperature (which we may assume to be zero) and the temperature of the other end is kept at $\phi(r)$ where r is distance from the axis of the cylinder. A steady state has been reached and the temperature is finite at all points in the cylinder. Take cylindrical coordinates (r, ϕ, z) with the z-axis along the axis of the cylinder and the origin at the centre of one end. The governing equation for the temperature T is

$$\frac{\partial^2 T}{\partial r^2} + \frac{1}{r}\frac{\partial T}{\partial r} + \frac{\partial^2 T}{\partial z^2} = 0.$$

Solve the problem in the case $\phi(r) = A$, constant.

46 A circular membrane of radius a with fixed circumference is made to vibrate. The displacement z satisfies the equation

$$\frac{\partial^2 z}{\partial t^2} = c^2\left\{\frac{\partial^2 z}{\partial r^2} + \frac{1}{r}\frac{\partial z}{\partial r} + \frac{1}{r^2}\frac{\partial^2 z}{\partial \theta^2}\right\}$$

To find the normal modes of vibration we try a solution of the form $z = R(r) \Theta(\theta) \cos(\omega t - \varepsilon)$. Find the solution which satisfies the condition $z = 0$ at $r = a$.

47 The differential equation satisfied by the displacement z in a vibrating membrane is

$$\frac{1}{r}\frac{\partial}{\partial r}\left(r\frac{\partial z}{\partial r}\right) + \frac{1}{r^2}\frac{\partial^2 z}{\partial \theta^2} = \frac{1}{c^2}\frac{\partial^2 z}{\partial t^2}$$

where (r, θ) are polar coordinates in the plane of the membrane, t is the time and c is a constant. Obtain a solution of this equation of the form $F(r) \cos n\theta \cos \omega t$, where $F(r)$ is a function of r only, n is an integer and ω is a constant.

If $z = 0$ when $r = a$ and when $r = b$, and is not identically zero, deduce that the admissible values of ω are given by

$$J_n\left(\frac{\omega a}{c}\right) Y_n\left(\frac{\omega b}{c}\right) = J_n\left(\frac{\omega b}{c}\right) Y_n\left(\frac{\omega a}{c}\right) \tag{LU}$$

48 The displacement y of a non-uniform taut string fixed at $x = 0$ and $x = l$ and whose density per unit length at distance x from the origin is

$$\rho(1 + kx)$$

where ρ and k are constants, is given by the equation

$$\rho(1 + kx)\frac{\partial^2 y}{\partial t^2} = T\frac{\partial^2 y}{\partial x^2}$$

T being the tension, supposed uniform, and t the time.

Assuming a periodic solution in t of the form

$$y = u(x) \sin pt$$

prove that p is given by the equation

$$J_{1/3}(pa) J_{-1/3}(pb) = J_{-1/3}(pa) J_{1/3}(pb)$$

where

$$a = \frac{2}{3}\frac{1}{k}\sqrt{\frac{\rho}{T}}, \qquad b = (1 + kl)^{3/2}\, a$$

6

FOURIER SERIES
APPROXIMATIONS

6.1 Introduction

We are often concerned with the approximation of one function by another function; the approximating functions are sines and cosines in the case of Fourier series. If our aim is to evaluate the given function at a given point to a certain number of decimal places then, provided the value of the approximating function agrees to the required number of decimal places we shall achieve our aim. We have to decide what constitutes a *good* approximation and whether the approximation we are using is, in fact, good in the sense of our criterion; furthermore, is it the *best* approximation in that sense?

Many phenomena that are studied in engineering are periodic in nature. For example the current and voltage in an alternating current circuit, the displacement, velocity and acceleration of a piston, and many parameters in a vibrating system are all periodic. To model such situations would suggest that sine and cosine functions, being the most well-known periodic functions, might well be used as the building blocks. We would hope that such modelling would give adequate representation over the whole cycle of periodicity rather than the local nature of a Taylor series representation.

Many wave phenomena are also periodic and it is well known, especially through acoustics, that a wave can in general be decomposed or analysed into several distinct waves of different frequencies. In the terminology of acoustics the wave with the lowest frequency is called the **fundamental** and the others are called **harmonics**. The plotting of amplitude versus frequency is called **spectral analysis**.

6.2 Approximation of a function by a trigonometric series

A series of the form,

$$\tfrac{1}{2} a_0 + a_1 \cos x + b_1 \sin x + a_2 \cos 2x + b_2 \sin 2x + a_3 \cos 3x + \ldots \qquad (6.1)$$

where the a_i and b_i are constants is called an **infinite trigonometric series**. It can also be written in the shortened form that follows

$$\frac{1}{2} a_0 + \sum_{k=1}^{\infty} (a_k \cos kx + b_k \sin kx) \tag{6.2}$$

The reason for writing the constant term as $\frac{1}{2} a_0$ will become apparent later.
Whether the series converges or not will depend on the value of x chosen and the coefficients a_k and b_k. Note that if it converges in any closed interval $[c, c + 2\pi]$, the periodic nature of the cosine and sine functions guarantees convergence for all values of x. Remember that a function $g(x)$ is **periodic** with period c if $g(x + c) = g(x)$ for all values of x. Such a function will also have period $2c, 3c, 4c, \ldots$. We usually choose the period as the **smallest** value of c for which $g(x + c) = g(x)$ is true for all x.

We shall now address the problem of approximating a function $f(x)$ by a finite trigonometric series

$$S_n(x) = \frac{a_0}{2} + \sum_{k=1}^{n} (a_k \cos kx + b_k \sin kx) \tag{6.3}$$

over the interval $(-\pi, \pi)$ and, more generally, the interval $(c, c + 2\pi)$. We shall in a later section extend the ideas to making an approximation over the interval $(c, c + L)$. We shall select the coefficients a_k, b_k by choosing that finite trigonometric series which is the *least squares* best approximation of $f(x)$ in the interval $(-\pi, \pi)$. The least squares criterion is interpreted as minimising

$$I_n = \int_{-\pi}^{\pi} \left\{ f(x) - \frac{1}{2} a_0 - \sum_{k=1}^{n} (a_k \cos kx + b_k \sin kx) \right\}^2 dx \tag{6.4}$$

We assume that $f(x)$ is integrable in $[-\pi, \pi]$.

To find the normal equations which will yield a solution for a_k, b_k we need to differentiate I_n partially with repect to a_k, b_k. Therefore

$$\frac{\partial I_n}{\partial a_0} = \frac{-1}{2} . 2 \int_{-\pi}^{\pi} \left\{ f(x) - \frac{a_0}{2} - \sum_{k=1}^{n} (a_k \cos kx + b_k \sin kx) \right\} dx$$

$$= - \left\{ \int_{-\pi}^{\pi} f(x) \, dx - \frac{1}{2} a_0 \int_{-\pi}^{\pi} dx - \sum_{k=1}^{n} \int_{-\pi}^{\pi} (a_k \cos kx + b_k \sin kx) \, dx \right\} \tag{6.5a}$$

For $r \neq 0$

$$\frac{\partial I_n}{\partial a_r} = -2 \int_{-\pi}^{\pi} \left\{ f(x) - \frac{a_0}{2} - \sum_{k=1}^{n} (a_k \cos kx + b_k \sin kx) \right\} . \cos rx \, dx$$

$$= -2 \left[\int_{-\pi}^{\pi} f(x) \cos rx \, dx - \frac{a_0}{2} \int_{-\pi}^{\pi} \cos rx \, dx \right.$$

$$\left. - \sum_{k=1}^{n} \left\{ \int_{-\pi}^{\pi} (a_k \cos kx \cos rx + b_k \sin kx \cos rx) \, dx \right\} \right] \qquad (6.5b)$$

Similarly,

$$\frac{\partial I_n}{\partial b_r} = -2 \left[\int_{-\pi}^{\pi} f(x) \sin rx \, dx - \frac{a_0}{2} \int_{-\pi}^{\pi} \sin rx \, dx \right.$$

$$\left. - \sum_{k=1}^{n} \left\{ \int_{-\pi}^{\pi} (a_k \cos kx \sin rx + b_k \sin kx \sin rx) \, dx \right\} \right] \qquad (6.5c)$$

Now we can effect a considerable simplification by noting the following standard results of integration where r and k are any positive integers. The results still hold if $\int_{-\pi}^{\pi}$ is replaced by $\int_{c}^{c+2\pi}$

(i) $\displaystyle\int_{-\pi}^{\pi} \cos rx \, dx = 0$ (ii) $\displaystyle\int_{-\pi}^{\pi} \sin rx \, dx = 0$

(iii) $\displaystyle\int_{-\pi}^{\pi} \sin rx \cos kx \, dx = 0, \ k \neq r$ (iv) $\displaystyle\int_{-\pi}^{\pi} \sin rx \cos rx \, dx = 0$

(v) $\displaystyle\int_{-\pi}^{\pi} \sin rx \sin kx \, dx = 0, \ k \neq r$ (vi) $\displaystyle\int_{-\pi}^{\pi} \sin^2 rx \, dx = \pi$

(vii) $\displaystyle\int_{-\pi}^{\pi} \cos rx \cos kx \, dx = 0, \ k \neq r$ (viii) $\displaystyle\int_{-\pi}^{\pi} \cos^2 rx \, dx = \pi$ $\qquad (6.6)$

Therefore,
$$\frac{\partial I_n}{\partial a_0} = -\left\{ \int_{-\pi}^{\pi} f(x)\, dx - \pi a_0 - 0 \right\} \tag{6.7a}$$

since all remaining integrals in (6.5a) are zero by (i) and (ii).

Likewise
$$\frac{\partial I_n}{\partial a_r} = -2\left\{ \int_{-\pi}^{\pi} f(x) \cos rx\, dx - 0 - \int_{-\pi}^{\pi} a_r \cos^2 rx\, dx \right\}$$

$$= -2\left\{ \int_{-\pi}^{\pi} f(x) \cos rx\, dx - a_r . \pi \right\} \quad \text{by (viii)} \tag{6.7b}$$

since all remaining integrals in (6.5b) are zero by (i), (iii), (iv), (vii). Similarly

$$\frac{\partial I_n}{\partial b_r} = -2\left\{ \int_{-\pi}^{\pi} f(x) \sin rx\, dx - b_r \pi \right\} \tag{6.7c}$$

since all remaining integrals in (6.5c) are zero by (ii), (iii), (iv), (v) and we have used (vi).

Now the least squares criterion requires $\dfrac{\partial I_n}{\partial a_0} = \dfrac{\partial I_n}{\partial a_r} = \dfrac{\partial I_n}{\partial b_r} = 0$. Therefore we obtain the following formulae for the coefficients

$$\left. \begin{array}{l} a_0 = \dfrac{1}{\pi} \displaystyle\int_{-\pi}^{\pi} f(x)\, dx \\[4mm] a_r = \dfrac{1}{\pi} \displaystyle\int_{-\pi}^{\pi} f(x) . \cos rx\, dx \qquad b_r = \dfrac{1}{\pi} \displaystyle\int_{-\pi}^{\pi} f(x) . \sin rx\, dx; \;\; r = 1, 2, 3, ... \end{array} \right\} \tag{6.8}$$

Now you see the reason for choosing the constant term in the trigonometric series to be $\frac{1}{2}a_0$: it allowed a systematic representation of the coefficients. Also note that $\frac{1}{2}a_0$ is the average value of $f(x)$ over a cycle. The coefficients defined above are the **Fourier coefficients** of $f(x)$.

Example

Find the Fourier coefficients for $f(x) = x + 2$, $-\pi \le x \le \pi$
Using the formulae (6.8) we obtain the following results.

$$a_0 = \frac{1}{\pi} \int_0^{\pi} (x + 2)\, dx = \frac{1}{\pi} \left[\frac{x^2}{2} + 2x \right]_0^{\pi} = 4\,1$$

$$a_r = \frac{1}{\pi} \int_{-\pi}^{\pi} (x + 2) \cos rx\, dx = \frac{1}{\pi} \left\{ \left[\frac{(x+2) \sin rx}{r} \right]_{-\pi}^{\pi} - \int_{-\pi}^{\pi} 1 \cdot \frac{\sin rx}{r}\, dx \right\}$$

$$= \frac{1}{\pi} \left\{ 0 - 0 + \left[\frac{\cos rx}{r^2} \right]_{-\pi}^{\pi} \right\} = 0$$

$$b_r = \frac{1}{\pi} \int_{-\pi}^{\pi} (x+2) \sin rx\, dx = \frac{1}{\pi} \left\{ \left[\frac{[(x+2)][-(\cos rx)]}{r} \right]_{-\pi}^{\pi} + \int_{-\pi}^{\pi} 1 \cdot \frac{\cos rx}{r}\, dx \right\}$$

$$= \frac{1}{\pi} \left\{ \frac{(-\pi + 2)}{r} \cos r\pi - \frac{(\pi + 2)}{r} \cos r\pi + \left[\frac{\sin rx}{r^2} \right]_{-\pi}^{\pi} \right\}$$

$$= \frac{1}{\pi} \left[\frac{-2\pi \cos r\pi}{r} \right] = \frac{-2}{r} \cos r\pi$$

Now $\cos \pi = -1$, $\cos 2\pi = 1$, $\cos 3\pi = -1$ etc. Hence, $b_1 = \frac{2}{1}$, $b_2 = \frac{-2}{2}$,

$b_3 = \frac{2}{3}$, $b_4 = \frac{-2}{4}$, $b_5 = \frac{2}{5}$ ...

Infinite trigonometric series

With Taylor's series under certain conditions, the more terms that are taken, the better the approximation. What is the effect of allowing n to tend to infinity in the trigonometric series approximation? For a fixed value of n, the Fourier coefficients are those which give the least squares best fit. Let us expand the integrand in the right-hand side of (6.4) and integrate formally the terms which are independent of $f(x)$. Then

$$I_n = \int_{-\pi}^{\pi} \{f(x)\}^2\, dx - a_0 \int_{-\pi}^{\pi} f(x)\, dx - 2 \sum_{k=1}^{n} \left\{ \int_{-\pi}^{\pi} f(x)(a_k \cos kx + b_k \sin kx)\, dx \right\}$$

$$+ a_0 \sum_{k=1}^{n} \left\{ \int_{-\pi}^{\pi} (a_k \cos kx + b_k \sin kx) \, dx \right\} + \frac{a_0^2 \pi}{2}$$

$$+ \sum_{k=1}^{n} \sum_{l=1}^{n} \left\{ \int_{-\pi}^{\pi} (a_k \cos kx + b_k \sin kx)(a_l \cos lx + b_l \sin lx) \, dx \right\}$$

By using the standard integration formulae and formulae (6.8) we obtain, after simplification,

$$I_n = \int_{-\pi}^{\pi} \left\{ f(x) \right\}^2 dx - \pi \left[\frac{a_0^2}{2} + \sum_{k=1}^{n} (a_k^2 + b_k^2) \right] \qquad (6.9)$$

Now I_n, being the integral of a squared function, can never be negative and

$$\sum_{k=1}^{n} (a_k^2 + b_k^2)$$ is an increasing function of n; hence I_n is a monotonically

decreasing function of n. Therefore the approximation of $f(x)$ by a finite trigonometric series with Fourier coefficients (finite Fourier series) improves with increasing n.

Note that the Fourier coefficients tend to zero with increasing n, provided $f(x)$ is an integrable function in $[-\pi, \pi]$.

We have not yet proved that the **Fourier series**

$$\frac{1}{2} a_0 + \sum_{k=1}^{\infty} (a_k \cos kx + b_k \sin kx) \qquad (6.10)$$

with the coefficients defined by (6.8) will converge for any value of x and even if it does converge, whether the result is $f(x)$. All we have so far demanded of $f(x)$ is that it be integrable in $[-\pi, \pi]$. Now we are in a certain difficulty in so far as the terms in (6.10) are periodic with period 2π. How then could we possibly represent a function such as $f(x)$? The answer is to strive only for a representation in the range $[-\pi, \pi]$ and forget about representation outside this range.

Consider Figure 6.1. In the first diagram we show the function $f(x) = x^2$ and in the second diagram we show the function

$$g(x) = x^2, \ -\pi < x < \pi$$
$$g(x + 2\pi) = g(x)$$

To an observer who is only allowed to view the functions in the range $(-\pi, \pi)$ there is no distinction between these functions. In this example $g(x)$ is continuous.

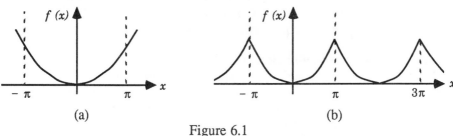

(a)   (b)

Figure 6.1

Now consider Figure 6.2 where the functions depicted are respectively $f(x) = x$ and $g(x) = x$, $-\pi < x < \pi$, $g(x + 2\pi) = g(x)$.

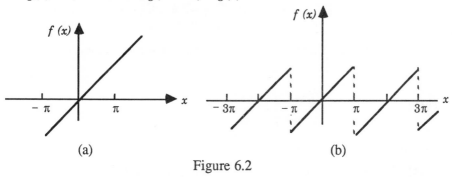

(a)   (b)

Figure 6.2

Again, the functions are indistinguishable in the range $(-\pi, \pi)$ but this time the function $g(x)$ is not continuous. However, at each point of discontinuity, it has a finite jump of -2π. These points are called points of **ordinary discontinuity**.

Dirichlet's Criterion for representation

A well-known sufficient condition for uniform convergence of the Fourier series is due to Dirichlet. It states

If $f(x)$ is a function, defined in $-\pi \leq x < \pi$ and extended outside this interval by $f(x + 2\pi) = f(x)$, having a finite number of points of ordinary discontinuity in the interval $[-\pi, \pi]$ and having a finite number of local maxima and minima there, then the series (6.10) with the coefficients given by (6.8) converges at every point x_0 of $[-\pi, \pi]$ to the value $[f(x_0 +) + f(x_0 -)]/2$.

$[f(x_0 +)$ means that $f(x)$ is evaluated at a value of x slightly greater than $x_0]$. Note that at points of continuity $f(x_0 +) = f(x_0 -)$ and therefore convergence is to simply $f(x_0)$, as we should hope. At the points of discontinuity the value converged to is the average of the values on either side. In comparison with the somewhat restrictive conditions a function has to satisfy to possess a Taylor series, a very wide class of functions do possess Fourier series which represent them. We can, of course, manufacture continuous functions which have infinitely many local maxima and minima in the interval $[-\pi, \pi]$ and hence do not necessarily permit a Fourier series representation, but most of the functions arising in engineering problems do satisfy the Dirichlet criterion.

Further notes on the Fourier coefficients

Note that if the Fourier series approximation including terms up to $\cos nx$ and $\sin nx$ has coefficients $a_0, a_1, ..., a_n, b_1, ..., b_n$, then to improve the approximation we merely need to calculate extra coefficients a_{n+1}, b_{n+1}; the existing coefficients are unchanged.

A set of continuous functions $\{f_1(x), f_2(x), ..., f_r(x), ...\}$ which do not vanish identically in the interval $[a, b]$ is **orthogonal** on the interval $[a, b]$ if

$$\int_a^b f_r(x) f_s(x)\, dx = 0, r \neq s \text{ and } \int_a^b \left\{f_r(x)\right\}^2 dx = d_r^2, \text{ where } d_r \neq 0, \text{ for all } r.$$

If $d_r = 1$ for all r the functions are said to be **orthonormal** on the interval $[a, b]$.

Therefore the functions $\dfrac{1}{\sqrt{2\pi}}, \dfrac{\cos x}{\sqrt{\pi}}, \dfrac{\sin x}{\sqrt{\pi}}, \dfrac{\cos 2x}{\sqrt{\pi}}, \dfrac{\sin 2x}{\sqrt{\pi}}, ...$ form an

orthonormal set on the interval $[-\pi, \pi]$.

The set of orthonormal functions $\phi_1(x), \phi_2(x), \phi_3(x), ...$ is said to be

complete if the conditions $\displaystyle\int_a^b \psi(x)\, \phi_1(x)\, dx = 0, \int_a^b \psi(x)\, \phi_2(x)\, dx = 0,$

$$\int_a^b \psi(x)\, \phi_3(x)\, dx = 0, ... \quad \text{imply that } \psi(x) \equiv 0 \text{ on } [a, b].$$

In general if we assume that a function $f(x)$ which is defined on the interval $[a, b]$ can be expanded in terms of a complete set of orthonormal functions

$\phi_1(x), \phi_2(x), \phi_3(x), ...$ in the form $f(x) = \displaystyle\sum_{k=1}^{\infty} c_k \phi_k(x)$ then the coefficients

can be found by the formulae $a_r = \displaystyle\int_a^b f(x)\, \phi_r(x)\, dx.$

6.3 Examples of Fourier Series

It is now time to analyse some standard wave forms to see the kind of representation that Fourier series provide.

Example 1   Square Wave

Figure 6.3 shows this wave form.

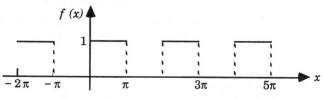

Figure 6.3

We can define this function by $f(x) = \begin{cases} 0, & -\pi < x < 0 \\ 1, & 0 \le x < \pi \end{cases}$ $\qquad f(x + 2\pi) = f(x)$

The function is periodic, with period 2π. It arises in practice if a repeated impulse is applied to a mechanical system.

Now the formulae (6.8) for coefficients involve integrations over the interval $[-\pi, \pi]$. We shall split this range into $[-\pi, 0]$ and $[0, \pi]$ to reflect the different definitions of $f(x)$. Hence

$$\pi a_0 = \int_{-\pi}^{\pi} f(x)\,dx = \int_{-\pi}^{0} 0.dx + \int_{0}^{\pi} 1.dx = \pi \text{ and so } a_0 = 1.$$

The fact that $f(0) = 1$ does not alter the value of $\int_{-\pi}^{0} f(x)\,dx$. Further,

$$\pi a_r = \int_{-\pi}^{\pi} f(x) \cos rx\,dx = \int_{-\pi}^{0} 0.dx + \int_{0}^{\pi} \cos rx\,dx = \left[\frac{\sin rx}{r}\right]_{0}^{\pi} = 0$$

Also, $b_r = \dfrac{1}{\pi r}(1 - \cos r\pi)$. [Verify this].

Now $\cos r\pi = 1$ for even r and -1 for odd r (sketch a graph of $\cos x$) and therefore $b_r = \dfrac{2}{r\pi}$ for odd values of r and $b_r = 0$ for even values of r. Then

$$f(x) = \frac{1}{2} + \sum_{r = 1,3,5...}^{\infty} \frac{2}{r\pi} \sin rx$$

i.e. $$f(x) = \frac{1}{2} + \frac{2}{\pi} \left[\frac{\sin x}{1} + \frac{\sin 3x}{3} + \frac{\sin 5x}{5} + \dots \right] \qquad (6.11)$$

In Figure 6.4 we sketch the first four partial sums (as far as the term in $\sin 5x$) on the graph of $f(x)$. Then, for completeness, we show the partial sum up to the term in $\sin 11x$. The graphs can be extended in either direction by periodicity.

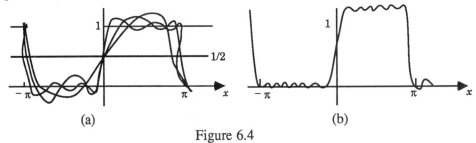

(a)   (b)

Figure 6.4

As we take more terms in the series the approximation improves in the following respects. First, the maximum amplitude of the oscillations decreases, meaning that no point is too badly approximated. Second, the frequency of the oscillations increases, meaning that more points (though still only a handful) are exactly represented. Furthermore, the representation stays reasonable over a greater part of the interval $[0, \pi]$ and of the interval $[-\pi, 0]$. However, there is still a poor representation near the points of discontinuity. Note that all partial

sum curves pass through the points $(0, \frac{1}{2})$, $(\pi, \frac{1}{2})$, $(-\pi, \frac{1}{2})$.

It is of little use obtaining a result such as (6.11) if we do not get a feel for the approximation by substituting values of x.

If we substitute $x = \pi/2$, the right-hand side of (6.11) becomes equal to

$\frac{1}{2} + \frac{2}{\pi} \left[1 - \frac{1}{3} + \frac{1}{5} \dots \right]$ and since Dirichlet's criterion is met, we are assured that

the series in square brackets will converge, in fact to $\pi/4$ (giving a total value of 1). What we are not assured is the rate of convergence of the series; in this instance the spin-off of a series for $\pi/4$ is devalued by the slowness of convergence.

Since we are dealing with a series of sine functions it does not follow that the smaller the value of x, the more rapid the convergence, as was the case with the Maclaurin's series. Since the Fourier coefficients are proportional to $1/r$ it means in physical terms that the square wave contains many high-frequency components and if an electronic device will not pass such components then the input will appear in a distorted form as output.

Notice that $\frac{1}{2}a_0 = 0.5$ is the average value of $f(x)$ over a cycle of length 2π.

Notice further that at $x = n\pi$, for integral n, the right-hand side of (6.11) has all its terms zero except the first and hence predicts the value 0.5 for $f(x)$. This value is midway between the values on either side of this discontinuity as predicted by Dirichlet's condition.

Example 2   Rectified cosine wave

Rectifiers are devices which convert alternating currents into direct currents. When the current flows in one direction, the rectifier will conduct, but when the current flows in the opposite direction, the rectifier will not conduct. If we combine two rectifiers we can produce full-wave rectification, in which the 'negative' half cycles are reversed. A sketch of the resulting waveform is given in Figure 6.5.

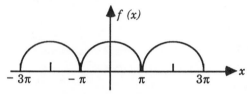

Figure 6.5

The function which represents this waveform is $f(x) = \cos \frac{1}{2}x$, $-\pi < x < \pi$

with $f(x + 2\pi) = f(x)$. The function we have chosen is of period 2π. (You can see that the fundamental frequency of this rectified wave is twice that of the

original cosine wave which had period 4π). The term $\frac{1}{2}a_0$ is the direct current

term since it is not affected by oscillations. It is given by

$$\pi a_0 = \int_{-\pi}^{\pi} \cos \frac{x}{2}\, dx = \left[2 \sin \frac{x}{2}\right]_{-\pi}^{\pi} = 4. \quad \text{Therefore } a_0 = 4/\pi.$$

Also $\pi a_r = \int_{-\pi}^{\pi} \cos \frac{x}{2} \cos rx\, dx = \frac{1}{2} \int_{-\pi}^{\pi} \left\{\cos \left(r + \frac{1}{2}\right)x + \cos \left(r - \frac{1}{2}\right)x\right\} dx$

$$= \frac{1}{2} \left[\frac{\sin \left(r + \frac{1}{2}\right)x}{\left(r + \frac{1}{2}\right)} + \frac{\sin \left(r - \frac{1}{2}\right)x}{\left(r - \frac{1}{2}\right)}\right]_{-\pi}^{\pi}$$

Hence $a_r = \frac{1}{\pi} \left[\dfrac{\sin \left(r + \frac{1}{2}\right)\pi}{\left(r + \frac{1}{2}\right)} + \dfrac{\sin \left(r - \frac{1}{2}\right)\pi}{\left(r - \frac{1}{2}\right)}\right]$

Now if r is odd, $\sin(r + \tfrac{1}{2})\pi$ and $\sin(r - \tfrac{1}{2})\pi$ are respectively -1 and 1 and

$$a_r = \frac{1}{\pi}\left\{\frac{-1}{(r+\frac{1}{2})} + \frac{1}{(r-\frac{1}{2})}\right\} = \frac{1}{\pi(r^2 - \frac{1}{4})} = \frac{4}{\pi(4r^2 - 1)}$$

If r is even, $\sin(r + \tfrac{1}{2})\pi$ and $\sin(r - \tfrac{1}{2})\pi$ are respectively 1 and -1.

$a_r = \dfrac{-4}{\pi(4r^2 - 1)}$. Similarly, $\pi b_r = 0$ for all r. [Verify this]. Hence

$$f(x) = \frac{4}{\pi}\left\{\frac{1}{2} + \frac{\cos x}{(4.1^2) - 1} - \frac{\cos 2x}{(4.2^2) - 1} + \frac{\cos 3x}{(4.3^2) - 1} - \ldots\right\}$$

Notice that in this example the Fourier coefficients decay like $(1/r^2)$ as opposed to the previous example where the coefficients decayed like $(1/r)$; in this latter case we expect a useful rapidly convergent approximation to $f(x)$.

Example 3   Saw-tooth wave

Figure 6.6 shows the graph of a waveform frequently occuring in electronics. It is defined by $f(t) = t/\pi$, $-\pi < t < \pi$ with $f(t + 2\pi) = f(t)$. Since we are dealing with a time variable problem we shall use t for the independent variable; do not be put off by this change.

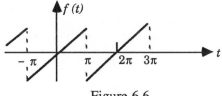

Figure 6.6

Let $f(t) = \dfrac{1}{2}a_0 + \displaystyle\sum_{k=1}^{\infty}(a_k \cos kt + b_k \sin kt)$. Then

$$\pi a_0 = \int_{-\pi}^{\pi} f(t)\,dt = \frac{1}{\pi}\int_{-\pi}^{\pi} t\,dt = 0$$

$$\pi a_r = \frac{1}{\pi}\int_{-\pi}^{\pi} t \cos rt\,dt = \frac{1}{\pi}\left[\frac{t \sin rt}{r}\right]_{-\pi}^{\pi} - \frac{1}{\pi}\int_{-\pi}^{\pi} 1.\frac{\sin rt}{r}\,dt$$

$$= 0 - 0 - \frac{1}{\pi} \left[\frac{-\cos rt}{r^2} \right]_{-\pi}^{\pi} = 0$$

Hence $a_r = 0$. Likewise $b_r = \dfrac{-2 \cos r\pi}{\pi r}$. If r is even, $\cos r\pi = 1$ and

$b_r = -2/\pi r$ whereas, if r is odd, $\cos r\pi = -1$ and $b_r = 2/\pi r$. Therefore

$$f(t) = \frac{2}{\pi} \left[\frac{\sin t}{1} - \frac{\sin 2t}{2} + \frac{\sin 3t}{3} - \cdots \right].$$

6.4 Odd and Even Functions; Half-Range Series

The rectified wave of Example 2 of the last section gave rise to a series without sine terms, whereas the saw-tooth wave gave rise to a series containing sine terms only. The rectified wave form is an even function of x and the saw-tooth wave form is an odd function of t. It can easily be shown that for an even function $f(x)$ the Fourier coefficients $b_r = 0$ for all r and as a bonus

$$a_0 = \frac{2}{\pi} \int_0^{\pi} f(x) \, dx, \quad a_r = \frac{2}{\pi} \int_0^{\pi} f(x) \cos rx \, dx$$

If $f(x)$ is an odd function, the Fourier coefficients $a_r = 0$ for all r and $a_0 = 0$.

Furthermore $b_r = \dfrac{2}{\pi} \displaystyle\int_0^{\pi} f(x) \sin rx \, dx$

Consequently we should always check at the outset whether a function is even or odd: we could save ourselves much calculation. To reinforce these ideas, let us examine two variants on the square wave of Example 1 of the last section.

Example 1 Anti-symmetric square wave

In Figure 6.7(a) we sketch the waveform $f(t) = \begin{cases} -1, & -\pi < t < 0 \\ 1, & 0 < t < \pi \end{cases}$ with

$f(t + 2\pi) = f(t)$. The function is odd and therefore $a_0 = a_r = 0$.

Now $b_r = \dfrac{2}{\pi} \displaystyle\int_0^{\pi} f(t) \sin rt \, dt$, therefore $\dfrac{\pi b_r}{2} = \displaystyle\int_0^{\pi} 1. \sin rt \, dt = \left[\dfrac{-\cos rt}{r} \right]_0^{\pi}$

i.e. $b_r = \dfrac{2}{r\pi}(1 - \cos r\pi)$. If r is even, $b_r = 0$ whereas if r is odd, $b_r = 4/r\pi$.

Therefore $f(t) = \dfrac{4}{\pi}\left[\dfrac{\sin t}{1} + \dfrac{\sin 3t}{3} + \dfrac{\sin 5t}{5} + ... \right]$

Compare this series with the one for Example 1 of the last section and comment on the similarities and differences.

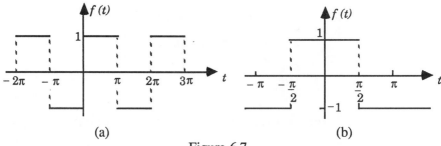

(a)   (b)

Figure 6.7

Example 2   Symmetric square wave

In Figure 6.7(b) the wave form is sketched whose definition is

$$f(t) = \begin{cases} -1, & -\pi < t < -\dfrac{\pi}{2} \\ 1, & -\dfrac{\pi}{2} < t < \dfrac{\pi}{2} \\ -1, & \dfrac{\pi}{2} < t < \pi \end{cases} \qquad f(t + 2\pi) = f(t).$$

The function is even and therefore $b_r = 0$, whilst

$$a_0 = \dfrac{2}{\pi}\int_0^\pi f(t)\,dt = \dfrac{2}{\pi}\int_0^{\pi/2} 1\,dt + \dfrac{2}{\pi}\int_{\pi/2}^\pi (-1)\,dt = 0$$

This we could have expected, since the average of $f(t)$ over a cycle is 0.

$$a_r = \dfrac{2}{\pi}\int_0^\pi f(t)\cos rt\,dt = \dfrac{2}{\pi}\int_0^{\pi/2}\cos rt\,dt + \dfrac{2}{\pi}\int_{\pi/2}^\pi (-\cos rt)\,dt$$

$$= \dfrac{2}{\pi}\left[\dfrac{\sin rt}{r}\right]_0^{\frac{\pi}{2}} + \dfrac{2}{\pi}\left[\dfrac{-\sin rt}{r}\right]_{\pi/2}^{\pi} = \dfrac{4\sin(r\pi/2)}{r\pi}$$

If r is even, $a_r = 0$. If $r = 1, 5, 9, ...$, $a_r = 4/r\pi$, whereas if $r = 3, 7, 11, ...$, $a_r = -4/r\pi$. Hence

$$f(t) = \frac{4}{\pi}\left[\frac{\cos t}{1} - \frac{\cos 3t}{3} + \frac{\cos 5t}{5} - \dots\right]$$

Note that we should be able to deduce this result from the result of the previous example by substituting $(t + \pi/2)$ for t. Try the substitution.

In this, as well as the last example, the even harmonics were zero. In the next section we investigate the underlying cause.

Half-range series

We can often make use of the oddness or evenness of a function giving rise to a sine series or a cosine series respectively. In the soution of some partial differential equations, the boundary conditions may restrict us to a series which contains only sine terms (see Section 7.4). We shall therefore need to investigate how to manufacture an odd function or an even function, given a function which may be neither.

Consider Figure 6.8. In the first diagram we show a function given in the range $[0, \pi]$. In the second diagram we have extended the domain of the function to $[-\pi, \pi]$ to create an even function which we then extend by periodicity. In the third diagram we have created an odd periodic function. Therefore the function defined on $[0, \pi]$ can be represented by either a cosine series or a sine series.

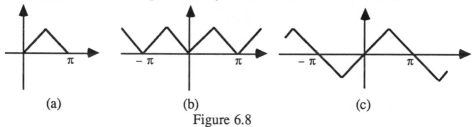

(a)   (b)   (c)

Figure 6.8

If the function of Figure 6.8(a) be $f(t) = \begin{cases} t & 0 < t < \frac{\pi}{2} \\ \pi - t, & \frac{\pi}{2} < t < \pi \end{cases}$ then for the function of Figure 6.8(b)

$$b_r = 0, \quad a_0 = \frac{2}{\pi}\int_0^{\pi/2} t\, dt + \frac{2}{\pi}\int_{\pi/2}^{\pi} (\pi - t)\, dt$$

and

$$a_r = \frac{2}{\pi}\int_0^{\pi/2} t \cos rt\, dt + \frac{2}{\pi}\int_{\pi/2}^{\pi} (\pi - t) \cos rt\, dt$$

This gives rise to the following series

$$f(t) = \frac{\pi}{4} - \frac{2}{\pi} \left[\frac{\cos 2t}{1^2} + \frac{\cos 6t}{3^2} + \frac{\cos 10t}{5^2} + ... \right] \qquad (6.12a)$$

For the function of Figure 6.8(c) $a_0 = 0$, $a_r = 0$, and the series obtained is,

$$f(t) = \frac{4}{\pi} \left[\frac{\sin\ t}{1^2} - \frac{\sin 3t}{3^2} + \frac{\sin\ 5t}{5^2} - ... \right] \qquad (6.12b)$$

For purposes of comparison we truncated the two series to the terms shown and estimated $f(t)$ for $t = 0, \pi/8, \pi/4, 3\pi/8, \pi/2$. Refer to Table 6.1.

Table 6.1

t	$f(t)$	Approximation (6.12a) Estimate (4 d.p.)	Error	Approximation (6.12b) Estimate (4 d.p.)	Error
0	0	0.0526	0.0526	0	0
$\pi/8$	0.3927	0.4033	0.0106	0.4036	0.0109
$\pi/4$	0.7854	0.7854	0	0.7643	0.0211
$3\pi/8$	1.1781	1.1675	–0.0106	1.2110	0.0329
$\pi/2$	1.5708	1.5182	–0.0526	1.4656	–0.1052

6.5 Further Features of Fourier Series

Consider the usual series representation

$$f(x) = \frac{1}{2} a_0 + \sum_{k=1}^{\infty} (a_k \cos kx + b_k \sin kx) \qquad (6.2)$$

Suppose we replace x by $(x + \pi)$, then

$$f(x + \pi) = \frac{1}{2} a_0 + \sum_{k=1}^{\infty} (a_k \cos kx \cos k\pi + b_k \sin kx \cos k\pi)$$

Now, if k is even $\cos k\pi = 1$ whereas if k is odd, $\cos k\pi = -1$.

The two terms $a_k \cos kx \cos k\pi$ and $b_k \sin kx \cos k\pi$ are equal to their counterparts in (6.2) when k is even but are the negative of their counterparts if k is odd. If we require $f(x + \pi) = f(x)$ then the odd harmonics vanish so that

$$f(x + \pi) = f(x) = \frac{1}{2} a_0 + \sum_{k=2,4,6...}^{\infty} (a_k \cos kx + b_k \sin kx) \qquad (6.13)$$

In this case $f(x)$ is of period π ; it is also of period 2π.

Example

Consider the function shown in Figure 6.9(a). It is specified either by

$$f(x) = \begin{cases} x & , & 0 < x < \pi \\ x - \pi & , & \pi < x < 2\pi \end{cases} \qquad f(x + 2\pi) = f(x)$$

or by $f(x) = x$, $0 < x < \pi$; $f(x + \pi) = f(x)$.

We shall adapt the formulae (6.8) to extend over the interval $[0, 2\pi]$ instead of $[-\pi, \pi]$. We find

$$\pi a_0 = \int\limits_0^{2\pi} f(x)\,dx = \int\limits_\pi^{2\pi} (x - \pi)\,dx + \int\limits_0^\pi x\,dx = \frac{\pi^2}{2} + \frac{\pi^2}{2} = \pi^2$$

Hence $a_0 = \pi$. Similarly, $a_r = 0$ and $b_r = -\frac{1}{r}(1 + \cos r\pi)$

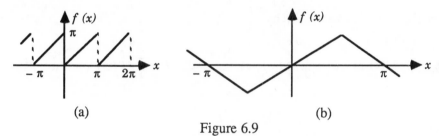

(a)   (b)

Figure 6.9

Therefore, $b_r = 0$ for r odd and $b_r = -2/r$ for r even. Hence

$$f(x) = \frac{\pi}{2} - 2\left[\frac{\sin 2x}{2} + \frac{\sin 4x}{4} + \frac{\sin 6x}{6} + \dots\right]$$

and the odd harmonics have vanished as promised.

The function graphed in Figure 6.9(b) can be defined by

$$f(x) = x, \frac{-\pi}{2} < x < \frac{\pi}{2} \; ; f(x + \pi) = -f(x)$$

By the standard method we can derive the Fourier series as

$$f(x) = \frac{4}{\pi}\left[\frac{\sin x}{1} - \frac{\sin 3x}{3} + \frac{\sin 5x}{5} - \dots\right]$$

and the even harmonics have vanished.

This function satisfies the condition $f(x - \pi) = -f(x)$.

We could have calculated the non-zero coefficients by $2\int\limits_0^\pi$ instead of $\int\limits_{-\pi}^\pi$ as indeed

we could have done for the previous example.

Just as any function can be written as the sum of an odd and an even function, any function can be written as the sum of the function with odd harmonics only and a function with even harmonics only. Now we can begin to see a way of producing a Fourier series with fewer terms and hence a greater rate of convergence.

Consider the function $f(x) = x$, $0 < x < \pi/2$. We can first extend the function to $f(x) = x$, $-\pi/2 < x < 0$, i.e. making it odd, and then extend it by periodicity to the function shown in Figure 6.9(b). We shall then lose all but the sine terms; of these we shall lose either the odd or the even ones.

General period of a function

Suppose $f(x)$ has a period T and suppose its basic domain of definition is $[a, a + T]$. Then it will be capable of a Fourier series representation under the Dirichlet conditions as

$$f(x) = \frac{1}{2} a_0 + \sum_{k=1}^{\infty} \left(a_k \cos \frac{2\pi kx}{T} + b_k \sin \frac{2\pi kx}{T} \right) \tag{6.14}$$

where the coefficients are given by

$$a_0 = \frac{2}{T} \int_a^{a+T} f(x)\, dx \quad a_r = \frac{2}{T} \int_a^{a+T} f(x) \cos \frac{2\pi rx}{T}\, dx \quad b_r = \frac{2}{T} \int_a^{a+T} f(x) \sin \frac{2\pi rx}{T}\, dx$$

$$\tag{6.15}$$

Example 1

The function $f(x) = x$ is to be represented by a Fourier sine series in the interval $[0,4]$.

We first extend the function by defining $f(x) = 8 - x$ in $[4,8]$ and then create an odd function in $[-8,8]$. We then have $f(x + 16) = f(x)$; [c.f. Figure 6.9(b)] and we get a Fourier series representation by taking our basic interval as $[-8,8]$. Since the function is odd, we need only consider sine terms. Then the representation

$$f(x) = \frac{1}{2} a_0 + \sum_{k=1}^{\infty} \left(a_k \cos \frac{2\pi kx}{16} + b_k \sin \frac{2\pi kx}{16} \right)$$

has $a_0 = a_r = 0$ and $b_r = \frac{2}{16} \int_{-8}^{8} f(x) \sin \frac{2\pi rx}{16}\, dx$

$$= \frac{4}{16} \left\{ \int_0^4 x \sin \frac{\pi rx}{8}\, dx + \int_4^8 (8 - x) \sin \frac{\pi rx}{8}\, dx \right\}$$

Hence $b_r = \frac{32}{\pi^2 r^2} \sin \frac{\pi r}{2}$.

Clearly, $b_r = 0$ if r is even. It follows that

$$f(x) = \frac{32}{\pi^2} \left\{ \frac{\sin \frac{\pi x}{8}}{1^2} - \frac{\sin \frac{3\pi x}{8}}{3^2} + \frac{\sin \frac{5\pi x}{8}}{5^2} - \ldots \right\}$$

Suppose we had tried to represent $f(x)$ by a Fourier series based on $[-4, 4]$ simply by making the extension $f(x) = x$ on $[-4, 0]$ to make it odd. Then we should have found

$$f(x) = \sum_{k=1}^{\infty} b_k \sin \frac{2\pi kx}{8} \quad \text{where } b_r = 2.\frac{2}{8} \int_0^4 x \sin \frac{2\pi rx}{8} \, dx = -\frac{8}{\pi r} \cos r\pi$$

Therefore $f(x) = \frac{8}{\pi} \left[\frac{\sin \pi x/4}{1} - \frac{\sin 2\pi x/4}{2} + \frac{\sin 3\pi x/4}{3} - \ldots \right]$.

Compare for yourselves the series after four terms each for $x = 0$ (0.5) 4

Example 2

A uniform horizontal beam of length l, freely supported at its ends carries a point load W at a distance a from one end. Show that the deflection is the same as would be produced by a non-uniformly distributed load $w(x)$ per unit length where

$$w(x) = \frac{2W}{l} \left[\sin \frac{\pi a}{l} . \sin \frac{\pi x}{l} + \sin \frac{2\pi a}{l} . \sin \frac{2\pi x}{l} + \ldots \right]$$

Find the deflected profile.

The approximation we make is that the point load can be replaced by a uniformly distributed load W/ε over a small distance ε as shown in Figure 6.10.

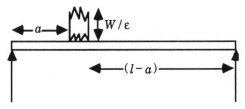

Figure 6.10

We seek a sine series representation and therefore the function we choose is

$$f(x) = \begin{cases} 0, & 0 < x < a \\ W/\varepsilon, & a < x < a + \varepsilon \\ 0, & a + \varepsilon < x < l \end{cases} \quad f(x) \text{ odd and } f(x + 2l) = f(x)$$

Then $f(x) = \sum_{r=1}^{\infty} b_r \sin \frac{r\pi x}{l}, \; b_r = \frac{2}{l} \int_0^l f(x) \sin \frac{r\pi x}{l} \, dx$

i.e. $b_r = \frac{2W}{l\varepsilon} \int_a^{a+\varepsilon} \sin \frac{r\pi x}{l} \, dx = \frac{2W}{l\varepsilon} \left[-\frac{l}{r\pi} \cos \frac{r\pi x}{l} \right]_a^{a+\varepsilon}$

$\approx \frac{4W}{r\pi\varepsilon} \sin \frac{r\pi a}{l} \cdot \frac{r\pi\varepsilon}{2l}$ if ε is small i.e. $b_r \approx \frac{2W}{l} \sin \frac{r\pi a}{l}$

Hence the result follows.

To obtain the deflected profile we use the formula $EI \, y^{iv}(x) = w(x)$. Integrating four times and using the boundary conditions applicable to this problem we obtain the deflected profile

$$y = \frac{2Wl^3}{\pi^4 EI} \left[\sin \frac{\pi a}{l} \cdot \sin \frac{\pi x}{l} + \frac{1}{2^4} \sin \frac{2\pi a}{l} \cdot \sin \frac{2\pi x}{l} + \ldots \right]$$

Gibbs Phenomenon

In Figure 6.4 we saw how the Fourier series representation of a square wave exhibited the phenomenon of breaking away from the function at points of ordinary discontinuity. Since in practice we only ever take a finite number of terms in the series approximation, it is important to know how many terms to take in order to overcome this poorness of approximation. It is interesting to note that the first observation of this phenomenon was reported by Michelson, who had constructed a machine to analyse numerically a waveform into its harmonics. Gibbs then investigated the phenomenon mathematically.

An infinite Fourier series will converge slowly near a discontinuity. The high-frequency oscillations of the truncated series around the true function value will always be present, but their amplitudes are usually small enough not to cause concern. Near a discontinuity, they become noticeable and unfortunately we cannot eliminate them by taking more terms of the series. We can merely reduce the range of x over which they are noticeable; we cannot reduce the amplitude at the discontinuity.

To investigate the phenomenon, we examine the error involved in approximating the square wave

$$f(x) = \begin{cases} -1, & -\pi < x < 0 \\ 1, & 0 < x < \pi \end{cases} \qquad f(x + 2\pi) = f(x)$$

Let the partial sum up to the nth harmonic be

$$S_n(x) = \frac{1}{2} a_0 + \sum_{k=1}^{n} (a_k \cos kx + b_k \sin kx)$$

It can be shown after much algebra that

$$2\pi S_n(x) = \int_{-x}^{x} \frac{\sin (n + \frac{1}{2})\theta}{\sin \theta/2} \, d\theta - \int_{\pi - x}^{\pi + x} \frac{\sin (n + \frac{1}{2})\theta}{\sin \theta/2} \, d\theta$$

We have expressed $S_n(x)$ in terms of two integrals, one centred on 0 and the other centred on π: the two points of ordinary discontinuity.

The second integral is evaluated over a region where the size of the integrand is approximately 1. The first integrand has a value at $\theta = 0$ of $(2n + 1)$, as we can discover from L'Hopital's rule. Therefore, for values of n which are not small (and this will be the case for the situation we are studying) we can neglect the second integral and use as our approximation

$$S_n(x) \approx \frac{1}{2\pi} \int_{-x}^{x} \frac{\sin (n + \frac{1}{2})\theta}{\sin \theta/2} \, d\theta = \frac{1}{\pi} \int_{0}^{x} \frac{\sin (n + \frac{1}{2})\theta}{\sin \theta/2} \, d\theta$$

since the integrand is even. Further, since we are interested in values of θ near the origin we approximate $\sin \theta/2$ by $\theta/2$ and put $m = (n + \frac{1}{2})$ to obtain

$$S_n(x) \approx \frac{2}{\pi} \int_{0}^{x} \frac{\sin m\theta}{\theta} \, d\theta$$

We wrote a computer program and obtained a table of values for $S_n(x)$ from which we produced the graph shown in Figure 6.11(a).

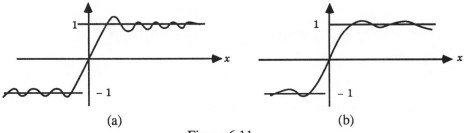

(a)   (b)

Figure 6.11

We see that as x increases from 0, the partial sum overshoots the value 1 after only a small interval of x. It reaches a maximum value and then oscillates about the value 1 with decaying amplitude; similar remarks apply for negative values of x. It can be shown that this first maximum has a value of about 1.18 irrespective of the value of n. Its position does change, however, moving nearer $x = 0$ as n increases.

Several methods of overcoming this problem have been proposed. The one we mention here is the **Lanczos smoothing factor**. The basic idea is to replace the approximate value $S_n(x) \approx \dfrac{1}{2} a_0 + \displaystyle\sum_{k=1}^{n} (a_k \cos kx + b_k \sin kx)$ by the

average of $S_n(x)$ between $(x - \pi/n)$ and $(x + \pi/n)$. It can be shown that this produces the revised approximation

$$\frac{1}{2} a_0 + \sum_{k=1}^{n} \frac{\sin (\pi k/n)}{\pi k/n} (a_k \cos kx + b_k \sin kx) \tag{6.16}$$

We have multiplied the oscillatory terms by the sigma-factors $\sigma_k = \dfrac{\sin (\pi k/n)}{\pi k/n}$

For consistency we can take $\sigma_0 = 1$. The effect of introducing the sigma factors is shown in Figure 6.11(b) where we see the approximations up to $\cos 20x$ and $\sin 20x$. The price we pay for reducing the maximum amplitude is that the rate of rise to this first maximum is slowed by half.

Parseval's result

Assume that $f(x)$ has a Fourier series representation (6.2). Parseval's result is that for a suitably integrable function $f(x)$,

$$\frac{1}{2\pi} \int_{-\pi}^{\pi} \{f(x)\}^2 \, \mathrm{d}x = \frac{1}{4} a_0^2 + \frac{1}{2} \sum_{k=1}^{\infty} (a_k^2 + b_k^2) \tag{6.17}$$

The left-hand side is the **power** in the wave form given by $f(x)$.

Example 1

Find the Fourier series expansion for the function $f(x)$ defined by

$f(x) = \begin{cases} -x, & -\pi < x < 0 \\ x, & 0 < x < \pi \end{cases}$ $f(x + 2\pi) = f(x)$ and deduce the power in the

wave form $g(x) = x^2$, $-\pi < x < \pi$, $g(x + 2\pi) = g(x)$.

The first part we shall leave to you (including a sketch of $f(x)$) The Fourier

coefficients are $a_0 = \pi$, $b_r = 0$, $a_r = \dfrac{2}{\pi r^2} [\cos r\pi - 1]$. Now $g(x) = [f(x)]^2$

and from Parseval's result

$$\frac{1}{2\pi} \int_{-\pi}^{\pi} x^2 \, \mathrm{d}x = \frac{1}{4} \cdot \pi^2 + \frac{8}{\pi^2} \sum_{k=1}^{\infty} \frac{1}{(2k-1)^4} \quad \text{,which is the power.}$$

Since the left-hand side can be integrated directly to produce the result $\frac{\pi^2}{3}$ we may deduce the result that

$$\frac{\pi^4}{96} = \frac{1}{1^4} + \frac{1}{3^4} + \frac{1}{5^4} + \dots$$

Example 2

The root-mean-square (RMS) value of $f(x)$ of period T is defined by

$$(\text{RMS value})^2 = \frac{1}{T} \int_0^T [f(x)]^2 \, dx$$

Find the RMS value of the wave $e = \sum_{k=1}^{\infty} E_k \sin(kwx + \alpha_k)$

Using the result that $\frac{1}{2} a_0$ is the mean value of $f(x)$ over one-cycle, denoted by $\bar{f}$, say, we have, from the generalised form of Parseval's result the relationship

$$(\text{RMS value})^2 = \bar{f} + \frac{1}{2} \sum_{k=1}^{\infty} (a_k^2 + b_k^2)$$

For the given wave,

$$e = \sum_{k=1}^{\infty} \{E_k \sin \alpha_k \cos kwx + E_k \cos \alpha_k \sin kwx\}$$

The period $T = 2\pi/w$ and therefore, since $\bar{f} = 0$

$$(\text{RMS value})^2 = \frac{1}{2} \sum_{k=1}^{\infty} [(E_k \sin \alpha_k)^2 + (E_k \cos \alpha_k)^2] = \frac{1}{2} \sum_{k=1}^{\infty} E_k^2$$

Integration of Fourier series

Since integration is a smoothing process, we expect that integration of a Fourier series is permissible. We quote the result that if $f(x)$ satisfies the Dirichlet condition then we can integrate its Fourier series term by term over the range $[a, x] \subset [-\pi, \pi]$ to obtain a series which converges to the integral of the original function.

That is to say

$$\int_a^x f(x)\, dx = \frac{a_0}{2}\, [x-a] + \sum_{k=1}^{\infty} \frac{1}{k}\, [b_k(\cos ka - \cos kx) + a_k\, (\sin kx - \sin ka)]$$

Note that the coefficients have an additional factor of k in the denominator and the new series should converge more rapidly than the original series. However, unless $a_0 = 0$ the new series will not be a Fourier series although the series for

$$\int_a^x f(x)\, dx - a_0\, [x-a]/2 \quad \textit{will be a Fourier Series.}$$

Example

Find the Fourier series of $f(x) = \begin{cases} -1, & -\pi < x < 0 \\ 1, & 0 < x < \pi \end{cases} \quad f(x + 2\pi) = f(x)$

and deduce the Fourier series for the function
$$g(x) = |x|,\ -\pi < x < \pi$$
$$g(x + 2\pi) = g(x)$$

The Fourier series for $f(x)$ can be found to be $\dfrac{4}{\pi} \displaystyle\sum_{k=1,3,5,\ldots}^{\infty} \dfrac{\sin kx}{k}$

Here $a_0 = 0$ since $f(x)$ is odd; also $a_r = 0$, $b_r = \dfrac{4}{r\pi}$ if r is odd and, $b_r = 0$ if r is even. Now

$$g(x) = \int_0^x f(x)\, dx = \sum_{k=1,3,5,\ldots}^{\infty} \frac{1}{k}\cdot\frac{4}{\pi}\cdot\frac{1}{k}\, (1 - \cos kx)$$

$$= \frac{4}{\pi} \sum_{k=1,3,5,\ldots}^{\infty} \frac{1}{k^2} - \frac{4}{\pi} \sum_{k=1,3,5,\ldots}^{\infty} \frac{\cos kx}{k^2}$$

Now, if we put $x = \pi$, the right-hand side becomes $2\cdot\dfrac{4}{\pi} \displaystyle\sum_{k=1,3,5,\ldots}^{\infty} \dfrac{1}{k^2}$, since

all the cosine terms are -1. Since $g(x)$ is continuous at $x = \pi$ and $g(\pi) = \pi$, it

follows that $\qquad\qquad \dfrac{8}{\pi} \displaystyle\sum_{k=1,3,5,\ldots}^{\infty} \dfrac{1}{k^2} = \pi$

and therefore $\qquad g(x) = \dfrac{\pi}{2} - \dfrac{4}{\pi} \displaystyle\sum_{k\,=\,1,3,5,...}^{\infty} \dfrac{\cos kx}{k^2}$

Differentiation of Fourier series

Since differentiation tends to destroy smooth properties of a function, we cannot expect the term-by-term differentiation of the Fourier series for $f(x)$ to converge uniformly to $f'(x)$. Now let

$$f(x) = \frac{1}{2} a_0 + \sum_{k=1}^{\infty} (a_k \cos kx + b_k \sin kx),$$

as usual and suppose $f'(x)$ can have a Fourier series representation

$$f'(x) = \frac{1}{2} A_0 + \sum_{k=1}^{\infty} (A_k \cos kx + B_k \sin kx)$$

Then $\qquad A_0 = \dfrac{1}{\pi} \displaystyle\int_{-\pi}^{\pi} f'(x)\mathrm{d}x = \dfrac{1}{\pi} [f(\pi) - f(-\pi)]$

$$A_r = \frac{1}{\pi} \int_{-\pi}^{\pi} f'(x) \cos rx \, \mathrm{d}x = \frac{1}{\pi} [f(\pi) - f(-\pi)] \cos r\pi + rb_r$$

Similarly $B_r = -ra_r$

Term-by-term differentiation yields $\displaystyle\sum_{k=1}^{\infty} (-ka_k \sin kx + kb_k \cos kx)$ and this series is the same as the Fourier series of $f'(x)$ if $f(\pi) = f(-\pi)$ i.e. if the periodic extension of $f(x)$ is continuous at the points $x = n\pi$.

Example 1

Consider the function $f(x) = x$, $-\pi < x < \pi$, $f(x + 2\pi) = f(x)$. Its

Fourier series is $\qquad 2 \displaystyle\sum_{k=1}^{\infty} (-1)^{k+1} \dfrac{1}{k} \sin kx$

Differentiating this term by term gives the series $2 \displaystyle\sum_{k=1}^{\infty} (-1)^{k+1} \cos kx$

which does not converge for any value of x. Note that there is a discontinuity of $f(x)$ at $x = n\pi$.

Example 2

Consider the function $f(x) = |x|$, $-\pi < x < \pi$, $f(x + 2\pi) = f(x)$, which is continuous at $x = n\pi$. Its Fourier series is

$$f(x) = \frac{\pi}{2} - \frac{4}{\pi} \sum_{k = 1, 3, 5 \ldots}^{\infty} \frac{1}{k^2} \cos kx$$

Differentiation of this series term-by-term produces

$$\frac{4}{\pi} \sum_{k = 1,3,5,\ldots}^{\infty} \frac{1}{k} \sin kx$$

which you can show is the Fourier series for the function

$$g(x) = \begin{cases} -1, & -\pi < x < 0 \\ 1, & 0 < x < \pi \end{cases} \quad g(x + 2\pi) = g(x)$$

Amplitude spectrum

The plot of the amplitude of the Fourier coefficients against the frequencies of the components they represent is called an **amplitude spectrum**.

For the last example considered, the spectra of $f(x)$ and $g(x)$ are shown in Figure 6.12(a) and (b) repectively.

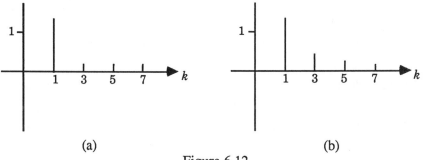

(a)   (b)

Figure 6.12

Addendum

The sending of signals without distortion is an important function of many electronic devices. The amplification of a waveform can be studied by Fourier analysis. The original wave is analysed into its harmonics and the amplification of each harmonic calculated from the characteristics of the device (it may, for example, distort very high frequency waves). The Fourier analysis will tell us how much of each harmonic is present in the original wave and hence the degree of distortion expected. The range of freedom from such distortion is called the **bandwidth** of the device. We need to know how many harmonics can be 'lost' before the quality of reproduction suffers markedly. The effects of filters can also be studied by first analysing the input wave into its harmonics.

6.6 Trigonometric series approximation of discrete data

Suppose that $f(x)$ is not specified by a formula but by a table of observed values. How then can we carry out our Fourier analysis? This problem arises in many contexts, an important one being in analysing the motion of the sea by taking sample readings over a period of time.

First consider the problem of fitting a finite trigonometric series to a function $f(x)$ specified by $(2n + 1)$ values: $f(x_0)$, $f(x_1)$, $f(x_2)$, ..., $f(x_{2n})$. (If the number of data points is $2n$, we in effect omit the last sine term). We shall merely state the relevant results and leave you either to carry out the proofs yourselves or consult suitable books.

The $(2n + 1)$ points will determine uniquely the coefficients in the series

$$\frac{1}{2} a_0 + \sum_{k=1}^{n} \left(a_k \cos \frac{2\pi kx}{(2n+1)} + b_k \sin \frac{2\pi kx}{(2n+1)} \right) \tag{6.18}$$

We have the following properties of orthogonality, given in the set of equations (6.19) below

$$\sum_{i=0}^{2n} \sin \frac{2\pi j x_i}{(2n+1)} \cos \frac{2\pi k x_i}{(2n+1)} = 0$$

$$\sum_{i=0}^{2n} \sin \frac{2\pi j x_i}{(2n+1)} \sin \frac{2\pi k x_i}{(2n+1)} = \begin{cases} 0 \text{ if } j \neq k \\ (2n+1)/2 \text{ if } j = k \neq 0 \end{cases} \tag{6.19}$$

$$\sum_{i=0}^{2n} \cos \frac{2\pi j x_i}{(2n+1)} \cos \frac{2\pi k x_i}{(2n+1)} = \begin{cases} 0 & \text{if } j \neq k \\ (2n+1)/2 & \text{if } j = k, \text{ neither is } 0 \text{ or } (2n+1) \\ (2n+1) & \text{if } j = k, \text{ is either } 0 \text{ or } (2n+1) \end{cases}$$

We can then proceed to determine the coefficients by the formulae

$$\left. \begin{aligned} a_k &= \frac{2}{(2n+1)} \sum_{i=0}^{2n} f(x_i) \cos \frac{2\pi k x_i}{(2n+1)} \quad, k = 0, 1, 2, ..., n \\ b_k &= \frac{2}{(2n+1)} \sum_{i=0}^{2n} f(x_i) \sin \frac{2\pi k x_i}{(2n+1)} \quad, k = 1, 2, ..., n \end{aligned} \right\} \tag{6.20}$$

Example
The following data is given. Fit a trigonometric approximation and estimate $f(2.5)$. Note $n = 2$.

x	0	1	2	3	4
$f(x)$	0	2	4	2	0

We find from (6.20) that $a_0 = \dfrac{2}{5}$ $[0 + 2 + 4 + 2 + 0] = 3.2$

$$a_1 = \frac{2}{5}\left[0 \times \cos\frac{2\pi.0}{5} + 2\cos\frac{2\pi.1}{5} + 4\cos\frac{2\pi.2}{5} + 2\cos\frac{2\pi.3}{5} + 0 \times \cos\frac{2\pi.4}{5}\right]$$

$\qquad = -1.6944$ (4 d.p.)

$$a_2 = \frac{2}{5}\left[0 \times \cos\frac{2\pi.2.0}{5} + 2\cos\frac{2\pi.2.1}{5} + 4\cos\frac{2\pi.2.2}{5} + 2\cos\frac{2\pi.2.3}{5} + 0 \times \cos\frac{2\pi.2.}{5}\right.$$

$\qquad = 0.0944$ (4 d.p.)

Similarly $b_1 = 1.2310$ (4 d.p.) and $b_2 = -0.2906$ (4 d.p.)

Then $f(x) \approx 1.6 - 1.6944 \cos\dfrac{2\pi x}{5} + 0.0944 \cos\dfrac{4\pi x}{5} + 1.2310 \sin\dfrac{2\pi x}{5}$

$\qquad\qquad - 0.2906 \sin\dfrac{4\pi x}{5}$

Hence $f(2.5) = 3.389$.

Least squares approximation

Suppose we wish to fit to a set of data a trigonometric series which has fewer coefficients than the number of data points. We shall then have an infinite number of series from which to choose and we select one of these by the method of least squares. For $(2n + 1)$ data points and a series with $(2m + 1)$ coefficients, where $m \leq n$ we seek to minimise

$$E = \sum_{i=0}^{2n} \left[f(x_i) - S_m(x_i)\right]^2 \tag{6.21}$$

where

$$S_m(x) = \frac{1}{2}\alpha_0 + \sum_{k=1}^{m}\left[\alpha_k \cos\frac{2\pi kx}{(2n+1)} + \beta_k \sin\frac{2\pi kx}{(2n+1)}\right] \tag{6.22}$$

The coefficients α_k, β_k which give rise to the minimum value of E are just those in (6.20). In other words, the Fourier coefficients are fixed, irrespective of the value of $m \leq n$. If we fit a trigonometric series with more terms we do not alter these earlier coefficients; if we have determined coefficients up to a_4, b_4 and we decide to take two more terms in the series approximation then we need only determine a_5 and b_5, since the other coefficients will remain at their chosen values. The same result held for a formula specification of $f(x)$ in Section 6.2.

This useful result is a direct consequence of orthogonality, as expressed in (6.19). We do not find this true for least squares polynomial fitting, where the selection of a higher degree polynomial necessitates the determination of all the coefficients from afresh.

Example

Fit a least squares approximation of the form $f(x) \approx b_1 \sin x + b_3 \sin 3x + b_5 \sin 5x$ to the data

x	0	$\pi/6$	$\pi/3$	$\pi/2$	$2\pi/3$	$5\pi/6$	π
$f(x)$	0	2.5	4	4.5	4	2.5	0

This is a slight variant on the material discussed in the text, but there are fewer coefficients than data points and the principles still apply, so

$$b_1 = \frac{2}{7}\left[0 \times \sin\left(\frac{2\pi}{7} . 0\right) + 2.5 \times \sin\left(\frac{2\pi}{7} . \frac{\pi}{6}\right) + 4 \times \sin\left(\frac{2\pi}{7} . \frac{\pi}{3}\right) + \ldots\right]$$

$$= 4.112 \text{ (3 d.p.)}$$

$$b_3 = \frac{2}{7}\left[0 \times \sin\left(\frac{2\pi.3}{7} . 0\right) + 2.5 \times \sin\left(\frac{2\pi}{7} .3. \frac{\pi}{6}\right) + 4 \times \sin\left(\frac{2\pi}{7} .3. \frac{\pi}{3}\right) + \ldots\right]$$

$$= -0.263 \text{ (3 d.p.)}$$

Similarly $b_5 = -0.235$ (3 d.p.)

Hence, $f(x) \approx 4.112 \sin x - 0.263 \sin 3x - 0.235 \sin 5x$

Data smoothing

Suppose we have some numerical data which may have random 'noise' in the sense that it respresents the true values of a function with random errors superimposed. If we further suppose that the true function is relatively smooth, whereas the errors are not, then we may make a start towards separating the two; because the true function is smooth, its Fourier coefficients will decay rapidly, whereas the error, being unsmooth, will have slowly decreasing Fourier coefficients. The higher frequency harmonics of the series representing the total function will be almost entirely due to error. We shall, of course, have some error contribution in the lower frequency harmonics.

Example

The data from an experiment is as shown below.

Assume that the first and last values are indeed zero and extend the function as an odd function. Smooth the data using trigonometric series and calculate the root-mean-square error of the given data and the smoothed data, on the assumption that the true function was $f(x) = x(2 - x)$ and that random errors were superimposed.

x	0	0.25	0.50	0.75	1.00	1.25	1.50	1.75	2.00
$f(x)$	0.000	0.407	0.823	0.841	1.061	0.946	0.845	0.348	0.000

Since we are manufacturing an odd function we need only compute the coefficients b_k, using the formulae (6.20). Having extended the function to an odd function of period 4, we note that the a_k are zero and the coefficients

$$b_k = \frac{1}{4} \sum_{j=1}^{7} f(x_j) \sin \frac{2\pi k x_j}{8}$$

Table 6.2 shows the smoothed values and the values of $f(x) = x(2 - x)$. (all values are quoted to 3 d.p.).

Table 6.2

Given values	0.000	0.407	0.823	0.841	1.061	0.946	0.845	0.348	0.000
Smoothed values	0.000	0.452	0.750	0.910	1.009	1.000	0.777	0.377	0.000
'True' values	0.000	0.438	0.750	0.938	1.000	0.938	0.750	0.438	0.000

The subject of data smoothing is quite extensive and we have only touched on it here. We merely remark that in order to get a series which should converge rapidly (so that a greater separation between true values and noise can be achieved) we first transform $f(x)$ into $g(x) = f(x) - (\alpha + \beta x)$ where α and β can be found from the boundary conditions $g(x_0) = 0 = g(x_{2n})$.

We can then extend $g(x)$ to an odd function. The resulting function will have no discontinuities in either itself or in its derived function. Therefore we determine the coefficients of the sine series in the usual way.

We must be careful, however, not to oversmooth the values and destroy the qualitative nature of the function.

Numerical Harmonic Analysis

We conclude this section with an example on fitting a Fourier series to a table of data by evaluating those terms which make a significant contribution. The data is the observed values of the current I flowing in an electronic device, the observations taken at regular intervals of time. We have chosen the times to be multiples of $\pi/6$ merely for numerical convenience. The data is as follows

t	0	$\pi/6$	$\pi/3$	$\pi/2$	$2\pi/3$	$5\pi/6$	π	$7\pi/6$	$4\pi/3$	$3\pi/2$	$5\pi/3$	$11\pi/6$	2π
$I(t)$	0	4.6	11.0	17.8	21.6	22.8	19.8	9.6	0	0	0	0	0

Now we assume a Fourier series representation

$$I(t) = \frac{1}{2} a_0 + \sum_{k=1}^{\infty} (a_k \cos kt + b_k \sin kt)$$

Then $a_0 = \dfrac{1}{\pi} \displaystyle\int_0^{2\pi} I(t)\, dt = 2$ (mean value of $I(t)$ over $[0, 2\pi]$)

$$a_r = \frac{1}{\pi} \int_0^{2\pi} I(t) \cos rt \, dt = 2 \ (\text{mean value of } I(t) \cos rt \text{ over } [0, 2\pi])$$

Also, $b_r = 2$ (mean value of $I(t) \sin rt$ over $[0, 2\pi]$)

We find mean values by summing the contributions from the twelve points $0(\pi/6) 11\pi/6$ and divide by 12. However, since we would then multiply the result by 2 we shall in fact obtain the total contribution and divide by 6. We display in Table 6.3 the result of the calculations for a_0, a_1, b_1 and summarise the results for higher coefficients. Results for coefficients are quoted to 1 d.p.

Hence $a_0 = 17.9$, $a_1 = -8.2$, $b_1 = 9.2$. Similarly, we can find that $a_2 = 0.7$, $b_2 = -2.8$, $a_3 = -1.5$, $b_3 = 0.0$, $a_4 = 0.5$, $b_4 = 0.3$, $a_5 = -0.2$, $b_5 = -0.4$.

In general, there is the result that with r intervals it is worth carrying this kind of analysis only as far as the $\left(\dfrac{r}{2} - 1\right)^{\text{th}}$ harmonic. In this example $r = 12$ and hence we need go no further than the 5th harmonic.

Table 6.3

t	$I(t)$	$\cos t$	$I \cos t$	$\sin t$	$I \sin t$
0	0	1	0	0	0
$\pi/6$	4.6	0.867	3.98	0.5	2.30
$\pi/3$	11.0	0.5	5.50	0.867	9.52
$\pi/2$	17.8	0	0	1	17.80
$2\pi/3$	21.6	−0.5	−10.8	0.867	18.70
$5\pi/6$	22.8	−0.867	−19.74	0.5	11.4
π	19.8	−1	−19.8	0	0
$7\pi/6$	9.6	−0.867	−8.33	−0.5	−4.8
$4\pi/3$	0	−0.5	0	−0.867	0
$3\pi/2$	0	0	0	−1	0
$5\pi/3$	0	0.5	0	−0.867	0
$11\pi/6$	0	0.867	0	−0.5	0
Σ	107.2		−49.18		54.92
$\Sigma/6$	17.9		−8.2		9.2

Problems

Section 6.3

1   For each of the functions below, defined in the range $[-\pi, \pi]$ and assumed to have period 2π, find their Fourier series.

   (a) $f(x) = 2, -\pi < x < 0; f(x) = 1, 0 < x < \pi$

       (What is the value of the sum of the series when $x = -\pi, 0, \pi$? Comment.)

   (b) $f(x) = 0, -\pi < x < 0; f(x) = x, 0 < x < \pi$

   (c) $f(x) = x^2, -\pi < x < \pi$

   (d) $f(x) = |\sin x|, -\pi < x < \pi$

   (e) $f(x) = -0.5 , -\pi < x < 0; f(x) = 0.5 , 0 < x < \pi$

       (What is the value of the sum of the series when $x = -\pi, 0, \pi$? Comment.)

   (f) $f(x) = 1, -\pi < x < 0; f(x) = x, 0 < x < \pi$

   (g) $f(x) = 0, -\pi < x < 0; f(x) = 1, 0 < x < \pi/2; f(x) = -1, \pi/2 < x < \pi$

2   A function $f(x)$ is of period 2π and is defined in the interval $0 \le x \le 2\pi$ in the form $f(x) = \sin x, 0 \le x \le \pi; f(x) = 0, \pi \le x \le 2\pi$. Show that

$$f(x) = \frac{1}{\pi} + \frac{1}{2} \sin x - \frac{2}{\pi} \sum_{n=1}^{\infty} \frac{\cos 2nx}{4n^2 - 1} .$$

Deduce the sum of the infinite series $\dfrac{1}{1.3} - \dfrac{1}{3.5} + \dfrac{1}{5.7} \, ...$ (LU)

3   A function $f(x)$ is defined in the range $(0, \pi)$ as follows: $f(x) = x$ for $0 \le x \le \frac{1}{4} \pi$,

$f(x) = \frac{1}{4} \pi$ for $\frac{1}{4} \pi \le x < \frac{3}{4} \pi, f(x) = \pi - x$ for $3\pi/4 < x < \pi$. If, in addition,

$f(\pi + x) = -f(x)$ for all values of x, expand $f(x)$ as a Fourier series, giving the first 3 non-zero terms and the general term.   (LU)

4   If $f(t)$ is a periodic function of period 2π and $f(t) = t/\pi$ for $0 < t < \pi$,

$f(t) = (2\pi - t)/\pi$ for $\pi < t < 2\pi$, show that $f(t) = \dfrac{1}{2} - \dfrac{4}{\pi^2} \sum_{0}^{\infty} \dfrac{\cos (2m + 1) t}{(2m + 1)^2} .$

Show also, that if ω is not an integer,

$$y = \frac{1}{2\omega^2} (1 - \cos \omega t) - \frac{4}{\pi^2} \sum_{m=0}^{\infty} \frac{\cos (2m + 1) t - \cos \omega t}{(2m + 1)^2 [\omega^2 - (2m + 1)^2]}$$

satisfies the equation $d^2 y/dt^2 + \omega^2 y = f(t)$ with the initial conditions $y = dy/dt = 0$ when $t = 0$.   (LU)

5   If $f(x)$ is an even periodic function with period 2π defined by $f(x) = x^2$ for $0 \le x \le \pi$, sketch its graph in the range $-3\pi \le x \le 3\pi$. Show that the Fourier series for $f(x)$ is

$$f(x) = \frac{\pi^2}{3} + 4 \sum_{n=1}^{\infty} \frac{(-1)^n \cos nx}{n^2}$$

Obtain a Fourier sine series, valid in the range $0 < x < \pi$, for the function $f(x)$. (LU)

Section 6.4

6 Which of the following functions are odd, even or neither?
   (a) $\cos x - \cos 5x$,   (b) $\sin x - 5 \cos 5x$ (c) $2 + x$ (d) $x/\{(x + 1)(x - 1)\}$,
   (e) $x \sin x$ (f) $\sin x + \cos x$.

7 Find the Fourier series for the function

$$f(x) = 0, -\pi < x < -\frac{1}{2}\pi; \ f(x) = \cos x, -\frac{1}{2}\pi \leq x \leq \frac{1}{2}\pi; \ f(x) = 0, \frac{1}{2}\pi < x < \pi$$

8 Find the Fourier sine series and the Fourier cosine series for each of the following functions and graph the two periodic extensions for each case.
   (a) $f(x) = 2, 0 < x < \pi$;   (b) $f(x) = x^2, 0 < x < \pi$;
   (c) $f(x) = x + 1, 0 < x < \pi$.

9 Given that $f(x) = 1 + \cos x, 0 < x < \pi$, find a half-range Fourier sine series for $f(x)$.

   Find the sum of the series when $x = 3\pi/2$ and show that $\frac{\pi}{4} = 1 - \frac{1}{3} + \frac{1}{5} - \frac{1}{7} + \frac{1}{9} - \ldots$

10 Prove that for $0 < x < \pi$, $\sin x = \frac{2}{\pi} - \frac{4}{\pi} \left[\frac{\cos 2x}{2^2 - 1} + \frac{\cos 4x}{4^2 - 1} + \frac{\cos 6x}{6^2 - 1} + \ldots \right]$.

   Sketch for $-2\pi \leq x \leq 3\pi$ the graph of the function represented by the right-hand side of this equation.

   Find the sum of the series $\frac{1}{1.3} - \frac{1}{3.5} + \frac{1}{5.7} - \frac{1}{7.9} + \ldots$ (LU)

11 Show that the Fourier cosine series for the function $f(x) = x, 0 \leq x \leq \pi$, is given by

   $\frac{\pi}{2} - \frac{4}{\pi} \sum_{0}^{\infty} \frac{\cos (2n + 1)x}{(2n + 1)^2}$. Express $f(x)$ as a Fourier sine series. Sketch the graph of
   each of these series for $-2\pi < x < 3\pi$. (LU)

12 Obtain the Fourier expansion of an even function $f(x)$ defined by
   $f(x) = x$, for $0 \leq x < \pi/3; f(x) = 0$, for $\pi/3 \leq x \leq 2\pi/3$;
   $f(x) = x - 2\pi/3$, for $2\pi/3 < x \leq \pi$,
   and simplify the coefficients of the first four terms. (LU)

Section 6.5

13 Find the Fourier series for the following functions; sketch their graphs
   (a) $f(x) = x, -1 < x \leq 0; f(x) = x + 2, 0 < x \leq 1$
   (b) $f(x) = -1, -2 < x \leq -1; f(x) = x, -1 < x < 1; f(x) = 1, 1 < x < 2$

(c) $f(x) = \dfrac{2}{3} x, 0 < x < \dfrac{1}{3} \pi; f(x) = \dfrac{1}{3} (\pi - x), \dfrac{\pi}{3} < x < \pi$

(d) $f(x) = \dfrac{4ax}{l}, 0 < x < \dfrac{l}{4}; \quad f(x) = \dfrac{4a}{l} (\dfrac{1}{2} l - x), \dfrac{l}{4} < x < \dfrac{3l}{4};$

$f(x) = \dfrac{4a}{l}(x - l), \dfrac{3l}{4} < x < l$

(e) $f(x) = ax/b, 0 < x < b; f(x) = a(l - x)/(l - b), b < x < l.$

14   Find the Fourier series for the functions whose graphs are shown.

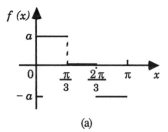 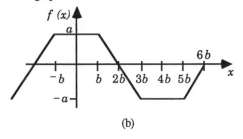

                   (a)   (b)

15   Sketch the graph of each of the following functions and find their Fourier series. Assume periodicity outside the quoted range.

(a) $f(x) = \begin{cases} 8, & 0 < x < 2 \\ -8, & 2 < x < 4 \end{cases}$ (b) $f(x) = \begin{cases} -x, & -4 \le x \le 0 \\ x, & 0 \le x \le 4 \end{cases}$

(c) $f(x) = 4x, 0 < x < 10$ (d) $f(x) = \begin{cases} 2x, & 0 \le x < 3 \\ 0, & -3 < x < 0 \end{cases}$

16   Prove that if $-\pi < x < \pi,$

$$\cosh ax = \frac{2a}{\pi} \left\{ \frac{1}{2a^2} + \sum_{n=1}^{\infty} (-1)^n \frac{1}{(n^2 + a^2)} \cos nx \right\} \sinh a\pi$$

and deduce that, for the same range of values of $x,$

$$\sinh ax = \frac{2}{\pi} \left\{ \sum_{n=1}^{\infty} (-1)^{n-1} \frac{n}{(n^2 + a^2)} \sin nx \right\} \sinh a\pi$$

17   (a)   Show that, for $-\pi \le x \le \pi,$

$$x \cos x = -\frac{1}{2} \sin x + 2 \left\{ \frac{2}{1.3} \sin 2x - \frac{3}{2.4} \sin 3x + \frac{4}{3.5} \sin 4x - ... \right\}$$

and deduce that, for $-\pi \le x \le \pi$

$$x \sin x = 1 - \frac{1}{2} \cos x - 2 \left\{ \frac{\cos 2x}{1.3} - \frac{\cos 3x}{2.4} + \frac{\cos 4x}{3.5} - ... \right\}$$

(b) Show that for $0 \le x \le \pi$, $x(\pi - x) = \dfrac{8}{\pi} \left[\dfrac{\sin x}{1^3} + \dfrac{\sin 3x}{3^3} + \dfrac{\sin 5x}{5^3} + \ldots \right]$

By differentiating this last result, show that for $0 \le x \le \pi$

$$x = \frac{1}{2} \pi - \frac{4}{\pi} \left[\frac{\cos x}{1^2} + \frac{\cos 3x}{3^2} + \frac{\cos 5x}{5^2} + \ldots \right]$$

18 Use Parseval's result and the Fourier series for $f(x) = x$ in $-\pi < x < \pi$ to deduce that

$$\sum_{m=1}^{\infty} \frac{1}{m^2} = \frac{\pi^2}{6}$$

19 (a) From the cosine series for $f(x) = x$, $0 < x < \pi$ deduce that

$$\sum_{m=1}^{\infty} \frac{1}{(2m-1)^4} = \frac{\pi^4}{96}$$

(b) From the sine series for $f(x) = 1$, $0 < x < \pi$ deduce that

$$\sum_{m=1}^{\infty} \frac{1}{(2m-1)^2} = \frac{\pi^2}{8}$$

Section 6.6

20 Find a Fourier series, as far as the third harmonic, to represent the periodic function $f(x)$ given by the values below.

x	0°	30°	60°	90°	120°	150°	180°	210°	240°	270°	300°	330°
f	−7.24	−11.32	17.5	18.26	21.72	23.48	16.8	6.68	−13.52	−27.88	−36.28	−28.5

21 Repeat Problem 20 for the f–values below:
(a) 1.0, 2.4, 4.0, 4.8, 3.2, 4.3, 4.3, 3.1, −2.0, −2.0, −0.8, 0.6
(b) 0, 1.0, 1.9, 2.8, 3.2, 3.4, 2.6, 1.8, 1.4, 0.8, 0.5, 0.2
(c) 0, 2.1, 2.8, 3.2, 2.1, 0, 0, 0, 0, 0, 1.7, 1.7

22 Find a Fourier series, as far as the harmonic requested for the data below.

(a) x	0	0.5	1.0	1.5	2.0	2.5	3.0	3.5	4.0	4.5	(4th)
f	3.0	0.3	−0.3	−1.4	−1.5	−1.0	0.1	1.4	3.5	4.6	

(b) x	0	0.2	0.4	0.6	0.8	1.0	1.2	1.4	1.6	1.8	2.0	(5th)
f	0	0.8	2.8	5.6	2.4	0	−2.4	−5.6	−2.8	−0.8	0	

7

PARTIAL DIFFERENTIAL
EQUATIONS

7.1 Introduction

Many engineering systems can be modelled by means of partial differential equations. In this chapter we shall concentrate on three basic kinds of linear partial differential equations which can each be used as the model for a wide class of physical phenomena. We shall examine, in the main, numerical and analytical solutions, but in some cases the use of analogues is helpful and we shall mention these for completeness. It will be impossible in a short space to describe all the major applications of our three types of equation, but we hope that through the text and the problems we have given you a reasonably wide compass. We shall sometimes use the abbreviation p.d.e.

7.2 Case Study: Steady state temperature distribution in a plate

We wish to find the temperature distribution in a rectangular metal plate under certain conditions. The plate in Figure 7.1(a) is covered on its top and bottom faces by layers of thermally insulating material so that heat is constrained to flow mainly in the x- and y-directions. Along the edges of the plate various conditions are applied – a particular case is shown in Figure 7.1(b). These conditions, being applied at the boundaries of the region of interest, are known as **boundary conditions**.

Assumptions
We now make assumptions about the physical situation in order to formulate a reasonably simple mathematical model.

(i) The metal is *uniform* in the sense that its thermal conductivity is the same at all points of the plate.
(ii) The plate is sufficiently *thin* so that we may neglect any heat flow in the directions perpendicular to its faces.
(iii) The temperature distribution is in the *steady state*, i.e. the temperature at any point in the plate does not depend on time.

Model

We are now entitled to assume that the model plate is infinitesimally thin and that the temperature function θ depends only on x and y, i.e. $\theta = \theta(x,y)$. We have chosen the origin of coordinates at one corner for convenience. Let the dimensions of the plate be as shown in Figure 7.1(c), so that the plate may be defined as $0 \leq x \leq a, 0 \leq y \leq b$. We now analyse the situation in a small rectangular region ABCD of the plate, whose position is as shown; its centre is the point (x_0, y_0).

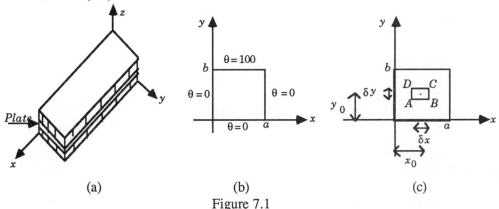

(a)   (b)   (c)

Figure 7.1

Consider what happens at the edge DA of this sub-region. The temperature will change along DA, since the y-coordinate varies, but it will change only with y. We shall ultimately take limits as δx and δy tend to zero, so we may approximate the derivative $\partial\theta/\partial x \equiv \theta_x$ at all points of the edge DA by its value at the mid-point of that edge, viz. $(x_0 - \frac{1}{2}\delta x, y_0)$. Then the heat flowing across DA to the right is approximately given by

$$H_{DA} \cong -k\delta y\theta_x(x_0 - \tfrac{1}{2}\delta x, y_0)$$

This is a mathematical statement of **Fourier's Law** which states that the flow of heat is proportional to the local temperature gradient. The constant of proportionality k is related to the thermal conductivity of the plate material. Now the flow of heat across CB is

$$H_{CB} \cong -k\delta y\theta_x(x_0 + \tfrac{1}{2}\delta x, y_0)$$

Notice that if θ decreases as x increases then $\partial\theta/\partial x$ will be negative and we shall have a positive flow of heat to the right.

Precisely analogous arguments show that the flows of heat across AB and DC are given, respectively, by

$$H_{AB} \cong -k\delta x\theta_y(x_0, y_0 - \tfrac{1}{2}\delta y) \text{ and } H_{DC} \cong -k\delta x\theta_y(x_0, y_0 + \tfrac{1}{2}\delta y)$$

Since we are considering the steady-state, there is no accumulation of heat in the region ABCD and the input of heat flow must equal the output. We shall not lose any generality if we assume that heat flows in through AB and DA and out through CB and DC.

Then $H_{DA} - H_{CB} + H_{AB} - H_{DC} = 0$

i.e. $-k \{ \delta y \ [\theta_x(x_0 - \tfrac{1}{2}\delta x, y_0) - \theta_x(x_0 + \tfrac{1}{2}\delta x, y_0)]$

$+ \delta x \ [\theta_y(x_0, y_0 - \tfrac{1}{2}\delta y) - \theta_y(x_0, y_0 + \tfrac{1}{2}\delta y)] \ \} \cong 0$

i.e. $+k\delta x\delta y \left\{ \left[\dfrac{\theta_x(x_0 + \tfrac{1}{2}\delta x, y_0) - \theta_x(x_0 - \tfrac{1}{2}\delta x, y_0)}{\delta x} \right] \right.$

$$\left. + \left[\frac{\theta_y(x_0, y_0 + \tfrac{1}{2}\delta y) - \theta_y(x_0, y_0 - \tfrac{1}{2}\delta y)}{\delta y} \right] \right\} \approx 0 \qquad (7.1)$$

We now let the area of the rectangle ABCD tend to zero by letting δx and δy tend to zero simultaneously. This will have the following effects

(i) $\left[\dfrac{\theta_x(x_0 + \tfrac{1}{2}\delta x, y_0) - \theta_x(x_0 - \tfrac{1}{2}\delta x, y_0)}{\delta x} \right] \rightarrow \left. \dfrac{\partial \theta_x}{\partial x} \right|(x_0, y_0) = \theta_{xx}(x_0, y_0)$

(ii)   The other term in square brackets will tend to the value $\theta_{yy}(x_0, y_0)$
(iii)   The approximation becomes an equality.

Of course, as equation (7.1) stands we shall get the result $0 = 0$; therefore we first divide both sides by $k\delta x\delta y$ *before* taking limits. As a result of letting δx and δy tend simultaneously to zero we obtain **Laplace's Equation** in two dimensions

$$\frac{\partial^2 \theta}{\partial x^2} + \frac{\partial^2 \theta}{\partial y^2} = 0 \qquad (7.2)$$

Boundary Conditions

Equation (7.2) possesses infinitely many solutions. It is necessary to specify extra conditions to cut the possibilities down to a *unique* solution. Too many conditions may not permit *any* solution and too few will not fix a unique solution. These conditions must be of the right kind. We shall study these **boundary conditions** more fully in the next section, but for the moment we shall specify simple conditions as shown in Figure 7.1(b). Three edges are maintained at zero temperature, the fourth is maintained at a temperature of 100°C. Roughly speaking, we need two conditions on the temperature for given values of x and two for given values of y. This follows because, in order to solve (7.2) analytically we need two integrations with respect to x and two integrations with respect to y. Each of these will introduce an arbitrary function with respect to the other variable and we need effectively to supply limits of integration. The conditions can be stated mathematically as

$\theta = 0$ when $x = 0$,   for $0 \leq y < b$ (7.3a)

$\theta = 0$ when $y = 0$,   for $0 \leq x \leq a$ (7.3b)

$\theta = 0$ when $x = a$,   for $0 \leq y < b$ (7.3c)

$\theta = 100$ when $y = b$,   for $0 \leq x \leq a$ (7.3d)

We may write these conditions in a different way; for example (7.3a) becomes $\theta(0,y) = 0$ for $0 \leq y < b$.

Notice that there is a discontinuity in the temperature function at the points $(0,b)$ and (a,b): this is the price we pay for taking a very simple condition along the edge DC.

The equation (7.2) together with the boundary conditions (7.3) forms the mathematical model of the problem. We now examine three methods of solution.

Analogue method

Laplace's equation is extremely widespread in its appearance in mathematical models. It applies to problems which can be expressed in terms of a **potential function**; for example, electrostatics, magnetostatics, temperature problems and problems in stress analysis, seepage and irrotational motion of fluids. Because of the same equation governing phenomena in electrostatics, we can produce an analogous model using voltages by specifying the appropriate geometry and boundary conditions. Reading the voltages at specified points gives us a picture of the overall distribution of voltage which will be the distribution of temperature in our model.

Take a sheet of conducting paper cut to the appropriate rectangular shape. We put a conducting wire along the top edge and cover it with conducting paint. We put a second wire along the remaining three edges and cover it with paint. Then the wires are connected to a variable voltage supply so that we may regard the top edge as being at 100λ volts and the other three edges at 0 volts; λ is a scaling factor. We now select a suitable temperature, say 80°C and set the variable scale of voltage to 80λ. We then take a conducting pencil connected to the output via a galvanometer. If we press hard on the conducting pencil at a point where the potential is 80λ we shall then find no deflection of the galvanometer needle. If we make several sets of exploratory moves, we can find a set of points at which the potential is the same: if we join these points we have a curve of equal potential which corresponds to an isotherm for 80°C. Proceeding in a similar way we may construct further isotherms. In Figure 7.2 we represent schematically the apparatus, while in Figure 7.3 we show a typical set of results; in this latter figure we have constructed a **flow net** for the left-hand part by adding the orthogonal lines of heat flow. Because of the temperature discontinuities at the top corners we have all the isotherms entering and leaving these corners.

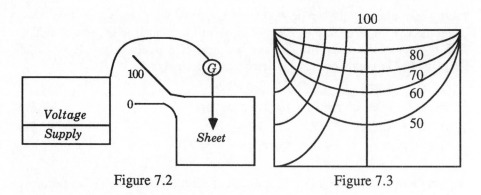

Figure 7.2   Figure 7.3

Analytical solution – Separation of variables method

Analytical techniques successfully employed produce a formula for θ which will be our solution. How useful this formula is remains to be seen. The technique most widely employed in such problems is the method of *separation of variables*. This is dealt with more fully in Section 7.4, but we outline its main principles here.

The approach used is a variant of a main line of attack in mathematics: reduce the problem at hand to one (or more) which you *can* solve. In this instance, whilst we know how to solve *ordinary* differential equations (particularly if they are linear with constant coefficients) we have no technique as yet for solving *partial* differential equations. We aim to separate the independent variables x and y so that we obtain *two* ordinary differential equations. We accomplish this aim by assuming that we can express $\theta(x,y)$ as the product of a function of x only and a function of y only. For example, $\theta(x,y) = 3 \cos 2x\, e^{-2y}$ is of the form we seek, whereas $\theta(x,y) = 4 \ln(x + 2y)$ is not; the form $\theta(x,y) = 2 \sin (x + 2y)$ *is* a possibility since it can be expanded into $\theta(x,y) = 2 \sin x \cos 2y + 2 \cos x \sin 2y$ and is a *linear combination* of acceptable forms. Of course, we have not shown that *any* of the three forms given actually satisfies Laplace's equation. (In fact, only the first form does.) What we do is to obtain the most general solution of Laplace's equation which stands a chance of satisfying the boundary conditions and then apply the boundary conditions one at a time, whittling down the possibilities until, if the problem is properly posed, we end up with a unique solution to the problem. As we shall see in Section 7.4, any function of the form

$$\theta(x,y) = (A \cos kx + B \sin kx)(C \cosh ky + D \sinh ky)$$

will satisfy Laplace's equation and also lead to satisfaction of the boundary conditions. There are four independent constants to be determined and we have four conditions to apply. It turns out, having allowed for the linearity of the equation, that the unique solution to the problem is

$$\theta(x,y) = \frac{400}{\pi} \sum_{n = 1,3,5,\ldots}^{\infty} \frac{\sin (n\pi x/a) \sinh (n\pi y/a)}{n \sinh (n\pi b/a)} \qquad (7.4)$$

However we certainly cannot immediately draw a graphical respresentation of $\theta(x,y)$ nor is it easy to calculate the values of θ at given points (x,y). We programmed a digital computer to calculate the value of θ at 9 internal points of the slab, equally spaced, so that we could get a reasonable idea of the temperature distribution. We took $a=b=1$ for simplicity. The series converged fairly quickly at each point to get the required accuracy of 1 d.p. The results are shown in Figure 7.4. It is important to remark that the problem is symmetric about the line $x = a/2$ and therefore we need only calculate the temperature at 6 internal points. We then programmed the computer to deal with 121 equally spaced internal points and the results of this allowed us to sketch the isotherms as in Figure 7.5(a).

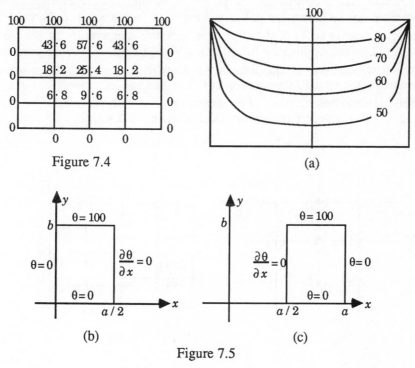

Figure 7.4   (a)

(b)   (c)

Figure 7.5

It is worth remarking that the symmetry of the problem means that the slopes of the isotherms are parallel to the x-axis at $x = a/2$ and this in turn means that the slab has two distinct regions of heat flow. If we make the line $x = a/2$ an insulating line then we can deal with the problem where no heat flows across a boundary. The two possible cases are shown in Figure 7.5(b) and (c).

The analytical method can be made to yield both qualitative and quantitative information. We have taken a geometrically simple region and straightforward boundary conditions and the resulting formula for θ was even so an unpleasant entity. If, for example, the geometry of the region is more awkward then the analytical method may become virtually impossible.

Numerical approach

In such a case we can resort to a numerical method of attack. We divide the region of the slab by a mesh and calculate the temperatures at the nodes of the mesh. This was what we effectively did in Figure 7.4. In the numerical approach, however, we replace the differential equation by a finite difference approximation. We shall find in Section 7.7 that an approximation which is $0(h^2)$, where h is the spacing of the mesh nodes in either the x- or the y-direction, is given by

$$\theta(x_r+h, y_s)+\theta(x_r-h, y_s)+\theta(x_r, y_s+h)+\theta(x_r, y_s-h)-4\theta(x_r, y_s) = 0 \qquad (7.5)$$

In Figure 7.6(a) we show a selection of mesh points around the point (x_r, y_s) and we note that (7.5) can be interpreted as stating that the average of the temperatures at the four mesh points nearest to (x_r, y_s) is equal to the value of the temperature at (x_r, y_s). Let $\theta_{r,s}$ represent the temperature at the point (x_r, y_s). Then the application of equation (7.5) to the 9 internal points of Figure 7.4 will provide 9 equations in 9 unknowns $\theta_{r,s}$. However, we appeal to symmetry to reduce this number to 6. In general, equation (7.5) can be simplified near boundaries. For example, if $r = 1$, then $\theta_{r-1,s} = 0$ for $s = 1, 2, 3$. The case of most interest occurs when $r = 2$ since

$$\theta_{r+1,s} = \theta_{r-1,s} \qquad \text{for} \quad s = 1, 2, 3. \qquad (7.6)$$

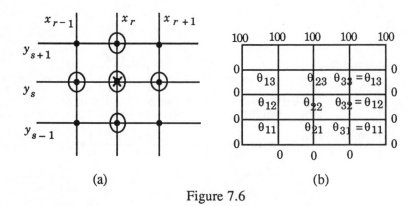

(a)   (b)

Figure 7.6

In Figure 7.6(b) we show the six unknown temperatures and leave you to show that the six equations in matrix form are

$$\begin{bmatrix} -4 & 1 & 1 & 0 & 0 & 0 \\ 2 & -4 & 0 & 1 & 0 & 0 \\ 1 & 0 & -4 & 1 & 1 & 0 \\ 0 & 1 & 2 & -4 & 0 & 1 \\ 0 & 0 & 1 & 0 & -4 & 1 \\ 0 & 0 & 0 & 1 & 2 & -4 \end{bmatrix} \begin{bmatrix} \theta_{11} \\ \theta_{21} \\ \theta_{12} \\ \theta_{22} \\ \theta_{13} \\ \theta_{23} \end{bmatrix} = \begin{bmatrix} 0 \\ 0 \\ 0 \\ 0 \\ -100 \\ -100 \end{bmatrix} \qquad (7.7)$$

We can solve these equations by Gauss elimination since the **sparseness** (i.e.

presence of many zeros) is not destroyed by the method. This is so because the coefficient matrix is banded. The solution we obtained was, to 1 d.p.,

$$\theta_{11} = 7.1; \ \theta_{21} = 9.8; \ \theta_{12} = 18.7; \ \theta_{22} = 25.0; \ \theta_{13} = 42.9; \ \theta_{23} = 52.2$$

This agrees with the analytical results only tolerably. We need a more sophisticated numerical approach using a finer mesh.

7.3 Some Basic Ideas

The **order** of a partial differential equation is the order of the highest partial derivative it contains. By an **analytical solution** to a partial differential equation, we mean a function of the independent variables which satisfies identically the equation at every point in a domain of the independent variables. For example, $\theta = A \cos kx \sinh ky$ is a solution of equation (7.2) and so is $\theta = B \sin kx \cosh ky$ for all (x,y).

In general, a partial differential equation of order n has a solution which contains at most n arbitrary functions. It may be, however, that we need to express the general solution as a sum of such solutions. This follows from the linearity of the equation: if θ_1 and θ_2 are solutions, so is $a\theta_1 + b\theta_2$ where a and b are constants. We use the ideas of complementary function and particular integral. If θ_1 is the complementary function of the associated homogeneous equation and θ_2 is a particular integral of the linear p.d.e. then $\theta = \theta_1 + \theta_2$ is a general solution of the full equation. The general solution can be particularised to a *unique* solution if appropriate extra conditions are provided. These are generally classed as **boundary conditions**; where time, $t,$ is one of the independent variables and we specify a configuration at $t = 0$, we refer to it as an **initial condition**. The kind of boundary condition we need to specify depends on the nature of the equation. We shall give specific examples in later sections, but we can make some general remarks here.

First, we classify the problems modelled by partial differential equations. We have already remarked that Laplace's equation arises in many different engineering contexts. Although the symbols may be different and the nomenclature may vary, the underling *structure* is the same.

Equilibrium problems relate to steady-state conditions. The problems which fall into this category include steady-state temperature distributions, steady flows of electric current, equilibrium stress situations and steady ideal fluid flows. We seek configurations of the system studied, for example, displacements, temperatures and velocities.

These problems are boundary-value problems. Correspondingly, we need to specify conditions which exist along the entire boundary. We may specify the value of the problem variable, for example temperature, at each point on the boundary. Alternatively we may specify the normal derivative of the variable at some points on the boundary and the problem variable at the others. See Figure 7.7(a). Notice that the boundary conditions have a tremendous influence on the solution. They have been likened to a jury demanding that the solution satisfies the equation at all points in the closed region, and, simultaneously, satisfies the conditions prescribed at all points on the boundary.

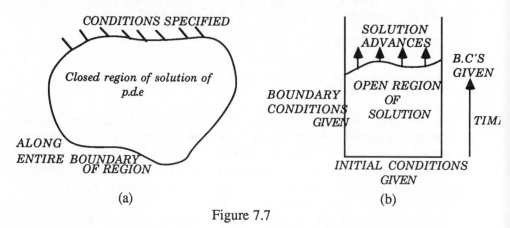

(a)   (b)

Figure 7.7

The two equations of this class which we shall study in depth are **Laplace's equation** which in two-dimensional cartesian coordinates is

$$\frac{\partial^2 \theta}{\partial x^2} + \frac{\partial^2 \theta}{\partial y^2} = 0 \qquad (7.2)$$

and **Poisson's equation**, an example of which is

$$\frac{\partial^2 \phi}{\partial x^2} + \frac{\partial^2 \phi}{\partial y^2} = f(x,y) \qquad (7.8)$$

The function $\theta(x,y)$ in (7.2) could be gravitational potential, electrostatic potential, magnetic potential, velocity potential in irrotational fluid flows as well as steady-state temperature in a uniform solid. The function $\phi(x,y)$ in (7.8) could be gravitational potential in a region where $f(x,y)$ is proportional to density of the material, or electrostatic potential in a region where $f(x,y)$ is proportional to charge distribution, or a measure of shear stress in a long bar where $f(x,y)$ is constant. Laplace's equation is a special case of Poisson's equation.

We may also mention the **biharmonic equation**

$$\frac{\partial^4 \phi}{\partial x^4} + 2\frac{\partial^4 \phi}{\partial x^2 \partial y^2} + \frac{\partial^4 \phi}{\partial y^4} = 0 \qquad (7.9)$$

The numerical approach is to use finite difference techniques to transform the differential system into a set of linear simultaneous algebraic equations; the solution will be found at suitable points.

Propagation problems are **initial-value problems**. Here we need to specify conditions for initial time as well as on the space boundaries Such problems may be concerned with unsteady-state and transient phenomena. Knowledge of a system in an initial state is used to predict its behaviour at later times. The kinds of problem studied include propagation of heat, of displacements and stresses in elastic structures, of pressure waves in air and studies of evaporation. Figure 7.7(b) shows schematically the solution marching out into the future, guided along by the space boundary conditions. The region of solution is therefore an open one.

An equation we shall study is the one-dimensional **diffusion equation**

$$k^2 \frac{\partial^2 \phi}{\partial x^2} = \frac{\partial \phi}{\partial t} \tag{7.10}$$

which has the two-dimensional extension

$$k^2 \left[\frac{\partial^2 \phi}{\partial x^2} + \frac{\partial^2 \phi}{\partial y^2} \right] = \frac{\partial \phi}{\partial t} \tag{7.11}$$

The diffusion can be of heat or mass and (7.10) could model the voltage (or current) in an electrical transmission line where both inductance and leakage are negligible. In both equations k is a constant.

We shall also mention the one-dimensional **wave equation**

$$c^2 \frac{\partial^2 \phi}{\partial x^2} = \frac{\partial^2 \phi}{\partial t^2} \tag{7.12}$$

which has a two-dimensional extension

$$c^2 \left[\frac{\partial^2 \phi}{\partial x^2} + \frac{\partial^2 \phi}{\partial y^2} \right] = \frac{\partial^2 \phi}{\partial t^2} \tag{7.13}$$

The function ϕ could be a component of displacement in a vibrating system, the velocity potential of a gas in acoustic theory or each component of the electric or magnetic vector in the electromagnetic theory of light.

Eigenvalue problems are extensions of equilibrium problems with a leaning towards initial value problems. In addition to the steady-state configuration or mode, it is required to find critical values (eigenvalues) of a scalar problem parameter. The kinds of problem studied include vibrations at natural frequencies, resonance in electrical circuits and acoustics, and the buckling of structures. It is often only the relative displacements or relative amplitudes in a particular mode which can be found.

Some further remarks are in order. Since eigenvalue problems can be regarded as almost equilibrium problems or almost propagation problems, equations under either of these latter categories *may* be models for the former class. The nature of the problem will allow us to decide if it can be classed as an eigenvalue problem.

Alternative classification

For development of solutions we shall find it convenient to classify the model equations by an alternative scheme. The most general linear partial differential equation of the second order with two independent variables is

$$A \frac{\partial^2 \phi}{\partial x^2} + B \frac{\partial^2 \phi}{\partial x \partial y} + C \frac{\partial^2 \phi}{\partial y^2} + D \frac{\partial \phi}{\partial x} + E \frac{\partial \phi}{\partial y} + F\phi + G = 0 \quad (7.14)$$

where A, B, C, D, E, F and G are functions of x and y, including constants.

By analogy with the general equation for a conic: $ax^2 + bxy + cy^2 + dx + ey + f = 0$ we have the following classification

(i)   If $B^2 < 4AC$,   the equation is **elliptic**.
(ii) If $B^2 > 4AC$,   the equation is **hyperbolic**.
(iii) If $B^2 = 4AC$,   the equation is **parabolic**.

Examples

(i)   Laplace's equation (7.2) is a special case of (7.14) with $A = 1, C = 1$ and all other constants zero. $B^2 < 4AC$ in this case, hence the equation is elliptic.
(ii) The one-dimensional wave equation (7.12) has $A = c^2, C = -1, B = 0$ and hence $B^2 > 4AC$ to show that the equation is hyperbolic.
(iii) The diffusion equation (7.10) has $A = k^2, E = -1$ and all other constants zero. Hence $B^2 - 4AC = 0$ and the equation is parabolic.

Notice that the extensions to more variables, (7.13) and (7.11), are hyperbolic and parabolic respectively. Notice too how often in these extensions the expression

$$\frac{\partial^2 \phi}{\partial x^2} + \frac{\partial^2 \phi}{\partial y^2}$$ occurs. We may recognise $\left[\frac{\partial^2}{\partial x^2} + \frac{\partial^2}{\partial y^2}\right]$ as the two-dimensional

case of the **Laplace operator**, ∇^2.

7.4 Separation of Variables Method

Essentially, the aim of this method is to reduce the one *partial* differential equation to two or more *ordinary* differential equations, each one involving only one of the independent problem variables. This will be accomplished by separating these variables from the very beginning. It is easiest to demonstrate this method by applying it to an example. We shall choose for this purpose Laplace's equation respresenting the steady-state temperature distribution of section 7.2. For mental refreshment we present the equation and boundary conditions. The equation is

$$\frac{\partial^2 \theta}{\partial x^2} + \frac{\partial^2 \theta}{\partial y^2} = 0 \qquad (7.2)$$

and the boundary conditions are

$\theta = 0$ when $x = 0$; $\theta = 0$ when $x = a$; $\theta = 0$ when $y = 0$; $\theta = 100$ when $y = b$

(7.3)

The underlying vital assumption is that we *can* find a form for the solution $\theta(x,y)$ as the product of a function of x and a function of y. Therefore, we assume that $\theta = X(x) \cdot Y(y)$. What we shall do is to cheat a little by producing for illustration a form for θ which does satisfy equation (7.2); we shall develop the first stage of the method for this particular θ in parallel with the development for the general expression for θ.

$\theta \quad = X(x)\, Y(y)$	$\theta \quad = \cos 2x \, \sinh 2y$
$\dfrac{\partial \theta}{\partial x} = X'(x)\, Y(y)$	$\dfrac{\partial \theta}{\partial x} = -2 \sin 2x \, \sinh 2y$
$\dfrac{\partial^2 \theta}{\partial x^2} = X''(x)\, Y(y)$	$\dfrac{\partial^2 \theta}{\partial x^2} = -4 \cos 2x \, \sinh 2y$
$\dfrac{\partial^2 \theta}{\partial y^2} = X(x)\, Y''(y)$	$\dfrac{\partial^2 \theta}{\partial y^2} = \cos 2x \, . \, 4 \sinh 2y$
$\dfrac{\partial^2 \theta}{\partial x^2} + \dfrac{\partial^2 \theta}{\partial y^2} = X''Y + XY''$	$\dfrac{\partial^2 \theta}{\partial x^2} + \dfrac{\partial^2 \theta}{\partial y^2} = (-4 \cos 2x)(\sinh 2y)$ $\qquad\qquad + (\cos 2x)(4 \sinh 2y) = 0$

Hence $X''Y = -XY''$ $\qquad\qquad$ $-4 \cos 2x \, \sinh 2y = \cos 2x \, (-4 \sinh 2y)$

and $\dfrac{X''}{X} = \dfrac{-Y''}{Y}$ $\qquad\qquad$ $\dfrac{-4 \cos 2x}{\cos 2x} = \dfrac{-4 \sinh 2y}{\sinh 2y}$

(We are justified in dividing both sides of the equation by XY unless either $X \equiv 0$ or $Y \equiv 0$; if either of these cases occurs then $\theta \equiv 0$ and this, whilst being a solution of (7.2) does not satisfy the fourth boundary condition.)

You will notice that with the particular form of θ we chose, both sides of the last equation reduce to a constant. In fact, looking at the general case, we see that this must always be so. For the left-hand side (X''/X) being the ratio of two functions of x, is, at worst, also a function of x. Similarly the right-hand side (Y''/Y) must be independent of x, containing at most y. However, the right-hand side and the left-hand side *must* always be in balance and if the right-hand side does not respond to changes in x then nor does the left-hand side. Since this latter cannot respond to changes in y, it must be a constant, as must the

right-hand side. Therefore we may conclude that $\dfrac{X''}{X} = \dfrac{-Y''}{Y} = \text{constant}$.

The question that now arises is whether to take the constant to be positive or negative (why not zero?). Consider the two equations
$$Z'' = k^2 Z \qquad \text{and} \qquad Z'' = -k^2 Z$$

The general solutions are respectively
$$Z = A' \cosh kz + B' \sinh kz \text{ or } Z = Ae^{kz} + Be^{-kz}$$
and $\qquad\qquad Z = C \cos kz + D \sin kz$
where A, B, A', B', C and D are constants.

It would seem, therefore, that the choice of a positive constant, k^2 will give rise to the equations $X = Ae^{kx} + Be^{-kx}$ and $Y = C \cos ky + D \sin ky$; ($k^2$ is chosen since it is always positive for k real); the choice of a negative constant, $-k^2$, will produce
$$X = C \cos kx + D \sin kx \text{ and } Y = Ae^{ky} + Be^{-ky}$$
Which sign we choose for the constant will be governed by the need to satisfy the boundary conditions, which determine the form of the solution. In Figure 7.8 we graph the six basic functions under consideration.

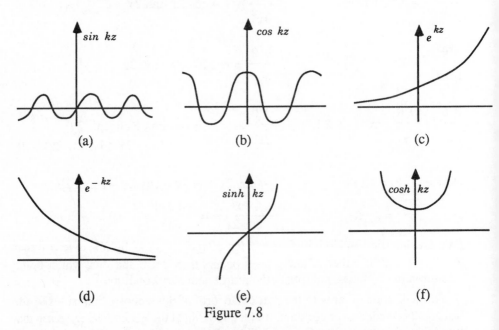

Figure 7.8

In some problems, physical considerations will help the selection; for example, in a cooling rod, the only suitable choice for time dependence is case (d). In our current example, the possible form which suggests itself is a sine-like function for

X since the first two boundary conditions require a function zero at $x = 0$ and at a later point. Let us choose a negative constant and hope this will see us through. Therefore, we take $X''/X = -Y''/Y = -k^2$ and we can write

$$\theta(x,y) \equiv X(x).\ Y(y) = (C \cos kx + D \sin kx)(Ae^{ky} + Be^{-ky}) \tag{7.15}$$

All expressions of the form (7.15) for θ will satisfy Laplace's equation (7.2); different combinations of values A, B, C and D give different particular solutions. We may need one of these particular solutions or a linear combination of them. Now apply the first boundary condition; $\theta = 0$ when $x = 0$, for $0 \le y < b$. Then $0 = (C.1 + D.0)(Ae^{ky} + Be^{-ky})$.

The second bracket cannot be zero for all y in the interval $(0, b)$ unless A and B are both zero; this would give $\theta \equiv 0$ and we therefore reject the possibility. We must have $C = 0$ so that $\theta = D \sin kx\ (Ae^{ky} + Be^{-ky})$

In practice, however, we would not distinguish between, say, $\theta = 2 \sin 3x(4e^{3y} + 5e^{-3y})$, $\theta = 5 \sin 3x\ (1.6\ e^{3y} + 2e^{-3y})$ and $\theta = \sin 3x(8e^{3y} + 10e^{-3y})$. We shall use this last form. It makes sense to absorb D into the constants A and B, so that from now on A replaces AD and B replaces BD. We then have $\theta = \sin kx\ (Ae^{ky} + Be^{-ky})$.

The third boundary condition requires that $\theta = 0$ when $y = 0$ for all x in $(0,a)$; this implies that $0 = \sin kx(A.1 + B.1)$. Since $\sin kx \equiv 0$ in $(0,a)$ would give $\theta \equiv 0$ we require that $A + B = 0$ i.e. $B = -A$ so that $\theta = A \sin kx\ (e^{ky} - e^{-ky})$ or $\theta = A \sin kx \sinh ky$.

Next we apply the second boundary condition, that $\theta = 0$ when $x = a$ for all y in $(0,b)$ i.e. $0 = A \sin ka \sinh ky$. You should be able to reason that if either $A = 0$ or $\sinh ky \equiv 0$ then $\theta \equiv 0$, therefore $\sin ka = 0$, i.e. we are restricted in the values of k that we can use to those which make $\sin ka = 0$. But $\sin z = 0$ at $z = n\pi$, n integer. Hence $k = n\pi/a$, n integer.

If n were zero then k would be zero and the corresponding value of θ would be zero. The effect of choosing k to be negative is to give us no new information; thus, $A \sin \dfrac{3\pi}{a}x \sinh \dfrac{3\pi}{a}y$ and $A \sin \left[\dfrac{-3\pi}{a}\right] x \sinh \left[\dfrac{-3\pi}{a}\right] y$

are in reality the same expression. Therefore, we restrict k to the values $n\pi/a$, where $n = 1, 2, 3, \ldots$

Each of the functions, $A_1 \sin \dfrac{\pi x}{a} \sinh \dfrac{\pi y}{a}$, $A_2 \sin \dfrac{2\pi x}{a} \sinh \dfrac{2\pi y}{a}$,

$A_3 \sin \dfrac{3\pi x}{a} \sinh \dfrac{3\pi y}{a}$ etc. satisfies equation (7.2) and the first three boundary conditions.

The most general form for θ before we fit the fourth boundary condition is as follows

$$\theta = \sum_{n=1}^{\infty} A_n \sin \frac{n\pi x}{a} \sinh \frac{n\pi y}{a} \qquad (7.16)$$

We have chosen different values for the constants $A_1, A_2, A_3,...$ to be as general as possible and used the linearity of the Laplace equation to superimpose solutions.

If we try to fit the fourth boundary condition we arrive at the result

$$100 = \sum_{n=1}^{\infty} A_n \sin \frac{n\pi x}{a} \sinh \frac{n\pi b}{a} = \sum_{n=1}^{\infty} F_n \sin \frac{n\pi x}{a}$$

where we have to determine the coefficients F_n. We have the Fourier series problem of trying to represent the function $f(x) = 100$, $0 \le x \le a$ by a series of sine terms. From the ideas in Section 6.5 we know that the coefficients are given by the formula

$$F_n = \frac{2}{a} \int_0^a 100 \sin \frac{n\pi x}{a}\, dx = \frac{200}{a} \left[-\frac{a}{n\pi} \cos \frac{n\pi x}{a} \right]_0^a = \frac{200}{n\pi} \left[1 - \cos n\pi \right]$$

Hence $F_n = 0$ for even n and $F_n = \dfrac{400}{n\pi}$ for odd n. Therefore $A_n = 0$ for even

n and $A_n = \dfrac{400}{n\pi} \cdot \dfrac{1}{\sinh (n\pi b/a)}$ for odd n. Substituting for A_n in (7.16), we

get finally $\theta = \dfrac{400}{\pi} \displaystyle\sum_{n=1,3,5,...}^{\infty} \dfrac{1}{n} \dfrac{\sin (n\pi x/a) \sinh (n\pi y/a)}{\sinh (n\pi b/a)}$ $\qquad (7.4)$

which is the result we quoted in Section 7.2.

7.5 Origin of some partial differential equations

In this section we develop the models for some physical problems and derive their solutions by separating the variables. We believe that the construction of these models is an important part of the process and we emphasise this for the first three examples.

Example 1 Torsion of uniform prism with rectangular section

We wish to calculate the amount of twist produced when a structural member sustains torsional loads; also we wish to calculate the shear stresses set up. We take the simple problem of a solid bar of uniform section twisted by a torque applied to one end, whilst the other end is fixed to prevent rotation. For simplicity we shall consider a rectangular cross-section, but we will impose this particular condition at a later stage. In Figure 7.9 we see the schematic representation of the process.

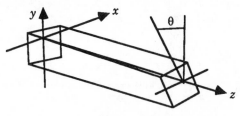

Figure 7.9

The free end rotates about the z-axis in such a way that a cross-section distant z from the fixed end rotates through an angle $\theta = \alpha z$, where α is a twist per unit length, assumed constant. Any point in a given section will be displaced in the x- and y-directions by amounts $d_x = -\alpha yz$, $d_y = \alpha xz$ respectively; these results follow from considering the displacement, d_θ in a tangential direction and resolving this in the x- and y-directions.

To complete the set up we *assume* that the displacement in the z-direction is a function of x and y only. If we write $d_z = \psi(x,y)$ then $\psi(x,y)$ is called the **warping function**. A section which was originally plane will take up a shape defined by $\psi(x,y)$. The *assumption* implicitly made is that the ends of the bar are free to warp.

We may obtain the strains by differentiating the displacement. For example,

$e_x = \dfrac{\partial}{\partial x}(d_x) = 0$; similarly, $e_y = 0 = e_z$. Since all the direct strains are zero,

it follows that throughout the bar, direct stresses are zero. The shear

displacements can be found to be $e_{xy} = \dfrac{1}{2}\left[\dfrac{\partial}{\partial x}(d_x) + \dfrac{\partial}{\partial y}(d_y)\right] = 0$,

$e_{zx} = \dfrac{\alpha}{2}(\psi_x - y)$, $e_{zy} = \dfrac{\alpha}{2}(\psi_y + x)$. The non-vanishing shear stresses

are $\sigma_{zx} = G\alpha(\psi_x - y)$ and $\sigma_{zy} = G\alpha(\psi_y + x)$, where G is the shear modulus for the bar. If we consider the three stress equilibrium equations, the

one which is not obviously satisfied is $\dfrac{\partial}{\partial z}\sigma_{zz} + \dfrac{\partial}{\partial x}\sigma_{zx} + \dfrac{\partial}{\partial y}\sigma_{zy} = 0$.

However, we can ensure that this equation always holds by introducing a **stress function** ϕ which gives rise to the shear stresses by the equations

$\sigma_{zx} = G\alpha\,\dfrac{\partial\phi}{\partial y}$, $\sigma_{zy} = -G\alpha\,\dfrac{\partial\phi}{\partial x}$. Therefore from the two expressions for the stress component σ_{zx} we obtain

$$\frac{\partial \phi}{\partial y} = \frac{\partial \psi}{\partial x} - y \tag{7.17a}$$

Similarly

$$\frac{\partial \phi}{\partial x} = -\frac{\partial \psi}{\partial y} - x \tag{7.17b}$$

Now

$$\frac{\partial^2 \phi}{\partial y^2} = \frac{\partial^2 \psi}{\partial y \partial x} - 1 \text{ and } \frac{\partial^2 \phi}{\partial x^2} = -\frac{\partial^2 \psi}{\partial x \partial y} - 1$$

On addition we obtain, for the stress function $\phi(x,y)$ the equation

$$\frac{\partial^2 \phi}{\partial x^2} + \frac{\partial^2 \phi}{\partial y^2} = -2 \tag{7.18}$$

This is a form of Poisson's Equation.

We may also obtain the following equation in $\psi(x,y)$

$$\frac{\partial^2 \psi}{\partial x^2} + \frac{\partial^2 \psi}{\partial y^2} = 0 \tag{7.2}$$

which is Laplace's equation.

We need to find boundary conditions for ϕ. Let the rectangular cross-section be $-a \leq x \leq a$, $-b \leq y \leq b$. $\sigma_{zy} = 0$ on the faces $x = \pm a$ and that the shear stress $\sigma_{zx} = 0$ on the faces $y = \pm b$ leads to the condition that ϕ be constant along the entire boundary. For convenience, we choose this constant value to be zero.

Having found a unique solution for ϕ we make use of (7.17a) and (7.17b) to obtain a solution for ψ. The method of separation of variables gives $\phi = X(x) \cdot Y(y)$ and leads to the equation $X''/X = -2 - Y''/Y =$ constant. To satisfy the conditions at $x = \pm a$ we must take the constant to be negative, i.e. $-k^2$. This gives $X = C \cos kx + D \sin kx$ and also we obtain $Y'' = (k^2 - 2) Y$. See whether you can follow the argument through to produce the result

$$\phi = b^2 - y^2 - \frac{32b^2}{\pi^3} \sum_{n = 1,3,5,\dots}^{\infty} (-1)^{(n-1)/2} \cdot \frac{1}{n^3} \text{ sech } \frac{n\pi a}{2b} \cosh \frac{n\pi x}{2b} \cos \frac{n\pi y}{2b}$$

$$(7.19)$$

Note that the terms of this series alternate in sign.

Example 2   Heat flow in one dimension

Suppose that we have a long thin bar of length l which is aligned along the x-axis. We wish to determine the temperature distribution $\theta(x,t)$ in the bar. We make the assumptions that the bar is insulated along its sides and that heat flows in the x-direction only. In Figure 7.10 we examine a section of the bar at a distance x_0 from one end.

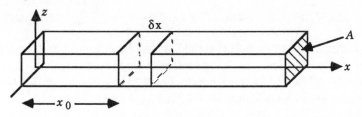

Figure 7.10

We make use of the following **empirical** laws of heat flow:

(i)   The amount of heat in a body is proportional to its mass and to its temperature.
(ii)   Heat flows from a point at a higher temperature to one at a lower temperature.
(iii) The rate of flow of heat through a plane surface is proportional to the area of the surface and to the rate of change of temperature with respect to distance in a direction perpendicular to the plane.

The rate of inflow of heat to the section is $-KA\theta_x(x_0, t)$ where K is the thermal conductivity of the material of the bar and A is the (constant) cross-section area of the bar. (Why the negative sign?) Similarly, the rate of outflow of heat is $-KA\theta_x(x_0 + \delta x, t)$. From law (i) we obtain the rate of build-up of heat in the section as $c\rho A\delta x\theta_t(x_0, t)$ where ρ is the density of the material, and c is its specific heat. Let $k = K/c\rho$ be the **diffusivity** of the material. We may equate the rate of build-up of heat to the rate of inflow to the section less the rate of outflow and, by taking the limiting case as $\delta x \to 0$, we eventually obtain the diffusion equation

$$\frac{\partial \theta}{\partial t} = k \frac{\partial^2 \theta}{\partial x^2} \tag{7.20}$$

Suitable boundary conditions might be

(i) $\theta = 100$ at $x = 0$ for all times $t > 0$
(ii) $\theta = 50$ at $x = l$ for all times $t > 0$
(iii) The initial temperature distribution is specified, e.g. $\theta = 100$ at $t = 0$ for all points in the bar.

(In this case we may imagine that one end of the bar is suddenly cooled to 50°C.)

Now the problem is exactly parallel to one where $\theta = 50$ at $x = 0$; $\theta = 0$ at $x = l$ and $\theta = 50$ at $t = 0$. This equation may be slightly easier to solve with these boundary conditions and we may then add 50 to recover the solution of the original problem.

Let $\theta = X(x)\,T(t)$.   Then $\partial\theta/\partial t = XT'$ and $\partial^2\theta/\partial x^2 = X''T$.

Substituting into (7.20) we obtain $XT' = kX''T$ or $\dfrac{X''}{X} = \dfrac{1}{k}\dfrac{T'}{T} = \text{constant}$.

Since the temperature of a point on the bar may be expected to decrease with the passage of time we must choose a negative value for the constant, $-\lambda^2$ say, so that we obtain $T' = -\lambda^2 kT$ and hence $T = e^{-\lambda^2 kt}$. Then it follows that $X = C\cos\lambda x + D\sin\lambda x$ and that $\theta = (C\cos\lambda x + D\sin\lambda x)\,e^{-\lambda^2 kt}$.

Applying the condition that $\theta = 50$ at $x = 0$ we find that $50 = (C.1 + D.0)e^{-\lambda^2 kt}$. Immediately, we have run into trouble. We overcome this difficulty by thinking about the physical problem. If we maintain the end $x = 0$ at 50 and the end $x = l$ at 0 then there will be a **steady state** reached eventually (in theory as $t \to \infty$). Superimposed on this is a **transient** solution which decays to zero.

We may then write $\theta(x,t) = \theta_s(x) + \theta_T(x,t)$, where $\theta_s(x)$ is the steady-state part of the solution, which is independent of time, and $\theta_T(x,t)$ the transient part.

Since $\theta_s(x)$ is independent of time $\dfrac{\partial\theta_s}{\partial t} = 0$ and from (7.20) it must satisfy

$k\,\dfrac{d^2\theta_s}{dx^2} = 0$, i.e. $\theta_s(x) = \alpha x + \beta$. Now $\theta_s(0) = 50$ and $\theta_s(l) = 0$, therefore,

$\theta_s(x) = 50 - 50x/l$.

Consequently, $\theta_T(0,t) = \theta(0,t) - \theta_s(0) = 0$, $\theta_T(l,t) = \theta(l,t) - \theta_s(l) = 0$ and $\theta_T(x,0) = \theta(x,0) - \theta_s(x) = 50 - 50 + 50x/l = 50x/l$.

At this stage we note that $\dfrac{\partial\theta}{\partial t} = \dfrac{\partial\theta_s}{\partial t} + \dfrac{\partial\theta_T}{\partial t} = \dfrac{\partial\theta_T}{\partial t}$ and that $\partial^2\theta/\partial x^2 = \partial^2\theta_s/\partial x^2 + \partial^2\theta_T/\partial x^2$. Hence both θ_s and θ_T satisfy (7.20) separately.

The most general solution possible which satisfies the first two boundary conditions is

$$\theta_T = \sum_{n=1}^{\infty} D_n \sin\frac{n\pi x}{l}\, e^{-n^2\pi^2 kt/l^2} \qquad \text{(We ignore } n \le 0.\ \text{Why?)}$$

The last boundary condition is $\theta_T = 50x/l$ at $t = 0$; therefore

$$\frac{50x}{l} = \sum_{n=1}^{\infty} D_n \sin\frac{n\pi x}{l}$$

Using the ideas applied to the steady-state temperature distribution problem we produce the formula

$$D_n = \frac{2}{l} \int_0^l \frac{50x}{l} \sin \frac{n\pi x}{l} \, dx = \frac{-100}{n\pi} (-1)^n$$

We now add on $\theta_s(x)$ and the 50°C we removed earlier on.

Finally we have the solution for $\theta(x, t)$ given by

$$\theta = 100 - \frac{50x}{l} - \frac{100}{\pi} \sum_{n=1}^{\infty} \frac{(-1)^n}{n} \sin \frac{n\pi x}{l} . e^{-n^2 \pi^2 kt/l^2} \qquad (7.21)$$

It is worth noting the presence of $-n^2$ in the index of the exponential; this should indicate that the terms will decay to zero reasonably rapidly and hence only a few terms may be needed to obtain a good approximation to the value of $\theta(x, t)$ being calculated. We programmed a computer to calculate $\theta(x, t)$ for suitable values of x and t and in Figure 7.11 we show the temperature profile $\theta(x)$ for selected values of t; k and l were taken to be 1. (What do you think will be the effects of changing k and/or l?)

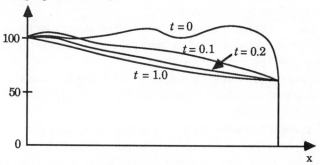

Figure 7.11

Example 3   Vibration of a string

We consider the small transverse vibrations of a stretched string. We know that this is an artificial example but we decided to discuss it not because of its historical importance but because it is easy to visualise the solution in terms of the physical problem.

In Figure 7.12(a) we see the string, initially of length l, fixed at both ends, subject to a constant tension T and in a state of vibration.

We wish to find the displacement function $y(x, t)$.   The following assumptions are made.

(i)   The string is perfectly flexible and therefore cannot resist bending moments.

(ii)   The displacements y are small compared to l.

(iii) At any point of time the slope of the deflected profile, $\dfrac{dy}{dx}$, is small.

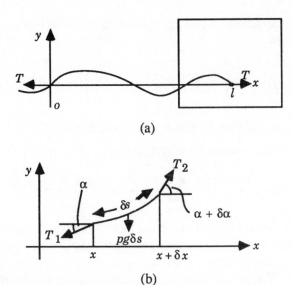

(a)

(b)

Figure 7.12

(iv)   The tension T remains constant **at all** times and for all points on the string.
(v)   The mass of the string can be neglected.
(vi)   The horizontal displacement of a point on the string can be neglected in comparison to the vertical displacement y, i.e. the motion is **transverse**.
(vii) The motion takes place in the x-y plane.

Note that these assumptions are consistent. Inconsistency would arise, if, for example, we required horizontal displacements to be negligible *and* did not require assumption (ii) to hold.

In Figure 7.12(b) we analyse the forces acting on a small part of the string. We let the section be of length δs and it can be considered to be a straight line; ρ is the density of the string per unit length. Since we assume no displacement in the x-direction, the resolute of the forces in the x-direction must be zero. Hence

$$T_1 \cos \alpha = T_2 \cos (\alpha + \delta\alpha) = T, \text{ a constant} \qquad (7.22)$$

Using Newton's second law of motion in the y-direction, we have

$$\rho\delta s \frac{\partial^2 y}{\partial t^2} = T_2 \sin(\alpha + \delta\alpha) - T_1 \sin \alpha - \rho g \delta s \qquad (7.23)$$

Since we are dealing only with small vibrations, $dy/dx = \tan \alpha$ will be small and

$$\sin \alpha = \frac{dy}{dx} \left/ \left\{ 1 + \left[\frac{dy}{dx}\right]^2 \right\}^{\frac{1}{2}} \right. \cong \frac{dy}{dx} = \tan \alpha \text{ and } \delta s \cong \delta x.$$

Hence approximately we have

$$\rho \delta x \cdot \frac{\partial^2 y}{\partial t^2} = T \tan(\alpha + \delta\alpha) - T \tan \alpha - \rho g \delta x$$

$$= T \left[\frac{\partial y}{\partial x} \bigg|_{x+\delta x} - \frac{\partial y}{\partial x} \bigg|_x \right] - \rho g \delta x$$

Dividing by $\rho \delta x$ and neglecting the gravity term we obtain

$$\frac{\partial^2 y}{\partial t^2} = \frac{T}{\rho} \left[\frac{\dfrac{\partial y}{\partial x} \bigg|_{x+\delta x} - \dfrac{\partial y}{\partial x} \bigg|_x}{\delta x} \right]$$

Now we take limits as $\delta x \to 0$ and arrive at the equation

$$\frac{\partial^2 y}{\partial t^2} = \frac{T}{\rho} \frac{\partial^2 y}{\partial x^2}$$

Since T/ρ has the dimensions of velocity squared we write it as c^2: it can be shown that this is the speed at which the disturbance or wave travels. We finally produce the one-dimensional wave equation

$$\frac{\partial^2 y}{\partial t^2} = c^2 \frac{\partial^2 y}{\partial x^2} \tag{7.24}$$

The boundary conditions for one version of the vibrating string problem are:
(i) $y(0,t) = 0$ for $t \geq 0$ (ii) $y(l,t) = 0$ for $t \geq 0$
(iii) $y(x,0) = f(x)$ for $0 \leq x \leq l$, i.e. the initial profile of the string

(iv) $\dfrac{\partial y}{\partial t} = 0$ at $t = 0$ for $0 \leq x \leq l$, i.e. the string is released from rest.

We seek a solution in the form $y = X(x) . T(t)$.

Separating the variables leads to the result $T''/T = c^2 X''/X = $ constant. Since we are dealing with a vibration, observation suggests that we expect a periodic behaviour in the time variable, i.e. we expect $T = C \cos pt + D \sin pt$.

This implies that we might choose the separation constant to be $-k^2 c^2$ and therefore we obtain the equation $X'' = -k^2 X$ so that $X = A \cos kx + B \sin kx$ and $y(x,t) = (A \cos kx + B \sin kx)(C \cos kct + D \sin kct)$.

The solution satisfying (i) and (ii) is of the form

$$y_n(x, t) = \sin \frac{n\pi x}{l} \left[C_n \cos \frac{n\pi ct}{l} + D_n \sin \frac{n\pi ct}{l} \right] \tag{7.25}$$

where we have attached suffices to the constants C and D for ease of calculation. We call such functions as (7.25) **eigenfunctions** and the values

$\lambda_n = \dfrac{n\pi c}{l}$ the **eigenvalues** of the string. Each solution $y_n(x, t)$ represents a

periodic motion with a frequency $(nc/2l)$ called the nth **normal mode**. The case $n = 1$ is the **fundamental** mode of vibration. In Figure 7.13 we graph the first three modes of vibration. The **nodes** for each mode are those points which do not move during the vibration. Together with $x = 0$ and $x = l$ they are equally spaced at $x = l/n$, $2l/n$, ..., $(n - 1)l/n$. In Figure 7.14 we show the third normal mode at different times.

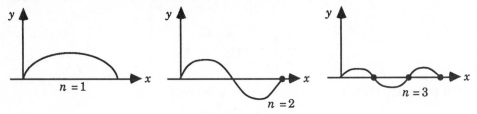

Figure 7.13

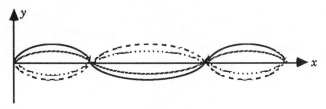

Figure 7.14

Returning to (7.25) we need to produce the most general solution which satisfies the first two boundary conditions. We do this by summing the basic solutions to obtain

$$y(x,t) = \sum_{n=1}^{\infty} \sin \frac{n\pi x}{l} \left[C_n \cos \frac{n\pi ct}{l} + D_n \sin \frac{n\pi ct}{l} \right]$$

Now $\partial y/\partial t = 0$ when $t = 0$; therefore we first find $\partial y/\partial t$;

$$\frac{\partial y}{\partial t} = \sum_{n=1}^{\infty} \sin \frac{n\pi x}{l} \left[-\frac{n\pi c}{l} C_n \sin \frac{n\pi ct}{l} + \frac{n\pi c}{l} D_n \cos \frac{n\pi ct}{l} \right]$$

Applying this boundary condition we have

$$0 = \sum_{n=1}^{\infty} \sin \frac{n\pi x}{l} \frac{n\pi c}{l} \cdot D_n$$

We need to take $D_n = 0$, for all n, and we are now able to state that

$$y(x,t) = \sum_{n=1}^{\infty} C_n \sin\frac{n\pi x}{l} \cos\frac{n\pi ct}{l} \tag{7.26}$$

The last boundary condition prescribes a form $f(x)$ for $y(x,t)$ at $t = 0$. Then (7.26) becomes

$$y(x,0) = f(x) = \sum_{n=1}^{\infty} C_n \sin\frac{n\pi x}{l}$$

This is a Fourier series problem and we manufacture the function shown in Figure 7.15.

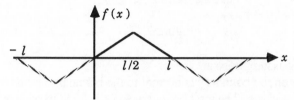

Figure 7.15

Hence $C_{2n} = 0$ for even n,

$$C_{4n+1} = \frac{8}{(4n+1)^2\,\pi^2}, \quad C_{4n+3} = \frac{-8}{(4n+3)^2\,\pi^2}.$$

We finally obtain the solution

$$y(x,t) = \frac{8}{\pi^2}\left[\frac{1}{1^2}\sin\frac{\pi x}{l}\cos\frac{\pi ct}{l} - \frac{1}{3^2}\sin\frac{3\pi x}{l}\cos\frac{3\pi ct}{l} + \frac{1}{5^2}\sin\frac{5\pi x}{l}\cos\frac{5\pi ct}{l} - \cdots\right]$$

$$\tag{7.27}$$

We programmed a digital computer to calculate values of y for suitable values of x at a number of times t. We used the results to produce the graphs of Figure 7.16, which show the progressive profiles $y(x)$ for selected times in a half-cycle of oscillations.

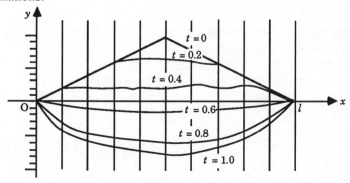

Figure 7.16

Example 4 Vibrating membrane

In Figure 7.17 we depict a rectangular membrane that is tightly stretched and held firmly by a rigid rectangular frame.

When the membrane is disturbed, we seek the deflection $u(x, y, t)$ normal to the x-y plane.

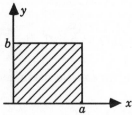

Figure 7.17

The following assumptions are made.

(i)   The force on the boundary is normal to the boundary and its value per unit length is constant. The total boundary force is large in comparison with the weight of the membrane.

(ii)   The membrane is thin and cannot resist bending moments.

(iii)   The deflections are small compared to the dimensions a and b.

(iv)   The displacements in the x- and y- directions are small compared with the deflection u.

(v)   The slopes on any particular deflected profile are small.

The governing partial differential equation is

$$\frac{\partial^2 u}{\partial t^2} = c^2 \left[\frac{\partial^2 u}{\partial x^2} + \frac{\partial^2 u}{\partial y^2} \right] \tag{7.28}$$

where c is a parameter akin to that for a vibrating string. Suitable boundary conditions are

(i) $u(0, y, t) = 0;$ $0 \le y \le b,$ $t > 0$

(ii) $u(a, y, t) = 0;$ $0 \le y \le b,$ $t > 0$

(iii) $u(x, 0, t) = 0;$ $0 \le x \le a,$ $t > 0$

(iv) $u(x, b, t) = 0;$ $0 \le x \le a,$ $t > 0$

If we displace the membrane initially to a profile $u = f(x,y)$ and release it from rest, then we have two initial conditions

(v) $u(x, y, 0) = f(x, y);$ $0 \le x \le a,$ $0 \le y \le b$

(vi) $u_t(x, y, 0) = 0;$ $0 \le x \le a,$ $0 \le y \le b$

The solution may be written

$$u = \frac{64a^2b^2}{\pi^6} \sum_{n=1,3,5,..}^{\infty} \sum_{m=1,3,5,..}^{\infty} \frac{1}{n^3 m^3} \sin \frac{n\pi x}{a} \sin \frac{m\pi y}{b} \times \cos c \left[\frac{n^2 \pi^2}{a^2} + \frac{m^2 \pi^2}{b^2} \right]^{\frac{1}{2}} t$$

$$\tag{7.31}$$

Example 5 Flow of electricity in a cable

We consider finally an example from electrical engineering. Suppose we have a transmission line in which we assume that the resistance, inductance and capacitance vary linearly with x, the distance from the source of electricity, measured along the line. We shall need the following quantities:

$e(x, t)$ = the potential at a point on the cable; $i(x, t)$ = the current there;
R = the resistance of the cable per unit length; L = inductance per unit length;
G = conductance to ground per unit length; C = capacitance to ground per unit length.

In Figure 7.18(a) we show schematically the system and in Figure 7.18(b) we show the equivalent circuit for the portion PQ. We allow the cable to leak to earth.

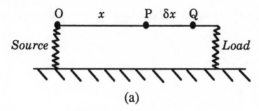

(a)

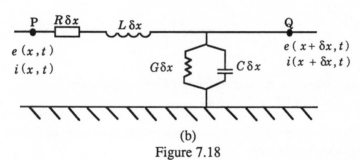

(b)

Figure 7.18

Applying Kirchhoff's second law to PQ, the drop in potential along the section PQ is given by

$$e(x, t) - e(x + \delta x, t) = iR\delta x + \frac{\partial i}{\partial t} L \, \delta x$$

Dividing by δx and letting $\delta x \to 0$ we obtain

$$\frac{\partial e}{\partial x} = -Ri - L\frac{\partial i}{\partial t} \tag{7.32}$$

Applying Kirchhoff's first law to the equivalent circuit

$$i(x, t) = i(x + \delta x, t) + C\delta x \frac{\partial e}{\partial t} + G\delta x e(x, t)$$

Dividing by δx and letting $\delta x \to 0$ we obtain the equation

$$\frac{\partial i}{\partial x} = -Ge - C\frac{\partial e}{\partial t} \qquad (7.33)$$

We differentiate (7.32) partially with respect to x to obtain

$$\frac{\partial^2 e}{\partial x^2} = -R\frac{\partial i}{\partial x} - L\frac{\partial^2 i}{\partial x \partial t} \qquad (7.34)$$

We differentiate (7.33) partially with respect to t to obtain

$$\frac{\partial^2 i}{\partial x \partial t} = -G\frac{\partial e}{\partial t} - C\frac{\partial^2 e}{\partial t^2} \qquad (7.35)$$

We now substitute (7.33) and (7.35) into (7.34) and produce the following equation for e

$$\frac{\partial^2 e}{\partial x^2} = LC\frac{\partial^2 e}{\partial t^2} + (RC + LG)\frac{\partial e}{\partial t} + RGe \qquad (7.36)$$

Similarly,

$$\frac{\partial^2 i}{\partial x^2} = LC\frac{\partial^2 i}{\partial t^2} + (RC + LG)\frac{\partial i}{\partial t} + RGi \qquad (7.37)$$

These last two equations are called the **telephone equations**.

We may consider three special cases.

(i)   If $G = L = 0$, i.e. we may neglect leakage and inductance, which is a reasonable assumption for a submarine cable, then we have the simplified **telegraph equations**

$$\frac{\partial^2 e}{\partial x^2} = RC\frac{\partial e}{\partial t} \quad \text{and} \quad \frac{\partial^2 i}{\partial x^2} = RC\frac{\partial i}{\partial t} \qquad (7.38)$$

(ii)   If $R = G = 0$ as is reasonable to assume for high frequencies, we obtain

the equations $\dfrac{\partial^2 e}{\partial x^2} = LC\dfrac{\partial^2 e}{\partial t^2}$ and $\dfrac{\partial^2 i}{\partial x^2} = LC\dfrac{\partial^2 i}{\partial t^2}$ which are forms of the

one-dimensional wave equation with wave speed $1/\sqrt{LC}$.

(iii) If $L = C = 0$ which is the case for very low frequencies and steady-state

conditions, we obtain the equations $\dfrac{\partial^2 e}{\partial x^2} = RGe$ and $\dfrac{\partial^2 i}{\partial x^2} = RGi$.

Transformation of equations to non-dimensional form

Consider the equation $\dfrac{\partial \theta}{\partial t} = k \dfrac{\partial^2 \theta}{\partial x^2}$ where k is constant. Suppose we have a reference value θ_0 at time $t = 0$; this may be a typical value, a maximum or minimum value or a constant initial temperature. If l is a suitable reference length in the problem then the transformations $X = x/l$, $\Theta = \theta/\theta_0$ will reduce the equation to the form $\partial\Theta/\partial t = \partial^2\Theta/\partial X^2$ where Θ and X are non-dimensional variables.

The need for a numerical approach

Sometimes the problem at hand cannot be solved by the separation of variables method, especially if the boundary is an awkward one geometrically. Even when such a solution is possible, a glance at a formula such as (7.4) makes one doubt whether there is much to commend it. The formula is an extremely tangled one to use to calculate values of θ and we would be well advised to attempt a numerical solution. In the next few sections we examine some numerical methods of solution of the well-known equations. We should remark perhaps that the equations we have solved by separation of variables are of importance in models of engineering systems.

7.6 Parabolic equations: Finite difference methods

The one-dimensional heat conduction equation is taken in the form $\dfrac{\partial \theta}{\partial t} = \dfrac{\partial^2 \theta}{\partial x^2}$,

with the boundary conditions: $\theta = 0$ when $x = 0$ and when $x = 1$ for $t \geq 0$ and $\theta = 100$ at $t = 0$ for $0 < x < 1$.

A simple explicit method

If finite differences are used as approximations to derivatives we can use central differences in space (they are the most accurate) but since we are propagating forwards in time we use a forward time difference. We therefore take as an approximation to the differential equation the difference equation

$$\frac{\theta_{i,j+1} - \theta_{i,j}}{k} = \frac{\theta_{i+1,j} - 2\theta_{i,j} + \theta_{i-1,j}}{h^2} \tag{7.39}$$

Reference to Figure 7.19(a) shows the relative positions of the four mesh values of θ which are involved. The fact that there is only one value at the further forward **time level** $(j + 1)$ indicates that the scheme (7.39) is an **explicit** one. Note that the scheme is valid for mesh points in the interior of the domain of the solution. The values on the boundary are specified.

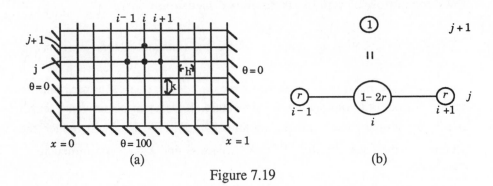

Figure 7.19

We may rearrange the scheme to give explicitly a formula for $\theta_{i, j+1}$. It is

$$\theta_{i, j+1} = r\theta_{i-1, j} + (1 - 2r)\theta_{i, j} + r\theta_{i+1, j} \qquad (7.40)$$

where
$$r = k/h^2 \qquad (7.41)$$

Figure 7.19(b) shows the **computational molecule** which summarises the scheme. It depicts the relative positions of the component mesh values and their weightings in the difference scheme.

Note that we have chosen the simplest forward difference approximation to $\partial\theta/\partial t$ and the simplest central difference approximation to $\partial^2\theta/\partial x^2$. The truncation errors in these approximations are $O(k)$ and $O(h^2)$ respectively.

Before embarking on the solution we need to specify the parameter r. From (7.41) we see that the same value of r can arise from many combinations of k and h. Suppose we decide to divide the interval $0 \le x \le 1$ into 10 equal parts making $h = 0.1$. We should remember that the problem, and hence the solution, is symmetric about $x = 0.5$. We shall therefore only need to compute θ for $x = 0.1(0.1)0.5$. What value do we choose for k? Bearing in mind the error magnitudes for the finite difference approximations, it would seem that the smaller we choose k the better. However, the smaller we choose k, the more steps are needed to reach a given time. We need to strike a compromise. Therefore, for simplicity, let us choose $r = 1$, i.e. $k = 0.01$. Then the formula for $\theta_{i, j+1}$ becomes

$$\theta_{i, j+1} = \theta_{i-1, j} - \theta_{i, j} + \theta_{i+1, j} \qquad (7.42)$$

We now obtain the solution for the first few time steps and show the results of selected time steps in Table 7.1.

For comparison, we show in Table 7.2 the corresponding results from the analytical solution

$$\theta = \frac{400}{\pi} \sum_{n = 1,3,5..}^{\infty} \frac{1}{n} e^{-n^2\pi^2 t} \sin n\pi x \qquad (7.43)$$

which you can derive by separation of variables.

To demonstrate briefly the method we work through a few calculations. We know the solution at $j = 0$, $i = 0, 1, \ldots, 10$. We also know that $\theta_{0, 1} = 0$ from the boundary condition. Then $\theta_{1,1} = \theta_{0,0} - \theta_{1,0} + \theta_{2,0} = 0 - 100 + 100 = 0$, $\theta_{2,1} = \theta_{1,0} - \theta_{2,0} + \theta_{3,0} = 100 - 100 + 100 = 100$. Similarly, $\theta_{3,1} = 100$ etc. Moving to the next time level, $\theta_{1,2} = \theta_{0,1} - \theta_{1,1} + \theta_{2,1} = 0 - 0 + 100 = 100$, $\theta_{2,2} = \theta_{1,1} - \theta_{2,1} + \theta_{3,1} = 0 - 100 + 100$.

Table 7.1 Finite difference solution, $r = 1$

x t	0 $(i=0)$	0.1 $(i=1)$	0.2 $(i=2)$	0.3 $(i=3)$	0.4 $(i=4)$	0.5 $(i=5)$
0	0	100	100	100	100	100
0.01	0	0	100	100	100	100
0.02	0	100	0	100	100	100
0.03	0	−100	200	0	100	100
0.04	0	300	−300	300	0	100
0.05	0	−600	900	−600	400	−100

Table 7.2 Analytical solution (1 d.p.)

x t	0	0.1	0.2	0.3	0.4	0.5
0	0	100	100	100	100	100
0.001	0	97.5	100	100	100	100
0.002	0	88.6	99.8	100	100	100
0.005	0	68.3	97.1	99.7	100	100
0.01	0	52.0	84.4	98.7	99.7	99.9
0.015	0	43.6	75.2	92.3	99.6	99.9
0.02	0	38.3	68.3	86.8	99.4	99.9
0.025	0	34.5	62.9	81.9	94.6	99.5
0.03	0	31.7	58.5	77.5	90.1	94.7
0.04	0	27.3	51.6	69.8	81.6	85.8
0.05	0	24.4	46.2	63.0	73.9	77.7
0.10	0	14.7	27.9	38.4	45.1	47.5

In Figure 7.20 we display temperature profiles for $t = 0.01, 0.05$, and 0.1.

Comparison with Table 7.2 is hardly necessary; the results of Table 7.1 are rubbish. They may satisfy (7.42) and the boundary conditions, but they cannot be said to satisfy the differential equation. It would seem as though the time step is too coarse, too insensitive, to adjust the changes occurring in the temperatures.

The first reaction might well be to reduce the time step. Let us take $k = 0.005$ i.e. $r = 0.5$. Then (7.40) reduces to

$$\theta_{i,j+1} = \tfrac{1}{2}\theta_{i-1,j} + \tfrac{1}{2}\theta_{i+1,j} \qquad (7.44)$$

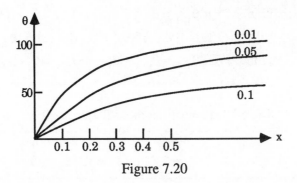

Figure 7.20

In Table 7.3 we show the results of the first few time steps. This is a better state of affairs, but there are still some blemishes, particularly in the early stages. Notice that at $t = 0.05$ the values for $x = 0.1$, 0.3 and 0.5 are relatively closer to analytical values than those for $x = 0.2$ and 0.4. If we had taken the results for $t = 0.055$, we should have found the reverse was the case. To see whether the situation can be improved markedly we decrease k to 0.001 and hence r to 0.1. Table 7.4, shows the finite difference solution, where we have worked to 3 s.f.

Table 7.3 Finite difference solution $r = 1/2$

x t	0	0.1	0.2	0.3	0.4	0.5
0	0	100	100	100	100	100
0.005	0	50	100	100	100	100
0.01	0	50	75	100	100	100
0.015	0	37.5	75	87.5	100	100
0.02	0	37.5	62.5	87.5	93.8	100
0.025	0	31.3	62.5	78.1	93.8	93.8
0.05	0	24.4	44.0	63.5	70.3	78.1

Table 7.4 Finite difference solution ($r = 0.1$)

x t	0	0.1	0.2	0.3	0.4	0.5
0	0	100	100	100	100	100
0.001	0	90	100	100	100	100
0.002	0	82	99	100	100	100
0.005	0	65.7	93.4	99.3	100	100
0.01	0	51.1	82.8	95.7	99.2	99.8
0.02	0	38.0	67.6	85.9	94.5	96.9
0.05	0	24.3	45.9	62.6	73.1	76.7
0.10	0	14.6	27.8	38.2	44.9	47.2

With $r = 0.1$ the formula (7.40) reduces to

$$\theta_{i,j+1} = \frac{1}{10} \left(\theta_{i-1,j} + 8\theta_{i,j} + \theta_{i+1,j} \right) \tag{7.45}$$

We see that this time the finite difference solution is good. However, we have the drawback that we need ten times as many steps to reach, say, $x = 0.10$. What we do not yet know is which values of r will give reasonable or good accuracy or which values of r cause problems such as those in Table 7.1. Is it in fact, simply the value of r which determines the nature of the results? We notice that the initial temperature distribution had discontinuities at $x = 0$ and $x = 1$. For small values of time, the finite difference solution is poor near $x = 0$. As the value of t increases so the accuracy improves. This suggests that a method which collected information from other than the first row and the first column near $x = 0$, $t = 0$ would be better; such an approach is to use an implicit method such as the Crank-Nicolson method discussed later. However it must be pointed out that the discontinuity will always cause trouble, no matter which finite difference method is used.

Convergence, Stability and Compatibility

In order that a finite difference scheme should give solutions which are reasonably accurate approximations to the true solution of the appropriate partial differential equation we require three conditions to be met: convergence, stability and compatibility. We briefly examine the ideas behind each of these in turn. Unless stated otherwise, these remarks apply to partial differential equations quite generally.

Convergence

When we solve the finite difference equations, we should remember that even if we were able to find an exact solution, we will not necessarily have solved the partial differential equation exactly.

Let $u(x, t)$ represent the **exact** solution to a partial differential equation and $u^*(x, t)$ the **exact** solution to the corresponding finite difference equation. We define the **convergence** of the finite-difference solution as occurring when $u^*(x_0, t_0) \rightarrow u(x_0, t_0)$ as $\delta x = h$ and $\delta t = k$ both tend to zero, for all x_0 and t_0 in the solution domain.

Some authors call the difference $(u - u^*)$ the **discretization error**, some refer to it as the **truncation error**. At any mesh point, the size of the discretization error depends partly on the grid spacing and partly on the accuracy of the finite difference approximations to the derivatives. If we attempt to reduce the discretization error by reducing h and k we shall increase the number of equations to be solved, and we then encounter problems of computer storage space and computing time required. It is, in general, very difficult to obtain criteria for convergence.

However, for the parabolic equation we are considering, it can be shown that the finite-difference solution converges to the true solution of the p.d.e. as

$h \rightarrow 0$ *provided* $r \leq \frac{1}{2}$.

Stability

When we solve the finite difference equations, we cannot generally work with exact arithmetic. Therefore, we will introduce **round-off error** at each stage. As we proceed along a given time-level, and then from one time-level to the next, we shall accumulate this error. We define the finite difference equations to be **stable** when the accumulating effect of round-off error can be neglected, i.e. when this error tends to damp out.

We remark that the trouble with the case $r = 1$ was that the solution was not convergent; stability did not enter into the story, since we worked with exact arithmetic.

There are two well-known methods for determining a stability criterion. The Fourier series method expresses an initial row of errors as a finite Fourier series, with as many terms in the series as there are mesh points in each row. The series is usually formulated in exponential form and the effect is examined for a single term propagating from one time level to the next. The overall effect is determined by superimposition. However, the Fourier series method neglects boundary conditions.

A second method involves the use of matrices and eigenvalues; it is more difficult to apply, but it has the advantage of taking into account the boundary conditions. The discussion of both these methods is beyond the scope of this book; however, it can be shown that either method produces the criterion for the heat conduction equation

$$r \le \tfrac{1}{2}. \tag{7.46}$$

In this instance, the stability criterion is the same as the convergence criterion.

Compatibility

In an attempt to produce a finite-difference scheme that is stable and effective, it is possible to end up with one which has a solution that, under certain combinations of k and h, will not converge to the solution of the original differential equation as $h \to 0$, but rather converges to the solution of a different differential equation. The scheme is then said to be **incompatible**. The explicit scheme (7.39) is compatible with the equation $\partial\theta/\partial t = \partial^2\theta/\partial x^2$.

Crank-Nicolson implicit scheme

We would like to replace the explicit scheme by a method which is convergent and stable for all finite values of r and yet which does not involve an inordinate amount of calculation. The **Crank-Nicolson method** replaces $\partial^2\theta/\partial x^2$ by the average of its finite-difference approximation on the current time level and the previous time level. Then the finite-difference scheme becomes

$$\frac{\theta_{i,j+1} - \theta_{i,j}}{k} = \frac{1}{2}\left\{ \frac{\theta_{i-1,j+1} - 2\theta_{i,j+1} + \theta_{i+1,j+1}}{h^2} + \frac{\theta_{i-1,j} - 2\theta_{i,j} + \theta_{i+1,j}}{h^2} \right\}$$

This can be arranged to give

$$-r\theta_{i-1,j+1} + (2 + 2r)\theta_{i,j+1} - r\theta_{i+1,j+1} = r\theta_{i-1,j} + (2 - 2r)\theta_{i,j} + r\theta_{i+1,j} \tag{7.47}$$

where $r = k/h^2$, as before.

If we have mesh points x_0 to x_n for a given time level, then there will be $(n-1)$ simultaneous equations for the $(n-1)$ unknowns.

In Figure 7.21(a) we show the computational molecule. In Figure 7.21(b) we represent the mesh near the point of interest.

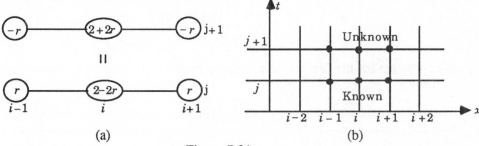

(a)

(b)

Figure 7.21

Example

We shall now apply the Crank-Nicolson method to the problem we have been studying, namely $\partial\theta/\partial t = \partial^2\theta/\partial x^2$, where $\theta = 0$ at $x = 0$ and $x = 1$ for all $t \geq 0$ and where $\theta = 100$ at $t = 0$ for $0 < x < 1$.

Let us choose $h = 0.1$ as before. If we choose $r = 1$, it has the advantage that we shall have simple coefficients to handle and, in particular, the middle term on the right-hand side of (7.47) disappears. We have

$$-\theta_{i-1, j+1} + 4\theta_{i, j+1} - \theta_{i+1, j+1} = \theta_{i-1, j} + \theta_{i+1, j} \qquad (7.48)$$

In our problem, we have symmetry about $x = 0.5$; hence $\theta_{4, j} = \theta_{6, j}$ etc. For simplicity we shall write $\theta_{i, 1}$ as θ_i.

In matrix form, we may write the five equations for the five unknowns θ_1, θ_2, θ_3, θ_4 and θ_5 at $j = 1$ as follows.

$$\begin{bmatrix} 4 & -1 & 0 & 0 & 0 \\ -1 & 4 & -1 & 0 & 0 \\ 0 & -1 & 4 & -1 & 0 \\ 0 & 0 & -1 & 4 & -1 \\ 0 & 0 & 0 & -2 & 4 \end{bmatrix} \begin{bmatrix} \theta_1 \\ \theta_2 \\ \theta_3 \\ \theta_4 \\ \theta_5 \end{bmatrix} = \begin{bmatrix} 0 + 100 \\ 100 + 100 \\ 100 + 100 \\ 100 + 100 \\ 100 + 100 \end{bmatrix} \qquad (7.49)$$

The basic pattern of coefficients is $-1\ 4\ -1$, but there are two exceptions. For $j = 1$, $\theta_{i-1, j} = \theta_{0, j} = 0$ by the boundary conditions. For $j = 5$, $\theta_{i+1, j} = \theta_{6, j} = \theta_{4, j} = \theta_{i-1, j}$ by symmetry and hence we get the pattern $-2\ 4$.

We saw in Section 1.5 how to solve such a tridiagonal matrix set of equations. The solution is, using the label $\theta_{i, 1}$ for θ_i, $\theta_{1, 1} = 46.4$, $\theta_{2, 1} = 85.6$, $\theta_{3, 1} = 96.1$, $\theta_{4, 1} = 98.9$, $\theta_{5, 1} = 99.4$. Then we may compute the next time step. Putting now $\theta_{i, 2} = \theta_i$ we have the equations

$$
\begin{bmatrix}
4 & -1 & 0 & 0 & 0 \\
-1 & 4 & -1 & 0 & 0 \\
0 & -1 & 4 & -1 & 0 \\
0 & 0 & -1 & 4 & -1 \\
0 & 0 & 0 & -2 & 4
\end{bmatrix}
\begin{bmatrix}
\theta_1 \\
\theta_2 \\
\theta_3 \\
\theta_4 \\
\theta_5
\end{bmatrix}
=
\begin{bmatrix}
0 + 85.6 \\
46.4 + 96.1 \\
85.6 + 98.9 \\
96.1 + 99.4 \\
98.9 + 98.9
\end{bmatrix}
\tag{7.50}
$$

The solution for the first few time steps is shown in Table 7.5. Compare with Table 7.2.

Table 7.5

x	0	0.1	0.2	0.3	0.4	0.5
t						
0	0	100	100	100	100	100
0.01	0	46.4	85.6	96.1	98.9	99.4
0.02	0	38.1	66.8	86.5	94.7	96.8
0.03	0	31.2	58.2	76.9	87.9	91.3
0.04	0	27.4	51.2	69.4	80.4	84.2
0.05	0	24.3	46.0	62.8	73.3	76.9
0.10	0	14.7	27.9	38.4	45.1	47.4

In Table 7.6 we compare the Crank-Nicolson estimated values with the analytical solution for $x = 0.5$ You can see that the Crank-Nicolson solution is more accurate than the explicit method. We emphasise that the Crank-Nicolson method is stable for all values of r.

Table 7.6

t	Analytical solution (3 d.p.)	Crank-Nicolson solution (7.48)	Percentage error	Explicit method $r = 0.1$	Percentage error
0.01	98.7	96.1	-2.6	95.7	-3.0
0.02	86.8	86.5	-0.03	85.9	-1.0
0.05	63.0	62.8	-0.3	62.6	-0.6
0.10	38.4	38.4	0	38.2	-0.5

Relative computing effort

We briefly compare the amounts of computing effort required by the explicit and implicit methods. As is customary, we consider only multiplications and divisions. Let us suppose that there are $(N - 1)$ internal mesh points for each time level and further suppose that there is no symmetry to employ. The explicit method requires $2(N-1)$ multiplications. The Crank-Nicolson method, solved via a Gauss elimination adapted to a tridiagonal matrix system requires $3(N-1)$ multiplications, a relatively minor increase, bearing in mind the lack of restriction on r. If we solve the Crank-Nicolson scheme by using the Gauss-Seidel technique, each iteration step takes $2(N-1)$ multiplications and this may be too time-consuming to be preferred to the Gauss elimination approach.

7.7 Elliptic Equations

In this section we consider further finite difference approaches and exploit further the idea of analogies. First we make a few general remarks on improving the accuracy of finite difference solutions. The most obvious way of proceeding is to make the mesh finer by reducing the mesh size. However, for elliptic problems the number of equations to be solved is inversely proportional to the square of the mesh length; an alternative method is therefore desirable. A second possibility is to use more accurate finite-difference formulae for the second partial derivatives $\partial^2\theta/\partial x^2$ and $\partial^2\theta/\partial y^2$.

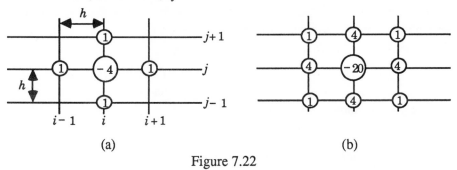

(a)   (b)

Figure 7.22

In Figure 7.22(a) we show the five points employed by the formulae that we used. The approximations were obtained by using Taylor's series.

$$\theta_{i,j+1} = \theta_{i,j} + h\theta_x\big|_{(i,j)} + \frac{h^2}{2!}\theta_{xx}\big|_{(i,j)} + \frac{h^3}{3!}\theta_{xxx}\big|_{(i,j)} + 0(h^4) \qquad (7.51)$$

$$\theta_{i,j-1} = \theta_{i,j} - h\theta_x\big|_{(i,j)} + \frac{h^2}{2!}\theta_{xx}\big|_{(i,j)} - \frac{h^3}{3!}\theta_{xxx}\big|_{(i,j)} + 0(h^4) \qquad (7.52)$$

Adding these two equations and then dividing by h^2 gives respectively

$$\theta_{i,j+1} + \theta_{i,j-1} - 2\theta_{i,j} = 2\frac{h^2}{2!}\theta_{xx}\big|_{(i,j)} + 0(h^4) \qquad (7.53)$$

$$\theta_{xx}\big|_{(i,j)} = \frac{\theta_{i,j-1} - 2\theta_{i,j} + \theta_{i,j+1}}{h^2} + 0(h^2) \qquad (7.54)$$

Similarly we may obtain

$$\theta_{yy} \mid_{(i,j)} = \frac{\theta_{i-1,j} - 2\theta_{i,j} + \theta_{i+1,j}}{h^2} + 0\,(h^2) \tag{7.55}$$

Therefore the approximation to Laplace's equation we used, viz.

$$\theta_{i-1,j} + \theta_{i+1,j} + \theta_{i,j-1} + \theta_{i,j+1} = 4\theta_{i,j}$$

has error $O(h^2)$. The weights are shown in Figure 7.22(a).

Suppose we now take all nine points shown in Figure 7.22(b) into account. We have, for example, to consider

$$\theta_{i+1,\,j+1} = \theta_{i,j} + h\theta_x \mid_{(i,j)} + h\theta_y \mid_{(i,j)} + \frac{h^2}{2}\theta_{xx} \mid_{(i,j)}$$

$$+ h^2\theta_{xy} \mid_{(i,j)} + \frac{h^2}{2}\theta_{yy} \mid_{(i,j)} + \frac{h^3}{3!}\,\{\theta_{xxx} \mid_{(i,j)}$$

$$+ 3\theta_{xxy} \mid_{(i,j)} + 3\theta_{xyy} \mid_{(i,j)} + \theta_{yyy} \mid_{(i,j)}\,\} + O(h^4) \tag{7.56}$$

In fact we take the expansions as far as terms $O(h^5)$ and we may derive the nine-point approximation to Poisson's equation $\nabla^2\theta = K$, K constant, which has a truncation error $O(h^6)$. This approximation is

$$\theta_{i-1,\,j-1} + \theta_{i+1,\,j+1} + \theta_{i+1,\,j-1} + \theta_{i-1,\,j+1} + 4\,[\theta_{i-1,\,j} + \theta_{i+1,\,j}$$

$$+ \theta_{i,\,j-1} + \theta_{i,\,j+1}] - 20\theta_{i,\,j} = 6h^2\,K \tag{7.57}$$

The weights are shown in Figure 7.22(b).

Gauss-Seidel method

The Gauss-Seidel method applied to Laplace's equation gives the iterative formula

$$\theta_{i,\,j,\,n+1} = \tfrac{1}{4}\,\{\theta_{i-1,\,j,\,n+1} + \theta_{i,\,j-1,\,n+1} + \theta_{i+1,\,j,\,n} + \theta_{i,\,j+1,\,n}\} \tag{7.58}$$

We define the residual at the point (i,j) as

$$R_{i,j} = \theta_{i-1,\,j,\,n+1} + \theta_{i,\,j-1,\,n+1} + \theta_{i+1,\,j,\,n} + \theta_{i,\,j+1,\,n} - 4\theta_{i,\,j,\,n} \tag{7.59}$$

The Gauss-Seidel method is known as the (unextrapolated) **Liebmann method**.

We choose the starting values by first estimating $\theta_{2,\,2,\,0}$ from the nine-point formula; the other values may be guessed by linear interpolation. Since the method converges only slowly, it is best suited to a digital computer.

The method of relaxation

This is a method mainly of historical interest, since the widespread use of digital computers has effectively made it redundant. However, we briefly examine it so that we may set the scene for the next subsection.

Essentially, the method calculates the residual at each mesh point and then selects the largest of these. Since this occurs at the point where the initial approximation is worst we take steps to improve the situation here before any other points. Altering the function value at this point has the effect of changing the residuals at neighbouring points as well at the point under consideration. If we use the five-point formula then, if the largest residual at a point is –40, adding 10 to that value of θ will reduce the residual to zero and add 10 to the four nearest residuals. Then one looks for the next largest residual and so on. However, **relaxation** has been developed to an art and residuals are not always relaxed exactly to zero. Sometimes the residual is **under-relaxed**, sometimes **over-relaxed**. An experienced operator would get a feel for how to proceed at each step. The method is therefore not suitable to programming for a computer. A common device is to multiply all mesh values by 10 when residuals get small, in order to keep to integer arithmetic.

Successive Over-Relaxation

This method, also known as **S.O.R.** or the **Extrapolated Liebmann method** seeks to accelerate the convergence of Liebmann's method and arrive at a technique which combines rapid convergence with the systematic approach needed for digital computation.

For the Gauss-Seidel method we can write the iteration formula as

$$\theta_{i,j,n+1} = \theta_{i,j,n} + \frac{1}{4} R_{i,j} \qquad (7.60a)$$

In the S.O.R. method we change $\theta_{i,j}$ by a larger amount than $\frac{1}{4} R_{i,j}$ – hence the term **over-relaxation**. The iterative formula we use is

$$\theta_{i,j,n+1} = \theta_{i,j,n} + \frac{\omega}{4} R_{i,j} \qquad (7.60b)$$

where ω is a positive constant between 1 and 2.

But what value should we choose for ω? One case which has managed to yield a simple rule for determining ω is a rectangular region with m and n mesh divisions on adjacent sides. It has been proved that the value of ω can be found by solving the equation

$$\left[\cos \frac{\pi}{m} + \cos \frac{\pi}{n} \right]^2 \omega^2 - 16\omega + 16 = 0 \qquad (7.60c)$$

General remarks on elliptic equations

Boundary conditions specify at each point of the boundary curve C *either* the value of the independent variable *or* its derivative normal to the boundary. They may not specify both at the *same* point.

Further, a result which any finite difference scheme should model is that the solution of Laplace's equation has neither its maximum nor its minimum value on points *inside* the boundary curve.

Experimental Analogies

In Section 7.2 we saw that since electrical potential satisfies Laplace's equation we were able to set up an **electrical analogue** to the steady-state temperature distribution problem. Of course, the method can apply to the solution of Laplace's equation in any context, placing a known potential at some points of the boundary and insulating other points (so that the derivative there normal to the boundary is zero).

A second analogy that is used is the **membrane analogy**. A thin rubber sheet or a soap film can be stretched over a frame which is shaped in such a way that the three-dimensional surface formed by the membrane has contours which in plan view correspond to streamlines of a flow pattern. Should we require the pattern of flow round an immersed body then the membrane can be attached to the body so that it meets it at a constant height. The contours may be traced by a movable point gauge in the case of a rubber membrane or by reflection of light rays in the case of the soap film.

A third analogy, which may be used in the study of irrotational flows, is the **viscous flow analogy**. Viscous fluid is made to flow between horizontal parallel plates separated by a small distance. Suitable obstacles may be placed between the plates. Dye is fed continuously at regular intervals across the boundary at which fluid enters the plates (which are usually sealed along two edges to form a cell) and it traces out the corresponding flow pattern. Near the boundaries of the flow, the analogy becomes somewhat innacurate, but reducing the spacing of the plates will alleviate this to an extent.

7.8 Characteristics

Consider the equation

$$\frac{\partial^2 \phi}{\partial x^2} + 4 \frac{\partial^2 \phi}{\partial x \partial y} + 3 \frac{\partial^2 \phi}{\partial y^2} = 0 \tag{7.61}$$

Let $\phi_1(x, y)$ and $\phi_2(x, y)$ be two solutions of this equation. Then suppose we define $\phi^*(x, y) = \alpha \phi_1(x, y) + \beta \phi_2(x, y)$, where α and β are constants. We can show that $\phi^*(x, y)$ is also a solution of (7.61) by direct substitution. This demonstrates the linearity of the equation. We can write it as either

$$\left[\frac{\partial}{\partial x} + \frac{\partial}{\partial y} \right] \left[\frac{\partial}{\partial x} + 3 \frac{\partial}{\partial y} \right] \phi = 0 \tag{7.62a}$$

or

$$\left[\frac{\partial}{\partial x} + 3 \frac{\partial}{\partial y} \right] \left[\frac{\partial}{\partial x} + \frac{\partial}{\partial y} \right] \phi = 0 \tag{7.62b}$$

Now it can be shown that the general solution of the equation

$$\left[\frac{\partial}{\partial x} + 3 \frac{\partial}{\partial y} \right] \phi = \frac{\partial \phi}{\partial x} + 3 \frac{\partial \phi}{\partial y} = 0$$

is $\phi_1 = f(y - 3x)$ for a suitably differentiable function f. It follows from (7.62a) that this form of ϕ is a solution of (7.61). Likewise, starting from the

equation $\left[\dfrac{\partial}{\partial x} + \dfrac{\partial}{\partial y} \right] \phi = 0$ whose general solution is $\phi_2 = g(y - x)$ for a

suitably differentiable g, we find that ϕ_2 is another solution of (7.61). Since arbitrary constants may be absorbed into f and g, it follows that the general solution of (7.61) can be written as

$$\phi = f(y - 3x) + g(y - x) \tag{7.63}$$

Change of variable

We can transform the independent variables x and y into u and v where $u = x - y$, $v = 3x - y$. Applying the chain rule for partial derivatives eventually leads to the result

$$4 \frac{\partial^2 \phi}{\partial u \partial v} = 0 \tag{7.64}$$

Integrating with respect to v gives $\dfrac{\partial \phi}{\partial u} = h(v)$ and integrating this last equation

with respect to u provides $\phi = f(v) + g(u)$ where $f'(v) = h(v)$. Hence we recover solution (7.63).

The curves $3x - y = $ constant and $x - y = $ constant are called **characteristics** of the equation. Along them the solution takes on a particularly simple form. The form (7.64) is called the **canonical form** of the equation (7.61). It can be shown that along the curves $3x - y = $ constant, the value of

$\left[\dfrac{\partial \phi}{\partial x} + \dfrac{\partial \phi}{\partial y} \right]$ is constant and along the curves $x - y = $ constant, the value of

$\left[\dfrac{\partial \phi}{\partial x} + 3 \dfrac{\partial \phi}{\partial y} \right]$ is constant.

Classification via characteristics

The general second-order quasi-linear partial differential equation with two independent variables, whose coefficients may be functions of x and/or y, is

$$A \frac{\partial^2 \phi}{\partial x^2} + B \frac{\partial^2 \phi}{\partial x \partial y} + C \frac{\partial^2 \phi}{\partial y^2} + D \frac{\partial \phi}{\partial x} + E \frac{\partial \phi}{\partial y} + F\phi + G = 0 \tag{7.65}$$

We shall confine our attention to the case $D = E = F = 0$, although our remarks will generalise. If we employ the notation which is customary, viz. $p = \partial\phi/\partial x$, $q = \partial\phi/\partial y$, $r = \partial^2\phi/\partial x^2$, $s = \partial^2\phi/\partial x\partial y$, $t = \partial^2\phi/\partial y^2$ then the simplified form of (7.65) may be written as

$$Ar + Bs + Ct = -G \tag{7.66}$$

Now $\quad dp = \dfrac{\partial p}{\partial x}\, dx + \dfrac{\partial p}{\partial y}\, dy = rdx + sdy = rdx + sdy + t.0 \tag{7.67a}$

and $\quad dq = sdx + tdy = r.0 + sdx + tdy \tag{7.67b}$

Suppose that the values of $p = \partial\phi/\partial x$ and $q = \partial\phi/\partial y$ are given on some curve in the x - y plane. From equations (7.66) and (7.67) we can deduce the values of r, s and t on the curve provided that

$$\Delta = \begin{vmatrix} dx & dy & 0 \\ 0 & dx & dy \\ A & B & C \end{vmatrix} \neq 0 \tag{7.68}$$

The requirement $\Delta = 0$ leads to the equation

$$A\left[\frac{dy}{dx}\right]^2 - B\left[\frac{dy}{dx}\right] + C = 0 \tag{7.69}$$

This quadratic equation will have two roots if $B^2 - 4AC > 0$; there are two real characteristics and the original equation is said to be **hyperbolic**. If $B^2 = 4AC$ then there is only one characteristic direction and the equation is **parabolic**. Finally, if $B^2 < 4AC$ the equation is **elliptic**; there are no real characteristic curves.

Hyperbolic equations may be transformed to $\dfrac{\partial^2\phi}{\partial u \partial v} = F\left(\phi,\, u,\, v,\, \dfrac{\partial\phi}{\partial u},\, \dfrac{\partial\phi}{\partial v}\right).$

The one-dimensional wave equation $c^2\partial^2\phi/\partial x^2 - \partial^2\phi/\partial t^2 = 0$ is hyperbolic. It can be reduced to $\partial^2\phi/\partial u \partial v = 0$ where $u = x - ct$, $v = x + ct$ with general solution $\phi = f(x - ct) + g(x + ct)$.

Note that a p.d.e. may be elliptic in one region of the plane, parabolic in another, and hyperbolic in yet another region; see Problem 46.

The method of characteristics

For reasons which will become clear, this method works only for hyperbolic equations.

We work through one example to show the ideas. Consider the equation

$$\frac{\partial^2\phi}{\partial x^2} - 2\frac{\partial^2\phi}{\partial x \partial y} - 3\frac{\partial^2\phi}{\partial y^2} = -1 \tag{7.70}$$

Suppose we are given that at $y = 0$, $\phi = x = \dfrac{\partial\phi}{\partial y}$ for $0 < x < 1$.

Then if we refer to the more general case shown in Figure 7.23 we see that from each point two characteristic curves may be drawn. A characteristic of one class from P will intersect one of the other class from Q in a point S. P, Q, R are points on the first level and S and T are points on the second level. It is true in general that as we proceed from one level to the next, the number of points available will become smaller. Returning to our example the characteristics are given by

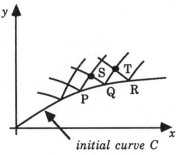

initial curve C

Figure 7.23

$$\left[\frac{dy}{dx}\right]^2 - 2\frac{dy}{dx} - 3 = 0 \text{ and are a pair of straight lines, viz. } \frac{dy}{dx} = 3, \frac{dy}{dx} = -1.$$

Families of these lines emanating from the initial curve $y = 0$ at $x = 0(0.1)1.0$ are shown in Figure 7.24. In general, the pattern of characteristics will not be so simple.

Now we start from the knowledge at $y = 0$, $x = 0(0.1)1.0 - 11$ points $-$ and use this to advance to the 10 points on level 2 then to the 9 points on level 3. Consider $P = (0.4, 0)$ and $Q = (0.5, 0)$ and let characteristics from those points intersect at R. Then we find that the coordinates of R are $(0.425, 0.075)$

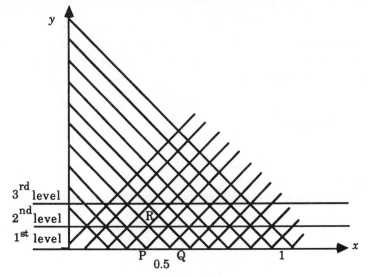

Figure 7.24

Note that we know that at $y = v$, $\partial\phi/\partial x = 1$ so that on the initial curve, ϕ, $\partial\phi/\partial x$ and $\partial\phi/\partial y$ are known.

It can be shown, although the proof is beyond the scope of this book, that we can find the values of p and q at R and the knowledge of these allows us to find the values of ϕ at R. We can therefore find all values at the points on the second level and from these proceed to the third level.

Note that if the initial curve is itself a characteristic we will be in trouble. There will be a unique solution only along characteristic curves of the same family.

In physical problems modelled by hyperbolic equations it may be that there are discontinuities in the initial conditions. These discontinuities are propagated along the characteristics as discontinuities in the solution. An example is that of shock waves.

Problems

Section 7.2

Apply the methods used on the case study to the following physical situations. The analytical solution is provided for you.

1   A thin square plate $0 \leq x \leq a$, $0 \leq y \leq a$ has its faces insulated. The edges $x = 0$, $y = a$ are maintained at zero temperature. The edge $y = 0$ is insulated and the edge $x = a$ is kept at a constant temperature 50°C. (An insulated edge implies that the normal derivative of temperature is zero there.) The steady-state temperature is given by

$$\theta(x,y) = \frac{200}{\pi} \sum_{n=0}^{\infty} \frac{(-1)^n}{2n+1} \frac{\sinh \dfrac{(2n+1)\pi x}{2a}}{\sinh \dfrac{(2n+1)\pi}{2}} \cos \frac{(2n+1)\pi y}{2a}$$

2 A rectangular plate $0 \leq x \leq 20$, $0 \leq y \leq 10$ has the edges $x = 0$, $x = 20$, $y = 0$ maintained at zero temperature whilst the temperature of the edge $y = 10$ is given by $\theta = 20x - x^2$. The steady-state temperature is given by

$$\theta(x,y) = \frac{3200}{\pi^3} \sum_{n=1}^{\infty} \frac{\sin \dfrac{(2n-1)\pi x}{20} \sinh \dfrac{(2n-1)\pi y}{20}}{(2n-1)^3 \sinh \dfrac{1}{2}(2n-1)\pi}$$

Section 7.3

3 Classify each of the following equations as elliptic, hyperbolic or parabolic.

(a) $\dfrac{\partial^2 u}{\partial x^2} - \dfrac{\partial^2 u}{\partial y^2} = 0$ (b) $\dfrac{\partial u}{\partial x} + \dfrac{\partial^2 u}{\partial x \partial y} = 8$

(c) $\dfrac{\partial^2 u}{\partial x^2} - 2\dfrac{\partial^2 u}{\partial x \partial y} + 2\dfrac{\partial^2 u}{\partial y^2} = x + 3y$

(d) $\dfrac{\partial^2 u}{\partial x^2} + 3\dfrac{\partial^2 u}{\partial x \partial y} + 4\dfrac{\partial^2 u}{\partial y^2} + 5\dfrac{\partial u}{\partial x} - 2\dfrac{\partial u}{\partial y} + 4u = 2x - 6y$

(e) $\dfrac{\partial^2 u}{\partial x^2} - 7\dfrac{\partial^2 u}{\partial x \partial y} + \dfrac{\partial^2 u}{\partial y^2} = 0$ (f) $\dfrac{\partial^2 u}{\partial x^2} + \dfrac{\partial^2 u}{\partial x \partial y} - 6\dfrac{\partial^2 u}{\partial y^2} = 0$

(g) $\dfrac{\partial^2 u}{\partial x^2} + 6\dfrac{\partial^2 u}{\partial x \partial y} + 9\dfrac{\partial^2 u}{\partial y^2} = 0$

Section 7.4

4   Obtain the analytical formulae given in Problems 1 and 2 of Section 7.2.

5   Find a solution of Laplace's equation which satisfies the conditions
(i) $\theta \to 0$ as $y \to \infty$;   (ii) $\theta = 0$ when $x = 0$ and $x = 1$;
(iii) $\theta = \sin \pi x + \sin 2\pi x$ when $y = 0$.

6   Obtain a solution of the equation $\dfrac{\partial^2 V}{\partial x^2} + 9\dfrac{\partial^2 V}{\partial t^2} = 0$ which satisfies the conditions

(i) V is periodic in t;   (ii) $V = 0$ when $t = 0$;   (iii) $\dfrac{\partial V}{\partial x} = 0$ when $x = 0$;

(iv) $V = 4 \cosh 15x$ when $t = \pi/2$.

7   Consider the rectangular plate $0 \leq x \leq a, 0 \leq y \leq b$.   The temperature distributions along the edges are given by $\theta(0,y) = f_1(y)$; $\theta(a,y) = f_2(y)$; $\theta(x,0) = f_3(x)$; $\theta(x,b) = f_4(x)$.   Consider four solutions of Laplace's equation as $\theta_1, \theta_2, \theta_3, \theta_4$ which satisfy the conditions respectively

$\theta_1 (0,y) = f_1(y)$, $\theta_1 = 0$ along the three other edges
$\theta_2 (a,y) = f_2(y)$, $\theta_2 = 0$ along the three other edges
$\theta_3 (x,0) = f_3(x)$, $\theta_3 = 0$ along the three other edges
$\theta_4 (x,b) = f_4(x)$, $\theta_4 = 0$ along the three other edges

Show that the steady-state solution to the original problem can be written as $\theta(x,y) = \theta_1 + \theta_2 + \theta_3 + \theta_4$.   Interpret this result.

8   Show that the solution of Laplace's equation which satisfies
(i) $\theta \to 0$ as $y \to \infty$ for $-a < x < a$;   (ii) $\partial\theta/\partial x = 0$ when $x = \pm a$;
(iii) $\theta = -T_0$ when $y = 0, -a < x < 0$;   (iv) $\theta = T_0$ when $y = 0, 0 < x < a$
is given by the formula

$$\theta = \dfrac{T_0}{\pi} \sum_{n=0}^{\infty} \dfrac{4}{(2n+1)} \exp\left\{-(n+\tfrac{1}{2})\pi y/a\right\} \sin\left\{(n+\tfrac{1}{2})\pi x/a\right\}$$ (LU)

9 The voltage V at any point (x, y) of a rectangular metal plate satisfies the equation

$$\frac{\partial^2 V}{\partial x^2} + \frac{\partial^2 V}{\partial y^2} = 0$$

Along the sides of the plate it is given by

$V = 0$ for $x = 0$,	$0 \le y \le b$	$V = 0$ for $y = 0$,	$0 \le x \le a$
$V = 0$ for $x = a$,	$0 \le y \le b$	$V = f(x)$ for $y = b$,	$0 \le x \le a$

Here a and b are the lengths of the sides and f is some known function. Show that

$$V = \sum_{n=1}^{\infty} A_n \sin\left[\frac{n\pi x}{a}\right] \sinh\left[\frac{n\pi y}{a}\right]$$

where A_n, $n = 1, 2, 3, ...$, are constants. Determine the constants A_n for the case when

$$f(x) = \begin{cases} x, & 0 \le x \le \frac{1}{2}a \\ a - x, & \frac{1}{2}a \le x \le a \end{cases} \tag{EC}$$

Section 7.5

10 The ends A and B of a rod 50cm. long have their temperatures kept at 0°C and 100°C, respectively until the temperatures are indistinguishable from those for the steady state. At some time after this, there is a sudden change in the end temperatures. Find the temperature of any point in the rod, $\theta(x, t)$ as a function of x (cm.), the distance from one end A, and t (sec) the time elapsed after the sudden change for each of the following cases.

   (a) The temperature of B is suddenly reduced to 0°C., and kept so, while that of A is kept at 0°C.;

   (b) The temperature of B is kept at 100°C., while that of A is suddenly increased to 100°C., and maintained at this value;

   (c) The temperature of B is suddenly reduced to 50°C., and kept at this value while that of A is kept at 0°C.;

   (d) The temperature of A is suddenly raised to 25°C., while that of B is suddenly reduced to 75°C., and then these temperatures are maintained;

   (e) The temperature of A is suddenly raised to 50°C., while that of B is suddenly raised to 150°C., and then these temperatures are maintained.

11 Find the solution of the one-dimensional diffusion equation $\dfrac{\partial^2 u}{\partial x^2} = \dfrac{1}{k}\dfrac{\partial u}{\partial t}$, where k is a constant, satisfying the boundary conditions $u(0,t) = u(\pi,t) = 0$, $t \ge 0, u(x,0) = f(x)$, $0 \le x \le \pi$ where

$$f(x) = x \text{ for } 0 \le x \le \frac{\pi}{2} \text{ and } f(x) = \pi - x \text{ for } \frac{\pi}{2} \le x \le \pi \tag{EC}$$

12 A thin metal bar has a length L which lies between $x = 0$ and $x = L$. Assuming a linear law of heat transfer between the surface of the bar and its surroundings, the partial differential equation satisfied by the temperature $u(x, t)$ in the bar is

$$\frac{\partial u}{\partial t} = a^2 \frac{\partial^2 u}{\partial x^2} - hu \quad (0 \le x \le L, t \ge 0)$$

where a and h are positive constants.

Assuming that the initial temperature distribution is given by $f(x)$ and that thereafter the ends $x = 0$ and $x = L$ are kept at zero temperature, find the general solution for $u(x, t)$ using the method of separation of variables.

Find the particular solution where $f(x) = x(L - x)$ (EC)

13 In the equation $\frac{\partial u}{\partial t} = h^2 \frac{\partial^2 u}{\partial x^2} - k(u - u_0)$ the change of variable $u = u_0 + e^{-kt}v$ is introduced, where h, k and u_0 are constants: show that the equation in v becomes

$\frac{\partial v}{\partial t} = h^2 \frac{\partial^2 v}{\partial x^2}$. Hence given boundary conditions $u(0, t) = u_0 = u(1, t)$ and assuming a solution of the form $u(x, t) = u_0 + X(x) e^{-(k+m^2)t}$, m real, obtain a general solution of the given equation. If the initial condition is $u(x, 0) = u_0 + x(1 - x)$, determine the solution completely. (EC)

14 The temperature $\theta(x, t)$ in a bar of length a is given by $\frac{\partial \theta}{\partial t} = k \frac{\partial^2 \theta}{\partial x^2}$ where k is the constant of diffusivity.

If the ends $x = 0$ and $x = a$ of the bar are maintained at zero temperature and if the initial temperature distribution is: $\theta = \theta_0 x(a - x)$ where θ_0 is constant, show that

$$\theta(x, t) = \frac{8\theta_0 a^2}{\pi^3} \sum_{n}^{\infty} \frac{1}{n^3} \sin \frac{n\pi x}{a} \exp\left[-\frac{kn^2\pi^2 t}{a^2}\right]$$

the summation extending over $n = 1, 3, 5, 7, \dots$ only. (EC)

15 In Terzaghi's theory of one-dimensional consolidation, the equation for the excess pore water

pressure, u, is $\frac{\partial u}{\partial t} = c_v \frac{\partial^2 u}{\partial z^2}$ where c_v is the (constant) coefficient of consolidation.

A soil sample of thickness $2H$ is tested and allowed to drain at both the upper and lower surfaces: the initial distribution of u is a constant u_i. Write down the three boundary conditions for this case and solve the equations. Defining the degree of consolidation, U,

as $\frac{u_i - u}{u_i}$ find $\overline{U}$, the average degree of consolidation over the sample layer, and sketch the

graph of $\overline{U}$ against t. Explain where the degree of consolidation is a minimum for all time and verify analytically that this is so.

16 A thin rectangular slab of cross-section area A, of thickness t is painted on the edges so that the liquid content of the slab can evaporate only through its faces. Assume that liquid evaporates *immediately* on reaching a face. The constant concentration of liquid at each face is called the **equilibrium concentration**. Assume that the initial concentration is uniform throughout the interior of the slab but falls immediately to the equilibrium value at the surface. What further assumptions are necessary to show that the excess of concentration over the equilibrium value satisfies the one-dimensional diffusion equation? Suggest suitable boundary conditions and hence find the concentration as a function of position and time.

17 The current i in a cable satisfies the equation $\dfrac{\partial^2 i}{\partial x^2} = \dfrac{2}{k}\dfrac{\partial i}{\partial t} + i$, ($k$ constant.)

By assuming a solution of the type $i = XT$, where X is a function x alone and T is a function of t alone, show that if $i = 0$ when $x = l$ and $\partial i/\partial x = -ae^{-kt}$ when $x = 0$,

the current is given by $i = ae^{-kt}\dfrac{\sin(l - x)}{\cos l}$. $\hspace{2cm}$ (LU)

18 In a telegraph wire the resistance R and the leakage conductance G are very large in comparison with the capacitance and inductance. The wire is of length l and the voltage at the sending end is V_0 (constant). If the other end of the wire is earthed, show that the voltage V and the current i at distance x from the sending end are given approximately by

$$V = V_0\,\frac{\sinh a(l-x)}{\sinh al}, \quad i = V_0\left[\frac{G}{R}\right]^{1/2}\frac{\cosh a(l-x)}{\sinh al}$$

where $a = (RG)^{1/2}$. $\hspace{5cm}$ (LU)

19 A distortionless transmission line ($RC = LG$) of length l is initially charged to unit potential and the end $x = l$ is insulated. If, at time $t = 0$, the end $x = 0$ is earthed, show that the potential at time t and distance x from the earthed end is given by

$$\frac{4}{\pi}e^{-kt}\sum_{r=0}^{\infty}\frac{1}{2r+1}\sin\frac{(2r+1)\pi x}{2l}\cos\frac{(2r+1)\pi ut}{2l},$$

where $u = (LC)^{-1/2}$ and $k = R/L$. $\hspace{3cm}$ (LU)

20 A light elastic string of length a has its ends $x = 0$ and $x = a$ fixed. The point where $x = a/3$ is drawn aside a small distance h and released from rest at time $t = 0$. Assume that the displacement $y(x, t)$ of the string satisfies the one-dimensional wave equation

$\dfrac{\partial^2 y}{\partial x^2} = \dfrac{1}{c^2}\dfrac{\partial^2 y}{\partial t^2}$ where c is a constant. By separating the variables show that

$$y(x,t) = \frac{9h}{\pi^2}\sum_{n=1}^{\infty}\frac{1}{n^2}\sin\frac{n\pi}{3}\sin\frac{n\pi x}{a}\cos\frac{n\pi ct}{a} \hspace{2cm}\text{(EC)}$$

21 Show that the Fourier sine series for the function $f(x) = kx$ for $0 \le x < 1$ and $f(x) = -kx + 2k$ for $1 < x \le 2$ is

$$\frac{8k}{\pi^2} \sum_{n=1}^{\infty} \frac{1}{n^2} \sin \frac{n\pi}{2} \sin \frac{n\pi x}{2}.$$

Use the method of separation of variables to solve the differential equation

$$\frac{\partial^2 y}{\partial t^2} = a^2 \frac{\partial^2 y}{\partial x^2} \quad (0 < x < 2, t > 0)$$ satisfying the boundary conditions

(i) $y(0,t) = 0$, $y(2,t) = 0$ for all t

(ii) $\dfrac{\partial y}{\partial t} = 0$ at $t = 0$ for all x

(iii) $y(x, 0) = f(x)$ as defined earlier.
Obtain the solution in the form

$$y(x, t) = \frac{8k}{\pi^2} \sum_{m=1}^{\infty} \frac{(-1)^{m+1}}{(2m-1)^2} \sin \frac{(2m-1)\pi x}{2} \cos \frac{(2m-1)\pi a t}{2} \qquad \text{(EC)}$$

22 (i) Show that the general form $y(x,t) = F(x - ct) + G(x + ct)$ satisfies the one-dimensional wave equation $\dfrac{\partial^2 y}{\partial x^2} = \dfrac{1}{c^2} \dfrac{\partial^2 y}{\partial t^2}$

If, at $t = 0$, $y(x, 0)$ is given by $\alpha(x)$ and $\dfrac{\partial y(x, 0)}{\partial t}$ by $\beta(x)$, for all values of x, show that

$$y(x, t) = \frac{1}{2}[\alpha(x - ct) + \alpha(x + ct)] + \frac{1}{2c} \int_{x-ct}^{x+ct} \beta(u)du.$$

(ii) Find the solution of the differential equation $\dfrac{\partial^2 F}{\partial x^2} + \dfrac{\partial^2 F}{\partial y^2} + 5F = 0$ in the form

$F(x,y) = f(x)g(y)$ satisfying the following conditions:
(a) F is periodic in x;
(b) $F = 0$ when $x = 0$ for all values of y;
(c) $F = 3e^{-2y}$ when $x = \pi/6$ for all values of y. (LU)

23 Solve the one-dimensional wave equation for a string of length 1 with zero initial velocity and a profile given by $f(x)$ where
(i) $f(x) = A \sin 2\pi x$; (ii) $f(x) = Ax(1 - x^2)$

(iii) $f(x) = \begin{cases} x & , & 0 < x < 1/4 \\ 1/2 - x & , & 1/4 < x \le 1/2 \\ 0 & , & 1/2 \le x \le 1 \end{cases}$ (iv) $f(x) = Ax^2(1-x^2)$.

24 The small vertical vibrations of a uniform beam of length l are given by the equation

$$EI \frac{\partial^4 y}{\partial x^4} + \rho A \frac{\partial^2 y}{\partial t^2} = 0$$ where EI is the flexural rigidity of the beam of cross-sectional

area A; the uniform density of the beam is ρ. Find the general solution of this equation. (This beam is referred to in Problems 25, 26, 27).

25 If the beam is simply supported at both ends find the fundamental mode of vibration.

26 If the beam is cantilevered at $x = 0$ show that the period of oscillation is $2\pi/\omega$ where $\rho A \omega^2 = EIa^4$ and $\cosh al \cos al = -1$.

27 If the beam is clamped horizontally at both ends show that the period of oscillation is $2\pi/\omega$ where $\rho A \omega^2 = EIa^4$ and $\cosh al \cos al = 1$. Find graphically the smallest root of this equation and hence the fundamental period. Show that for vibrations of higher frequency ω

is given by $$\omega = \left[\frac{EI}{\rho A} \right]^{\frac{1}{2}} \left[\frac{(2n+1)\pi}{2l} \right]^2 \quad n = 2, 3, 4, \ldots$$

28 A long rectangular wave-guide has its axis parallel to the z-axis, its walls being formed by the four planes $x = 0$, $x = a$, $y = 0$, $y = b$. Associated with a transverse magnetic wave travelling in the guide is a function ϕ which vanishes on the four walls of the guide,

and satisfies the equation $\dfrac{\partial^2 \phi}{\partial x^2} + \dfrac{\partial^2 \phi}{\partial y^2} + \dfrac{\partial^2 \phi}{\partial z^2} = \dfrac{1}{c^2} \dfrac{\partial^2 \phi}{\partial t^2}$ inside the guide. In order that a

wave of frequency f should be transmitted, ϕ must be of the form $U(x, y) \cos (\beta z - 2\pi ft)$, where β is real. Show that if $U(x, y)$ has the form $X(x) Y(y)$, then the only possible solutions of the problem are $\phi = A \sin (m\pi x/a)$ $\sin (n\pi y/b) \cos (\beta z - 2\pi ft)$, where A is a constant, and m, n are integers such that $m^2/a^2 + n^2/b^2 = 4f^2/c^2 - \beta^2/\pi^2$. Deduce that if $a = b$, waves given by a solution

of the above form must have a frequency not less than $c/(a\sqrt{2})$.   (LU)

29 Show that the transmission line equation $\dfrac{\partial^2 e}{\partial x^2} = LC \dfrac{\partial^2 e}{\partial t^2} + (RC + GL) \dfrac{\partial e}{\partial t} + RGe$

with boundary conditions $e(0, t) = E_0 \cos \omega t$ (E_0, ω constant) and $e(x, t)$ bounded as $x \to \infty$ or $t \to \infty$, where $e(x, t)$ is the voltage at distance x and time t and R, L, G, C are constants of the line, does not have a separable solution of the form $e(x, t) = X(x).T(t)$. Further, show that there exists a solution of form $e(x, t) = E_0 e^{-ax} \cos (\omega t + bx)$ provided that $a^2 - b^2 = RG - LC\omega^2$ and $2ab = -(RC + GL)\omega$. Hence obtain the parameters a and b.

                                                                                        (EC)

30 Show that the equation $\dfrac{\partial^2 V}{\partial r^2} + \dfrac{1}{r} \dfrac{\partial V}{\partial r} + \dfrac{1}{r^2} \dfrac{\partial^2 V}{\partial \theta^2} = 0$ has a solution of the form

$$V = \left[Ar^n + \dfrac{B}{r^n} \right] f(\theta)$$ where $f(\theta)$ is a trigonometric function of θ only with period $2\pi/n$.

Find $f(\theta)$.
Find V for the case $n = 2$, given that $V \to 0$ as $r \to \infty$, $V = 0$ when $\theta = 0$, $V = 1$ when $r = 1$ and $\theta = \pi/4$.   (LU)

31 (i)   The temperature V at any point of a circular metal plate with radius a satisfies, when the steady state has been reached, the partial differential equation

$$r^2 \dfrac{\partial^2 V}{\partial r^2} + r \dfrac{\partial V}{\partial r} + \dfrac{\partial^2 V}{\partial \theta^2} = 0, \text{ where } r, \theta \text{ are polar coordinates and the centre of the}$$

plate is the pole. Find a solution of the equation which is bounded at the centre of the plate and is such that $V = a^2(1 - \cos 2\theta)$ whenever $r = a$.

(ii)   Obtain, by the method of separation of variables, a solution of the equation

$$x \dfrac{\partial f}{\partial x} + x^3 \dfrac{\partial f}{\partial t} = 2f. \text{ Determine the function } f(x, t) \text{ if } f(1, t) = e^{-t}.$$ (LU)

Section 7.6

32 Use the explicit method to solve the equation $u_{xx} = u_t$ where $u(0, t) = u(20, t) = 200$,

$$u(x, 0) = \begin{cases} 100 - 10x & 0 \le x \le 10 \\ -200 + 20x, & 10 \le x \le 20 \end{cases}$$

Use $h = \dfrac{1}{20}$ and choose k so that $r = 0.2$. Continue the solution for 4 time steps. Then

write a computer program to carry out 100 time steps.

33 Repeat Problem 32 for $r = 0.1, 0.4, 0.8, 1.0$.

34 Repeat Problem 32 for the new boundary condition $u(12,0) = 200$. Comment.

35 Repeat Problem 32 with $u(0, t) = 0$ and initial condition $u(x, 0) = \begin{cases} 0, & 0 \le x \le 10 \\ 20x - 200, & 10 \le x \le 20 \end{cases}$

36 Repeat Problem 32 for conditions $u(0, t) = 0$, $u(20, t) = 0$,

$$u(x, 0) = \begin{cases} 50x & , 0 \le x \le 4 \\ 400 - 50x & , 4 \le x \le 8 \\ 0 & , x \ge 8 \end{cases}$$

37 Repeat Problems 32 to 36 using the Crank-Nicolson method with $r = 0.8, 1.0, 2.0, 5.0$. Use a computer program to help you in your calculations.

Section 7.7

38 A square plate is heated along its edges to temperatures which vary in a way indicated in the diagram. The mesh is square. We require to find the temperatures at points A, B, C and D. First guess the temperatures T_A, T_B, T_C and T_D by inspection. Use the four-point formula (7.40) for Laplace's equation, to find a new estimate for T_A; then find new values for T_B, T_C and T_D in that order. Repeat the cycle until the temperatures are obtained to the nearest degree. Write and run a computer program to continue the iterations until the temperatures are obtained to a prescribed accuracy. Use the computer program of Problem 38 or a suitable modification to solve Problems 39 to 43 which all relate to Laplace's equation with suitable boundary conditions.

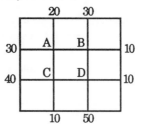

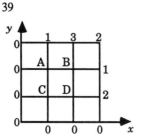   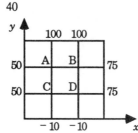   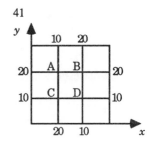

42 A rectangular area is bounded by the lines $x = 0$, $x = 4$, $y = 0$, $y = 5$. A function V

satisfies the equation $\dfrac{\partial^2 V}{\partial x^2} + \dfrac{\partial^2 V}{\partial y^2} = 0$ within the area, and $V = 5(2 - x)^2 + y^2$ on the

boundary. Using a mesh of unit squares, find the values of V at the mesh points.   (LU)

43 A square plate is bounded by the lines $x = \pm 2$; $y = \pm 2$. The temperature θ of the plate

obeys Laplace's equation $\dfrac{\partial^2 \theta}{\partial x^2} + \dfrac{\partial^2 \theta}{\partial y^2} = 0$. The boundary temperatures are on the line

$y = 2$, $\theta = -80$; on the lines $x = \pm 2$, $\theta = -40y$;
one the line $y = -2$, $\theta = 240 + 80x$ for $-2 \le x \le 0$.
$\quad\quad\quad\quad\quad\quad\quad\theta = 240 - 80x$ for $0 \le x \le 2$
Find the temperatures at the nine points (x_i, y_i) where $i = -1, 0, 1, j = -1, 0, 1$.   (LU)

44 A long square pipe with a square hole carries a hot fluid. The bottom half of the outside of the pipe is at a temperature of 0°C. The top surface of the pipe is at 100°C. Assume that the temperature varies linearly along the top half of each vertical side. The temperature of

the fluid inside the pipe is 200°C. The inside dimension of the pipe is 4cm and the outside dimension is 10cm. Choose a suitable mesh size and find the temperature at mesh points.

45 Repeat Problems 39, 40 and 41 using the method of successive over-relaxation.

Section 7.8

46 For what regions are the equations below elliptic, parabolic, hyperbolic?

(a) $x \dfrac{\partial^2 u}{\partial x^2} + y \dfrac{\partial^2 u}{\partial y^2} + 3y^2 \dfrac{\partial u}{\partial x} = 0$

(b) $x^2 \dfrac{\partial^2 u}{\partial x^2} + 2xy \dfrac{\partial^2 u}{\partial x \partial y} + y^2 \dfrac{\partial^2 u}{\partial y^2} = 0$

(c) $(x^2 - 1) \dfrac{\partial^2 u}{\partial x^2} + 2xy \dfrac{\partial^2 u}{\partial x \partial y} + (y^2 - 1) \dfrac{\partial^2 u}{\partial y^2} = x \dfrac{\partial u}{\partial x} + y \dfrac{\partial u}{\partial y}$

(d) $\dfrac{\partial^2 u}{\partial x^2} + x \dfrac{\partial^2 u}{\partial x \partial y} + y \dfrac{\partial^2 u}{\partial y^2} - xy \dfrac{\partial u}{\partial x} = 0$

47 The canonical form of a hyperbolic equation given by (7.65) is

$\dfrac{\partial^2 u}{\partial \xi \partial \eta} = d \dfrac{\partial u}{\partial \xi} + e \dfrac{\partial u}{\partial \eta} + fu$ where d, e and f are real constants and $\xi = \lambda_1 x + y$,

$\eta = \lambda_2 x + y$; λ_1 and λ_2 are the roots of the equation $A\lambda^2 + B\lambda + C = 0$.

For a parabolic equation given by (7.65) the transformation $\xi = \lambda x + y$, $\eta = y$ where λ is the repeated root of $A\lambda^2 + B\lambda + C = 0$, will produce the canonical form

$\dfrac{\partial^2 u}{\partial \eta^2} = d \dfrac{\partial u}{\partial \xi} + e \dfrac{\partial u}{\partial \eta} + fu$.

For an elliptic equation given by (7.65) the transformation $\xi = ax + y$, $\eta = bx$ where $a \pm bi$ are the roots of the equation $A\lambda^2 + B\lambda + C = 0$, will produce the canonical

form $\dfrac{\partial^2 u}{\partial \xi^2} + \dfrac{\partial^2 u}{\partial \eta^2} = d \dfrac{\partial u}{\partial \xi} + e \dfrac{\partial u}{\partial \eta} + fu$.

Verify that the transformation stated will produce the given canonical forms.

48 Transform the equations below to canonical form

(i) $\dfrac{\partial^2 u}{\partial x^2} - 5 \dfrac{\partial^2 u}{\partial x \partial y} + 6 \dfrac{\partial^2 u}{\partial y^2} = 0$ (ii) $\dfrac{\partial^2 u}{\partial x^2} - 2 \dfrac{\partial^2 u}{\partial x \partial y} - 8 \dfrac{\partial^2 u}{\partial y^2} + 9 \dfrac{\partial u}{\partial x} = 0$

(iii) $\dfrac{\partial^2 u}{\partial x^2} - 4 \dfrac{\partial^2 u}{\partial x \partial y} + 4 \dfrac{\partial^2 u}{\partial y^2} = 0$ (iv) $\dfrac{\partial^2 u}{\partial x^2} + 2 \dfrac{\partial^2 u}{\partial x \partial y} + \dfrac{\partial^2 u}{\partial y^2} + 3 \dfrac{\partial u}{\partial x} + 9u = 0$

49   Find suitable constants a and b so that, under the transformation of co-ordinates

$u = x + ay$, $v = x + by$, the differential expression $2 \dfrac{\partial^2 z}{\partial x^2} - 3 \dfrac{\partial^2 z}{\partial x \partial y} + \dfrac{\partial^2 z}{\partial y^2}$ becomes

$\dfrac{\partial^2 z}{\partial u \partial v}$. Hence derive the general solution of the equation $2 \dfrac{\partial^2 z}{\partial x^2} - 3 \dfrac{\partial^2 z}{\partial x \partial y} + \dfrac{\partial^2 z}{\partial y^2} = 0$. (EC)

50   If in the equation $\dfrac{\partial u}{\partial t} = h^2 \dfrac{\partial^2 u}{\partial x^2} - k(u - u_0)$ the change of variable $u = u_0 + e^{-kt}v$ is

introduced, where h, k and u_0 are constants, show that the equation in v becomes

$\dfrac{\partial v}{\partial t} = h^2 \dfrac{\partial^2 v}{\partial x^2}$.   Hence, given boundary conditions $u(0, t) = u_0 = u(1, t)$ and

assuming a solution of the form $u(x, t) = u_0 + X(x). e^{-(k+m^2)t}$, m real, obtain a
general solution of the given equation.   If the initial condition is
$u(x,0) = u_0 + x(1 - x)$, determine the solution completely.   (EC)

51   A function $y(x, t)$ satisfies the differential equation $\dfrac{\partial^2 y}{\partial x^2} = \dfrac{\partial^2 y}{\partial t^2}$ subject to the conditions

(a) $y = 0$, which implies $\dfrac{\partial y}{\partial t} = 0$ when $x = 0$ and $x = 10$ for all $t \geq 0$;

(b) $\dfrac{\partial y}{\partial t} = 0$ when $t = 0$ for $0 \leq x \leq 10$;

(c) when $t = 0$, $\dfrac{\partial y}{\partial x}$ has values given by the table

x	0	1	2	3	4	5	6	7	8	9	10
$\dfrac{\partial y}{\partial x}$	2.00	1.90	1.62	1.18	0.62	0.00	−0.62	−1.18	−1.62	−1.90	−2.00

Using the method of characteristics determine the values of $\dfrac{\partial y}{\partial t}$ when $t = 4$ for $x = 1, 5, 8$

explaining clearly the steps in the solution.   (LU)

8

INTEGRAL TRANSFORMS

8.1 Introduction

In this chapter we study examples of transforms. An integral transform of a function $f(x)$, which is defined for all x in the interval $a \leq x \leq b$, is a function

$$F(s) = \int_a^b f(x)\, k(x, s)\, \mathrm{d}x \qquad (8.1)$$

where $k(x, s)$ is a function known as the **kernel** of the transform.

We can regard the process of obtaining the transform as performing an **operation** on f. Sometimes $F(s)$ is said is to be defined on the image-space or s–space. (Note that some authors use p instead of s.)

Much of the application of integral transforms is for the purpose of solving differential equations. The method is depicted in Figure 8.1.

The idea is that the equation for $F(s)$ is easier to solve than the original model equation; in the case of the original being an ordinary differential equation, we aim to produce an algebraic equation for $F(s)$, whereas a partial differential equation can be converted to an ordinary differential equation. (The method can be applied to a model involving simultaneous differential equations.)

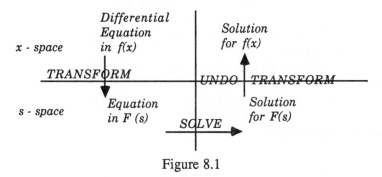

Figure 8.1

For the approach to be useful, we require that the process of taking a transform is a one-to-one operation, i.e. for each function $f(x)$ there is a unique transform $F(s)$ and, conversely, for each transform $F(s)$ there is a unique function $f(x)$ upon inverting the process. We may note in passing that the transform operation is a linear one as a consequence of the definition (8.1).

In Chapter 24 of *Engineering Mathematics* we introduced the Laplace

Transform and applied it to some simple ordinary differential equations. We begin this chapter by considering further applications where the forcing term is discontinuous or periodic.

In Chapter 6 of this book, we introduced the Fourier series representation of a function and applied the technique to the solution of partial differential equations in Chapter 7. We had to assume that the function represented was defined on a finite interval and then we were able to extend artificially its domain to make it periodic. In the latter part of this chapter we introduce the idea of a **Finite Fourier Transform** which for the moment can be regarded as a way of transforming the partial differential equation in anticipation of the Fourier series representation. Then we develop the limiting case as the domain of definition becomes infinite (or semi-infinite) and produce the **Infinite Fourier Transforms**.

It is perhaps more an art than a science to decide which transform method to employ on a partial differential equation. Much depends on the boundary conditions and the expected nature of the solution. We shall study the heat conduction problem with differing boundary conditions to help illustrate the contrast between the effectiveness of various transforms.

We can analyse waveforms by Fourier Transforms in an analogous method to Fourier series analysis. We end the chapter by looking at this and related transforms, viz. z-transforms and discrete Fourier transforms.

8.2 Basic results on the Laplace Transform

Some important results on Laplace transforms were derived in *Engineering Mathematics*; in this section we produce some more. We assume that the Laplace transform of $f(t)$ exists and is defined by

$$\mathcal{L}\{f(t)\} \equiv F(s) = \int_0^\infty e^{-st} f(t) \, dt \qquad (8.2)$$

Since the Laplace transform is usually applied to problems where time is the independent variable, we work with $f(t)$.

First we provide a resumé of some results referred to above.

(i) **First Shift Theorem**

This represents a shift in s-space. Let a be any constant, then

$$\mathcal{L}\{e^{at} f(t)\} = F(s - a) \qquad (8.3)$$

(ii) **Multiplication by t^n**

If n is any positive integer, $\mathcal{L}\{t^n f(t)\} = (-1)^n \dfrac{d^n F(s)}{ds^n}$ (8.4)

(iii) **Division by t** $\mathcal{L}\left\{\dfrac{f(t)}{t}\right\} = \int_s^\infty F(s) \, ds$ (8.5)

(iv) **Transform of Integral of** $f(t)$ $\mathcal{L}\left\{\displaystyle\int_0^t f(u)\, du\right\} = \dfrac{1}{s} F(s)$ (8.6)

(v) **Transform of derivatives**

If $f(t)$ is continuous and $\lim\limits_{t \to 0} f(t) = f(0)$ then

$$\mathcal{L}\{f'(t)\} = sF(s) - f(0) \qquad (8.7a)$$

If, further, $f'(t)$ is continuous and $\lim\limits_{t \to 0} f'(t) = f'(0)$ then

$$\mathcal{L}\{f''(t)\} = s^2 F(s) - sf(0) - f'(0) \qquad (8.7b)$$

These results generalise to the result that if $f(t)$ and all its derivatives up to and including order $(n-1)$ are continuous then

$$\mathcal{L}\{f^{(n)}(t)\} = s^n F(s) - s^{n-1} f(0) - s^{n-2} f'(0) - \ldots - f^{(n-1)}(0) \qquad (8.7c)$$

(vi) **Change of scale**

If a is any positive constant then

$$\mathcal{L}\{f(at)\} = \frac{1}{a} F\left[\frac{s}{a}\right] \qquad (8.8)$$

(vii) **Parametric differentiation**

If a is some parameter involved in the basic function then

$$\mathcal{L}\left\{\frac{\partial}{\partial a} f(t, a)\right\} = \frac{\partial}{\partial a} [F(s, a)] \qquad (8.9)$$

The above results allow us to extend the table of standard transforms. A further result we may quote is that $\lim\limits_{s \to \infty} F(s) = 0$.

Initial– and Final–Value Theorems

Sometimes we want to know, not the solution of a differential equation but either the initial or, more usually, the final value of the dependent variable. We have two useful results. Provided the limits exist then

(i) $\lim\limits_{s \to \infty} sF(s) = \lim\limits_{t \to 0} f(t)$ (8.10)

(ii) $\lim\limits_{s \to 0} sF(s) = \lim\limits_{t \to \infty} f(t)$ (8.11)

Example

A mass m falls through a viscous liquid. The equation of motion of the mass in terms of its velocity $v(t)$ is $m \dfrac{dv}{dt} = -kv + mg$ where k is the viscous damping coefficient of the liquid. Taking Laplace transforms, we obtain

$$m[sV(s) - v(0)] = -kV(s) + \frac{mg}{s} \quad \text{so that} \quad V(s) = \frac{mg}{s(ms + k)} + \frac{mv(0)}{(ms + k)} \ .$$

Then the terminal velocity of the mass is given by

$$\lim_{t \to \infty} v(t) = \lim_{s \to 0} sV(s) = \frac{mg}{k}$$

Step response

The Laplace Transform of $U(t - a)$ is found directly from the definition to be

$$\int_0^\infty e^{-st} U(t - a) \, dt = \int_a^\infty e^{-st} \, dt = \frac{1}{s} e^{-as} \tag{8.12}$$

Obviously the transform of $U(t)$ is $\dfrac{1}{s}$.

The response of a system to excitation by a unit step function is called its **indicial response** or **step response**.

Example

A light beam is freely supported at its ends by point supports at the same horizontal level. It carries a uniform load over the middle third of its length. Find the deflected profile of the beam.

If the length of the beam is l and the uniform load is w/unit length then the differential equation governing the beam is $y^{(iv)}(x) = \dfrac{f(x)}{EI}$ where y is the deflection of a point on the beam a distance x from the left-hand end and $f(x)$ is the load distribution. The end conditions are $y = 0$ at $x = 0$ and at $x = l$ (zero deflections at the ends) and $y'' = 0$ at $x = 0$ and at $x = l$ (zero moment at the ends). Note that x is the independent variable here instead of the t we have used in previous theory. You can check that

$$f(x) = w[U(x - l/3) - U(x - 2l/3)]$$

and that the transformed equation is

$$EI[s^4 Y(s) - s^2 y'(0) - y'''(0)] = w \left[\frac{e^{-sl/3} - e^{-2sl/3}}{s} \right]$$

where $y'(0)$ and $y'''(0)$ are unknowns.

Then
$$Y(s) = \frac{w}{EI} \left[\frac{e^{-sl/3} - e^{-2sl/3}}{s^5} \right] + \frac{y'(0)}{s^2} + \frac{y'''(0)}{s^4} \tag{8.13}$$

Second shift theorem

If $\quad \mathcal{L}\{f(t)\} = F(s) \quad$ then $\quad \mathcal{L}\{U(t - a) f(t - a)\} = e^{-as} F(s) \tag{8.14}$

This represents a shift in the t-direction.

Example
We are now in a position to complete the previous example. Since we know
$\mathcal{L}\{x^4\} = 4!/s^5$ it follows that $e^{-sl/3}/s^5$ must be the transform of a function
similar   to $U(x - l/3)(x - l/3)^4$.   In fact it is the transform of
$U(x- l/3)(x - l/3)^4/4!$ as you can check.
   Hence we may invert (8.13) to obtain

$$y(x) = \frac{w}{24EI} [(x - l/3)^4 \, U(x - l/3) - (x - 2l/3)^4 \, U(x - 2l/3)] +$$

$$y'(0)x + y'''(0) \, x^3/6$$

Now $y(l) = 0$ so that

$$0 = \frac{w}{24EI} \left(\left[\frac{2l}{3}\right]^4 - \left[\frac{l}{3}\right]^4 \right) + y'(0)l + y'''(0) \frac{l^3}{6} \qquad \text{(8.15a)}$$

Furthermore $y''(l) = 0$ so that

$$0 = \frac{w}{24EI} \left(12\left[\frac{2l}{3}\right]^2 - 12\left[\frac{l}{3}\right]^2 \right) + y'''(0)l \qquad \text{(8.15b)}$$

From (8.15a) and (8.15b) we obtain

$$y'''(0) = -\frac{wl}{6EI} \text{ and } y'(0) = \frac{13}{648}\frac{wl^3}{EI}$$

We can alternatively express the solution via three formulae:

(i)   for $0 \le x \le l/3$, $y(x) = \dfrac{13}{648}\dfrac{wl^3 x}{EI} - \dfrac{wlx^3}{36EI}$

(ii)   for $l/3 \le x \le 2l/3$, $y(x) = \dfrac{w}{24EI} (x - l/3)^4 + \dfrac{13wl^3 x}{648EI} - \dfrac{wlx^3}{36EI}$

(iii) for $2l/3 \le x \le l$, $y(x) = \dfrac{w}{24EI} \left(\left[x - \dfrac{l}{3}\right]^4 - \left[x - \dfrac{2l}{3}\right]^4 \right) + \dfrac{13wl^3 x}{648EI} - \dfrac{wlx^3}{36EI}$

8.3 Further results on Laplace Transforms

(a) Unit Impulse Function
   When a large force acts for a very short time as for example in a shock test, or
when a shock voltage passes through an electrical system we can idealise the

situation by assuming all the effect to occur at a point in time. We can also idealise in space as with point loads on a structure. Figure 8.2(a) shows a rectangular pulse of finite duration and of unit strength (measured by the area under the graph). If we keep the strength constant and reduce the duration of the pulse we obtain a function similar to the one depicted in Figure 8.2(b). In the limit as $\varepsilon \to 0$ we obtain the idealised **Dirac delta function** $\delta(t-a)$, defined by

$$\delta(t-a) = \lim_{\varepsilon \to 0} \left[\frac{1}{\varepsilon}\,(U(t) - U(t-a)) \right] \tag{8.16}$$

Note that $\delta(t-a) = 0$ for all $t \ne a$ and that $\displaystyle\int_{-\infty}^{\infty} \delta(t-a)\mathrm{d}t = 1$.

(Note also the special case when $a = 0$.)

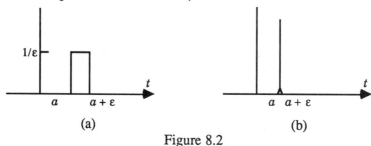

(a)   (b)

Figure 8.2

For example, if an impulse voltage of strength E is applied to an electrical circuit at $t = a$ then we represent it as $E\delta(t-a)$. We have the result that if $b < a < c$ and if $f(t)$ is integrable over the interval then

$$\int_{b}^{c} f(t)\,\delta(t-a)\,\mathrm{d}t = f(a) \tag{8.17}$$

This result follows from the definition of a definite integral. The Laplace transform of the delta function is given by

$$\mathcal{L}\{\delta(t-a)\} = \int_{0}^{\infty} \mathrm{e}^{-st}\,\delta(t-a)\,\mathrm{d}t = \mathrm{e}^{-sa} \tag{8.18a}$$

from result (8.17). Note the special case

$$\mathcal{L}\{\delta(t)\} = 1 \tag{8.18b}$$

An important result is provided by the following equation which you can verify for yourself.

$$\mathcal{L}\{f(t)\,\delta(t-a)\} = \int_{0}^{\infty} \mathrm{e}^{-st}\,f(t)\,\delta(t-a)\,\mathrm{d}t = f(a)\,\mathrm{e}^{-sa}$$

Example 1

(i) $\quad \mathcal{L}\{t^2\,\delta(t-3)\} = 3^2\,e^{-3s} = 9e^{-3s}$

(ii) $\quad \mathcal{L}\{\sin t\delta(t-\pi/2)\} = \sin\dfrac{\pi}{2}\,e^{-\pi s/2} = e^{-\pi s/2}$

We defined the step function $U(t-a)$ to be of value 1 when $t=a$. If we modify the definition slightly so that $U(t-a)=0$ when $t=a$ then we will not markedly alter any results so far obtained. However, if we denote this new function $U^*(t-a)$ then it can be shown that

$$\delta(t-a) = \frac{d}{dt}\,U^*(t-a) \qquad (8.19)$$

Some authors work with $U^*(t-a)$ in preference to $U(t-a)$.

Example 2

The system shown at rest in Figure 8.3 is given an impulsive blow $F\delta(t)$.

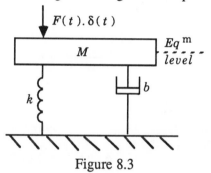

Figure 8.3

The equation of motion for the mass is $M\ddot{x} = F\delta(t) - kx - b\dot{x}$ and the initial

conditions are $x = \dot{x} = 0$ at $t = 0$. The transformed equation is

$$M[s^2 X(s)] = F - kX(s) - b[sX(s)] => X(s) = \frac{F}{(Ms^2 + bs + k)}$$

This is precisely the same result as if we had removed the impulsive blow and

given the system an initial velocity $\dot{x}(0) = F/m$, as we might have expected.

(b) Periodic function

Consider a function of period T defined on the interval $[0, T]$ by $f(t)$. Then $f(t + T) = f(t)$ and

$$F(s) = \frac{1}{(1 - e^{-sT})}\int_0^T e^{-st} f(t)\,dt \qquad (8.20)$$

Example

The square wave pulse is given by

$$f(t) = U(t) - U(t-a), \quad 0 \le t < a \qquad\qquad f(t+2a) = f(t), \; t \ge 0$$
$$= -U(t-a) + U((t-2a), \quad a \le t \le 2a$$

A domestic example of this general waveform is the pressure wave generated near the valve when a tap is turned off suddenly: the phenomenon of water hammer. Let us now derive its Laplace transform as a periodic function. From (8.20) this is

$$\frac{1}{(1-e^{-2sa})}\left\{ \int_0^a e^{-st}.1\, dt + \int_a^{2a} e^{-st}(-1)\, dt \right\} = \frac{1}{(1-e^{-2sa})}\left\{ \left[\frac{e^{-st}}{-s}\right]_0^a + \left[\frac{e^{-st}}{s}\right]_a^{2a} \right\}$$

$$= \frac{1}{(1-e^{-2sa})}\left\{ \frac{1}{s}(1-e^{-sa}) + \frac{1}{s}(e^{-2sa}-e^{-sa}) \right\}$$

$$= \frac{1}{(1-e^{-2sa})}\frac{1}{s}(1-e^{-sa})^2 = \frac{1}{s}\cdot\frac{(1-e^{-sa})}{(1+e^{-sa})}$$

(c) Convolution Theorem

We shall examine more closely the concept of convolution in Section 8.7. For the moment we shall regard it as a means of finding the inverse transform when the transform $F(s)$ is a product of the transforms of known functions. For

example, if $F(s) = \dfrac{2}{s^3} \cdot \dfrac{a}{(s^2+a^2)}$ then we recognise $\dfrac{2}{s^3}$ as the transform of t^2

and $a/(s^2+a^2)$ as the transform of $\sin at$. We make use of the following **convolution theorem**:

If $H(s)$ is the Laplace transform of $h(t)$ and $G(s)$ is the transform of $g(t)$ then $H(s) \cdot G(s)$ is the Laplace transform of

$$\int_0^t h(u)\, g(t-u)\, du = \int_0^t h(t-u)\, g(u)\, du \qquad\qquad (8.21)$$

Either integral is called the convolution of $h(t)$ and $g(t)$, and written $h * g$.

With reference to our example, $h(t) = t^2$ and $g(t) = \sin at$ so that $F(s)$ is

the transform of $I = \displaystyle\int_0^t u^2 \sin a(t-u)du$. We evaluate this integral by parts twice to obtain $I = \dfrac{t^2}{a} - \dfrac{2}{a^3}(1 - \cos at)$. You can verify directly that this is the

function whose transform is $\dfrac{2}{s^3}\dfrac{a}{(s^2 + a^2)}$.

Example

Find the function whose transform is $s^2/(s^2 + 4)^2$.

We see that the given expression is the square of $s/(s^2 + 4)$ which is the transform of $\cos 2t$. Consequently both $h(t)$ and $g(t)$ are $\cos 2t$. By the convolution theorem the function we seek is given by

$$f(t) = \int_0^t \cos 2u \cdot \cos 2(t - u)\, du = \frac{1}{2}\int_0^t [\cos 2t + \cos(4u - 2t)]\, du$$

$$= \frac{1}{2}[u \cos 2t + \frac{1}{4}\sin(4u - 2t)]_0^t$$

$$= \frac{1}{2}(t \cos 2t + \frac{1}{4}\sin 2t + \frac{1}{4}\sin 2t) = \frac{1}{2}t \cos 2t + \frac{1}{4}\sin 2t$$

For the moment we may note that if the impulsive response $h(t)$ of a system is known, then the response to any arbitrary excitation $g(t)$ is given by the convolution $h * g$.

Transfer functions

If a linear system has all initial conditions zero then the **transfer function** of the system is defined to be the ratio of the Laplace transform of the output variable to the Laplace transform of the input variable. For example, the transformed equation for a damped spring-mass system with zero initial conditions may be written

$$Ms^2 Y(s) + ksY(s) + n^2 Y(s) = F(s) \tag{8.22}$$

The transfer function of the system is

$$G(s) = \frac{Y(s)}{F(s)} = \frac{1}{Ms^2 + ks + n^2} \tag{8.23}$$

In general, the transfer function tells us how much of the input variable is transferred to the output variable. Often the transfer function is a ratio of polynomials in s. The values of s which make the denominator zero are called the **poles** of the function. Since the denominator is evolved from the input variable, the poles give information as to the natural motions that the system may possess. The poles may be complex numbers; in addition we might need to resort to numerical methods for their location.

Inversion of the Transform

A vital stage in the process of solving a differential equation by Laplace transforms is that of inverting the transform to obtain the solution. So far we have not considered this stage other than to refer to standard results or show ways in which the transform can be related to one or more standard transforms. In practice, it might not be always that straightforward. The rigorous definition of the Laplace transform allows the parameter s to be a complex number with a positive real part. The appropriate inversion can, in general be achieved via integration in the complex plane; this matter is taken up in Chapter 10.

8.4 Finite Fourier Transforms

Let us suppose that a function $f(x)$ is defined over the interval $0 \le x \le l$ and that it satisfies the Dirichlet conditions (see page 183). We can convert it into an odd function by defining $f(-x) = -f(x)$ and then we can extend its domain by periodicity. Then we can represent $f(x)$ by a half-range sine series as follows:

$$f(x) = \sum_{s=1}^{\infty} b_s \sin\left[\frac{s\pi x}{l}\right] \tag{8.24a}$$

where the coefficients are given by

$$b_s = \int_0^l f(x)\left(\frac{2}{l}\sin\left[\frac{s\pi x}{l}\right]\right) dx \tag{8.24b}$$

Expressed in this form you can see that b_s is the effect of applying an integral

transform to $f(x)$. The kernel of the transform is $\dfrac{2}{l}\sin\left[\dfrac{s\pi x}{l}\right]$. Formula

(8.24b) defines this **finite Fourier sine transform** and, in a sense, formula (8.24a) acts as an **inversion formula**.

Equally well, we can convert the original function $f(x)$ into an even function and then define the **finite Fourier cosine transform** and its inversion by the following formulae

$$f(x) = \tfrac{1}{2}a_0 + \sum_{s=1}^{\infty} a_s \cos\left[\frac{s\pi x}{l}\right] \tag{8.25a}$$

where the coefficients are given by

$$a_s = \int_0^l f(x) \cdot \left(\frac{2}{l}\cos\left[\frac{s\pi x}{l}\right]\right) dx \tag{8.25b}$$

If $f(x)$ is neither even nor odd and defined on the interval $[-l, l]$ then we may use the complex form of Fourier series representation and define the **finite Fourier transform** by

$$f(x) = \sum_{s=-\infty}^{\infty} c_s \exp(is\pi x/l) \qquad (8.26a)$$

where

$$c_s = \int_{-l}^{l} f(x) \left(\frac{1}{2l} \exp(-is\pi x/l) \right) dx \qquad (8.26b)$$

We shall shortly give two examples of how to use Fourier transforms to solve partial differential equations which we would have solved by separation of variables. First we need the transform of the second derivative of a function, since this is the derivative occurring most frequently in such equations. Note that we use x as the variable in our definition of the transforms.

Let us denote the finite Fourier sine transform of $f(x)$ by $\bar{f}_s(s)$ or $\bar{f}_s$. Then the finite Fourier sine transform of $\partial^2 f/\partial x^2$ is

$$\int_0^l \frac{\partial^2 f}{\partial x^2} \frac{2}{l} \sin\left[\frac{s\pi x}{l}\right] dx = \int_0^l \frac{2}{l} \sin\left[\frac{s\pi x}{l}\right] \cdot \frac{\partial^2 f}{\partial x^2} dx$$

Integrating by parts, we obtain

$$\left[\frac{2}{l} \sin\left[\frac{s\pi x}{l}\right] \cdot \frac{\partial f}{\partial x} \right]_0^l - \int_0^l \frac{2s\pi}{l^2} \cos\left[\frac{s\pi x}{l}\right] \frac{\partial f}{\partial x} dx$$

$$= 0 - \frac{2s\pi}{l^2} \int_0^l \cos\left[\frac{s\pi x}{l}\right] \frac{\partial f}{\partial x} dx$$

$$= \frac{-2s\pi}{l^2} \left\{ \left[\cos\left[\frac{s\pi x}{l}\right] f(x) \right]_0^l + \frac{s\pi}{l} \int_0^l \sin\left[\frac{s\pi x}{l}\right] f(x) \, dx \right\}$$

$$= \frac{-2s\pi}{l^2} [(-1)^s f(l) - f(0)] - \frac{s^2 \pi^2}{l^2} \bar{f}_s \qquad (8.27)$$

If we know $f(l)$ and $f(0)$ then we can specify this transform completely.

If the finite Fourier cosine transform of $f(x)$ be $\bar{f}_c$ then the same transform of

$$\frac{\partial^2 f}{\partial x^2} \text{ is } \qquad \frac{2}{l} \left[(-1)^s \left(\frac{\partial f}{\partial x}\right)_{x=l} - \left(\frac{\partial f}{\partial x}\right)_{x=0} \right] - \frac{s^2 \pi^2}{l^2} \bar{f}_c \qquad (8.28)$$

If the value of $\dfrac{\partial f}{\partial x}$ is specified at $x = 0$ and $x = l$ then this transform is determined completely.

Before we apply the transforms to differential equations we should note that the factor $(2/l)$ is a slight nuisance and we could re-define our transforms to put it in the inversion formulae. Also, note that the boundary conditions on the problem will select the transform to be used.

Example 1

The equation of heat conduction for a bar whose sides are insulated and which is of length l is

$$\frac{\partial \theta}{\partial t} = K \frac{\partial^2 \theta}{\partial x^2} \tag{8.29}$$

If the bar is insulated across the end faces $x = 0$ and $x = l$ we have the boundary conditions $\partial \theta / \partial x = 0$ at $x = 0$ and $x = l$

Finally, let $\theta = 100x(l - x)$ at $t = 0$ for $0 < x < l$ $\tag{8.30}$

Since $\partial \theta / \partial x$ is specified on the boundaries, a cosine transform is required to eliminate the x – dependence. We multiply both sides of (8.29) by $\cos\left[\dfrac{s\pi x}{l}\right]$

and integrate over the interval $[0, l]$ to obtain

$$\int_0^l \cos\left[\frac{s\pi x}{l}\right] \frac{\partial \theta}{\partial t}\ dx = K \int_0^l \cos\left[\frac{s\pi x}{l}\right] \frac{\partial^2 \theta}{\partial x^2}\ dx \tag{8.31}$$

If we let $$\overline{\theta}_c = \int_0^l \theta(x,\, t) \cos\left[\frac{s\pi x}{l}\right] dx$$

then the left-hand side of (8.31) is simply $\dfrac{\partial \overline{\theta}_c}{\partial t}$. The right-hand side becomes

$$K\left[\cos\left[\frac{s\pi x}{l}\right] \frac{\partial \theta}{\partial x}\right]_0^l + \frac{Ks\pi}{l} \int_0^l \sin\left[\frac{s\pi x}{l}\right] \cdot \frac{\partial \theta}{\partial x}\ dx$$

$$= K[0 - 0] + \frac{Ks\pi}{l} \int_0^l \sin\left[\frac{s\pi x}{l}\right] \frac{\partial \theta}{\partial x}\ dx$$

$$= \frac{Ks\pi}{l} \left[\sin\left[\frac{s\pi x}{l}\right] \cdot \theta(x, t) \right]_0^l - \frac{K s^2 \pi^2}{l^2} \int_0^l \cos\left[\frac{s\pi x}{l}\right] \theta(x, t) \, dx$$

$$= -\frac{K s^2 \pi^2}{l^2} \overline{\theta}_c$$

(Note the comparison with (8.28).)

The partial differential equation (8.29) has been reduced to an ordinary differential equation

$$\frac{\partial \overline{\theta}_c}{\partial t} = -\frac{K s^2 \pi^2}{l^2} \overline{\theta}_c$$

although we retain the partial derivative sign since, strictly, $\overline{\theta}_c = \overline{\theta}_c (s, t)$. This has the general solution

$$\overline{\theta}_c = A e^{-Ks^2\pi^2 t/l^2}$$

where the constant A is the value of $\overline{\theta}_c$ at $t = 0$

i.e. $$A = \int_0^l \theta(x, 0) \cos\left[\frac{s\pi x}{l}\right] dx = \int_0^l 100x(l - x) \cos\left[\frac{s\pi x}{l}\right] dx$$

Eventually, $$A = \frac{-100l^3}{s^2 \pi^2} (1 + \cos s\pi) = \begin{cases} 0 & \text{if } s \text{ is odd} \\ \dfrac{-200l^3}{s^2 \pi^2} & \text{if } s \text{ is even} \end{cases}$$

Now the inversion formula would give us

$$\theta(x, t) = \frac{1}{2} \cdot \frac{2}{l} a_0 + \frac{2}{l} \sum_{s=1}^{\infty} a_s \cos\left[\frac{s\pi x}{l}\right]$$

where $$a_s = \int_0^l \theta(x, t) \cos\left[\frac{s\pi x}{l}\right] dx = \overline{\theta}_c (s, t)$$

Note that the factor $\dfrac{2}{l}$ now appears in the inversion formula and that

$$a_0 = \int_0^l \theta(x, 0)dx = 100 \int_0^l x(l-x)dx = 100l^3/6 = \bar{\theta}_c (0, t)$$

Therefore

$$\theta(x, t) = \frac{1}{l} \bar{\theta}_c (0, t) + \frac{2}{l} \sum_{s=1}^{\infty} \bar{\theta}_c (s, t) \cos\left[\frac{s\pi x}{l}\right]$$

$$= 100l^2 \left\{ \frac{1}{6} - \sum_{s=2,4,6...}^{\infty} \frac{4}{s^2\pi^2} \cos\left[\frac{s\pi x}{l}\right] e^{-Ks^2\pi^2 t/l^2} \right\} \qquad (8.32)$$

Example 2

Consider the one-dimensional wave equation for a vibrating string of length l, fixed at its ends and released from rest with an initial profile $y = f(x)$.
The governing equation is

$$\frac{1}{c^2} \frac{\partial^2 y}{\partial t^2} = \frac{\partial^2 y}{\partial x^2} \qquad (8.33)$$

with boundary conditions and initial conditions
$$y(0, t) = y(l, t) = 0$$

$$y = f(x) \quad \text{and} \quad \frac{\partial y}{\partial t} = 0 \quad \text{at} \quad t = 0 \qquad (8.34)$$

We choose the sine transform to eliminate x since y itself is specified at the boundaries.

We multiply (8.33) by $\sin\left[\frac{s\pi x}{l}\right]$ and integrate over the interval $[0, l]$ to obtain

$$\frac{1}{c^2} \int_0^l \sin\left[\frac{s\pi x}{l}\right] \frac{\partial^2 y}{\partial t^2} dx = \int_0^l \sin\left[\frac{s\pi x}{l}\right] \frac{\partial^2 y}{\partial x^2} dx \qquad (8.35)$$

Let $\bar{y}_s(s, t) = \int_0^l \sin\left[\frac{s\pi x}{l}\right] y(x, t) \, dx$.

Then the left-hand side of (8.35) is $\dfrac{1}{c^2} \dfrac{\partial^2 \bar{y}_s}{\partial t^2}$.

After integrating by parts, the right-hand side becomes, $-\dfrac{s^2\pi^2}{l^2}\,\bar{y}_s$.

Hence we have

$$\frac{1}{c^2}\frac{\partial^2 \bar{y}_s}{\partial t^2} = -\frac{s^2\pi^2}{l^2}\,\bar{y}_s$$

and this has the general solution

$$\bar{y}_s(s,\,t) = A_s \cos\left[\frac{s\pi ct}{l}\right] + B_s \sin\left[\frac{s\pi ct}{l}\right]$$

where A_s and B_s can be determined from conditions (8.34).

Hence $\quad \bar{y}_s(x,\,0) = \displaystyle\int_0^l f(x) \sin\left[\frac{s\pi x}{l}\right] dx = A_s$ and $\dfrac{\partial \bar{y}_s}{\partial t} = 0 = B_s$

Therefore, $\quad \bar{y}_s(s,\,t) = \cos\left[\dfrac{s\pi ct}{l}\right] \displaystyle\int_0^l f(x) \sin\left[\dfrac{s\pi x}{l}\right] dx$

Inverting the transform we obtain

$$y(x,\,t) = \frac{2}{l}\sum_{s=1}^{\infty}\left\{\sin\left[\frac{s\pi x}{l}\right]\cos\left[\frac{s\pi ct}{l}\right]\right\}\int_0^l f(u)\sin\left[\frac{s\pi u}{l}\right] du \qquad (8.36)$$

where we have changed the variable of integration from x to u.

The diagram below shows the process we have followed.

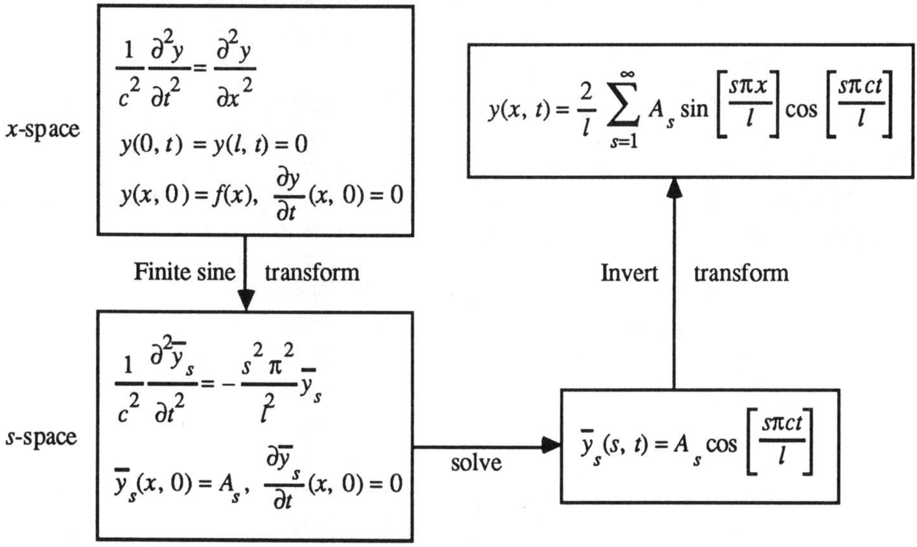

x-space:
$$\frac{1}{c^2}\frac{\partial^2 y}{\partial t^2} = \frac{\partial^2 y}{\partial x^2}$$
$$y(0,\,t) = y(l,\,t) = 0$$
$$y(x,\,0) = f(x),\ \frac{\partial y}{\partial t}(x,\,0) = 0$$

$$y(x,\,t) = \frac{2}{l}\sum_{s=1}^{\infty} A_s \sin\left[\frac{s\pi x}{l}\right]\cos\left[\frac{s\pi ct}{l}\right]$$

Finite sine transform

Invert transform

s-space:
$$\frac{1}{c^2}\frac{\partial^2 \bar{y}_s}{\partial t^2} = -\frac{s^2\pi^2}{l^2}\bar{y}_s$$
$$\bar{y}_s(x,\,0) = A_s,\ \frac{\partial \bar{y}_s}{\partial t}(x,\,0) = 0$$

solve

$$\bar{y}_s(s,\,t) = A_s \cos\left[\frac{s\pi ct}{l}\right]$$

Figure 8.4

Notes on the transforms

We cannot exclude odd order derivatives by the Fourier transforms and therefore in the heat conduction problem we would naturally eliminate x. In the case of the wave equation the boundary conditions on x were more suited to transformations. The Fourier transformation is less flexible than the Laplace transform but it does produce a quicker result when it is applicable.

For an equation involving three independent variables we can apply transforms repeatedly, but we shall not take the matter further here.

It may be of help to summarise the last problem in diagram form as shown in Figure 8.4.

8.5 Infinite Fourier Transforms

In some physical problems we consider one or more space dimensions to be infinite or semi-infinite. We may have a bar which is very long in comparison to its width and depth or, in order to neglect edge effects we may wish to choose as a model a plate which is infinite in extent. In such cases we want to consider the domain of the transform to extend to infinity in both directions. Furthermore, the finite transform, or equivalently the Fourier series required a waveform which was effectively periodic or which was defined on a finite interval. If, for example, we wish to study transient waveforms then the finite transforms will not suffice. In this section we examine infinite transforms as applied to partial differential equations. In Section 8.7 we shall study the interpretation of transforms of waveforms.

Fourier's Integral Formula and the Fourier Transforms

We shall merely quote this result, which can be regarded as the starting point for the transforms. It states that for a function $f(x)$ which satisfies Dirichlet's conditions (page 183) we may write

$$f(x) = \frac{1}{\pi} \int_0^\infty \left\{ \int_{-\infty}^\infty f(u) \cos [s(x - u)] \, du \right\} ds \qquad (8.37)$$

Here u is a dummy variable of integration.

(The proof follows by considering the limiting case of a Fourier series representation.) An alternative formulation is

$$f(x) = \frac{1}{2\pi} \int_{-\infty}^\infty e^{isx} \left\{ \int_{-\infty}^\infty f(u) \, e^{-isu} \, du \right\} ds$$

$$= \frac{1}{\sqrt{2\pi}} \int_{-\infty}^\infty e^{isx} \left\{ \frac{1}{\sqrt{2\pi}} \int_{-\infty}^\infty f(u) \, e^{-isu} \, du \right\} ds \qquad (8.38)$$

Equate the real part of the right-hand side to $f(x)$.

This last symmetrical form suggests that we define the **ordinary Fourier Transform** by

$$F(s) = \frac{1}{\sqrt{2\pi}} \int_{-\infty}^{\infty} e^{-isu} f(u) \, du \qquad (8.39a)$$

with inversion formula

$$f(x) = \frac{1}{\sqrt{2\pi}} \int_{-\infty}^{\infty} e^{isx} F(s) \, ds \qquad (8.39b)$$

If $f(x)$ is defined on the interval $0 \le x$ we may extend the domain by creating it an even function. Then, returning to (8.37) we find that

$$f(x) = \frac{1}{\pi} \int_{0}^{\infty} \left\{ \int_{-\infty}^{0} f(u) \cos [s (x - u)] \, du + \int_{0}^{\infty} f(u) \cos [s (x - u)] \, du \right\} \, ds$$

$$= \frac{1}{\pi} \int_{0}^{\infty} \left\{ \int_{0}^{\infty} f(u) \cos [s (x + u)] \, du + \int_{0}^{\infty} f(u) \cos [s (x - u)] \, du \right\} \, ds \; ^\dagger$$

$$= \frac{1}{\pi} \int_{0}^{\infty} \left\{ \int_{0}^{\infty} 2f(u) \cos sx \cos su \, du \right\} \, ds$$

Hence we may define the **Fourier cosine transform** by

$$F_c(s) = \sqrt{\frac{2}{\pi}} \int_{0}^{\infty} f(u) \cos su \, du \qquad (8.40a)$$

with the inversion formula

$$f(x) = \sqrt{\frac{2}{\pi}} \int_{0}^{\infty} F_c(s) \cos sx \, ds \qquad (8.40b)$$

In a similar way we define the **Fourier sine transform** by

$$F_s(s) = \sqrt{\frac{2}{\pi}} \int_{0}^{\infty} f(u) \sin su \, du \qquad (8.41a)$$

with the inversion formula

---

† Put $u = -u$ and $f(-u) = f(u)$

$$f(x) = \sqrt{\frac{2}{\pi}} \int_0^\infty F_s(s) \sin sx \, ds \qquad (8.41b)$$

Example 1

(i)   We find the ordinary Fourier transform of the rectangular pulse shown in Figure 8.5(a) whose definition is

$$f(t) = \begin{cases} \dfrac{1}{2T} & -T < t < T \\[2mm] 0 & \text{otherwise} \end{cases}$$

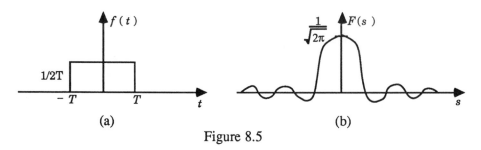

(a)   (b)

Figure 8.5

The transform is, from (8.39a)

$$F(s) = \frac{1}{\sqrt{2\pi}} \int_{-T}^{T} \frac{1}{2T} e^{-isu} \, du = \frac{1}{\sqrt{2\pi}} \cdot \frac{1}{2T} \left[\frac{e^{-isu}}{-is} \right]_{-T}^{T}$$

$$= \frac{1}{\sqrt{2\pi}} \cdot \frac{1}{Ts} \frac{1}{2i} (e^{isT} - e^{-isT}) = \frac{1}{\sqrt{2\pi}} \cdot \frac{\sin sT}{sT}$$

Note that as $s \to 0$, $F(s) \to 1/\sqrt{2\pi}$; the graph of $F(s)$ is presented in Figure 8.5(b). If $f(t)$ represents a rectangular pulse then $F(s)$ represents a frequency spectrum. Conversely, if $F(s)$ represents a frequency distribution then $f(t)$ represents the signal which produced it. In some applications, ω is used as the variable of the transform instead of s.

(ii)   Since $f(t)$ is symmetrical we could have found its Fourier cosine transform. It is

$$\sqrt{\frac{2}{\pi}} \int_0^T \frac{1}{2T} \cos su \, du = \sqrt{\frac{2}{\pi}} \frac{1}{2T} \left[\frac{\sin su}{s} \right]_0^T = \frac{1}{\sqrt{2\pi}\, T} \frac{\sin sT}{s}$$

as before.

(iii) To complete the picture, we shall now find the Fourier sine transform of this function. It is

$$\sqrt{\frac{2}{\pi}} \int_0^T \frac{1}{2T} \sin su \, du = \frac{1}{\sqrt{2\pi}} \cdot \frac{1}{T} \left[\frac{\cos su}{-s} \right]_0^T = \frac{1}{\sqrt{2\pi}} \cdot \frac{1}{sT} (1 - \cos sT)$$

Example 2 Unit impulse

This can be regarded as the limit of the rectangular pulse as $T \to 0$. We can see that the spectrum is given by $F(s) = 1$.

Miscellaneous Results

(It is assumed that $f(x)$ has transforms $F(s)$, $F_s(s)$, $F_c(s)$.)

(i)   **Transforms of derivatives**

If $f(x) \to 0$ as $x \to \infty$ and $f'(x) \to 0$ as $x \to \infty$ then

$f'(x)$ has cosine transform $-\sqrt{\dfrac{2}{\pi}} f(0) + sF_s(s)$ (8.42a)

and $f'(x)$ has sine transform $-sF_c(s)$ (8.42b)

If, further, $f''(x) \to 0$ as $x \to \infty$ then

$f''(x)$ has cosine transform $-\sqrt{\dfrac{2}{\pi}} f'(0) - s^2F_c(s)$ (8.42c)

$f''(x)$ has sine transform $\sqrt{\dfrac{2}{\pi}} sf(0) - s^2F_s(s)$ (8.42d)

If $f(x) \to 0$ as $x \to \pm\infty$ and $f'(x) \to 0$ as $x \to \pm\infty$ then
$f'(x)$ has Fourier transform $-isF(s)$ (8.42e)

If, further, $f''(x) \to 0$ as $x \to \pm\infty$ then
$f''(x)$ has Fourier transform $(is)^2 F(s)$ (8.42f)

(ii)   **Scaling theorem**

For $\lambda > 0$, $f(\lambda x)$ has a Fourier transform equal to $\lambda^{-1} F(s/\lambda)$ with similar results holding for the other transforms.

(iii) **Shift theorem**

$e^{i\lambda x} f(x)$ has transform $F(s - \lambda)$ and $f(x - \lambda)$ has transform $e^{-i\lambda s}F(s)$

(iv)   **Derivatives of transforms**

Assuming that differentiation is possible, $F'(s)$ is the Fourier transform of $xf(x)$, $F'_c(s)$ is the cosine transform of $xf(x)$ and $F'_s(s)$ is the sine transform of $-x f(x)$.

(v)   Convolutions

The convolution of the functions g and h is defined as

$$f(x) = \int_{-\infty}^{\infty} g(u)\, h(x-u)\, du = \int_{-\infty}^{\infty} g(x-u)\, h(u)\, du$$

The Fourier transform of the convolution is the product of the transforms of $g(x)$ and $h(x)$.

The use of convolutions in solving problems with time-dependent boundary conditions allows reduction to problems with constant boundary conditions. We discuss convolutions more fully in Section 8.7.

Comparison of Fourier and Laplace transforms

Let us first investigate the direct transforms.

The Fourier transform, being defined by an integral with limits from $t = -\infty$ to $t = +\infty$, has the advantage of being applicable to functions involving negative values of the variable t. This may not seem to be a significant advantage if t represents time; very seldom do we use time functions known or existing at $t \to -\infty$. However, if t does not represent time but some other variable, distance for instance, it is quite conceivable for this variable to assume negative values and for a function to be non-zero for such negative values of its variable. Obviously the direct Laplace transform, where the integration is carried from $t = 0$ to $t = \infty$, of such a function does not exist. (Actually, an implicit assumption is made that $f(t) = 0$ for $t < 0$.) A fundamental difference therefore between the Laplace and Fourier transforms lies in the fact that the former is applicable only to functions whose variable is always positive and the latter to functions whose variable may include negative values.

The Laplace transform, however compensates for this disadvantage by being applicable to a greater variety of functions (all functions of exponential order). Consider the function $f(t) = e^{at}$ where $a > 0$. Its Fourier transform does not exist since $f(t) \to \infty$ as $t \to \infty$; the function to be integrated is $e^{at}\, e^{-i\omega t}$ $= e^{at}(\cos \omega t - i \sin \omega t)$ and the area under the curve $e^{at} \cos \omega t$ or $e^{at} \sin \omega t$ is not finite as $t \to \infty$. However, in the case of the Laplace transform $(s = \sigma + i\omega)$, the function to be integrated is $e^{at}[e^{-st} = e^{-\sigma t}(\cos \omega t - i \sin \omega t)]$ and the area under the curve $e^{(a-\sigma)t} \cos \omega t$ or $e^{(a-\sigma)t} \sin \omega t$ is finite if $\sigma > a$. Hence the function e^{at}, $(a > 0)$, has a Laplace transform but not a Fourier transform. We see that the Laplace transform has a convergence factor $e^{-\sigma t}$, which gives the advantage over the Fourier transform. It is evident why the Laplace transform cannot involve negative values of t: the factor e^{-at} contributes to convergence only for $t > 0$ and causes divergence for $t < 0$. We may conclude that the Laplace transform, because of its better convergence properties, is applicable to a greater variety of functions provided these functions are zero for negative values of the variable.

Let us now examine the inverse transforms. The inverse Fourier transform has the advantage of involving an integration with a real variable ω, whereas the inverse Laplace transform necessitates a contour integration in the complex

s–plane. From this point of view it may sometimes be easier to find the inverse Fourier transform than the inverse Laplace transform.

Where $f(t)$ is a time function, the Fourier transform $F(i\omega)$ represents the frequency spectrum of $f(t)$.

Because of the greater variety of functions to which the Laplace transform is applicable, it is used more extensively in engineering than the Fourier transform. Its primary purpose is to facilitate the solution and to give a better insight into differential equations involved in engineering. In circuit theory it gives rise to the concept of operational impedances and admittances.

Since $\mathcal{L}[\delta(t)] = 1$, the inverse Laplace transform of operational impedances and admittances as well as of transfer functions has the physical meaning of being the time response to a unit impulse excitation. If $G(s)$ is the transfer function of a linear system (ratio of the Laplace transforms of, let us say, the output to input voltages), then $\mathcal{L}^{-1}[G(s)]$ is the output for a unit impulse input.

8.6 Solution of heat conduction equation

We now use various Fourier transforms to solve the heat conduction equation for a semi-infinite bar $x \geq 0$ with differing boundary conditions. Basically if the range of definition of the problem is $-\infty < x < \infty$ then the ordinary transform is called for, whereas if the range is $0 < x < \infty$ then either sine or cosine transforms are used to eliminate the **even order derivative**.

The heat conduction equation is

$$\frac{\partial \theta}{\partial t} = K \frac{\partial^2 \theta}{\partial x^2} \tag{8.29}$$

Case I

The bar is lagged at $x = 0$ so that $\dfrac{\partial \theta}{\partial x} = 0$ there, and $\theta = \theta_0(x)$ at $t = 0$.

From result (8.42c) we see that knowledge of $\partial\theta/\partial x$ at $x = 0$ suggests the cosine transform to eliminate x from the problem. We therefore apply the transform to both sides of (8.29) to obtain

$$\sqrt{\frac{2}{\pi}} \int_0^\infty \frac{\partial \theta}{\partial t} \cos sx \, dx = K \sqrt{\frac{2}{\pi}} \int_0^\infty \frac{\partial^2 \theta}{\partial x^2} \cos sx \, dx$$

i.e.
$$\frac{\partial \Theta_c}{\partial t} = K \left\{ -\sqrt{\frac{2}{\pi}} \theta'(0) - s^2 \Theta_c \right\}$$

where $\Theta_c \equiv \Theta_c(s, t)$ is the cosine transform of $\theta(x, t)$.

We therefore have the following equation

$$\frac{\partial \Theta_c}{\partial t} = -Ks^2 \Theta_c \tag{8.43}$$

which has solution $\qquad \Theta_c = Ae^{-Ks^2 t} \tag{8.44}$

where A is a function of s which is effectively the value of Θ_c at $t = 0$ i.e. $\Theta_c(s, 0)$. Now

$$\Theta_c(s, 0) \quad = \sqrt{\frac{2}{\pi}} \int_0^\infty \theta(x, 0) \cos sx \, dx$$

$$= \int_0^\infty \theta_0(x) \cos sx \, dx = F(s), \qquad \text{say.}$$

Note that for $F(s)$ to exist $\theta_0(x)$ must $\to 0$ as $x \to \infty$.

Hence $\qquad \Theta_c = F(s)e^{-Ks^2 t} \tag{8.45}$

We invert both sides by (8.40b) and obtain

$$\theta(x, t) = \sqrt{\frac{2}{\pi}} \int_0^\infty F(s)e^{-Ks^2 t} \cos sx \, ds$$

To be specific, we let $\theta_0(x)$ be $100e^{-x}$ so that

$$F(s) = \sqrt{\frac{2}{\pi}} \int_0^\infty 100e^{-x} \cos sx \, dx = \frac{100}{(s^2 + 1)} \sqrt{\frac{2}{\pi}} \tag{8.46}$$

Hence $\qquad \theta(x, t) = \frac{200}{\pi} \int_0^\infty \frac{1}{(s^2 + 1)} \cdot e^{-Ks^2 t} \cos sx \, ds$

It is not always easy to evaluate such integrals analytically.

Case II

The initial temperature distribution in the bar is zero and the end $x = 0$ is maintained at a constant temperature θ_0. On this occasion, the boundary conditions suggest the use of a sine transform.

Applying this transform to both sides of (8.29) we obtain

$$\sqrt{\frac{2}{\pi}} \int_0^\infty \frac{\partial \theta}{\partial t} \sin sx \, dx = K \sqrt{\frac{2}{\pi}} \int_0^\infty \frac{\partial^2 \theta}{\partial x^2} \sin sx \, dx$$

i.e.
$$\frac{\partial \Theta_s}{\partial t} = K \left\{ \sqrt{\frac{2}{\pi}} \, s\theta(0) - s^2 \Theta_s \right\}$$

where $\Theta_s = \Theta_s(s, t)$ is the sine transform of $\theta(x, t)$.

Hence we have the equation

$$\frac{\partial \Theta_s}{\partial t} + Ks^2 \Theta_s = K \sqrt{\frac{2}{\pi}} \, s\theta_0$$

which has solution

$$\Theta_s = Ae^{-Ks^2t} + \sqrt{\frac{2}{\pi}} \frac{\theta_0}{s}$$

where A is a yet undetermined function of s. When $t = 0$, $\theta = 0$ and hence

$\Theta_s(s, 0) = 0$. It follows that $A = -\sqrt{\dfrac{2}{\pi}} \dfrac{\theta_0}{s}$ and therefore

$$\Theta_s = \sqrt{\frac{2}{\pi}} \frac{\theta_0}{s} (1 - e^{-Ks^2t}) \tag{8.47}$$

We invert the transform via (8.41b) to obtain

$$\theta(x, t) = \frac{2}{\pi} \theta_0 \int_0^\infty \frac{1}{s} (1 - e^{-Ks^2t}) \sin xs \, ds \tag{8.48}$$

$$= \frac{2\theta_0}{\pi} \int_0^\infty \frac{1}{s} \sin xs \, ds - \frac{2\theta_0}{\pi} \int_0^\infty \frac{1}{s} e^{-Ks^2t} \sin xs \, ds$$

Now, by result (A3) in the Appendix the first integral is equal to $\pi/2$. The second

integral can be reduced by putting $s = \dfrac{u}{\sqrt{Kt}}$; it then becomes

$$\int_0^\infty \frac{e^{-u^2} \sin\left(xu / \sqrt{(Kt)}\right)}{u} \, du$$

as you can check. Result (A6) tells us that this integral is equal to $\dfrac{\pi}{2} \, \mathrm{erf}\left(\dfrac{x}{2\sqrt{Kt}}\right)$

Hence
$$\theta(x, t) = \theta_0 \left(1 - \mathrm{erf}\left[\frac{x}{2\sqrt{Kt}} \right] \right) \tag{8.49}$$

The solution is sketched in Figure 8.6.

Figure 8.6

Case III

We now repeat Case II using Laplace transforms to eliminate the time dependence (it transpires that this is easier than eliminating x-dependence).

If we let $\Theta(x, s)$ be the transform of $\theta(x, t)$ then if we multiply both sides of (8.29) by e^{-st} and integrate from $t = 0$ to $t = \infty$ we shall obtain

$$\int_0^\infty \frac{\partial \theta}{\partial t} e^{-st} \, dt = K \int_0^\infty \frac{\partial^2 \theta}{\partial x^2} e^{-st} \, dt$$

i.e.
$$s\Theta(x, s) - \theta(x, 0) = K \frac{\partial^2 \Theta(x, s)}{\partial x^2}$$

i.e.
$$\frac{\partial^2 \Theta}{\partial x^2} = \frac{s}{K} \Theta \tag{8.50}$$

This equation has general solution

$$\theta(x, s) = A \exp\left(\sqrt{\frac{s}{K}} \, x \right) + B \exp\left(-\sqrt{\frac{s}{K}} \, x \right)$$

where A and B are constants to be determined.

The boundary conditions for (8.50) can be found from the boundary conditions of (8.29) which have not yet been employed.

Now $\theta(x, t) \to 0$ as $x \to \infty$ and therefore $\Theta(x, s) \to 0$ as $x \to \infty$. Hence A must be zero and

$$\Theta(x, s) = B \exp\left(-\sqrt{\frac{s}{K}} \, x \right)$$

Now
$$\Theta(0, s) = \int_0^\infty \theta_0\, e^{-st}\, dt = \frac{\theta_0}{s} = B$$

and so
$$\Theta(x, s) = \theta_0 \frac{1}{s} \exp\left(-\sqrt{\frac{s}{K}}\, x\right) \qquad (8.51)$$

Normally, to invert the transform would require integration in the complex plane, but on this occasion we can proceed by ingenious manipulation. This involves appealing to results already established, but we shall merely quote the result.

$$\Theta(x, s) = \mathcal{L}\left\{\theta_0\left(1 - \mathrm{erf}\left[\frac{x}{2\sqrt{Kt}}\right]\right)\right\} \qquad (8.52)$$

Hence we obtain the solution (8.49) again for $\theta(x, t)$.

We were lucky to be able to invert the transform by appealing to results already established. Furthermore, equation (8.50) was reasonably straightforward; had it been less so we might have needed transforms even to solve it.

Choice of Transform

It would seem that the choice of which transform to use is partly dictated by boundary conditions. Sometimes, however, the choice is not always so clear-cut and only by trying two or more approaches can we decide which suits a particular problem best.

8.7 More on Fourier Transforms

We can place physical interpretation on Fourier transforms, and for those workers whose concern with the transforms lies in studying waveforms, such physical interpretations allow a deeper insight into their subject. Here, we shall merely look at one or two ideas.

The Fourier transform of a waveform is its spectrum and vice-versa; hence we may regard each as equally measurable. An electrical waveform can be visualised on an oscilloscope whilst its spectrum can be analysed on a spectroscope. The waveform is a function of time and the spectrum is a function of frequency. The approach of the transform to a rectangular pulse, say, is to regard it as an isolated event whereas the Fourier series approach requires it to be repeated in both directions; therefore, the transform approach is the more general.

Where the system properties are linear, Fourier transforms may be directly applied when it is required to represent a variable of time as a variable of frequency or vice-versa.

Convolutions

One concept which is often treated only as an integral definition is that of convolution. Many problems in electricity or optics can be tackled by applying a

Fourier transform to a function, solving a governing equation (which may represent the action of some circuit element) and then inverting the solution. We therefore have two transforming processes: one into the intermediate system (circuit element) and one out again. Perhaps this could be accomplished at a stroke by one transformation. In this way the intermediate system is transformed and then applied to the original function. The resultant output is the integral of the product of each point of the original function with the transformed response.

The convolution is the result obtained by scanning one function with each point of the other and summing the products at each point. Note that the scanner may, in fact spread out the function.

System response and convolution integral

First we note that a function $x(t)$ can be represented as a continuum of impulses

$$x(t) = \int_{-\infty}^{\infty} x(\lambda)\, \delta(\lambda - t)\, d\lambda \tag{8.53}$$

The input $x(t)$ to a linear system produces an output $y(t)$ which we wish to find. We first find the response to elementary input signals and then use superposition to obtain the total response.

Suppose the response to $\delta(t)$ is known to be $h(t)$; then the response to $x(\lambda)\delta(t - \lambda)$ is $x(\lambda)\, h(t - \lambda)$ and the total response of the system is given by

$$y(t) = \int_{-\infty}^{\infty} x(\lambda)\, h(t - \lambda)\, d\lambda = x(t) * h(t) \tag{8.54}$$

This as we know is the **convolution integral**.

Example

The low pass filter shown in Figure 8.7 has a unit impulse response $h(t) = e^{-t}$. Sketch the time response of the filter to a unit step function of voltage.

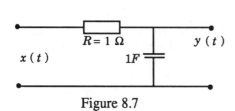

Figure 8.7   Figure 8.8

Solution using Laplace transforms

Since $x(t) = U(t)$ it follows that $X(s) = \dfrac{1}{s}$. Now

$$Y(s) = H(s)\, X(s) = \frac{1}{(s + 1)} \cdot \frac{1}{s} = \frac{1}{s} - \frac{1}{(s + 1)}$$

Therefore $\qquad\qquad y(t) = U(t) - U(t)\,e^{-t}$

We obtain the sketch of $y(t)$ shown in Figure 8.8.

Time domain solution

$$y(t) = \int\limits_{-\infty}^{\infty} x(\lambda)\,h(t-\lambda)\mathrm{d}\lambda = \int\limits_{-\infty}^{\infty} U(\lambda)\,e^{-(t-\lambda)}\,\mathrm{d}\lambda$$

Now convolution is the result of scanning one function with each point of the other and summing the products at each point. This is illustrated in Figure 8.9.

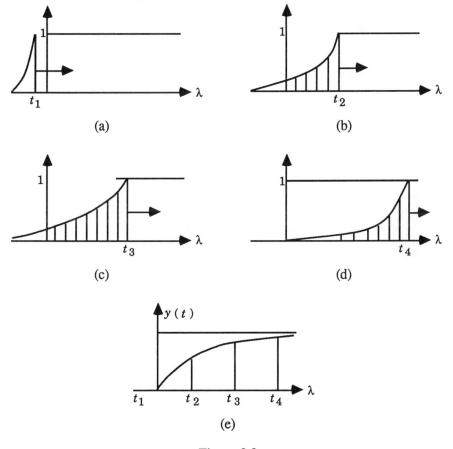

Figure 8.9

Figures (a) to (d) show how the value of the convolution integral is obtained by sliding one function over the other. The shaded area in each case represents the product for the particular t. Figure (e) shows the accumulated results.

Often in the analysis of systems and signals the Fourier transform is written $F(\omega)$ to emphasise the frequency domain, and the factor 2π is brought into the inversion integral.

We now present a table of some Fourier Transform pairs, Table 8.1.

Table 8.1

$$f(t) = \frac{1}{2\pi} \int_{-\infty}^{\infty} F(\omega)\, e^{j\omega t}\, d\omega \qquad\qquad F(\omega) = \int_{-\infty}^{\infty} f(t)\, e^{-j\omega t}\, dt$$

Rectangular pulse	$\begin{cases} 1 & ,\ |t| \le a \\ 0 & ,\ \text{elsewhere} \end{cases}$	$\dfrac{2 \sin \omega a}{\omega}$
Decay	$Ae^{-at}\, U(t)$	$\dfrac{A}{a + j\omega}$
Triangular pulse	$\begin{cases} A\left(1 - \dfrac{|t|}{a}\right), & |t| \le a \\ 0, & \text{elsewhere} \end{cases}$	$Aa\left[\dfrac{\sin (a\omega/2)}{a\omega/2}\right]^2$
Step function	$\begin{cases} A & ,\quad t > 0 \\ 0 & ,\quad \text{elsewhere} \end{cases}$	$A\left[\pi\delta(\omega) - j\,\dfrac{1}{\omega}\right]$
Constant	A	$2\pi A\delta(\omega)$

Table 8.2 on the following page shows some of the main properties of Fourier Transforms.

Example
(i)   Find the Fourier Transform of the pulse given by

$$f(t) = P_a(t) = \begin{cases} 1, & |t| \le a \\ 0, & |t| > a \end{cases}$$

Now $F(\omega) = \displaystyle\int_{-\infty}^{\infty} P_a(t)\, e^{-j\omega t}\, dt = \int_{-a}^{a} e^{-j\omega t}\, dt$

$$= \left[\frac{e^{-j\omega t}}{-j\omega}\right]_{-a}^{a} = -\frac{1}{j\omega}\{e^{-j\omega a} - e^{+j\omega a}\} = \frac{-2j \sin \omega a}{-j\omega} = \frac{2 \sin \omega a}{\omega}$$

By using the symmetry property we can show that if $f(t) = \dfrac{2 \sin at}{t}$ then

$F(\omega) = 2\pi\, P_a(\omega)$.

Table 8.2

Property	function $f(t)$	Transform $F(\omega)$
1 **Linearity**	$af_1(t) + bf_2(t)$	$aF_1(\omega) + bF_2(\omega)$
2 **Symmetry**	$F(t)$	$2\pi f(-\omega)$
3 **Time shift**	$f(t + t_0)$	$e^{j\omega t_0} F(\omega)$
4 **Scaling**	$f(at)$	$\dfrac{1}{|a|} F(\omega/a)$
5 **Frequency shift**	$e^{j\omega_0 t} f(t)$	$F(\omega + \omega_0)$
6 **Differentiation**	$f^{(n)}(t)$	$(j\omega)^n F(\omega)$
7 **Time convolution**	$f(t) * g(t)$	$F(\omega) G(\omega)$
8 **Frequency convolution**	$f(t) g(t)$	$\dfrac{1}{2\pi} F(\omega) * G(\omega)$
9 **Parseval's formula**	$\displaystyle\int_{-\infty}^{\infty} f^2(t)\, dt$	$\dfrac{1}{2\pi} \displaystyle\int_{-\infty}^{\infty} |F(\omega)|^2\, d\omega$

The last result relates energy in the time domain signal to energy in the frequency domain transform.

(ii)   If $f(t) = e^{-(t - t_0)} u(t - t_0)$ we need the shifting property and Table 8.1. From the table with $A = 1$ we know that the transform of $g(t) = e^{-at} U(t)$ is $1/(a + j\omega)$ so that the transform of $f(t)$ is $e^{-j\omega t_0}/(a + j\omega)$. Refer to Figure 8.10.

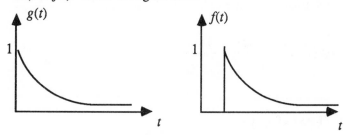

Figure 8.10

(iii)   The triangular pulse can be regarded as the time convolution of $P_a(t)$ with itself. Consider the following convolution and refer to Figure 8.11.

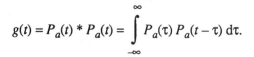

$$g(t) = P_a(t) * P_a(t) = \int\limits_{-\infty}^{\infty} P_a(\tau)\, P_a(t-\tau)\, d\tau.$$

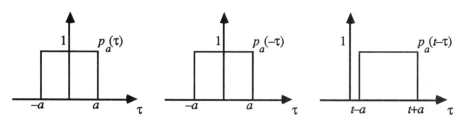

Figure 8.11

The functions do not overlap if $t < -2a$ or $t > 2a$. Consider as an example $t = a/2$ and refer to Figure 8.12.

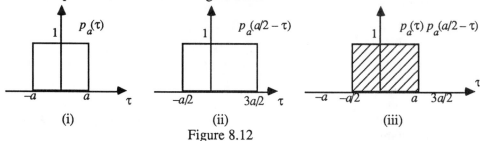

Figure 8.12

$g(a/2)$ is the shaded area in Figure 8.12(iii), which is $3a/2$. As t makes its way between $-2a$ and $2a$ it can be seen that

$$g(t) = \begin{cases} 2a & (1 - t/2a)\,, & 0 \le t \le 2a \\ 2a & (1 + t/2a)\,, & -2a \le t \le 0 \\ 0 & , & \text{elsewhere} \end{cases}$$

which is the triangular pulse of Table 8.1 with a replaced by $2a$.

$$\text{Hence } G(\omega) = \left[\frac{2 \sin \omega a}{\omega} \right]^2 = \frac{4 \sin^2 \omega a}{\omega^2}$$

8.8 Other Transforms

In this section we examine briefly two transforms which are used to handle linear discrete systems. The signal being studied is available as a sequence of values rather than as a continuous function.

Z–transforms

If we have a sequence of values $y(k)$ then the Z-transform of this sequence is given by

$$Y(z) = Z\{y(k)\} = \sum_{k=0}^{\infty} y(k) \, z^{-k} \tag{8.55}$$

Example 1   Let $\{y(k)\} = \{0, 2, 1, 0, 0, ...\}$

$$Y(z) = \frac{2}{z} + \frac{1}{z^2} = \frac{2z+1}{z^2}$$

As a function of the complex variable z, $Y(z)$ has a pole of order 2 at $z = 0$ and a zero at $z = -1/2$.

Example 2

Suppose that the function $f(t) = Ce^{-\lambda t}$ is sampled every T seconds. The sampled values are

$$C, \, Ce^{-\lambda T}, \, Ce^{-2\lambda T} \text{ etc.}$$

The Z transform of this sequence is

$$F(z) = C + Ce^{-\lambda T} z^{-1} + Ce^{-2\lambda T} z^{-2} + ...$$

which is a G.P. with common ratio $e^{-\lambda T} z^{-1}$.

Hence $F(z) = C/ (1 - e^{-\lambda T} z^{-1}) = Cz/(z - e^{-\lambda T})$.

Example 3 $y(k) = \delta(k)$; $Y(z) = y(0)z^{-0} = 1$
Similarly, if $y(k) = \delta(k - N)$, $Y(z) = z^{-N}$.

Some properties of the transform

1   Linearity
$$Z\{ay_1(k) + by_2(k)\} = aZ\{y_1(k)\} + bZ\{y_2(k)\}$$
2   Right Shift
$$Z\{y(k - n)\} = z^{-n} Z\{y(k)\} = z^{-n} Y(z)$$
3   Left Shift

$$Z\{y\,(k + n)\} = z^n Y(z) - \sum_{k=0}^{\infty} y(k) \, z^{n-k}$$

Example
Consider the system governed by the equation
$$y(k + 1) + 4y(k) = \delta(k + 1) \tag{8.56}$$
It is represented in Figure 8.13.
Taking the Z–transforms of the terms in (8.56) we obtain
$$Z\{y(k + 1)\} + 4Z\{y(k)\} = Z\{\delta(k + 1)\}$$
i.e. $$zY(z) - zy(0) + 4Y(z) = z$$

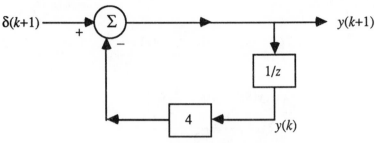

Figure 8.13

Hence
$$Y(z) = \frac{z + z\,y(0)}{z + 4} = \frac{z}{z + 4}$$

This follows because, putting $k = 0$ in (8.56) produces
$$y(1) + 4y(0) = \delta(1)$$
and since the input is applied one time unit ahead of $T = 0$ it is implicit that $y(0) = 0$. Note that by applying a right-shift of one time unit to (8.56) we obtain the difference equation

$$y(k) + 4y(k-1) = \delta(k) \qquad (8.57)$$

Applying the right-shift property then (8.57) becomes after transforming
$$Y(z) + 4z^{-1}\,Y(z) = 1$$
leading to the same formula for $Y(z)$.

4   Time scaling $Z\{a^k y(k)\} = Y(z/a)$

5   Periodic sequences
Let $y(k + N) = y(k)$ and $y_1(k)$ be the sequence in the first period.

Then $Z\{y(k)\} = \dfrac{z^N}{z^N - 1}\, Z\{y_1(k)\}$.

Example   Consider the sequence depicted in Figure 8.14

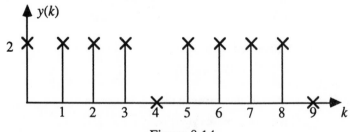

Figure 8.14

Here $N = 5$ and

$$Y(z) = \frac{z^5}{z^5 - 1} (2 + 2z^{-1} + 2z^{-2} + 2z^{-3})$$

$$= \frac{2z^2}{z^5 - 1} (z^3 + z^2 + z + 1)$$

6   **Multiplying by k** $Z\{ky(k)\} = -z \dfrac{d}{dz} Y(z)$

7   **Initial value**   If a sequence has zero value for $k < k_0$, then

$$y(k_0) = z^{k_0} \lim_{z \to \infty} Y(z)$$

8   **Final value**   If $y(k)$ exists as $k \to \infty$ then

$$\lim_{k \to \infty} y(k) = \lim_{z \to 1} \left[\left(1 - \frac{1}{z} \right) Y(z) \right]$$

9   **Convolution**   If $y(k) = h(k) * x(k) = \displaystyle\sum_{n=0}^{\infty} h(k-n)\, x(n)$ and

$h(k - n) = 0, \; n > k$ then $Y(z) = H(z) \cdot X(z)$

Example

The series LR circuit where $L = R = 1$ for simplicity and an applied voltage $v(k)$ produces an output voltage $v_0(k)$. The voltage $v_0(k)$ is governed by

$$\frac{dv_0(t)}{dt} + v_0(t) = v(t)$$

which in discrete form with $T = 1$ becomes

$$v_0(k) - v_0(k - 1) + v_0(k) = v(k)$$

and this is rearranged to

$$v_0(k) = (1/2)v(k) + (1/2)v_0(k - 1) \tag{8.58}$$

To determine the impulse response of the system we apply the input $v(k) = \delta(k)$. Then the response is found by transforming (8.58) to obtain

$$H(z) = (1/2) + (1/2)z^{-1} H(z)$$

from which

$$H(z) = \frac{1}{2} \frac{1}{1 - \frac{1}{2} z^{-1}} = \frac{1}{2} \left(1 + \frac{1}{2} z^{-1} + \frac{1}{2^2} z^{-2} + \dots \right)$$

Inverting the transform we find that $h(k)$ is given by the equation

$$h(k) = \frac{1}{2} \left(\frac{1}{2}\right)^k \quad , k \geq 0.$$

Now suppose that the input voltage is given by $v(k) = 1$, $k = 0, 1, 2, 3$ and zero elsewhere. Applying the convolution

$$v_0(k) = \sum_{n=0}^{k} h(k-n) \, v(n)$$

we evaluate the output response at successive intervals of time.

$$v_0(0) = h(0) \, v(0) = (1/2) \times 1 = 1/2$$
$$v_0(1) = h(1) \, v(0) + h(0) \, v(1) = (1/4) \times 1 + (1/2) \times 1 = 3/4$$
$$v_0(2) = h(2) \, v(0) + h(1) \, v(1) + h(0) \, v(2) = 7/8$$

You are advised to continue the evaluation for a few more steps and plot graphs of $v(t)$ and $v_0(k)$.

In Table 8.3 we show a few examples of Z transform pairs.

Table 8.3

$y(k)$	$Y(z)$	$y(k)$	$Y(z)$
$\delta(k)$	1	$\delta(k-m)$	z^{-m}
1 or $U(k)$	$z/(z-1)$	k	$z/(z-1)^2$
k^2	$z(z+1)/(z-1)^3$	a^k	$z/(z-a)$
$(k+1)a^k$	$z^2/(z-a)^2$	$\dfrac{(k+1)(k+2)}{2!} a^k$	$z^3/(z-a)^3$
ka^k	$az/(z-a)^2$	$a^k/k!$	$e^{a/z}$
$\sin k\omega T$	$\dfrac{z \sin \omega T}{z^2 - 2z \cos \omega T + 1}$	$\cos k\omega T$	$\dfrac{z(z - \cos \omega T)}{z^2 - 2z \cos \omega T + 1}$
e^{-akT}	$\dfrac{z}{z - e^{-kT}}$	ke^{-akT}	$\dfrac{z - e^{-aT}}{(z - e^{-aT})^2}$

Examples of Inverse Transforms

(i) $F(z) = \dfrac{1}{1 - 0.2z^{-1}} = 1 + 0.2z^{-1} + (0.2)^2 z^{-2} + (0.2)^3 z^{-3} + \ldots$

We read off the coefficients of z^{-k} directly to see that $f(k) = (0.2)^k$, $k \geq 0$

(ii) $F(z) = \dfrac{z^2 + 4z + 6}{(z-3)(z+3)(z+4)} = -\dfrac{1}{6} + \dfrac{3}{14}\dfrac{z}{z-3} + \dfrac{1}{6}\dfrac{z}{z+3} - \dfrac{3}{14}\dfrac{z}{z+4}$

where we have used partial fractions of the form $\dfrac{z}{z-a}$. From Table 8.3

we can write down

$$f(k) = -\frac{1}{6}\,\delta(k) + \frac{3}{14}\,3^k + \frac{1}{6}\,(-3)^k - \frac{3}{14}\,(-4)^k$$

(iii) $F(z) = \dfrac{z^2 - 16}{(z-2)(z-3)^3}$

$$= -\frac{8}{27} + \frac{6z}{z-2} - \frac{500}{81}\frac{z}{z-3} + \frac{97}{162}\frac{z^2}{(z-3)^2} - \frac{7}{54}\frac{z^3}{(z-3)^3}$$

From Table 8.3,

$$f(k) = -\frac{8}{27}\,\delta(k) + 6.2^k - \frac{500}{81}\,3^k + \frac{97}{162}\,(k+1)\,3^k - \frac{7}{54}\,\frac{(k+1)(k+2)\,3^k}{2}$$

(iv)   We return to the series LR circuit on page 297.
   Note that we can alternatively use the idea of the transfer function $H(z)$ for the system in a direct fashion.

Now $H(z) = \dfrac{1}{2}\dfrac{1}{1 - \dfrac{1}{2}z^{-1}} = \dfrac{1}{2}\dfrac{z}{z - \dfrac{1}{2}}$

Consider the step input $v(k) = u(k)$, its transform is

$$V(z) = U(z) = \frac{z}{z-1}$$

The response of the system to such an input is given in the z-domain by

$$V_0(z) = H(z)\,V(z) = \frac{1}{2}\frac{z}{z-\dfrac{1}{2}}\frac{z}{z-1} = \frac{z}{z-1} - \frac{1}{2}\frac{z}{z-\dfrac{1}{2}}$$

Hence $v_0(k) = 1 - \dfrac{1}{2}\left(\dfrac{1}{2}\right)^k$, $k \ge 0$ which agrees with the result earlier.

Solution of Difference Equations

Consider the difference equation
$$y(k) + 4y(k-1) = 7u(k)$$

where $y(-1) = 0$. Taking Z-transforms we obtain

$$Y(z) + 4z^{-1} Y(z) = 7z/(z-1)$$

Solving for $Y(z)$ we obtain

$$Y(z) = \frac{7z}{z-1} \cdot \frac{z}{z+4} = \frac{7}{5} \frac{z}{z-1} + \frac{28}{5} \frac{z}{z+4}$$

so that inverting the transforms we find

$$y(k) = 1.4u(k) + 5.6(-4)^k \qquad k \geq 0$$

as the required solution. Note that

$$y(k-1) = 1.4u\,(k-1) + 5.6\,(-4)^{k-1}$$

so that

$$y(k) + 4y(k-1) = 1.4u(k) + 5.6u(k-1) + 5.6(-4)^{k-1}(-4+4)$$

Now for $k \geq 1$ $u(k) = 1$, $u(k-1) = 1$ so that
$y(k) + 4y(k-1) = 7u(k)$ as required.
If $k = 0$,

$$y(0) + 4y(-1) = y(0)$$

and from the gereneral solution

$$y(0) = 1.4u(0) + 5.6 = 7.$$

Discrete Fourier Transforms

This technique is applied to signals which have been sampled. It is useful when describing phenomena which are related to a discrete time series.

Suppose we sample the signal at frequencies $\dfrac{2\pi}{\sqrt{T}}$ where T is the sampling interval. The Discrete Fourier Transform (DFT) of the sample values $f(kT)$, $k = 0, 1, \ldots (N-1)$ is given by

$$D\left\{f(kT)\right\} = \sum_{k=0}^{n-1} f(kT)e^{-j2\pi nkT/NT} = F\left(\frac{2\pi n}{NT}\right) \tag{8.59}$$

It can be shown that the DFT is the evaluation of the Z transform of the finite sequence $\{f(kT)\}$ at N points equally spaced along a unit circle in the z-plane. To invert the DFT we use

$$f(kT) = \frac{1}{N} \sum_{n=0}^{N-1} F\left(\frac{2\pi n}{NT}\right) e^{j2\pi nkT/NT}, \; k = 0, 1, \ldots, (N-1) \tag{8.60}$$

Among the properties of the DFT for a pair $F(k)$ and $F(n)$ we may mention

1 **Linearity** $D\left\{af_1(k) + bf_2(k)\right\} = aF_1(n) + bF_2(n)$

2 **Symmetry** $D\left\{\dfrac{1}{N} F(k)\right\} = f(-n)$

3 **Time shift** $D\left\{f(k-i)\right\} = F(n)\,e^{-jni}$

4 **Frequency shift** $f(k)\,e^{jki} = D^{-1}\left\{F(n-i)\right\}$

5 **Parseval's theorem** $\displaystyle\sum_{k=0}^{N-1} f^2(k) = \frac{1}{N} \sum_{n=0}^{N-1} |F(n)|^2$

A computational technique known as the Fast Fourier Transform (FFT) reduces considerably the number of operations required to evaluate the DFT. This technique has enhanced the usefulness of the DFT as a tool in the design of digital filters, for example.

Problems

Section 8.2

1   Find the Laplace transform of each of the following:

$e^{-t} \sin^2 t$; $t^2 \cos t$; $\sin kt \cos kt$; $\sin \omega t.t e^{-2t}$; $e^{-4t} \cosh 2t$;

$e^{2t}(t^2 - 5t + 6)$; $\dfrac{1}{t} \sinh t$.

2   The concentrations N_1 and N_2 of Iodine and Xenon in a nuclear reactor are given by the differential equations

$$\frac{dN_1}{dt} = a - \lambda_1 N_1, \quad \frac{dN_2}{dt} = \lambda_1 N_1 - \lambda_2 N_2 - bN_2$$

where a, b, λ_1 and λ_2 are constants. Use Laplace transforms to solve these equations subject to the initial conditions $N_1(0) = 0 = N_2(0)$, and obtain the steady-state values A_1 and A_2, say, of N_1 and N_2. If the reactor is then shut down the system becomes

$$\frac{dN_1}{dt} = -\lambda_1 N_1, \quad \frac{dN_2}{dt} = \lambda_1 N_1 - \lambda_2 N_2$$

subject now to the initial conditions $N_1(0) = A_1, N_2(0) = A_2$, where $t = 0$ now denotes the time of shut-down.

Solve these equations and show that the Xenon concentration builds up from its equilibrium value A_2 to a maximum concentration before decaying with time.   (EC)

3   The primary and secondary currents I_1 and I_2 respectively, in a pair of inductively coupled electrical circuits are given by the differential equations

$$L_1 \frac{dI_1}{dt} + R_1 I_1 + M \frac{dI_2}{dt} = E, \quad L_2 \frac{dI_2}{dt} + R_2 I_2 + M \frac{dI_1}{dt} = 0$$

in which L_1, R_1, L_2, R_2, M and E are all constant. If initially both I_1 and I_2 are zero use Laplace Transforms to show that the secondary current at time t is given by

$$I_2 = \frac{EM}{L_1 L_2 - M^2} \frac{e^{a_1 t} - e^{a_2 t}}{a_2 - a_1}$$

where a_1 and a_2 are the roots of the equation

$$(L_1 L_2 - M^2)a^2 + (L_1 R_2 + L_2 R_1)a + R_1 R_2 = 0$$

Give a rough sketch graph to show the general behaviour of $I_2(t)$, demonstrating in particular that I_2 varies linearly with t for small t. (You may assume that $L_1 L_2 > M$.)   (EC)

4   Prove that

(i) $\mathcal{L}\{(1/4)(1 - \cos 2t)\} = \dfrac{1}{s(s^2 + 4)}$ (ii) $\mathcal{L}\{U(t - a)\} = \dfrac{e^{-sa}}{s}$, a any positive constant

(iii) $\mathcal{L}\{U(t - a)f(t - a)\} = e^{-sa} F(s)$
Use the above results to solve the equation

$$\frac{d^2 y}{dt^2} + 4y = 3U(t - 2)$$

given that $y = 1$, $\dfrac{dy}{dt} = 0$ when $t = 0$.   (EC)

5   Evaluate

(i) $f(t) = L^{-1}\left\{\dfrac{1}{p(1 - e^{-p\pi})}\right\}$ (ii) $f(t) = L^{-1}\left\{\dfrac{p}{(1 + p^2)(1 - e^{-p\pi})}\right\}$

Sketch a graph of each function. Solve the differential equation

$$\frac{d^2 y}{dt^2} + y = f(t)$$

where $f(t) = n + 1$ for $n\pi < t < (n + 1)\pi$ and given that $y = 0 = \dfrac{dy}{dt}$ when $t = 0$.   (LU)

6   (i) Show that if $L\{f(t)\} = \bar{f}(p)$, and $g(t)\begin{cases} = f(t - a),\ t > a, \\ = 0,\quad\quad\ t < a, \end{cases}$ then $L\{g(t)\} = e^{-ap}\bar{f}(p)$

(ii) A function $f(t)$ is periodic, of period $\dfrac{2\pi}{b}$, and is defined as

$$f(t)\begin{cases} = 0, & 0 < t < c \\ = \sin b(t - c), & c < t < c + \dfrac{\pi}{b} \\ = 0, & c + \dfrac{\pi}{b} < t < c + \dfrac{2\pi}{b} \end{cases}$$

Obtain the Laplace transform of $f(t)$.   (LU)

7   If the periodic function $g(t)$ of period T is such that
$g(t) = c$ $0 < t < 1/2T$
$g(t) = 0$ $1/2T < t < T$,

show that $L\{g(t)\} = \dfrac{c}{p(1 + e^{-pT/2})}$

The variable θ satsifies the differential equation $\dfrac{d^2\theta}{dt^2} + \omega^2\theta = g(t)$ where $g(t)$ is defined above and at time $t = 0$, $\theta = 0$, $d\theta/dt = 0$. Show that at time $t = T(> 0)$, the value of θ

is $\dfrac{c}{\omega^2} (\cos \frac{1}{2}\omega T - \cos \omega T)$. (LU)

8 (i) If $f(t)$ takes the form
$$f(t) = 1 \qquad 0 < t < T$$
$$f(t) = -1 \qquad T < t < 2T$$

and is of period $2T$ show that $\bar{f}(p) = \dfrac{1}{p}$ tanh $(1/2)\, pT$.

(ii) If $f(t)$ takes the form
$$f(t) = \sin t \qquad 0 < t < \pi$$
$$f(t) = 0 \qquad \pi < t < 2\pi$$

and is of period 2π, show that $\bar{f}(p) = \dfrac{1}{(p^2 + 1)(1 - e^{-\pi p})}$ (LU)

9 The input θ_1 and output θ_0 of a servomechanism are related by the equation

$$\dfrac{d^2\theta_0}{dt^2} + 8\dfrac{d\theta_0}{dt} + 16\theta_0 = \theta_i, \quad t > 0 \text{ and initially } \theta_0 = 0 = \dfrac{d\theta_0}{dt}.$$

If $\theta_i = f(t)$, where $\quad f(t) = 1 - t, \qquad 0 < t < 1$
$$f(t) = 0, \qquad t > 1$$

show that $\mathcal{L}\{f(t)\} = \dfrac{p-1}{p^2} + \dfrac{1}{p^2} \cdot e^{-p}$

and hence obtain an expression for θ_0 in terms of t. (LU)

10 (i) Find $\mathcal{L}\{U(t - b)\}$

(ii) Find $\mathcal{L}\{(t - b)U(t - b)\}$ where $b > 0$

(iii) Show that $\mathcal{L}\{(t - b)^3\, U(t - b)\} = \dfrac{6e^{-pb}}{p^4}$

If $\dfrac{d^2y}{dx^2} = 1 + \dfrac{(x - (1/2)a)}{a} U(x - (1/2)a)$ for $0 < x < a$ and $y = 0$, $dy/dx = 0$ at

$x = 0$, find y when $x = a$. (LU)

11   A uniform light bar of length l has a load wx/unit length for $0 < x < l/2$ and $w(l - x)$ / unit length for $l/2 < x < l$. The bar is freely hinged at $x = 0$ and clamped horizontally at $x = l$. Find the deflection of the mid-point.

12   An electric circuit comprises an inductance of L henrys, in series with a capacitance C farads. At $t = 0$ an e.m.f. given by

$$E(t) = \begin{cases} E_0\, t/T, & 0 < t < T \\ 0 & t > T \end{cases}$$

is applied to the circuit. The initial current and the charge on the capacitor are zero. Find the charge for any time $t > 0$.

Section 8.3

13   (a)   Write down the Laplace transforms of

        (i) t (ii) te^{-t}

        and use the convolution theorem to show that the Inverse Laplace transform of

$$\frac{1}{s^2(s + 1)^2} \quad \text{is} \quad te^{-t} + 2e^{-t} + t - 2.$$

        Hence find the solution of the differential equation

$$\frac{d^2y}{dt^2} + 2\,\frac{dy}{dt} + y = t$$

        given that $\dfrac{dy}{dt} = 1$ and $y = 0$ when $t = 0$.

    (b)   State the convolution theorem for Laplace transforms. Hence write down the Laplace transform of

$$\int_0^t \sin(5u)\, f(t - u)\,du$$

        in terms of s and $F(s)$ where $F(s)$ is the Laplace transform of $f(t)$.
        Given the integral equation

$$f(t) = 2t + \int_0^t \sin(5u)\, f(t - u)\,du$$

        take the Laplace transform to find an equation for $F(s)$. Hence find $f(t)$.   (EC)

14   The equation representing the forced oscillations of a damped harmonic oscillator is

$$\frac{d^2x}{dt^2} + 2k\,\frac{dx}{dt} + n^2x = f(t) \text{ subject to } x = x_0 \text{ and } \frac{dx}{dt} = v_0 \text{ at } t = 0$$

$(x_0, v_0, n$ and k are constants with $k < n)$.
Show that $x(t) = x_1(t) + x_2(t)$ where

$$x_1(t) = x_0 e^{-kt} \cos \omega t + \frac{1}{\omega}\,(kx_0 + v_0)\,e^{-kt} \sin \omega t,$$

$$x_2(t) = \frac{1}{\omega} \int_0^t f(y)\, e^{-k(t-y)} \sin \omega(t-y)\,dy \quad \text{and} \quad \omega^2 = n^2 - k^2. \tag{EC}$$

15 Using the convolution theorem, evaluate

(a) $L^{-1}\left\{ \dfrac{1}{(s+1)(s^2+1)} \right\}$ (b) $L^{-1}\left\{ \dfrac{1}{(s-2)(s+1)^2} \right\}$ (EC)

16 (a) State the convolution theorem for Laplace transforms.

(b) If the Laplace transform of $f(t)$ is $F(s)$, use the convolution theorem to show that

$$\mathcal{L}^{-1}\left[\frac{1}{s} F(s) \right] = \int_0^t f(u)\,du$$

Deduce that

$$\mathcal{L}^{-1}\left[\frac{1}{s^2} F(s) \right] = \int_0^t \int_0^u f(v)\,dv\,du$$

and obtain a similar expression for $\mathcal{L}^{-1}\left[\dfrac{1}{s^3} F(s) \right]$

Hence, or otherwise, find the inverse Laplace transform of $\dfrac{1}{s^3(s+1)^2}$

(c) Use Laplace transform to solve the differential equation

$$\frac{d^2 y}{dt^2} + 2 \frac{dy}{dt} + y = t^2$$

given that $(dy/dt) = 3$ and $y = 4$ when $t = 0$.   (EC)

Section 8.4

17 The temperature distribution $\theta(x, t)$ in a uniform bar of length l is given by

$k \dfrac{\partial^2 \theta}{\partial x^2} = \dfrac{\partial \theta}{\partial t}$. The end $x = 0$ is given a temperature $\theta = f(t)$ and the other end

$x = l$ is kept at $\theta = 0$. If the initial temperature is zero throughout the bar, use a finite Fourier sine transform to show that θ at any point x at any time t is

$$\theta(x, t) = \frac{2\pi k}{l^2} \sum_{n=1}^{\infty} n \sin\left[\frac{n\pi x}{l} \right] \int_0^t f(t-u) e^{-kn^2\pi^2 u/l^2}\,du$$

18 The concentration $V(x, t)$ at position x and time t of a liquid diffusing along a channel

$0 < x < \alpha$ is given by $\dfrac{\partial V}{\partial t} = k \dfrac{\partial^2 V}{\partial x^2}$ where k is the constant diffusion coefficient. If

$\partial V/\partial x = 0$ at $x = 0$ and $x = \alpha$ and also $V = V_0 x/\alpha$ initially where V_0 is a constant, use a finite Fourier cosine transform to show that

$$V(x, t) = V_0/2 - (4V_0/\pi^2) \sum_n (1/n^2) \cos (n\pi x/\alpha) \exp (-kn^2\pi^2 t/\alpha^2) \text{ where the}$$

summation extends over $n = 1, 3, 5, \ldots$

19 A uniform metal sphere of radius a is heated to a constant temperature θ_0 throughout and is then immersed in a coolant liquid so that its surface is thereafter maintained at zero temperature. Assume that the temperature distribution $\theta(r, t)$ at radial distance r after a time t is given by the heat conduction equation $\dfrac{\partial^2 \theta}{\partial r^2} + \dfrac{2}{r}\dfrac{\partial \theta}{\partial r} = \dfrac{1}{k}\dfrac{\partial \theta}{\partial t}$. Substitute

$\theta(r, t) = r^{-1}f(r, t)$, then use a finite Fourier transform to solve the resulting equation for $f(r, t)$ and show that

$$\theta(r, t) = \frac{2\theta_0 a}{r} \sum_{n=1}^{\infty} \frac{(-1)^{n+1}}{n} \sin\left[\frac{n\pi r}{a}\right] e^{-kn^2\pi^2 t/a^2}$$

Section 8.5

20 Obtain

(i)   the Fourier transform of $f(x) = \begin{cases} \dfrac{1}{2a} & |x| \le b, \\ 0 & |x| > b \end{cases}$

(ii) the Fourier sine transform of $g(x) = e^{-\pi x}, x \ge 0$.
State the reciprocal relationship existing between the function $g(x)$ and its Fourier sine transform and use this relationship to evaluate

$$\int_0^\infty \frac{x \sin mx}{x^2 + \pi^2} \, dx \qquad\qquad\qquad \text{(LU)}$$

21 (i)   Obtain the Fourier sine transform $F_s(p)$ of the function $g(x)$ given by $g(x) = xe^{-|x|}$.

(ii) An even function $f(x)$ is given by $f(x) = \begin{cases} x^2 \text{ for } 0 \le x \le 2 \\ 0 \text{ for } x > 2. \end{cases}$ \qquad\qquad \text{(LU)}

22 Evaluate $F_c(s)$ when

(i) $f(x) = \begin{cases} \cos x, & 0 < x < \pi \\ 0 & , & x < \pi \end{cases}$

(ii) $f(x) = e^{-x}$
Assuming the reciprocal relation between a function and its Fourier cosine transform, use the second result to show that

$$\int_0^\infty \frac{\cos kx}{1 + x^2}\, dx = \frac{\pi}{2}\, e^{-k}.$$ (LU)

23 Find the cosine transform of $g(x)$ if $g(x) = 1$ for $0 < x < a$, $g(x) = 0$ for $x > a$.

Find the function whose sine transform is $\dfrac{\sin pa - pa \cos pa}{p^2}$ (LU)

24 Verify that the sine transform of $f''(x)$ is $pf(0) - p^2 F(p)$.
State the inversion formula for $f(x)$ in terms of $F(p)$. (LU)

Section 8.7

25 Find the Fourier transform of the following functions $f(t)$
(i) $p_2(t - 2)$ (ii) $p_a(t) + p_{2a}(t - 2)$ (iii) $e^{jt}\, p_a(t)$.

26 Sketch the functions given by the following formulae and find their Fourier transforms
(i) $f(t) = 4$, $0 \le t \le 1$, $f(t)$ odd and of period 2
(ii) $f(t) = t$, $-2 \le t \le 2$, $f(t)$ of period 4.

27 Find the Fourier transforms of the signals below; assume that $a > 0$
(i) $Ae^{-at} U(t)$ (ii) $Ae^{-a|t|}$ (iii) $Ae^{at} U(-t)$

28 Find the inverse Fourier transforms of the following:

(i) $\dfrac{4}{\omega} \sin 3\omega\, e^{j\omega}$ (ii) $\dfrac{3e^{j\omega}}{1 - j\omega}$ (iii) $2e^{-\omega^2}$

(iv) $\rho_a(\omega - \omega_1) + \rho_a(\omega + \omega_1)$ (v) $\dfrac{\sin 4\omega}{4\omega} e^{-3j\omega}$

29 (i) Use the signal $f(t) = \frac{1}{2} e^{-a|t|}$, $a > 0$ to deduce that $\displaystyle\int_{-\infty}^\infty \frac{dx}{\left(1 + x^2\right)^2} = \frac{\pi}{2}$

(ii) By means of the signal $f(t) = \begin{cases} e^{-t}, & 0 < t < 1 \\ 0, & \text{otherwise} \end{cases}$

and Parseval's theorem obtain the result that $\displaystyle\int_{-\infty}^\infty \frac{\cos \omega}{\left(1 + \omega^2\right)}\, d\omega = \frac{\pi}{e}$

30 The unit impulse response of a system is $e^{-t} \cos t\, U(t)$; find the unit step response of the system.

Section 8.8

31 Find the Z-transform for the sequences

(i) $f(k) = \begin{cases} (1/4)^k, & k = 0, 1, 2, \dots \\ 0, & k < 0 \end{cases}$

(ii) $f(k) = \begin{cases} 0, & k \leq 0 \\ -2, & k = 1 \\ a^k, & k = 2,3\dots \end{cases}$

32 Find the sequences $f(k)$ which transform to the following

(i) $\dfrac{z^2}{(z-1)(z-4)}$

(ii) $\dfrac{z^2 + 3z}{(z-2)^2}$

33 Find the Z-transforms of the following functions, each sampled every T seconds
(i) $\cos 3t \, U(t)$ (ii) $2^t \sin 3t \, U(t)$ (iii) $2^k \, U(t)$
(iv) $e^{-2t} \sin 2t \, U(t)$.

34 Prove properties 4 and 6 for the Z-transform.

35 Find the Z-transforms of the following sequences, where each is assumed to be zero for negative n.
(i) $(-1)^k$ (ii) $1/2^k$ (iii) k^3 (iv) $1/k!$
(v) $\cosh 2k$ (vi) ke^{-3k} (vii) $k^2 a^{k-1}$

(viii) $f(t) = 2t^2$, sampled at an interval of T.

36 Show that if $g(k) = f(0) + f(1) + \dots + f(k)$ then $G(z) = \left[\dfrac{z}{z-1} \right] F(z)$

and deduce the Z-transform of the sequence $g(k) = \begin{cases} 1 & n \text{ even} \\ 0 & n \text{ odd} \end{cases}$

37 Find the inverse transform of each of the following $F(z)$:

(i) $\dfrac{4z(z+1)}{(z-1)(4z^2 - 4z + 1)}$

(ii) $\dfrac{2z^3}{(z+2)(z+1)^2}$

(iii) $1 + \dfrac{1}{z} + \dfrac{2z}{(z-2)(z-1)}$

(iv) $\dfrac{5ez}{(ez-1)^2}$

(v) $\dfrac{225z^2 + 29.46z}{z^2 - 0.4z + 0.04}$

38 Solve the following difference equations:
(i) $y(k) - 2y(k-1) = k$; $y(-1) = 0$
(ii) $y(k) - 3y(k-1) = 6$; $y(-1) = 4$
(iii) $y(k) - 7y(k-1) + 10y(k-2) = 0$; $y(-2) = 5, y(-1) = 16$
(iv) $y(k) - 6y(k-1) + 9y(k-2) = 0$; $y(-2) = 0, y(-1) = 1$

39 Find the unit step response of the system $y(k) - \frac{1}{3}y(k-1) = x(k)$ and hence find its

response to the input $x(k) = \begin{cases} 0, & k < 0 \\ 2, & k = 0 \text{ or } 1 \\ 1, & k > 1 \end{cases}$, k integer (of course)

40 (a) Both directly and via Z-transforms find the convolutions

    (i) $3^k * 2^k$ (ii) $2^k * 3^k$ (iii) $(\frac{1}{2})^k * U(k)$

(b) If $x(k) = u(k) - u(k-4)$ and $h(k) = (\frac{1}{2})^k$ find the convolution $x(k) * h(k)$

(c) Find the impulse response of the system
$$y(k) - 0.3y(k-1) = x(k) + 0.7x(k-1)$$
and, via convolution, find its response to an input $(-1)^k U(k)$.

41 Let $f(k) = \{1, 0, 0, 1\}$. Verify the properties on pages 300 and 301 for its DFT.

9

INTEGRATION AND VECTOR
FIELD THEORY

9.1 Scalar and Vector Fields; Differentiation and Integration of Vectors

In many areas of applied mathematics the governing equation takes on different forms according to the geometry of the particular problem under consideration. For example, steady state temperature distribution in a long bar is governed by

$$\frac{d^2\theta}{dx^2} = 0, \text{ in a rectangular plate by } \frac{\partial^2\theta}{\partial x^2} + \frac{\partial^2\theta}{\partial y^2} = 0 \text{ and in a cylindrical conductor}$$

with radial flow by $\frac{\partial^2\theta}{\partial r^2} + \frac{1}{r}\frac{\partial\theta}{\partial r} = 0$. Yet the same physical process underlies

all three problems. In order to remove the effects of geometrical variation, and hence to produce an equation independent of coordinate systems we resort to **vector analysis.** The governing equation, valid for steady state heat conduction processes where thermal conductivity is constant with one, two or three space dimensions and any shape of conductor is

$$\nabla^2\theta = 0 \qquad (9.1)$$

This is familiar to us as **Laplace's equation** and takes on the form apposite to the appropriate geometry in each particular problem. The equation can be obtained for the general case using vector methods. The great power of vector analysis lies in its generality of approach.

In measurements of atmospheric pressure, balloons are used to record the pressure p at various points (x, y, z) in space. The totality of possible measurements $p(x, y, z)$ form a **scalar field**; regarded as a function, $p(x, y, z)$ is called a **scalar point function.** We could take a fixed value of the pressure, p_0 say and consider the set of all points in space at which the pressure is p_0; these form a surface, known as a **level surface** of pressure (though 'level' does not necessarily imply 'horizontal'). Through each point in space one and only one level surface will pass. Weather maps show isobars which are two-dimensional projections of these surfaces onto the plane of the map. We could also talk about the density $\rho(x, y, z)$ of the air as a scalar point function, as well as the temperature $\theta(x, y, z)$.

In general, the air will be in motion and the velocity of the air at each point will have a specific direction as well as magnitude, i.e. it will be a vector quantity. We may thus describe the velocity $v(x, y, z)$ as a **vector point function** which will give rise to a **vector field**.

So far we have ignored time changes in our discussion, i.e. we have assumed **time-independent** or **steady fields**. If we generalise our arguments then we must consider $p(x, y, z, t)$, $v(x, y, z, t)$ etc.

Differentiation of vectors

We are frequently concerned with the rate of change of a vector quantity in time or in space. We now examine the ways of calculating these rates of change.

Differentiation with respect to one scalar

If a vector quantity v depends upon one scalar variable t so that $v = v(t)$ then let v change to $v + \delta v$ whilst t changes to $t + \delta t$. If in the limit as $\delta t \to 0$, the ratio $\dfrac{\delta v}{\delta t}$ tends to a limiting value, this limiting value, denoted $\dfrac{dv}{dt}$, is the derivative of v with respect to t.

If in cartesian coordinates $v = v_1 \mathbf{i} + v_2 \mathbf{j} + v_3 \mathbf{k}$ then

$$\frac{dv}{dt} = \frac{dv_1}{dt}\mathbf{i} + \frac{dv_2}{dt}\mathbf{j} + \frac{dv_3}{dt}\mathbf{k} \tag{9.2}$$

since v_1, v_2, v_3 are functions of t only.

For example, if $v = t^2\mathbf{i} + \sqrt{t}\,\mathbf{j} - \sin t\,\mathbf{k}$ then $\dfrac{dv}{dt} = 2t\,\mathbf{i} + \dfrac{1}{2\sqrt{t}}\,\mathbf{j} - \cos t\,\mathbf{k}$

Formal rules for differentiation

Assume that $\mathbf{a}$, $\mathbf{b}$ and ϕ are functions of one scalar t and that the derivatives exist as necessary. Then the following rules hold as you can verify.

(i) $\dfrac{d\mathbf{a}}{dt} = 0$ if and only if $\mathbf{a}$ is a constant vector $\tag{9.3}$

(ii) $\dfrac{d}{dt}(\mathbf{a} + \mathbf{b}) = \dfrac{d}{dt}\mathbf{a} + \dfrac{d}{dt}\mathbf{b}$ $\tag{9.4}$

(iii) $\dfrac{d}{dt}(\phi\,\mathbf{a}) = \dfrac{d\phi}{dt}\mathbf{a} + \phi\dfrac{d\mathbf{a}}{dt}$ $\tag{9.5}$

(iv) $\dfrac{d}{dt}(\mathbf{a} \cdot \mathbf{b}) = \dfrac{d\mathbf{a}}{dt} \cdot \mathbf{b} + \mathbf{a} \cdot \dfrac{d\mathbf{b}}{dt}$ $\tag{9.6}$

(v) $\dfrac{d}{dt}(\mathbf{a} \wedge \mathbf{b}) = \dfrac{d\mathbf{a}}{dt} \wedge \mathbf{b} + \mathbf{a} \wedge \dfrac{d\mathbf{b}}{dt}$ $\tag{9.7}$

Other rules follow as one might expect. For example, if $\mathbf{a} = \mathbf{b}$ in (9.6) then

$\dfrac{d}{dt}(\mathbf{a} \cdot \mathbf{a}) = 2\mathbf{a} \cdot \dfrac{d\mathbf{a}}{dt}$ and if $|\mathbf{a}|$ is constant then $2\mathbf{a} \cdot \dfrac{d\mathbf{a}}{dt} = 0$ and $\mathbf{a}$ is perpendicular

to $\dfrac{d\mathbf{a}}{dt}$. In the special case where $\mathbf{a}$ is the **position vector**

$$\mathbf{r}(t) = x(t)\mathbf{i} + y(t)\mathbf{j} + z(t)\mathbf{k}$$

then $\dfrac{d\mathbf{r}}{dt}$ is the velocity vector $\mathbf{v}$ and we have the result that if a particle moves so

that its distance from the origin, $|\mathbf{r}|$, is constant then its velocity is in a direction perpendicular to the line joining it to the origin. What interpretation can you draw?

Example

A particle of mass m and charge q moves with velocity $\mathbf{v}$ in a uniform magnetic field $\mathbf{B}$. Let us choose cartesian axes so that the z-axis is parallel to $\mathbf{B}$, i.e. that $\mathbf{B} = (0, 0, B)$ and let $\mathbf{v} = (v_x, v_y, v_z)$. The equation of motion is

$$m\frac{d\mathbf{v}}{dt} = q(\mathbf{v} \wedge \mathbf{B}) \tag{9.8}$$

If we resolve (9.8) into three component equations and introduce the constant

$\omega = qB/m$ we find that $v_z = V$, a constant, and that $\ddot{v}_x = -\omega^2 v_x$,

$\ddot{v}_y = -\omega^2 v_y$, where the dots above the components indicate time differentiation. If we align the x- and y-axes so that the initial velocity, U, was along the latter and if the particle is at the origin at $t = 0$, then the position vector at any subsequent time is given by

$$\mathbf{r} = \left(\frac{U}{\omega} [1 - \cos \omega t], \quad \frac{U}{\omega} \sin \omega t, \; Vt \right) \tag{9.9}$$

If we eliminate t from the equations for x and y we obtain the projection of the path of the particle on the x-y plane. It is

$$(x - U/\omega)^2 + y^2 = U^2/\omega^2 \tag{9.10}$$

This is clearly a circle, centre $(U/\omega, 0)$ and radius U/ω; the path is therefore a helix, (Figure 9.1) which is wrapped round the cylinder whose equation is (9.10). The velocity vector is $\mathbf{v} = (U \sin \omega t, U \cos \omega t, V)$ and we see that the particle rotates uniformly about the axis of the cylinder whilst moving

uniformly parallel to this axis. Note that from (9.6) $\dfrac{d}{dt}(v^2) = 2\mathbf{v} \cdot \dfrac{d\mathbf{v}}{dt}$

$= 2\dfrac{q}{m}\mathbf{v} \cdot (\mathbf{v} \wedge \mathbf{B}) = 0$ and hence v^2 and the kinetic energy $\frac{1}{2}mv^2$ are constant.

Further, if the acceleration vector $\dfrac{d\mathbf{v}}{dt}$ be denoted $\mathbf{f}$, then $\mathbf{f}$ is perpendicular to $\mathbf{v}$ at all stages of the motion and is directed to the axis of the cylinder.

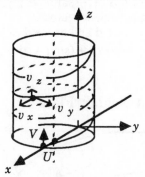

Figure 9.1

Partial differentiation

Now consider a vector $\mathbf{A}$ which is a function of x, y and z so that $\mathbf{A} = \mathbf{A}(x, y, z)$ and each component is a function of x, y and z. Then the partial derivatives of $\mathbf{A}$, provided that they exist, may be calculated in the usual way. For example, if $\mathbf{A} = 2xy^2\mathbf{i} - y^2z\mathbf{j} + xz^2\mathbf{k}$, then $\dfrac{\partial \mathbf{A}}{\partial x} = 2y^2\mathbf{i} + z^2\mathbf{k}$,

$\dfrac{\partial \mathbf{A}}{\partial y} = 4xy\mathbf{i} - 2yz\mathbf{j}$, $\dfrac{\partial \mathbf{A}}{\partial z} = -y^2\mathbf{j} + 2xz\mathbf{k}$ with results for higher derivatives following as is obvious. If, in turn, x, y and z are functions of one scalar variable t then the **total derivative**

$$\frac{d\mathbf{A}}{dt} = \frac{\partial \mathbf{A}}{\partial x}\frac{dx}{dt} + \frac{\partial \mathbf{A}}{\partial y}\frac{dy}{dt} + \frac{\partial \mathbf{A}}{\partial z}\frac{dz}{dt} \qquad (9.11)$$

The **total differential** of $\mathbf{A}$ is

$$d\mathbf{A} = \frac{\partial \mathbf{A}}{\partial x}\,dx + \frac{\partial \mathbf{A}}{\partial y}\,dy + \frac{\partial \mathbf{A}}{\partial z}\,dz \qquad (9.12)$$

The differential $d\mathbf{r} = (dx\mathbf{i} + dy\mathbf{j} + dz\mathbf{k})$ \qquad (9.13)

If we use the notation (in cartesian coordinates)

$$\nabla = \mathbf{i}\,\frac{\partial}{\partial x} + \mathbf{j}\,\frac{\partial}{\partial y} + \mathbf{k}\,\frac{\partial}{\partial z} \qquad (9.14)$$

then $\qquad\qquad d\mathbf{A} = (\nabla \,.\, d\mathbf{r})\,\mathbf{A}$ \qquad (9.15)

Ordinary integrals of vectors

Just as we can differentiate a position vector to obtain a velocity vector, we can reverse the process; corresponding to the arbitrary constant of scalar integration we have to add a time-independent arbitrary vector. We give one example as an illustration of the general process of integration.

If the velocity vector is $\mathbf{v} = 2 \sin t \, \mathbf{i} + (\cos t - 1)\mathbf{j} + 6t\mathbf{k}$ and $\mathbf{r} = \mathbf{0}$ at $t = 0$ we first find

$$\mathbf{r} = \int (2 \sin t \, \mathbf{i} + (\cos t - 1)\mathbf{j} + 6t\mathbf{k}) \, dt$$
$$= -2 \cos t \, \mathbf{i} + (\sin t - t) \, \mathbf{j} + 3t^2\mathbf{k} + \mathbf{d}$$

where $\mathbf{d}$ is an arbitrary time-independent vector.

Since $\mathbf{r} = \mathbf{0}$ when $t = 0$, $\mathbf{0} = -2\mathbf{i} + \mathbf{d}$ and hence $\mathbf{d} = 2\mathbf{i}$. Therefore

$$\mathbf{r} = (2 - 2 \cos t)\mathbf{i} + (\sin t - t)\mathbf{j} + 3t^2\mathbf{k}$$

9.2 The Gradient of a Scalar Field

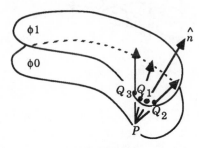

Figure 9.2

Let $\phi(x, y, z)$ constitute a scalar field. Then let us examine how to calculate the rate of change of ϕ with distance in any given direction. In Figure 9.2 we show two level surfaces corresponding to two constant values of ϕ, ϕ_0 and ϕ_1, say. Let $\phi(P) = \phi_0$ and $\phi(Q_1) = \phi(Q_2) = \phi(Q_3) = \phi_1$. If we consider the values of

$$\frac{\phi(Q) - \phi(P)}{PQ}$$ where Q represents points like Q_1, Q_2, Q_3 on the surface $\phi = \phi_1$

then these values depend only on the distance PQ; the ratio will have its maximum value when PQ is least, that is when PQ is normal to the level surface at P (why is

this?). In general we write $\dfrac{\phi(Q) - \phi(P)}{PQ}$ as $\dfrac{\delta\phi}{\delta s}$ and if we let $\delta s \to 0$ then if a

limit exists we denote it $\dfrac{\partial\phi}{\partial s}$ and call it a **directional derivative** of ϕ. The

maximum value is denoted $\dfrac{\partial\phi}{\partial n}$ in direction $\hat{\mathbf{n}}$, where $\hat{\mathbf{n}}$ is a unit vector normal to

the level surface at P. We call this vector the **gradient** of ϕ, written grad ϕ and

$$\text{grad } \phi = \frac{\partial\phi}{\partial n} \, \hat{\mathbf{n}} \qquad\qquad (9.16)$$

Let PQ_n be the normal at P to the surface on which P lies; Figure 9.3. Then

$$\frac{\partial \phi}{\partial s} = \lim_{\delta s \to 0} \left(\frac{\delta \phi}{\delta s} \right) = \lim_{\delta s \to 0} \left(\frac{\delta \phi}{\delta n} \cdot \frac{\delta n}{\delta s} \right) = \frac{\partial \phi}{\partial n} \cos \theta$$

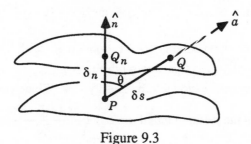

Figure 9.3

If $\hat{\mathbf{a}}$ is a unit vector in the direction of PQ then

$$\cos \theta = \hat{\mathbf{n}} \cdot \hat{\mathbf{a}} \quad \text{and} \quad \frac{\partial \phi}{\partial s} = \frac{\partial \phi}{\partial n} (\hat{\mathbf{n}} \cdot \hat{\mathbf{a}}) = \text{grad } \phi \cdot \hat{\mathbf{a}}$$

We see that $\frac{\partial \phi}{\partial s}$ is the resolved component of grad ϕ in the direction $\hat{\mathbf{a}}$. In

particular, if we take $\hat{\mathbf{a}}$ to be $\mathbf{i}$ then $\frac{\partial \phi}{\partial x} = \text{grad } \phi \cdot \mathbf{i}$. Similarly $\frac{\partial \phi}{\partial y} = \text{grad } \phi \cdot \mathbf{j}$

and $\frac{\partial \phi}{\partial z} = \text{grad } \phi \cdot \mathbf{k}$. It follows that in cartesian coordinates

$$\text{grad } \phi = \frac{\partial \phi}{\partial x} \mathbf{i} + \frac{\partial \phi}{\partial y} \mathbf{j} + \frac{\partial \phi}{\partial z} \mathbf{k} \tag{9.17}$$

Now we may via (9.14) identify grad ϕ with $\nabla \phi$. ∇ is a **vector operator** which associates with a scalar ϕ a unique vector $\nabla \phi$ or grad ϕ.

We should emphasise that result (9.16) is *independent of the coordinate system chosen* and this demonstrates the power of vector field theory. The expressions (9.14) and (9.17) are merely one particular form of (9.16).

Example 1

The temperature of a gas in a room is given by $T = T_0(x^2 - y^2 + xyz) + 273$. We wish to find the maximum rate of change of temperature at the point $(1, 1, 1)$ and the direction in which this occurs.

First we find grad $T = T_0[(2x + yz)\mathbf{i} + (xz - 2y)\mathbf{j} + xy\mathbf{k}]$ and then evaluate it at $(1, 1, 1)$ to obtain $T_0(3\mathbf{i} - \mathbf{j} + \mathbf{k})$. The magnitude of this vector is

$T_0\sqrt{11}$ which is the required maximum rate of change with direction given by

$$\hat{n} = \frac{3i - j + k}{\sqrt{11}}$$

Example 2

We wish to find a unit normal to the surface $x^2y + y^2z + z^2x = 5$ at the point $(2, 1, -1)$.

Let $\phi = x^2y + y^2z + z^2x$; then we are concerned with the level surface $\phi = 5$ which is the only one to pass through the point $(2, 1, -1)$. The value of grad ϕ at this point is easily shown to be $5i + 2j - 3k$. Since grad ϕ is normal to

the level surface then the required unit normal is $\pm \dfrac{5i + 2j - 3k}{\sqrt{38}}$.

Example 3

Find the rate of change of pressure (with distance) in the direction $2i + j - k$ at the point $(2, 3, 1)$ if pressure in a given region is $p = p_0 xyz^2$.

We require the directional derivative $\hat{a}$. grad ϕ where $\hat{a}$ is the unit vector $(2i + j - k) / \sqrt{6}$. The value of grad ϕ at $(2, 3, 1)$ is $p_0 (3i + 2j + 12k)$ and

the required result is $p_0 (2i + j - k) . (3i + 2j + 12k) / \sqrt{6} = -4p_0 / \sqrt{6}$.

Further results on gradients

(i)   Certain results follow fairly obviously from the definition of grad. For example, $\nabla(\phi_1 + \phi_2) = \nabla\phi_1 + \nabla\phi_2$

(ii)   If r is the position vector then $r = |r| = (x^2 + y^2 + z^2)^{1/2}$ and it is easily shown that

$$\nabla r = r/r = \hat{r} \qquad (9.18)$$

(iii) In general,

$$\nabla \{f(r)\} = \hat{r} f'(r) \qquad (9.19)$$

(iv)   If A is a constant vector then

$$\nabla(A . r) = A \qquad (9.20)$$

(v)   If B is an arbitrary vector then $(B . \nabla)\phi = B . (\nabla\phi)$. In particular

$$(B . \nabla)r = B . \hat{r} \qquad (9.21)$$

Example

Consider the transport of heat or mass by a fluid. Let this quantity be specified by $\phi(x, y, z, t) = \phi(r, t)$. Then

$$\delta\phi = \frac{\partial\phi}{\partial t}\,\delta t + \frac{\partial\phi}{\partial x}\,\delta x + \frac{\partial\phi}{\partial y}\,\delta y + \frac{\partial\phi}{\partial z}\,\delta z = \frac{\partial\phi}{\partial t}\,\delta t + (\delta\mathbf{r}\,.\,\nabla)\phi$$

The **total rate of change of** ϕ with time experienced by a particle of fluid as it moves (called differentiation following the fluid) is given by

$\dfrac{d\phi}{dt} = \dfrac{\partial\phi}{\partial t} + \left[\dfrac{d\mathbf{r}}{dt}\,.\,\nabla\right]\phi$. The local rate of change $\dfrac{\partial\phi}{\partial t}$ represents the

changes taking place at a fixed point in space and the convective rate of

change $\left[\dfrac{d\mathbf{r}}{dt}\,.\,\nabla\right]\phi$ represents the changes that take place independent of

time variation as the property ϕ is convected with the fluid at its speed $\dfrac{d\mathbf{r}}{dt}$.

(vi)   In cylindrical polar coordinates (r, θ, z)

$$\nabla V = \hat{\mathbf{r}}\,\frac{\partial V}{\partial r} + \hat{\theta}\,\frac{1}{r}\,\frac{\partial V}{\partial \theta} + \mathbf{k}\,\frac{\partial V}{\partial z} \qquad (9.22)$$

(vii)   In spherical polar coordinates (r, θ, ϕ)

$$\nabla V = \hat{\mathbf{r}}\,\frac{\partial V}{\partial r} + \hat{\theta}\,\frac{1}{r}\,\frac{\partial V}{\partial \theta} + \hat{\phi}\,\frac{1}{r\sin\theta}\,\frac{\partial V}{\partial \phi} \qquad (9.23)$$

9.3 Divergence of a Vector Field

In section 14.2 of *Engineering Mathematics* we considered the two-dimensional flow of a fluid into a cuboidal region. Let us now extend the argument to allow the flow to be three-dimensional. (Although we have taken fluid flow as our practical application the same argument would apply to any vector field.)

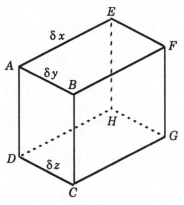

Figure 9.4

Figure 9.4 shows an infinitesimal element of volume with sides parallel to the coordinate axes, lying within the region of fluid flow. Its centre is at the point (x, y, z). Let $\mathbf{v}(x, y, z) = (v_1, v_2, v_3)$ be the time-independent velocity of the fluid and we consider the flux of $\mathbf{v}$ through the volume; (the flux is the volume of fluid crossing the faces of the box in unit time). We shall then let the volume shrink to the point (x, y, z) just as we did with the two-dimensional case and we arrive at the results that the net velocity flux out of the volume element is $\left(\dfrac{\partial v_1}{\partial x} + \dfrac{\partial v_2}{\partial y} + \dfrac{\partial v_3}{\partial z} \right) \delta x \, \delta y \, \delta z$ so that the amount of outward flux per unit volume, called the **divergence** of $\mathbf{v}$ is given by

$$\operatorname{div} \mathbf{v} = \frac{\partial v_1}{\partial x} + \frac{\partial v_2}{\partial y} + \frac{\partial v_3}{\partial z} \tag{9.24}$$

It would seem reasonable to represent div $\mathbf{v}$ by $\nabla . \mathbf{v}$ although we have only obtained a correspondence in cartesian coordinates; we shall accept the general result, i.e.

$$\nabla . \mathbf{v} = \operatorname{div} \mathbf{v} \tag{9.25}$$

Note that certain results follow fairly straightforwardly in cartesian coordinates but hold in any coordinate system; it is usual to verify them in cartesian coordinates. For example

(i) $\nabla.(\mathbf{A} + \mathbf{B}) = \nabla.\mathbf{A} + \nabla.\mathbf{B}$ (9.26)

(ii) $\nabla.(\phi\mathbf{A}) = (\nabla\phi).\,\mathbf{A} + \phi(\nabla.\,\mathbf{A})$ (9.27)

Notice in this last result that the expected rule for differentiation also has the feature that the scalar product is preserved in both terms on the right-hand side.

Example 1

$$\operatorname{div} \mathbf{r} = \nabla .\, (x\mathbf{i} + y\mathbf{j} + z\mathbf{k}) = \frac{\partial}{\partial x}\,(x) + \frac{\partial}{\partial y}\,(y) + \frac{\partial}{\partial z}\,(z) = 3$$

Example 2

The electrostatic field associated with an isolated charge e is given by $\mathbf{E} = -\operatorname{grad}\ \phi$ where $\phi = \dfrac{e}{r}$. $\mathbf{E} = -\nabla\,[\,e/(x^2 + y^2 + z^2)^{1/2}]$ and the symmetry involved will mean that we need only evaluate the first component E_x. In fact

$$E_x = -e\,\frac{\partial}{\partial x}\,(x^2 + y^2 + z^2)^{-1/2} = ex(x^2 + y^2 + z^2)^{-3/2} = ex/r^3$$

By using symmetry to obtain the other components we may deduce that

$$\mathbf{E} = \frac{e}{r^3}\,(x\mathbf{i} + y\mathbf{j} + z\mathbf{k}) = \frac{e\mathbf{r}}{r^3} = \frac{e\hat{\mathbf{r}}}{r^2}$$

Now
$$\text{div } \mathbf{E} = e\left[\frac{\partial}{\partial x}\left(\frac{x}{r^3}\right) + \frac{\partial}{\partial y}\left(\frac{y}{r^3}\right) + \frac{\partial}{\partial z}\left(\frac{z}{r^3}\right)\right]$$

But
$$\frac{\partial}{\partial x}\left(\frac{x}{r^3}\right) = \frac{1}{r^3} + x\frac{\partial}{\partial x}\left([x^2 + y^2 + z^2]^{-3/2}\right) = \frac{1}{r^3} - \frac{3x^2}{r^5}$$

Hence
$$\text{div } \mathbf{E} = \frac{3}{r^3} - \frac{3}{r^5}(x^2 + y^2 + z^2) = 0$$

A vector field whose divergence is zero is termed **solenoidal**. We may interpret this by saying that the lines of flow form closed curves.

Remarks on divergence

Returning to the fluid flow example, if the divergence is non-zero at a point in the fluid, it expresses the rate/unit volume at which fluid is flowing away from (if positive) or towards (if negative) the point. When the divergence is positive, then either the fluid is expanding and its density is falling with time or there is a **source** of fluid at the point where fluid is entering the field of flow. For negative divergence the opposite is true; that is either the fluid is contracting and the density rising with time at the point or there is a **sink** of fluid there at which fluid is leaving the field. In an electric field for example the existence of positive divergence at a point can indicate the presence of an electric pole at that point.

The case of zero divergence indicates that flux entering any element of space is exactly balanced by that leaving it. This implies that fluid is neither 'created' nor 'destroyed' there which implies that there is no source or sink of fluid at that point. If div $\mathbf{v} = 0$ everywhere in a region then no flux can have been generated within it and the lines of flow of the vector $\mathbf{v}$ must either form closed curves (the examples are magnetic field lines) or terminate on a boundary; an example is the electric field lines in a condenser.

9.4 Line Integrals

We concern ourselves in this section with a class of line integrals which arise frequently in practice; these are **tangential line integrals**. In Figure 9.5 let C be a curve drawn in a vector field and let δs be an element of arc along it, starting at a point P. Let $\mathbf{v}$ denote a vector from P making an angle θ with the length element. Then $\mathbf{v} \cdot \mathbf{dr} = v \cos \theta \, \delta r$ and if we sum such contributions from A to B along the curve we obtain in the limit as $\delta s \to 0$ the **line integral**

$$\int_A^B \mathbf{v} \cdot \mathbf{ds} = \int_A^B \mathbf{v} \cdot \mathbf{dr} = \int_A^B v \cos \theta \, dr \qquad (9.28)$$

where $\mathbf{dr}$ has been identified with $\mathbf{ds}$. (Why?).

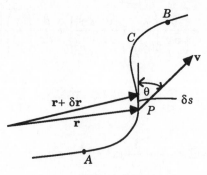

Figure 9.5

If **v** represents a mechanical force then (9.28) will represent the work done in displacing a particle along the curve C from A to B. If **v** is electric field strength then (9.28) represents the **potential difference** between A and B. Methods of evaluation are illustrated in the following examples.

Example 1

Given $\mathbf{v} = (2x + y)\mathbf{i} + (x - 3y)\mathbf{j}$, A = (0, 0), B = (1, 1) evaluate the integrals of **v.dr** from A to B (see Figure 9.6)

(i)   along the straight lines AD and DB
(ii) along the straight lines AE and EB
(iii) along the straight line AB
(iv) along the arc AB of the circle, centre D

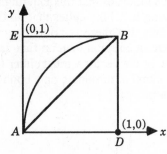

Figure 9.6

(i) $\displaystyle \int_{A}^{B} \mathbf{v}.\,d\mathbf{r} = \int_{AD} \mathbf{v}.\,d\mathbf{r} + \int_{DB} \mathbf{v}.\,d\mathbf{r}$

Along AD, $y = 0$, $d\mathbf{r} = dx\mathbf{i}$ and $\mathbf{v} = 2x\mathbf{i} + x\mathbf{j}$. Then $\mathbf{v}.d\mathbf{r} = 2x\,dx$ and the required integral becomes

$$\int_{0}^{1} 2x\,dx = 1$$

Along DB, $x = 1$, $d\mathbf{r} = dy\mathbf{j}$ and $\mathbf{v} = (2 + y)\mathbf{i} + (1 - 3y)\mathbf{j}$. Then $\mathbf{v}.\,d\mathbf{r} = (1 - 3y)\,dy$ and the integral becomes

$$\int_0^1 (1 - 3y)\,dy = 1 - \frac{3}{2} = -\frac{1}{2}$$

On addition we obtain the result $\dfrac{1}{2}$.

(ii)   Along AE, $x = 0$, $d\mathbf{r} = dy\mathbf{j}$, $\mathbf{v} = y\mathbf{i} - 3y\mathbf{j}$ so that $\mathbf{v}.\,d\mathbf{r} = -3y\,dy$ and

$$\int_0^1 -3y\,dy = -3/2.$$ Along EB, $y = 1$, $d\mathbf{r} = dx\mathbf{i}$, $\mathbf{v} = (2x + 1)\mathbf{i} + (x - 3)\mathbf{j}$

so that $\mathbf{v}.\,d\mathbf{r} = (2x + 1)\,dx$ and $\displaystyle\int_0^1 (2x + 1)\,dx = 2.$ On addition we obtain

the result $\dfrac{1}{2}$ as before.

(iii)   The equation of the line AB is $y = x$ and we may write $x = y = t$ and $dx = dy = dt$, $A \equiv t = 0$ and $B \equiv t = 1$. Then

$$\int_A^B \mathbf{v}.\,d\mathbf{r} = \int_A^B [(2x + y)\mathbf{i} + (x - 3y)\mathbf{j}].[dx\mathbf{i} + dy\mathbf{j}]$$

$$= \int_A^B [(2x + y)dx + (x - 3y)\,dy] = \int_0^1 (3t\,dt - 2t\,dt) = \int_0^1 t\,dt = \frac{1}{2}$$

(iv)   The circle in question has equation $(x - 1)^2 + y^2 = 1$. It may be represented parametrically by $x = 1 + \cos\theta$, $y = \sin\theta$. Then $A \equiv \theta = \pi$, $B \equiv \theta = \pi/2$, $dx = -\sin\theta\,d\theta$ and $dy = \cos\theta\,d\theta$. Also

$$\int_A^B \mathbf{v}.d\mathbf{r} = \int_A^B [(2x + y)\,dx + (x - 3y)\,dy]$$

$$= \int_\pi^{\pi/2} [-(2 + 2\cos\theta + \sin\theta)\sin\theta\,d\theta + (1 + \cos\theta - 3\sin\theta)\cos\theta\,d\theta]$$

$$= \int_{\pi}^{\pi/2} [-2 \sin \theta - 2 \cos \theta \sin \theta - \sin^2 \theta + \cos \theta + \cos^2 \theta - 3 \sin \theta \cos \theta] d\theta$$

$$= [2 \cos \theta + \frac{5}{4} \cos 2\theta + \frac{1}{2} \sin 2\theta + \sin \theta]_{\pi}^{\pi/2} = \frac{1}{2} \,.$$

Example 2

Repeat with paths (i), (ii) and (iii) for $\mathbf{v} = (1 + x^2 y)\mathbf{i} + 2xy\mathbf{j}$. Now

$$\int_A^B \mathbf{v} \cdot d\mathbf{r} = \int_A^B [(1 + x^2 y) \, dx + 2xy \, dy]$$

(i)   Along AD, $\mathbf{v} \cdot d\mathbf{r} = 1 \, dx$ and so $\int_{AD} \mathbf{v} \cdot d\mathbf{r} = \int_0^1 dx = 1$. Along DB, $\mathbf{v} \cdot d\mathbf{r} =$

$2y \, dy$ and so $\int_{DB} \mathbf{v} \cdot d\mathbf{r} = \int_0^1 2y \, dy = 1$. Adding, we get the result of 2.

(ii)   Along AE, $\mathbf{v} \cdot d\mathbf{r} = 0$ and so $\int_{AE} \mathbf{v} \cdot d\mathbf{r} = 0$. Along EB, $\mathbf{v} \cdot d\mathbf{r} = (1 + x^2) dx$

and so $\int_{EB} \mathbf{v} \cdot d\mathbf{r} = \int_0^1 (1 + x^2) dx = \frac{4}{3}$. Adding, we get the result of $\frac{4}{3}$.

(iii)   Along AB, $\mathbf{v} \cdot d\mathbf{r} = [(1 + t^3)\mathbf{i} + 2t^2\mathbf{j}] \cdot [dt\mathbf{i} + dt\mathbf{j}] = (1 + t^3 + 2t^2) dt$ so

that $\int_A^B \mathbf{v} \cdot d\mathbf{r} = \int_0^1 (1 + t^3 + 2t^2) \, dt = \frac{23}{12}$.

At once we have come across a conflict. Not only have we found that $\int_A^B \mathbf{v} \cdot d\mathbf{r}$ is

dependent on the path, as well as on the end points A and B but we now *cannot*

be *sure* that in Example 1 any other path would give the result $\frac{1}{2}$. We might have

just chosen the paths which give that result: it is unlikely but possible. We need to sort out whether an integral is path independent and if so *then* we can proceed to evaluate it by choosing any suitable path.

Consider Figure 9.7. Let us consider the result of adding $\int\limits_{AEB}$ v.dr to

$\int\limits_{BDA}$ v.dr. We denote this, $\oint$ v.dr and it is understood that such an integral is

called the **circulation**. The anti-clockwise direction is positive; now if $\int\limits_{A}^{B}$ v.dr is

path independent then $\int\limits_{BEA} = -\int\limits_{AEB} = -\int\limits_{ADB}$ and therefore $\oint$ v.dr = 0. If we

consider a particle moving around a closed curve in a force field, no net work

would be done and hence energy conserved. In general when $\oint$ v.dr = 0 we say

the field is **conservative**. In the case where $\int\limits_{A}^{B}$ v.dr is path dependent

$\oint$ v.dr ≠ 0 and the field is non-conservative.

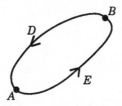

Figure 9.7

Condition for path independence

If $\int\limits_{A}^{B}$ v.dr is to be path independent then v.dr must be an exact differential of

some function ϕ i.e. v.dr = dϕ so that $\int\limits_{A}^{B}$ v.dr $= \int\limits_{A}^{B}$ dϕ = ϕ(B) − ϕ(A). But

d$\phi = \dfrac{\partial\phi}{\partial x}$ dx $+ \dfrac{\partial\phi}{\partial y}$ dy so that if v.dr = Pdx + Qdy where P and Q are functions of x and y then $P = \dfrac{\partial \phi}{\partial x}$ and $Q = \dfrac{\partial \phi}{\partial y}$ so that if $\dfrac{\partial^2 \phi}{\partial x \partial y} = \dfrac{\partial^2 \phi}{\partial y \partial x}$ as is highly likely then

$$\frac{\partial P}{\partial y} = \frac{\partial Q}{\partial x} \qquad (9.29)$$

By reversing the argument we may show that (9.29) is a necessary and sufficient condition for path independence.

In the case of Example 1, $P = 2x + y$ and $Q = x - 3y$ so that $\dfrac{\partial P}{\partial y} = 1 = \dfrac{\partial Q}{\partial x}$

and any path between A and B would have given the result of $\frac{1}{2}$. However in

Example 2, $P = 1 + x^2 y$ and $Q = 2xy$. Here $\dfrac{\partial P}{\partial y} = x^2 \neq 2y = \dfrac{\partial Q}{\partial x}$ and the

integral is path dependent.

Extension to three dimensions

If we extend our ideas to three dimensions we can place a conservative field in a new light. The actual calculation of an integral proceeds much as we might expect. Here

$$\int \mathbf{v}.d\mathbf{r} = \int (v_x \mathbf{i} + v_y \mathbf{j} + v_z \mathbf{k}).(dx\mathbf{i} + dy\mathbf{j} + dz\mathbf{k}) = \int (v_x dx + v_y dy + v_z dz)$$

If this integral is to be path independent, then

$$\mathbf{v}.d\mathbf{r} = v_x dx + v_y dy + v_z dz = d\phi$$

Now
$$d\phi = \frac{\partial \phi}{\partial x}\, dx + \frac{\partial \phi}{\partial y}\, dy + \frac{\partial \phi}{\partial z}\, dz = \nabla\phi.\, d\mathbf{r}$$

Hence $(\mathbf{v} - \nabla\phi).\, d\mathbf{r} = 0$ and we conclude that since $d\mathbf{r}$ is arbitrarily orientated and non-zero

$$\mathbf{v} = \nabla\phi \qquad (9.30)$$

that is, $\mathbf{v}$ is the gradient of some scalar ϕ, called the **scalar potential** of the conservative field. (Note that ϕ is determined apart from a constant and that in some branches of application a minus sign is introduced, thus $\mathbf{E} = -\text{grad } \phi$ where $\mathbf{E}$ is the electric field and ϕ the electrostatic potential.) You are probably familiar with gravitational potential where the gravity field may be approximately represented by $\mathbf{F} = -mg\mathbf{k}$ and the potential as mgz; $\nabla(mgz) = mg\mathbf{k}$.

The steps taken to obtain (9.30) are reversible and so a *necessary and*

sufficient condition that $\oint_C \mathbf{v}.d\mathbf{r}$ be zero is that $\mathbf{v} = \nabla\phi$, for some scalar ϕ.

Example

If $\mathbf{v} = (2x + y)\mathbf{i} + (x - 3y)\mathbf{j}$ then a suitable candidate for ϕ is

$\phi = x^2 + xy - \dfrac{3}{2} y^2$ as you can verify by forming $\nabla\phi$.

However it is a tedious task to try and find ϕ in cases where none exists and we look for a simple test on $\mathbf{v}$ itself to discover whether it is conservative. We effectively want a generalisation of condition (9.29).

9.5 Curl of a Vector Field

There are examples in the flow of fluids where although the fluid appears to be moving in straight lines, the individual particles are in fact rotating about their axes. Consider Figure 9.8. An incompressible fluid is in steady one-dimensional flow between two plane walls $y = 0$ and $y = 1$. The velocity vector is $\mathbf{v} = 4Uy(1 - y)\mathbf{i}$ where U is the velocity of the fluid along $y = 0.5$. If a small paddle wheel is placed in the fluid with its axis in the z-direction then it will spin as it moves downstream, if the velocity of the fluid hitting blade oF is different from that of the fluid hitting blade Go. It is observed that on the line $y = 0.5$ no spinning occurs; that the spinning is anticlockwise above the line $y = 0.5$ and clockwise below it; moreover, the angular velocity of the wheel increases the nearer it is placed to either wall.

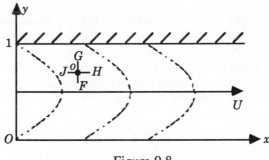

Figure 9.8

Consider a more general two-dimensional motion in which the velocity components at P are u and v, both functions of x and y. See Figure 9.9. In a small interval of time let the fluid particle at P move to P', by when its velocity components have become, to first order $u + (\partial u/\partial y)\delta y$ and $v + (\partial v/\partial x)\, \delta x$. The angular velocity of the line PQ is $[v + (\partial v/\partial x)\delta x - v]/\delta x$ which becomes $\dfrac{\partial v}{\partial x}$ in the limit as $\delta x \to 0$. Similarly we have a limiting angular velocity of the line PR of $-(\partial u/\partial y)$. Consequently, the rotation of the fluid element at P is the average of these, viz. $\dfrac{1}{2}\left(\dfrac{\partial v}{\partial x} - \dfrac{\partial u}{\partial y}\right) = \omega_z$.

Note that in our particular example $u = 4Uy(1-y)$ and $v = 0$ so that $\omega_z = \frac{1}{2}(0 - 4U[1-2y]) = -2U(1-2y)$. Along $y = \frac{1}{2}$ there is no angular velocity ; $y < \frac{1}{2}$ gives a negative velocity whilst $y > \frac{1}{2}$ gives a positive velocity; as $y \to 0$ or $y \to 1$ the *magnitude* of ω_z increases. This bears out the experimental observations.

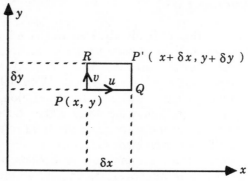

Figure 9.9

We have taken a special case of motion in the x-y plane and in a general motion the angular velocity could be about any axis. Let us associate with the angular velocity a vector whose magnitude is that of the angular velocity and whose direction is along the axis of rotation in the sense indicated by Figure 9.10 ; this, it can be shown, will mean that two such vectors can in fact be added by the parallelogram law – as should be the case.

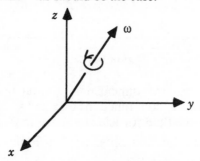

Figure 9.10

The quantity $\left(\dfrac{\partial v}{\partial x} - \dfrac{\partial u}{\partial y}\right)$ is reminiscent of the quantity $\left(\dfrac{\partial Q}{\partial x} - \dfrac{\partial P}{\partial y}\right)$ we met in the section on line integrals. However, angular velocity is a localised entity so let us consider the effect of finding the limit as $S \rightarrow 0$ of the ratio $\oint_C \mathbf{v} . \, d\mathbf{r}/S$,

where S is the plane area bounded by the curve C. Integrating around the element PQP'R of Figure (9.9) and neglecting velocity terms of second order, we obtain

$$\Gamma = \oint \mathbf{v}.d\mathbf{r} = u\delta x + [v + (\partial v/\partial x)\delta x]\delta y - [u + (\partial u/\partial y)\delta y]\delta x - v\delta y$$

$$= \left(\frac{\partial v}{\partial x} - \frac{\partial u}{\partial y}\right)\delta x \, \delta y$$

Then
$$\frac{\Gamma}{S} \approx \left(\frac{\partial v}{\partial x} - \frac{\partial u}{\partial y}\right)$$

and in the limit as $S \rightarrow 0$, the approximation, due to the neglected terms, becomes an exact relationship. Notice that this is *twice* the angular velocity.

We define the **curl** of a vector $\mathbf{v}$ by

$$\text{curl } \mathbf{v} = \lim_{S \to 0} \left\{ \oint \mathbf{v}.d\mathbf{r}/S \right\} \qquad (9.31)$$

where S and C have the meanings stated earlier.

In fluid dynamics, the curl of the velocity is called the **vorticity**. Note that the curl operator gives a vector result. The current flowing in a wire produces a magnetic field. We generalise this situation to say that a volume distribution of current $\mathbf{J}$ gives rise to a magnetic field strength $\mathbf{H}$ which is given by curl $\mathbf{H} = \mathbf{J}$.

Cartesian components of curl

Definition (9.31) is quite general and it means that curl $\mathbf{v}$ can in general be inclined to all three coordinate axes. By extending the argument earlier it can be shown that if the vector $\mathbf{v} = v_1 \mathbf{i} + v_2 \mathbf{j} + v_3 \mathbf{k}$ then

$$\text{curl } \mathbf{v} = \left(\frac{\partial v_3}{\partial y} - \frac{\partial v_2}{\partial z}\right)\mathbf{i} + \left(\frac{\partial v_1}{\partial z} - \frac{\partial v_3}{\partial x}\right)\mathbf{j} + \left(\frac{\partial v_2}{\partial x} - \frac{\partial v_1}{\partial y}\right)\mathbf{k} \qquad (9.32)$$

(We were earlier looking at the $\mathbf{k}$ component only.) More compactly we may write

$$\text{curl } \mathbf{v} = \begin{vmatrix} \mathbf{i} & \mathbf{j} & \mathbf{k} \\ \dfrac{\partial}{\partial x} & \dfrac{\partial}{\partial y} & \dfrac{\partial}{\partial z} \\ v_1 & v_2 & v_3 \end{vmatrix} = \nabla_\wedge \mathbf{v} \qquad (9.33)$$

both these forms following by analogy with the simple vector product. How nice it is to see the operation $\nabla_\wedge \mathbf{v}$ arising; it somehow brings us round full circle.

Note that the form $\nabla \wedge \mathbf{v}$ holds generally; the form for ∇ will be chosen to suit the particular coordinate system of any particular problem.

Conservative fields

If we recall the condition for a field in the x-y plane to be conservative then we recognise this as being that the z-component of curl $\mathbf{v}$ be zero; since the two other components do not exist this means that a necessary and sufficient condition for the field to be conservative is that its curl should vanish, i.e. curl $\mathbf{v} = 0$ if and only if $\mathbf{v}$ is conservative. *This result can be shown to hold quite generally.*

Example

Find curl $\mathbf{v} =$ if (i) $\mathbf{v} = x^2\mathbf{i} + x^2y\mathbf{j} + yz\mathbf{k}$ (ii) $\mathbf{v} = \text{grad } \phi$, for some scalar ϕ.

(i) $\quad \text{curl } \mathbf{v} = \mathbf{i}\left(\dfrac{\partial}{\partial y}\,[yz] - \dfrac{\partial}{\partial z}\,[x^2y]\right) + \mathbf{j}\left(\dfrac{\partial}{\partial z}\,[x^2] - \dfrac{\partial}{\partial x}\,[yz]\right)$

$$+ \mathbf{k}\left(\dfrac{\partial}{\partial x}\,[x^2y] - \dfrac{\partial}{\partial y}\,[x^2]\right) = z\mathbf{i} + 2xy\mathbf{k}$$

(ii) $\quad \text{curl } \mathbf{v} = \mathbf{i}\left(\dfrac{\partial}{\partial y}\left[\dfrac{\partial \phi}{\partial z}\right] - \dfrac{\partial}{\partial z}\left[\dfrac{\partial \phi}{\partial y}\right]\right) + \ldots$

$$= \mathbf{i}\left(\dfrac{\partial^2 \phi}{\partial y \partial z} - \dfrac{\partial^2 \phi}{\partial z \partial y}\right) + \mathbf{j}\left(\dfrac{\partial^2 \phi}{\partial z \partial x} - \dfrac{\partial^2 \phi}{\partial x \partial z}\right) + \mathbf{k}\left(\dfrac{\partial^2 \phi}{\partial x \partial y} - \dfrac{\partial^2 \phi}{\partial y \partial x}\right) = 0$$

Hence the curl of any gradient is zero, i.e. curl grad $\phi \equiv 0$. Note also that curl $\mathbf{r} = 0$. A vector whose curl is zero is termed **irrotational**.

Classification of Vector Fields

In practice we meet four kinds of vector fields classified as follows

(a) curl $\mathbf{v} = 0$ and div $\mathbf{v} = 0$. Here since curl $\mathbf{v} = 0$, then $\mathbf{v} = \text{grad } \phi$ and the second condition gives div grad $\phi = 0 = \nabla^2 \phi$ which is **Laplace's equation**. Typical fields of this kind occur in irrotational motion of incompressible fluids and in electric fields arising from static charges on boundary conductors.

(b) curl $\mathbf{v} = 0$ but div $\mathbf{v} \neq 0$. Again, $\mathbf{v} = \text{grad } \phi$ but div grad $\phi = \nabla^2 \phi \neq 0$ and we obtain **Poisson's equation** which arises for example in the electric fields of volume distribution of charges (electrons in a thermionic tube) and in the irrotational, compressible flow of a fluid.

(c) curl $\mathbf{v} \neq 0$ but div $\mathbf{v} = 0$. The field is rotational and no scalar potential exists. Such a field is typified by rotational, incompressible fluid flow or the magnetic field within a conductor carrying a current.

(d) curl $\mathbf{v} \neq 0$ and div $\mathbf{v} \neq 0$. This is the most general type of vector field and is found in, for example, the rotational motion of a compressible fluid. It can be shown that such a field is expressible as the sum of two fields, one of type (b) and one of type (c).

9.6 Vector Identities

We now present some results which are left to you to prove. We assume $\phi(x, y, z)$ and $\psi(x, y, z)$ are scalar point functions and $\mathbf{u}(x, y, z)$, $\mathbf{v}(x, y, z)$ are vector point functions. Note that these results are independent of coordinate systems; you will most easily prove them in Cartesians.

(i)   grad $(\phi\,\psi) = \psi$ grad $\phi + \phi$ grad ψ (9.34)

(ii) div $(\phi\,\mathbf{v}) = (\text{grad } \phi)\,.\,\mathbf{v} + \phi \text{ div } \mathbf{v}$ (9.35)

(iii) curl $(\phi\,\mathbf{v}) = (\text{grad } \phi) \wedge \mathbf{v} + \phi \text{ curl } \mathbf{v}$ (9.36)

(iv) div $(\mathbf{u} \wedge \mathbf{v}) = \mathbf{v}.\,\text{curl }\mathbf{u} - \mathbf{u}\,.\,\text{curl }\mathbf{v}$ (9.37)

(v) curl $(\mathbf{u} \wedge \mathbf{v}) = \mathbf{u} \text{ div } \mathbf{v} - (\mathbf{u}\,.\,\nabla)\mathbf{v} + (\mathbf{v}\,.\,\nabla)\mathbf{u} - \mathbf{v} \text{ div } \mathbf{u}$ (9.38)

(vi) grad $(\mathbf{u}\,.\,\mathbf{v}) = (\mathbf{v}\,.\,\nabla)\mathbf{u} + (\mathbf{u}\,.\,\nabla)\,\mathbf{v} + \mathbf{v} \wedge \text{curl }\mathbf{u} + \mathbf{u} \wedge \text{curl }\mathbf{v}$ (9.39)

Second order operators

We have seen that curl grad $\phi \equiv 0$; similarly, we can show that div curl $\mathbf{v} \equiv 0$.

The other second order operators which exist are grad div $\mathbf{v}$, div grad ϕ and curl curl $\mathbf{v}$. We examine these in turn.

(i)   If $\mathbf{v}$ represents electric force, div $\mathbf{v}$ is the space charge density and grad div $\mathbf{v}$ represents the greatest rate of increase of charge (in magnitude and direction) at any point. If $\mathbf{v}$ represents fluid velocity, grad div $\mathbf{v}$ represents the greatest rate of space increase of density.

(ii) div grad ϕ is written $\nabla^2\phi$ and ∇^2 is called **Laplace's operator**. In

cartesian coordinates $\nabla^2\phi = \dfrac{\partial^2\phi}{\partial x^2} + \dfrac{\partial^2\phi}{\partial y^2} + \dfrac{\partial^2\phi}{\partial z^2}$. $\nabla^2\phi = 0$ is, of course,

Laplace's equation. Note that if $\mathbf{A} = A_1\mathbf{i} + A_2\mathbf{j} + A_3\mathbf{k}$ then

$\nabla^2\mathbf{A} = \nabla^2 A_1\mathbf{i} + \nabla^2 A_2\mathbf{j} + \nabla^2 A_3\mathbf{k}$.

(iii) It can be shown that

$$\text{curl curl } \mathbf{v} = \text{grad div } \mathbf{v} - \nabla^2\mathbf{v} \qquad\qquad (9.40)$$

(In proving some of these vector identities we may note that if each side is a vector it often suffices to show that the $\mathbf{i}$-component of each side is the same and appeal to symmetry to establish the result.)

Example

If $\mathbf{E}$ represents electric field strength, $\mathbf{H}$ magnetic field strength and if we assume unit magnetic permeability and dielectric constant, charge-free and current-free region then Maxwell's equations for the electromagnetic field reduce to the form

$$\text{div } \mathbf{E} = 0 = \text{div } \mathbf{H}, \quad \text{curl } \mathbf{E} = -\frac{\partial \mathbf{H}}{\partial t}, \quad \text{curl } \mathbf{H} = \frac{\partial \mathbf{E}}{\partial t}$$

Now curl curl $\mathbf{H} = \text{curl } (\partial\mathbf{E}/\partial t) = \partial(\text{curl } \mathbf{E})/\partial t$ i.e.

$$\text{grad div } \mathbf{H} - \nabla^2\mathbf{H} = \frac{\partial}{\partial t}\left(-\frac{\partial \mathbf{H}}{\partial t}\right) = -\frac{\partial^2 \mathbf{H}}{\partial t^2}$$

But div $\mathbf{H} = 0$, therefore $\mathbf{H}$ satisfies the **wave equation**

$$\nabla^2\mathbf{H} = \partial^2\mathbf{H}/\partial t^2$$

and a similar equation is satisfied by $\mathbf{E}$.

9.7 Double Integration

Before proceeding further with vector analysis and the vector integral theorems we need to study, in this section, double integrals and in a later section the extension to triple integrals.

Example 1

A bar of length L with a cross-section shown in Figure 9.11 is subjected to a torque T applied at one end while the other end is prevented from rotating. The bar will attempt to rotate about the z – axis which passes through the centroid of each cross-section. If the angle of twist is θ/unit length of bar, then the torsional

rigidity is given by $C = \dfrac{2}{\theta} \displaystyle\iint_A \psi(x, y)\mathrm{d}A$ where A is the area of cross-

section, $\mathrm{d}A$ an element of that cross-section, $\psi(x, y)$ is a *stress function* which

satisfies $\dfrac{\partial^2 \psi}{\partial x^2} + \dfrac{\partial^2 \psi}{\partial y^2} = 2G\theta$ inside A and $\psi = 0$ on the perimeter of A; and G

is the shear modulus of the material comprising the bar. The symbolism $\displaystyle\iint_A$

means that the contributions $\psi(x, y)\ \mathrm{d}A$ are summed in a sense to be developed later; for the moment, we merely remark that in this summation we cover A.

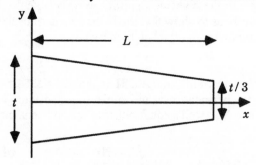

Figure 9.11

Example 2

The total strain energy in a plate is given by

$$U = \frac{1}{2} \iint\limits_{A} \left[\sigma_{xx} \frac{\partial u_x}{\partial x} + \sigma_{yy} \frac{\partial u_y}{\partial y} + \sigma_{xy} \left(\frac{\partial u_y}{\partial x} + \frac{\partial u_x}{\partial y} \right) \right] dA$$

where σ_{xx}, σ_{xy} and σ_{yy} are direct and shear stresses and u_x, u_y are displacements in the x- and y- directions.

Example 3

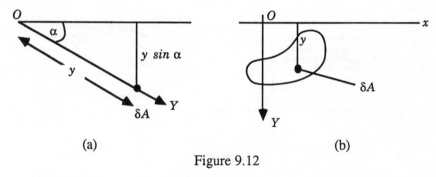

(a)   (b)

Figure 9.12

A plane area R is submerged in water (Figure 9.12). If the weight of unit volume of water be w at any point in the fluid then the total thrust on the plane area is $\iint\limits_{R} wy \sin \alpha \, dA$. We may want to find the coordinates of the centre of pressure.

In each of these summations we need a double summation to cover the area. We shall now develop some techniques of double integration.

First consider a rectangular region R shown in Figure 9.13(a).

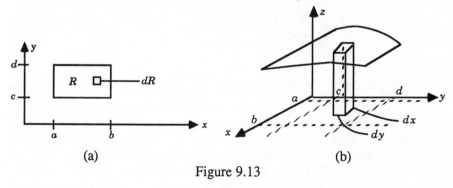

(a)   (b)

Figure 9.13

We have chosen positive coordinates merely for convenience. Every double integral $\iint\limits_{R} f(x, y)\, dR$ can be interpreted as a volume; the volume under a surface $z = f(x, y)$ where x and y are restricted to the domain R. In Figure 9.13(b) we show the building block from which we construct that volume. Its volume is $f(\xi, \eta)\, .\, dx\, .\, dy$, where (ξ, η) is a point somewhere in the base. Just as for an integral of a function of one variable, so here we can form an **upper sum** and a **lower sum**. Suppose we divide R into small rectangles of equal area as shown in Figure 9.14(a).

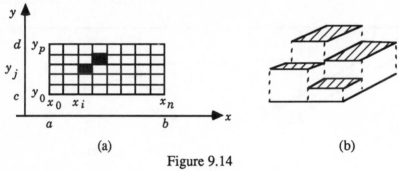

(a)   (b)

Figure 9.14

The rectangle whose right-hand boundary is $x = x_i$ and whose top boundary is $y = y_j$ may be called the i,jth rectangle. Let M_{ij} be the largest value of $f(x, y)$ where $x_{i-1} < x < x_i$, $y_{j-1} < y < y_j$ and let m_{ij} be the least value of $f(x, y)$ in that region. Then we form an upper sum

$$\overline{S}_{np} = \sum_{i=1}^{n} \sum_{j=1}^{p} M_{ij}(x_i - x_{i-1})(y_j - y_{j-1})$$

and a lower sum

$$\underline{S}_{np} = \sum_{i=1}^{n} \sum_{j=1}^{p} m_{ij}(x_i - x_{i-1})(y_j - y_{j-1})$$

Note that we could write the double sums as $\sum\limits_{j=1}^{p} \sum\limits_{i=1}^{n}$ just as easily. This would imply that we carried out the summation first along a column, then from row to row, instead of vice-versa. In Figure 9.14(b) we show schematically a few adjacent rectangles and corresponding upper sums for each rectangle. Suppose we divide the rectangular region R into smaller rectangles, thereby increasing their number. We carry out this process by saying that if the maximum diagonal of all rectangles in a particular subdivision of R be d, then we arrange that $d \rightarrow 0$. If the sequence of upper sums $\overline{S}_{np}$ tends to a finite number $\overline{S}$ and the sequence of lower sums $\underline{S}_{np}$ tends to a finite limit $\underline{S}$ *and* if the two limits are equal, then we say that $f(x, y)$ is **integrable** over R and the value of the

definite integral $\displaystyle\iint_R f(x, y) \, dR$ is the value of that common limit.

If the rectangle R has sides parallel to the coordinate axes then we can write

$$\iint_R f(x, y) \, dR \text{ as } \iint_R f(x, y) \, dx \, dy.$$

The way in which we formed the double sum suggests the way in which we can calculate the double integral. We can first integrate along a fixed value of y, with respect to x and then integrate with respect to y. Geometrically we may interpret this as shown in Figure 9.15(a).

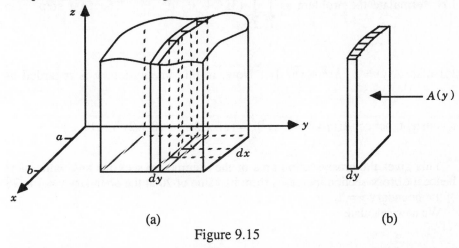

(a)   (b)

Figure 9.15

The building blocks are stacked up in the x-direction to form a slice as shown in Figure 9.15(b); this will have cross-section area $A(y)$, to emphasise that this area depends on the position of the slice, and thickness dy. We then stack together these slices in the y- direction to produce the volume required. We symbolise this process as

$$\int_c^d \int_a^b f(x, y) \, dx \, dy$$

where the interpretation is that the inner integration $\displaystyle\int_a^b f(x, y) \, dx$ is carried out first to produce a function of y and then the resulting function of y, $A(y)$, integrated as $\displaystyle\int_c^d A(y)\,dy$. Notice that, since R is rectangular, the limits of integration on x are the same for any value of y. We now work through three simple examples to fix ideas.

Example 1

Find $\displaystyle\iint_R (3x^2 + y^2)\,dR$ where R is the region $2 \le x \le 3$, $1 \le y \le 2$. We

first formulate the problem as $\displaystyle\int_1^2 \int_2^3 (3x^2 + y^2)\,dx\,dy$. The first step is to

calculate $A(y) = \displaystyle\int_2^3 (3x^2 + y^2)\,dx$. Now, in this integration, y is regarded as

a constant, hence $A(y) = [x^3 + xy^2]_2^3 = (27 + 3y^2) - (8 + 2y^2) = 19 + y^2$

This gives the cross-section area of the volume we seek for any value of y; hence the cross-section increases from its value of 20 at the boundary $y = 1$ to 23 at the boundary $y = 2$.

We now calculate

$$\int_1^2 (19 + y^2)\,dy = \left[19y + \frac{y^3}{3}\right]_1^2 = \left(38 + \frac{8}{3}\right) - \left(19 + \frac{1}{3}\right) = 21\frac{1}{3}$$

This example shows how we can transform a double integral into **repeated integration**.

Example 2

Find $I = \displaystyle\int_0^2 \int_1^4 (x + 2y)\,dx\,dy$. First, calculate

$$\int_1^4 (x + 2y)\,dx = \left[\frac{1}{2}x^2 + 2yx\right]_1^4 = (8 + 8y) - (\frac{1}{2} + 2y) = 7\frac{1}{2} + 6y$$

Then calculate

$$\int_0^2 (7\tfrac{1}{2} + 6y)dy = \left[7\tfrac{1}{2}\, y + 3y^2 \right]_0^2 = 15 + 12. \text{ Hence } I = 27$$

If we refer back to the definition of double integral, we said that the double summation was carried out first in the direction of x and then in the direction of y or vice-versa. Suppose we now reverse the order of integration i.e.

$$I = \int_1^4 \int_0^2 (x + 2y)\, dy\, dx. \text{ Refer to Figure 9.16}$$

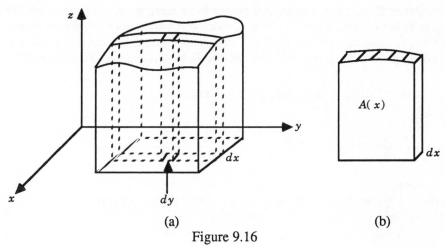

(a)   (b)

Figure 9.16

First we evaluate $\int_0^2 (x + 2y)\, dy = \left[xy + y^2 \right]_0^2 = 2x + 4$ and then we evaluate

$$\int_1^4 (2x + 4)\, dx = \left[x^2 + 4x \right]_1^4 = (16 + 16) - (1 + 4). \text{ Hence } I = 27, \text{ as before.}$$

Example 3

Find $I = \int_0^2 \int_1^4 xy^2\, dx\, dy$

In this example, we can separate the function xy^2 into the product of a function of x and a function of y. We can then write

$$I = \left(\int_0^2 y^2 \, dy \right) \left(\int_1^4 x \, dx \right) = \left[\frac{y^3}{3} \right]_0^2 \left[\frac{x^2}{2} \right]_1^4 = \frac{8}{3} \cdot \frac{15}{2} = 20$$

You can check for yourself that this is the same result as would have been obtained by the method of the previous examples. We have, in effect, reduced the problem to one of two separate integrations. We have not rigorously justified the methods used but we shall take the results on trust.

Extension of ideas to non-rectangular regions

Figure 9.17 shows a non-rectangular region D over which a double integral is to be defined. In order that an extension can be carried out we require that the region D have a boundary C which is a simple closed curve. We also assume that D can be enclosed in a circle of sufficiently large, but finite, radius: this avoids improper integrals. We can enclose D by a rectangle R and then, if we

wish to find $\iint_D f(x, y) \, dx \, dy$ we can define a function

$$g(x, y) = \begin{cases} f(x, \ y), & (x, \ y) \in D \\ 0, & (x, \ y) \notin D \end{cases}$$

so that $\iint_D f(x, y) \, dx \, dy = \iint_R g(x, y) \, dx \, dy$. See Figure 9.17(a)

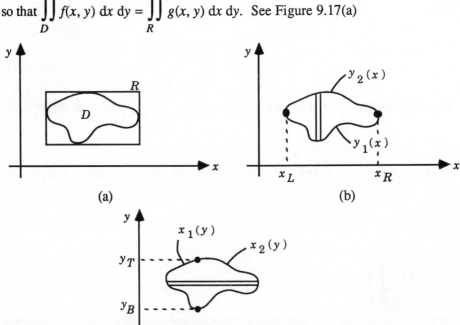

(a)

(b)

(c)

Figure 9.17

When we come to evaluate the double integral we shall first vary y and then vary x, as shown in Figure 9.17(b). Since D is restricted as shown above then there will be a least value of x, x_{min} say, and a greatest value x_{max}. As we traverse the boundary curve C from x_{min} to x_{max} we proceed along the upper part prescribed by a curve $y = y_2(x)$ and along the lower branch by a curve $y = y_1(x)$. A strip such as the one shown has upper and lower limits specified by $y_2(x)$ and $y_1(x)$ respectively. Hence

$$\iint_D f(x, y)\ dy\ dx = \int_{x_{min}}^{x_{max}} \int_{y_1(x)}^{y_2(x)} f(x, y)\ dy dx$$

Equally, since D is closed and bounded we can have a greatest y-value, y_{max} and a least y-value, y_{min} as shown in Figure 9.17(c). We shall then specify the limits on x by $x = x_1(y)$ and $x = x_2(y)$ respectively.

Example 1
Let D be the shaded region shown in Figure 9.18(a).

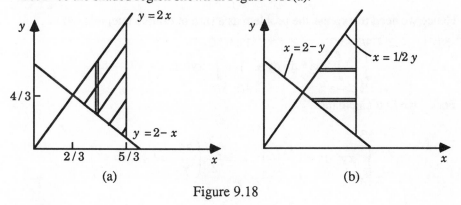

Figure 9.18

If we wish to integrate a function xy^2 over D then

$$I = \iint_D xy^2\ dy\ dx = \int_{2/3}^{5/3} \int_{2-x}^{2x} xy^2\ dy\ dx$$

The first step in the evaluation would be

$$\int_{2-x}^{2x} xy^2\ dy = \left[\frac{xy^3}{3}\right]_{2-x}^{2x} = \frac{x(8x^3)}{3} - \frac{x(2-x)^3}{3}$$

$$= \frac{8x^4 - 8x + 12x^2 - 6x^3 + x^4}{3} = 3x^4 - 2x^3 + 4x^2 - \frac{8}{3}x$$

Then we calculate

$$\int\limits_{2/3}^{5/3} (3x^4 - 2x^3 + 4x^2 - \frac{8}{3}x)\, dx = \left[\frac{3x^5}{5} - \frac{x^4}{2} + \frac{4x^3}{3} - \frac{4x^2}{3} \right]_{2/3}^{5/3}$$

$$= \left(\frac{5}{3}\right)^2 \left(\frac{3}{5} \cdot \frac{125}{27} - \frac{25}{18} + \frac{20}{9} - \frac{4}{3} \right) - \left(\frac{2}{3}\right)^2 \left(\frac{3}{5} \cdot \frac{8}{27} - \frac{4}{18} + \frac{8}{9} - \frac{4}{3} \right) = \frac{589}{90}$$

Bearing in mind that we can reverse the order of integration we may re-calculate

$\iint\limits_{D} xy^2 \, dy \, dx$. Reference to Figure 9.18(b) shows that whereas the right-hand

boundary of D is always prescribed by $x = \dfrac{5}{3}$ the left-hand boundary is given

by $x = \dfrac{1}{2} y$ for y in the range $\left[\dfrac{4}{3}, \dfrac{10}{3} \right]$ and by $x = 2 - y$ in the range $\left[\dfrac{1}{3}, \dfrac{4}{3} \right]$.

Hence we need to express the problem as a sum of double integrals, viz.

$$\int\limits_{1/3}^{4/3} \int\limits_{2-y}^{5/3} xy^2 \, dx \, dy + \int\limits_{4/3}^{10/3} \int\limits_{y/2}^{5/3} xy^2 \, dx \, dy = I_1 + I_2$$

For I_1, we first calculate

$$\int\limits_{2-y}^{5/3} xy^2 \, dx = \left[\frac{x^2 y^2}{2} \right]_{2-y}^{5/3} = \frac{y^2}{2} \left(\frac{25}{9} - 4 + 4y - y^2 \right)$$

$$= \frac{y^2}{2} \left(\frac{-11}{9} + 4y - y^2 \right)$$

and then we evaluate

$$\int\limits_{1/3}^{4/3} \left(-\frac{11}{18} y^2 + 2y^3 - \frac{1}{2} y^4 \right) dy = \left[-\frac{11}{54} y^3 + \frac{y^4}{2} - \frac{1}{10} y^5 \right]_{1/3}^{4/3}$$

$$= \frac{4941}{10 \times 3^6}$$

For I_2, we first calculate

$$\int\limits_{y/2}^{5/3} xy^2\,dx = \left[\frac{x^2 y^2}{2}\right]_{y/2}^{5/3} = \frac{y^2}{2}\left(\frac{25}{9} - \frac{y^2}{4}\right)$$

and then we evaluate

$$\int\limits_{4/3}^{10/3} \left(\frac{25}{18}y^2 - \frac{y^4}{8}\right) dy = \left[\frac{25}{54}y^3 - \frac{y^5}{40}\right]_{4/3}^{10/3} = \frac{42768}{10 \times 3^6}$$

Hence $I_1 + I_2 = \dfrac{47709}{10 \times 3^6} = \dfrac{589}{90}$ as before.

Notice that this way round required two double integrals and it would obviously be preferable to choose that order of integration which avoided this difficulty. The problem may be such, however, that one way round the first integral may be hard or even impossible to evaluate. If this should be the case, we have to reverse the order of integration even if it means performing two double integrals. However, it may not always be easy to spot which way round is best.

Example 2
Evaluate

$$I = \int\limits_{0}^{1} \int\limits_{x}^{\sqrt{2-x^2}} \frac{x}{\sqrt{x^2 + y^2}}\,dy\,dx$$

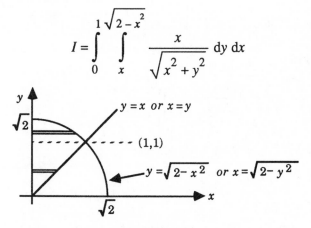

Figure 9.19

As the integral stands, the first step would be to calculate

$$\int\limits_{x}^{\sqrt{2-x^2}} \frac{x}{\sqrt{x^2 + y^2}}\,dy$$

Since x is constant we are able to obtain a solution by putting $y = x \cosh \theta$, but we are in for some tedious algebra. Let us reverse the order of integration; and first we shall sketch the region over which the integral is defined; see Figure 9.19. Notice the price we shall have to pay: there are two double integrals to evaluate.

$$I = \int_0^1 \int_0^y \frac{x}{\sqrt{x^2 + y^2}} \, dx \, dy + \int_1^{\sqrt{2}} \int_0^{\sqrt{2-y^2}} \frac{x}{\sqrt{x^2 + y^2}} \, dx \, dy = I_1 + I_2$$

Now the first step for I_1 is to evaluate

$$\int_0^y \frac{x}{\sqrt{x^2 + y^2}} \, dx = \left[(x^2 + y^2)^{1/2} \right]_0^y = \sqrt{2}y - y$$

Then the second step gives

$$I_1 = \int_0^1 (\sqrt{2}y - y) \, dy = \left[(\sqrt{2} - 1)y^2/2 \right]_0^1 = (\sqrt{2} - 1)/2$$

The first step for I_2 is to evaluate

$$\int_0^{\sqrt{2-y^2}} \frac{x}{\sqrt{x^2 + y^2}} \, dx = \left[(x^2 + y^2)^{1/2} \right]_0^{\sqrt{2-y^2}} = \sqrt{2} - y$$

and the second step gives

$$I_2 = \int_1^{\sqrt{2}} (\sqrt{2} - y) dy = \left[\sqrt{2}y - y^2/2 \right]_1^{\sqrt{2}} = 2 - 1 - \sqrt{2} + 1/2 = 3/2 - \sqrt{2}$$

Therefore,

$$I = I_1 + I_2 = (1 - \sqrt{2}/2)$$

9.8 Further Features of Double Integrals

We first quote, without proof, some properties of double integrals (we have tacitly assumed some of these). Suitable conditions on continuity can be assumed.

(i)   If c is a constant then $\iint_D c \, f(x,y) \, dx \, dy = c \iint_D f(x,y) \, dx \, dy$

(ii) $\displaystyle\iint\limits_{D} \{f(x, y) + g(x, y)\}\ dx\ dy = \iint\limits_{D} f(x, y)\ dx\ dy + \iint\limits_{D} g(x, y)\ dx\ dy$

(iii) If $D_1 \cap D_2 = \emptyset$ then

$$\int\limits_{D_1 \cup D_2}\!\!\int f(x, y)\ dx\ dy = \iint\limits_{D_1} f(x, y)\ dx\ dy + \iint\limits_{D_2} f(x, y)\ dx\ dy$$

(iv) If $\displaystyle\iint\limits_{D} \left[f(x, y)\right]^2 dx\ dy = 0$ and $D \neq \emptyset$ then $f(x, y) \equiv 0$ throughout D.

(v) If $m < f(x, y) < M$ for $(x, y) \in D$ and we denote $\displaystyle\iint\limits_{D} 1\ .\ dx\ dy$ by

A then $mA < \displaystyle\iint\limits_{D} f(x, y)\ dx\ dy < MA$

(vi) **Mean value theorem:** there is a point $(x_n, y_n) \in D$ such that

$$\iint\limits_{D} f(x, y)\ dx\ dy = f(x_n, y_n)\ .A \quad [A \text{ as defined in (v)}]$$

(vii) $\left| \displaystyle\iint\limits_{D} f(x, y)\ dx\ dy \right| \leq \iint\limits_{D} |f(x, y)|\ dx\ dy$

Application of double integrals

(i) **Area** If $f(x, y) \equiv 1$ in D then $\displaystyle\iint\limits_{D} f(x, y)\ dx\ dy$ is the area of D.

Note the relationship with single integration to find an area.

(ii) **Mass** If $\rho(x, y)$ is mass/unit area, i.e. a surface density, then

$M = \displaystyle\iint\limits_{D} \rho(x, y)\ dx\ dy$ is the mass of a lamina whose profile is D.

(iii) **Centre of mass** The coordinates of the centre of mass of the above lamina are given by

$$M\bar{x} = \iint\limits_{D} x\rho(x, y)\ dx\ dy \quad \text{and} \quad M\bar{y} = \iint\limits_{D} y\rho(x, y)\ dx\ dy$$

(iv) **Moments of inertia** The moment of inertia of the lamina about Ox is

$$\iint_D \rho(x, y) \cdot y^2 \, dx \, dy, \quad \text{about } Oy \text{ is } \iint_D \rho(x, y) x^2 \, dx \, dy \quad \text{and the}$$

product of inertia about both axes is $\iint_D \rho(x, y) \cdot xy \, dx \, dy$. The

moment of inertia about the axis Oz is $\iint_D (x^2 + y^2) \, \rho(x, y) \, dx \, dy$

$$= \iint_D x^2 \rho(x, y) \, dx \, dy + \iint_D y^2 \rho(x, y) \, dx \, dy.$$

(This is the **perpendicular axes theorem**.)

Example

Find the moments and product of inertia about the coordinate axes of the triangular lamina shown in Figure 9.20 where the mass distribution is given by $\rho(x, y) = xy$.

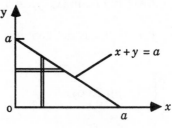

Figure 9.20

The moment of inertia about the axis Ox is $I_{xx} = \displaystyle\int_0^a \int_0^{a-y} xy \cdot y^2 \, dx \, dy$

(we have taken a horizontal strip whose distance from the x-axis is y). Now

$$\int_0^{a-y} xy^3 \, dx = \left[\frac{x^2}{2} y^3 \right]_0^{a-y} = \frac{(a-y)^2}{2} y^3$$

Hence

$$I_{xx} = \int_0^a \frac{(a-y)^2}{2} y^3 \, dy = \frac{1}{2} \int_0^a (a^2 y^3 - 2a y^4 + y^5) \, dy$$

$$= \frac{1}{2} \left[a^2 \frac{y^4}{4} - \frac{2}{5} a y^5 + \frac{y^6}{6} \right]_0^a = \frac{a^6}{2} \left(\frac{1}{4} - \frac{2}{5} + \frac{1}{6} \right) = \frac{a^6}{120}$$

Note that the moment of inertia about the axis Oy, $I_{yy} = \dfrac{a^6}{120}$ by symmetry. It is customary to quote moments of inertia in terms of the mass of the lamina,

$$M = \int_0^a \int_0^{a-y} xy \, dx \, dy = \int_0^a \frac{y(a-y)^2}{2} \, dy = \frac{a^4}{24}$$

Hence $I_{xx} = I_{yy} = \dfrac{Ma^2}{5}$

By the perpendicular axes theorem, $I_{zz} = \dfrac{Ma^2}{5} + \dfrac{Ma^2}{5} = \dfrac{2Ma^2}{5}$

The product of inertia

$$I_{xy} = \int_0^a \int_0^{a-y} (xy)(xy) \, dx dy = \int_0^a \frac{(a-y)^3}{3} y^2 \, dy$$

$$= \frac{1}{3} \int_0^a (a^3 y^2 - 3a^2 y^3 + 3ay^4 - y^5) \, dy = \frac{a^6}{180} = \frac{2Ma^2}{15}$$

Polar coordinates

Some double integrals may be best evaluated by transforming to a new system of coordinates. We merely introduce the idea of transforming to polar coordinates via a geometrical approach.

Figure 9.21(a) shows an element of area in cartesian coordinates bounded by lines $x = $ constant and $y = $ constant.

Figure 9.21(b) shows a corresponding element of area in polar coordinates, this time bounded by the curves $r = $ constant and $\theta = $ constant.

For completeness, we show in Figure 9.21(c) an element of area bounded by curves $\phi(x, y) = $ constant, $\psi(x, y) = $ constant; note that ϕ and ψ are orthogonal coordinates, i.e. curves of constant ψ and of constant ϕ intersect at right angles.

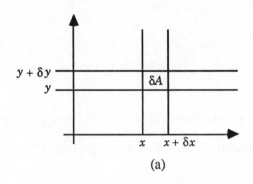

(a)

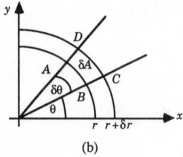

(b)

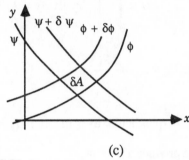

(c)

Figure 9.21

In polar coordinates, $\delta A \approx r\delta\theta\,\delta r$, since we can treat the element of area, δA, as approximately rectangular and $AB = r\delta\theta \approx DC$. This will allow us to state

that $$\iint_A f(x, y)\,\mathrm{d}x\,\mathrm{d}y = \iint_A g(r, \theta)\,r\mathrm{d}r\,\mathrm{d}\theta \tag{9.41}$$

where $g(r, \theta)$ is $f(x, y)$ expressed in polar coordinates via the equations

$$x = r \cos\theta, \, y = r \sin\theta \tag{9.42}$$

For example, $f(x,y) = xy^2 \Rightarrow r^3 \cos\theta \sin^2\theta = g(r, \theta)$.

Of course, we shall have to express the boundaries of A in terms of polar coordinates.

Example

Find the volume of the solid bounded above by the sphere $x^2 + y^2 + z^2 = a^2$ and below by the cone $z^2 = x^2 + y^2$. The solid is shown in Figure 9.22(a).

An element of volume whose base is centred at (x, y) in the plane $z = 0$ can be regarded as the difference between two volumes, one from the sphere down to $z = 0$ and the other from the cone down to $z = 0$. This element of volume is $\{(a^2 - x^2 - y^2)^{1/2} - (x^2 + y^2)^{1/2}\}\,\mathrm{d}x\,\mathrm{d}y$. We now have to define the region over which x and y are allowed to move. This is found by eliminating z between the equations of the two bounding surfaces, viz. $2(x^2 + y^2) = a^2$.

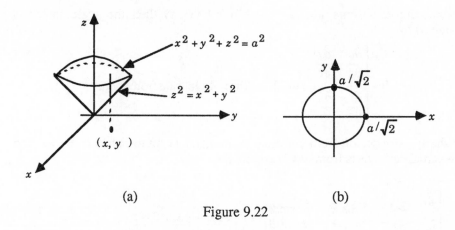

(a)   (b)

Figure 9.22

Then $D \equiv x^2 + y^2 < \dfrac{a^2}{2}$; see Figure 9.22(b). Hence the volume we seek is

$$V = \iint\limits_{D} \{(a^2 - x^2 - y^2)^{1/2} - (x^2 + y^2)^{1/2}\} \, dx \, dy$$

Converting to polar coordinates, we have $x^2 + y^2 = r^2$ and therefore

$$V = \int_0^{2\pi} \int_0^{a/\sqrt{2}} \{(a^2 - r^2)^{1/2} - r\} \, r \, dr \, d\theta$$

$$= \int_0^{2\pi} \left[\frac{-(a^2 - r^2)^{3/2}}{3} - \frac{r^3}{3} \right]_0^{a/\sqrt{2}} \, d\theta$$

$$= \int_0^{2\pi} \left(\frac{-a^3}{3.2\sqrt{2}} - \frac{a^3}{3.2\sqrt{2}} + \frac{a^3}{3} \right) \, d\theta$$

$$= \frac{a^3}{6} (2 - \sqrt{2}) \int_0^{2\pi} \, d\theta = \frac{\pi a^3}{3} (2 - \sqrt{2})$$

Notice that by exploiting the circular symmetry of the problem we were able to separate the integrals.

Some further remarks will be made on general transformation of coordinate systems. We have added a scale factor to convert $dr \, d\theta$ into an element of area ($dr \, d\theta$ has dimensions of length). It can be shown in general that if we convert from one coordinate system (u, v) to a second system (s, t), where the connecting equations are $u = u(s, t)$, $v = v(s, t)$ then the scale factor is provided by

$$J = \begin{vmatrix} \dfrac{\partial u}{\partial s} & \dfrac{\partial u}{\partial t} \\[2ex] \dfrac{\partial v}{\partial s} & \dfrac{\partial v}{\partial t} \end{vmatrix} \text{, the \textbf{Jacobian} of the transfromation.}$$

In the transformation from cartesian coordinates (x, y) to polars (r, θ) we find the partial derivatives from (9.42) and obtain

$$J = \begin{vmatrix} \dfrac{\partial x}{\partial r} & \dfrac{\partial x}{\partial \theta} \\[2ex] \dfrac{\partial y}{\partial r} & \dfrac{\partial y}{\partial \theta} \end{vmatrix} = \begin{vmatrix} \cos\theta & -r\sin\theta \\ \sin\theta & r\cos\theta \end{vmatrix} = r\cos^2\theta + r\sin^2\theta = r$$

Area of curved surfaces

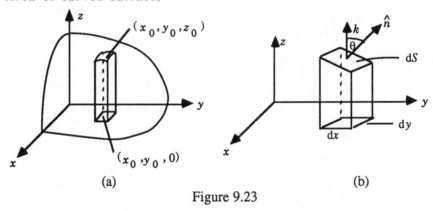

(a)   (b)

Figure 9.23

In Figure 9.23(a) we show a portion of a curved surface and we require to find its area. We show a small patch of the surface and one point on that patch; that point (x_0, y_0, z_0) is projected down onto the x - y plane. In the same way the small surface patch can be projected down onto the x - y plane. In Figure 9.23(b) we see the process in reverse: the element of area $dx\,dy$ in the x - y plane is projected back onto the curved surface to form an elementary area dS. If this elementary area is sufficiently small then it can be regarded as a plane. We know from earlier work that this plane has slopes given by $f_x(x_0, y_0)$ and $f_y(x_0, y_0)$ where the equation of the surface is expressed as $z = f(x, y)$. The equation of the tangent plane is

$$z = f(x_0, y_0) + (x - x_0)f_x(x_0, y_0) + (y - y_0)f_y(x_0, y_0)$$

or

$$[f(x_0, y_0) - x_0 f_x(x_0, y_0) - y_0 f_y(x_0, y_0)] + x f_x(x_0, y_0) + y f_y(x_0, y_0) - z = 0$$

The unit vector $\hat{n}$ is given by the following

$$\left(\frac{f_x\,(x_0,\,y_0)}{D}, \frac{f_y\,(x_0,\,y_0)}{D}, \frac{-1}{D} \right) \text{ where } D = \sqrt{[f_x\,(x_0,\,y_0)]^2 + [f_y\,(x_0,\,y_0)]^2 + 1}$$

The area of the surface patch $dS = dx\,dy/(\hat{\mathbf{n}} \cdot \mathbf{k}) = dx\,dy/\cos\theta$ where $\mathbf{k}$ is the unit vector in the z-direction. But $\cos\theta = 1/D$ and therefore $dS = dx\,dy \cdot D$

i.e. $$dS = dx\,dy\sqrt{[f_x\,(x_0,\,y_0)]^2 + [f_y\,(x_0,\,y_0)]^2 + 1} \qquad (9.43)$$

Example

Find the surface area of the hemispherical shell $x^2 + y^2 + z^2 = a^2$, $z > 0$. It should be clear that the projection of the hemispherical shell on the x - y plane is the disk $x^2 + y^2 \le a^2$. Then, from (9.43), the surface area is, by symmetry,

$$4\int_0^a \int_0^{\sqrt{a^2 - x^2}} \sqrt{[f_x\,(x_0,\,y_0)]^2 + [f_y\,(x_0,\,y_0)]^2 + 1}\ dy\,dx$$

But for the surface $z = (a^2 - x^2 - y^2)^{1/2} = f(x,\,y)$,

$$f_x = \frac{-x}{(a^2 - x^2 - y^2)^{1/2}}, \quad f_y = \frac{-y}{(a^2 - x^2 - y^2)^{1/2}}$$

Hence $$([f_x(x_0,\,y_0)]^2 + [f_y(x_0,\,y_0)]^2 + 1)^{1/2} = \left[\frac{x^2 + y^2 + (a^2 - x^2 - y^2)}{(a^2 - x^2 - y^2)} \right]^{1/2}$$

$$= a/(a^2 - x^2 - y^2)^{1/2}$$

Therefore the area we require, $$S = 4\int_0^a \int_0^{\sqrt{(a^2 - x^2)}} \frac{a}{(a^2 - x^2 - y^2)^{1/2}}\ dy\,dx$$

Converting this to polar coordinates, we obtain

$$S = 4\int_0^{\pi/2} \int_0^a \frac{a}{(a^2 - r^2)^{1/2}}\ r\,dr\,d\theta$$

i.e. $$S = 4a\left[\int_0^{\pi/2} d\theta \right]\left[\int_0^a \frac{r}{(a^2 - r^2)^{1/2}}\,dr \right] = 4a\,\frac{\pi}{2}\,.a = 2\pi a^2$$

9.9 Triple Integrals

The techniques obtained for double integrals may be extended to triple integrals. We restrict our attention to using triple integration to find volumes. The basic idea will be to take an element of volume $dxdydz$ in cartesian coordinates and perform three integrations in turn with respect to x, y and z.

Example 1
Find the volume of the solid bounded by the surfaces $z = x + 2$, $z = 0$, $y = x^2$ and $y = 2x + 3$.

The relevant solid is shown in Figure 9.24(a). The vertical walls are the parabolic cylinder $y = x^2$ and the plane $y = 2x + 3$. The base of the solid is shown in Figure 9.24(b) and is the projection on the x - y plane of any cross-section of the solid.

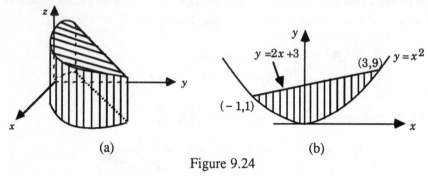

(a)   (b)

Figure 9.24

The volume V of the solid is found by integrating the element of volume $dxdydz$ over the region specified by $0 < z < x + 2$, $x^2 < y < 2x + 3$, $-1 < x < 3$. We could change the order of integration so as not to integrate with respect to x last, but this would involve a rewriting of the equations of the boundaries of the solid. Therefore

$$V = \int_{-1}^{3} \int_{x^2}^{2x+3} \int_{0}^{x+2} dzdydx = \int_{-1}^{3} \int_{x^2}^{2x+3} (x+2)\, dydx$$

$$= \int_{-1}^{3} (x+2)(2x+3-x^2)\, dx$$

$$= \int_{-1}^{3} (-x^3 + 7x + 6)dx = \left[-\frac{x^4}{4} + \frac{7x^2}{2} + 6x \right]_{-1}^{3}$$

$$= \left(-\frac{81}{4} + \frac{63}{2} + 18 \right) - \left(-\frac{1}{4} + \frac{7}{2} - 6 \right) = 32$$

We can usefully employ other coordinate systems. Two of the most common are cylindrical coordinates and spherical coordinates. In Figure 9.25(a) we show the element of volume in **cylindrical coordinates**; it is $r\,dr\,d\theta\,dz$. The transforming equations are

$$x = r\cos\theta,\ y = r\sin\theta,\ z = z.$$

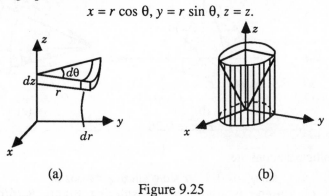

(a)   (b)

Figure 9.25

Example 2

Find the volume of the solid in the first octant bounded by the cone $z^2 = x^2 + y^2$ and the cylinder $x^2 + y^2 = a^2$; See Figure 9.25(b). In cylindrical polars, the cylinder has equation $r = a$ and the cone has the equation $z = r$. The solid can be defined by the conditions $0 < r < a$, $0 < \theta < \pi/2$, $0 < z < r$. Therefore the volume required,

$$V = \int_0^{\pi/2} \int_0^a \int_0^r r\,dz\,dr\,d\theta = \left(\int_0^{\pi/2} d\theta\right)\left(\int_0^a r^2 . dr\right) = \frac{\pi}{2} . \frac{a^3}{3} = \frac{\pi a^3}{6}$$

You might care to try this problem in cartesian coordinates. If we extend the idea of a Jacobian to three variables we might expect the scale factor to be

$$\begin{vmatrix} \dfrac{\partial x}{\partial r} & \dfrac{\partial x}{\partial \theta} & \dfrac{\partial x}{\partial z} \\[2mm] \dfrac{\partial y}{\partial r} & \dfrac{\partial y}{\partial \theta} & \dfrac{\partial y}{\partial z} \\[2mm] \dfrac{\partial z}{\partial r} & \dfrac{\partial z}{\partial \theta} & \dfrac{\partial z}{\partial z} \end{vmatrix} \text{ i.e. } \begin{vmatrix} \cos\theta & -r\sin\theta & 0 \\ \sin\theta & r\cos\theta & 0 \\ 0 & 0 & 1 \end{vmatrix} = r$$

Spherical coordinates are defined by the following rules. Choose a value of r. The condition $r = a$ fixes a spherical shell centred at the origin of radius a; a point in space must lie on one such shell and we fix its position on that shell by two angles. The angle ϕ measures declination from the vertical and the curve ϕ = constant is a cone; see Figure 9.26(a). Such a cone will intersect the spherical shell $r = a$ in a horizontal circle; see Figure 9.26(b). The angle θ is made by a line joining our point in space to the origin with a fixed line through the origin. In a sense, θ is a measure of longitude (though in our system $0 < \theta < 2\pi$) and ϕ is a kind of co-latitude, where $0 < \phi < \pi$.

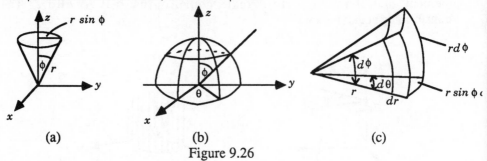

(a)   (b)   (c)

Figure 9.26

The transforming equations are

$$x = r \sin \phi \cos \theta, \; y = r \sin \phi \sin \theta, \; z = r \cos \phi \qquad (9.44)$$

The element of volume is found by considering Figure 9.26(c) to be $r^2 \sin \phi \, dr \, d\phi \, d\theta$.

Alternatively, we find the scale factor from the Jacobian

$$\begin{vmatrix} \dfrac{\partial x}{\partial r} & \dfrac{\partial x}{\partial \phi} & \dfrac{\partial x}{\partial \theta} \\[2mm] \dfrac{\partial y}{\partial r} & \dfrac{\partial y}{\partial \phi} & \dfrac{\partial y}{\partial \theta} \\[2mm] \dfrac{\partial z}{\partial r} & \dfrac{\partial z}{\partial \phi} & \dfrac{\partial z}{\partial \theta} \end{vmatrix} = \begin{vmatrix} \sin \phi \cos \theta & r \cos \phi \cos \theta & -r \sin \phi \sin \theta \\ \sin \phi \sin \theta & r \cos \phi \sin \theta & r \sin \phi \cos \theta \\ \cos \phi & -r \sin \phi & 0 \end{vmatrix}$$

$$= \cos \phi \begin{vmatrix} r \cos \phi \cos \theta & -r \sin \phi \sin \theta \\ r \cos \phi \sin \theta & r \sin \phi \cos \theta \end{vmatrix} + r \sin \phi \begin{vmatrix} \sin \phi \cos \theta & -r \sin \phi \sin \theta \\ \sin \phi \sin \theta & r \sin \phi \cos \theta \end{vmatrix}$$

$$= [\cos \phi] r^2 (\cos \phi \sin \phi \cos^2 \theta + \cos \phi \sin \phi \sin^2 \theta)$$
$$\qquad + r^2 \sin \phi [\sin^2 \phi \cos^2 \theta + \sin^2 \phi \sin^2 \theta]$$
$$= r^2 \cos^2 \phi \sin \phi + r^2 \sin^3 \phi = r^2 \sin \phi$$

Example

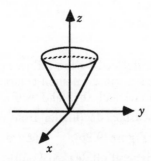

Figure 9.27

Find the volume of the solid bounded above by the sphere, $x^2 + y^2 + (z - a)^2 = a^2$ and below by the cone $z^2 = x^2 + y^2$. See Figure 9.27.

The cone has the equation $r^2 \cos^2 \phi = r^2 \sin^2 \phi$ i.e. $\tan \phi = 1$ and hence $\phi = \pi/4$. The sphere has the equation $r^2 \sin^2 \phi + (r \cos \phi - a)^2 = a^2$ i.e. $r^2 - 2ar \cos \phi + a^2 = a^2$ or $r = 2a \cos \phi$.

The solid is defined by the conditions $0 \leq \theta \leq 2\pi$, $0 \leq \phi \leq \pi/4$, $0 < r < 2a \cos \phi$. Using symmetry, the volume

$$V = 4 \int_0^{\frac{\pi}{2}} \int_0^{\frac{\pi}{4}} \int_0^{2a \cos \phi} r^2 \sin \phi \, dr d\phi d\theta$$

$$= 4 \left(\int_0^{\frac{\pi}{2}} d\theta \right) \int_0^{\frac{\pi}{4}} \sin \phi \left[r^3/3 \right]_0^{2a \cos \phi} d\phi$$

$$= 4 \frac{\pi}{2} \cdot 8 \int_0^{\frac{\pi}{4}} \frac{a^3}{3} \cos^3 \phi \sin \phi \, d\phi$$

$$= -\frac{16\pi a^3}{3} \left[\frac{\cos^4 \phi}{4} \right]_0^{\frac{\pi}{4}} = \frac{4a^3 \pi}{3} \left(1 - \frac{1}{4} \right) = \pi a^3$$

Since the volume of the spherical 'hat' is $\frac{2}{3} \pi a^3$ it follows that the cone

contributes $\frac{1}{3} \pi a^3$ to the total.

9.10 Green's theorem in the plane

In Figure 9.28 we have a closed curve which lies in the x-y plane and passes through points A, D, B and E. A and B are the points with the smallest and greatest x-coordinates on the curve. The lower portion of the curve, which passes through D, is specified by $y = y_1(x)$ and the upper portion by $y = y_2(x)$. Green's theorem states that if R is a closed region in the x-y plane bounded by a simple closed curve C and if P and Q are both continuous

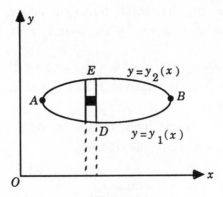

Figure 9.28

functions of x and y having continuous first derivatives in R then

$$\oint_C (P\mathrm{d}x + Q\mathrm{d}y) = \iint_R \left(\frac{\partial Q}{\partial x} - \frac{\partial P}{\partial y} \right) \mathrm{d}x\,\mathrm{d}y$$

The proof is outlined as follows.

$$\iint_R \frac{\partial P}{\partial y}\,\mathrm{d}x\,\mathrm{d}y = \int \mathrm{d}x \int \frac{\partial P}{\partial y}\,\mathrm{d}y \text{ between appropriate limits}$$

$$= \int \mathrm{d}x\,[P]_{y_1}^{y_2} = \int [P(x, y_2) - P(x, y_1)]\,\mathrm{d}x$$

$$= \int_{AEB} P\mathrm{d}x - \int_{ADB} P\mathrm{d}x = -\oint_C P\mathrm{d}x$$

Similarly

$$\iint_R \frac{\partial Q}{\partial x}\,\mathrm{d}x\,\mathrm{d}y = \oint_C Q\mathrm{d}y$$

and the result follows.

Example

Let us evaluate $\oint_C \mathbf{v}.\mathrm{d}\mathbf{r}$ where $\mathbf{v}$ is the vector of Example 2 (on page 322) and

C is the path ADBA (see Figure 9.6) by applying Green's theorem.

We calculate $I = \int_0^1 \mathrm{d}x \int_0^x (2y - x^2)\,\mathrm{d}y.$

Now $\displaystyle\int_0^x (2y - x^2)\, dy = \left[y^2 - x^2 y \right]_0^x = x^2 - x^3$

and $\displaystyle\int_0^1 (x^2 - x^3)\, dx = \left[\frac{x^3}{3} - \frac{x^4}{4} \right]_0^1 = \frac{1}{12}$

Note that, from parts (i) and (iii) of Example 2 we find that

$$\int_{ADBA} = \int_{ADB} + \int_{BA} = \int_{ADB} - \int_{AB} = 2 - \frac{23}{12} = \frac{1}{12}$$

In a sense, $\left(\dfrac{\partial Q}{\partial x} - \dfrac{\partial P}{\partial y} \right)$ represents some residual quantity picked up on going

once round the closed path C. If it is zero then the field is conservative.

9.11 Surface Integrals: Stokes' Theorem

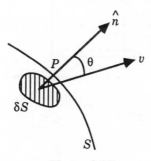

Figure 9.29

In the course of developing the concept of divergence, flux through a surface was considered. We now look at the flux through an arbitrary surface. In Figure 9.29 we consider a point P on a given surface S and an element of area δS surrounding P. The outward (positive) unit normal vector to the surface at P is denoted $\hat{\mathbf{n}}$. (If the surface is not closed we always define $\hat{\mathbf{n}}$ on the same side of the surface as P moves over the surface.) The vector element of area is defined as $\delta \mathbf{S} = \hat{\mathbf{n}}\, \delta S$. Let $\mathbf{v}$ be a vector field in which the surface lies. Then $\mathbf{v} \cdot \delta \mathbf{S} = \mathbf{v} \cdot \hat{\mathbf{n}}\, \delta S = |\mathbf{v}| \cos \theta \delta S$ represents the **flux** of $\mathbf{v}$ through the element of surface $\delta \mathbf{S}$. The total flux of $\mathbf{v}$ through the surface S is given by

$$\iint_S \mathbf{v} \cdot d\mathbf{S} \tag{9.45}$$

where the integral is taken over the surface S.

If $\mathbf{v}$ represents fluid velocity, then (9.45) represents the net amount of fluid crossing the surface in unit time. The same formula can represent current flow, magnetic flux, flow of heat etc.

Example

Evaluate (9.45) where $\mathbf{v} = z\mathbf{i} + x\mathbf{j} - 2y^2z\mathbf{k}$ and S is the curved surface of the cylinder $x^2 + y^2 = a^2$ included in the first octant and lying between $z = 0$ and $z = b$; see Figure 9.30. The steps are straightforward.

(i)   Find a unit normal $\hat{\mathbf{n}}$ to the surface.

(ii)   Form $\mathbf{v} \cdot \hat{\mathbf{n}}$.

(iii)   Project dS on to the appropriate plane (in this case the x-z plane will do, as would the y-z plane).

(iv)   Perform the double integral which results.

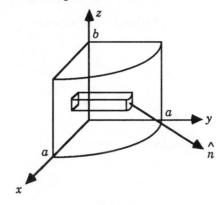

Figure 9.30

In our example, the surface may be written as $\phi(x, y, z) = x^2 + y^2 = a^2$ and a normal vector $\nabla\phi = 2x\mathbf{i} + 2y\mathbf{j}$.

(i)   A suitable unit normal is $\hat{\mathbf{n}} = (2x\mathbf{i} + 2y\mathbf{j}) / \sqrt{(2x)^2 + (2y)^2} = (x\mathbf{i} + y\mathbf{j})/a$.

(ii)   Then $\mathbf{v} \cdot \hat{\mathbf{n}} = (zx + xy)/a$.

(iii)   The projection of dS on the x-z plane is $dx\,dz/(\hat{\mathbf{n}} \cdot \mathbf{j})$, since $\mathbf{j}$ is a unit vector normal to the x-z plane, i.e. $dx\,dz(a/y)$

(iv)   If R is the projection of S on the x-z plane then

$$\iint_S \mathbf{v} \cdot d\mathbf{S} = \iint_R \frac{(zx + xy)}{a} \frac{dx\, dz\, a}{y}$$

$$= \int_0^b dz \int_0^a dx \left(\frac{zx}{\sqrt{a^2 - x^2}} + x \right)$$

$$= \int_0^b dz \left[-z\,(a^2 - x^2)^{1/2} + \frac{1}{2}\,x^2 \right]_0^a$$

$$= \int_0^b dz \left(za + \frac{1}{2}\,a^2 \right) = \frac{1}{2}\,ab(a + b).$$

Stokes' Theorem

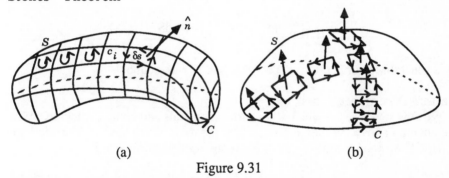

(a)   (b)

Figure 9.31

We now present a generalisation of Green's Theorem in the plane, which allows us to convert a line integral into a surface integral. We shall first state the theorem and then outline how it may be proved. Let an open surface S have a bounded edge C (as in Figure 9.31(a)); you may imagine a basin and its rim – although there is no restriction in the theorem on the edge C being planar. Let $\mathbf{F}$ be a vector field. Then Stokes' theorem states that *the total flux of curl* $\mathbf{F}$ *through the surface S is equal to the circulation of* $\mathbf{F}$ *round the circuit C* i.e.

$$\iint_S \text{curl } \mathbf{F} \cdot d\mathbf{S} = \oint_C \mathbf{F} \cdot d\mathbf{r} \tag{9.46}$$

The proof is effected by dividing S into elementary areas δS_i with boundaries C_i as shown in Figure 9.31(a). From the definition of curl $\mathbf{F}$,

$$\text{curl } \mathbf{F} \cdot \hat{\mathbf{n}}\, \delta S_i \approx \oint_{C_i} \mathbf{F} \cdot d\mathbf{r} \tag{9.47}$$

If we sum the left-hand side of (9.47) over all elementary areas we obtain the left-hand side of (9.46) after letting $\delta S_i \to 0$. If we sum the right-hand side of (9.47) and note as shown in Figure 9.31(b) that parts of boundaries C_i in common with neighbouring areas will lead to cancellation then the net sum is simply the right-hand side of (9.46), i.e. the integral around the external boundary C. Convince yourself by dividing S into 4 areas, say, and noting the cancellation. Hence the result may be established.

Note that if S lies in the $x - y$ plane, we recover Green's Theorem, since $dS \equiv dx\, dy\, \mathbf{k}$. Note further that if $\mathbf{F} = \text{grad } \phi$ then the left-hand side of (9.46) vanishes identically and we are left with a zero line integral as we should expect from a conservative field.

9.12 Gauss' Divergence Theorem

We first produce an alternative definition of divergence based on the ideas in the definition of curl. Let V be the volume enclosed by a surface S then if $\mathbf{A}$ be a vector field which operates throughout the volume V we have the definition

$$\text{div } \mathbf{A} = \lim_{V \to 0} \left\{ \frac{1}{V} \iint_S \mathbf{A} \cdot d\mathbf{S} \right\} \qquad (9.48)$$

This generalises our previous approach where V was a cuboid.

Gauss' Divergence Theorem

We now state a result which links surface integrals and volume integrals. With the notation above the theorem states that *the total divergence of* $\mathbf{A}$ *through the volume* V *is equal to the net flux of* div $\mathbf{A}$ *through the surface* S, i.e.

$$\iiint_V \text{div } \mathbf{A}\, dV = \iint_S \mathbf{A} \cdot d\mathbf{S} \qquad (9.49)$$

If we have an incompressible fluid and draw an imaginary volume in the field of flow, the theorem effectively states that as much fluid leaves through the surface as enters, unless there are sources or sinks inside the volume.

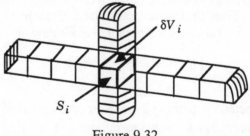

Figure 9.32

The proof of the theorem follows a similar line to that for Stokes' theorem. Refer to Figure 9.32. For the volume elements shown we have, using (9.48) that

$$\operatorname{div} \mathbf{A} \, \delta V_i \approx \iint_{S_i} \mathbf{A} . \, d\mathbf{S}$$

Summation of the left-hand side over all blocks produces the left-hand side of (9.49) after letting $\delta V_i \rightarrow 0$. Summing the right-hand side and allowing for cancellation at common faces leads to the right-hand side of (9.49); the cancellation is due to the fact that a flow outwards from one face of an elementary volume is a flow inwards (i.e. a negative outwards flow) to the same face of the neighbouring elementary volume.

Problems

Section 9.1

1 Describe the level surfaces of the scalar point functions

(i) $\phi = x + y + z$ (ii) $\phi = x^2 + y^2 - z^2$

2 An electrostatic field $\mathbf{F}$ is given by

$$\mathbf{F} = \frac{1}{[(x+1)^2 + y^2] [(x-1)^2 + y^2]} [2(x^2 - y^2 - 1)\mathbf{i} + 4xy\mathbf{j}]$$

Find the directions of this field at points
(i) on the x-axis (ii) on the y-axis (iii) on the lines $y = \pm x$
(iv) at the points $(\pm 1, 0)$.
Hence sketch the field.

3 Find $\dfrac{d\mathbf{A}}{dt}$, $\dfrac{d^2\mathbf{A}}{dt^2}$ for $\mathbf{A} = e^{-t} \mathbf{i} + 4 \cos 3t \mathbf{j} + 2 \sin t \mathbf{k}$

4 If $\mathbf{A} = (2x^2y - x^4)\mathbf{i} + y \sin x\mathbf{j} + \cos y\mathbf{k}$, find $\dfrac{\partial \mathbf{A}}{\partial x}$, $\dfrac{\partial \mathbf{A}}{\partial y}$

5 Show that a vector function $v(t)$ of the variable t

(i) will have a constant magnitude if and only if $\mathbf{v} . \dfrac{d\mathbf{v}}{dt} = 0$

(ii) will remain parallel to a fixed line if and only if $\mathbf{v} \wedge \dfrac{d\mathbf{v}}{dt} = 0$ (LU)

6 The position vector $\mathbf{r}$, measured from an origin O, of a particle satisfies the differential equation

$$\frac{d^2\mathbf{r}}{dt^2} + 2n \frac{d\mathbf{r}}{dt} + 2n^2\mathbf{r} = 0$$

where n is a positive constant and 0 is the zero vector. Find the values of the constant m

Advanced Engineering Mathematics

which make $\mathbf{A}e^{mt}$, where $\mathbf{A}$ is a constant vector, a solution of the equation, and deduce the general solution for $\mathbf{r}$. Find the particular solution if at $t = 0$ the particle starts from O with velocity $\mathbf{V}$. (LU)

Section 9.2

7 Find grad ϕ and $\dfrac{\partial \phi}{\partial n}$ for the following

(a) $\phi = x + y + z$ (b) $\phi = x^2 + y^2 - z^2$ (c) $\phi = 2xz^4 - x^2y$ (d) $\phi = 2z - x^3y$

8 Find the directional derivative of ϕ at point P in the direction $\mathbf{a}$
(i) $\phi = x^2 + y^2 + z^2$; $P = (1, 1, 2)$; $\mathbf{a} = \mathbf{i} - \mathbf{j}$
(ii) $\phi = 2x^2 - 4y^2 + z^2$; $P = (0, 1, 2)$; $\mathbf{a} = \mathbf{i} - 2\mathbf{j} + 3\mathbf{k}$
(iii) $\phi = xyz$; $P = (-1, 1, 2)$; $\mathbf{a} = \mathbf{i} + 2\mathbf{j} + \mathbf{k}$

9 Find ϕ if $\mathbf{v} = \nabla \phi$ (i) $\mathbf{v} = x\mathbf{i} + y\mathbf{j}$ (ii) $\mathbf{v} = 3x\mathbf{i} - 2y\mathbf{j} + z\mathbf{k}$

10 Find a unit normal vector to the surface at P.
(i) $2x + y + z = 2$; $P = (2, 1, -3)$ (ii) $x^2 + y^2 + 3z^2 = 8$; $P = (2, 2, 0)$.

11 A scalar point function ϕ is given by $\phi = \dfrac{2}{x^2 + y^2 + z^2}$. Describe the level surface which

passes through the point $P = (6, 2, 3)$. What is the unit normal to this surface at P and what is the normal rate of change of ϕ across the surface?

12 Show that if r is the distance of a point from the origin, $\nabla r^3 = 3r\mathbf{r}$ and $\nabla r^m = mr^{m-2}\mathbf{r}$.

13 (a) The electrostatic field $\mathbf{E}$ is derived from the potential ϕ by the equation $\mathbf{E} = -\,\text{grad }\phi$.
For a dipole of strength m pointing along the y-axis at the origin

$$\phi = \frac{m}{4\pi\varepsilon} \frac{y}{(x^2 + y^2 + z^2)^{3/2}} \quad \text{where } \varepsilon \text{ is a constant.}$$

Obtain $\mathbf{E}$ at the point $(6, 2, 3)$. (EC)

Section 9.3
14 Find the divergence of the following
(i) $3x\mathbf{i} + y\mathbf{j} + 2z\mathbf{k}$ (ii) $x^2\mathbf{i} + y^2\mathbf{j} + z^2\mathbf{k}$ (iii) $xy\mathbf{i} + yz\mathbf{j} + zx\mathbf{k}$
(iv) $(x\mathbf{i} + y\mathbf{j} + z\mathbf{k})/r^{3/2}$

15 Verify (9.26) and (9.27).

Section 9.4

16 Evaluate $\displaystyle\oint_C \frac{(x^3\,dy - y^3\,dx)}{x^2 + y^2}$

(i) when C is the boundary of the rectangle whose sides lie along the lines $x = \pm a$, $y = \pm b$,
(ii) when C is the circle $x^2 + y^2 = k^2$,

(iii) the triangle with vertices at $(0, 0)$, (c, c), $(0, c)$.   (LU)

17 Evaluate $\oint_C \{(y - \cos x)\, dx + \sin x\, dy\}$

where C is the perimeter of the triangle formed by the lines $y = 2x/\pi$, $x = \pi/2$, $y = 0$.

                                                                            (LU)

18 (a) Show that the line integral $\int \dfrac{yz}{x^2 + y^2}\, ds$ taken along the arc of the curve in space

given by the equations $x = 3at$, $y = 3at^2$, $z = 2a(1 + t^3)$ between the points for which $t = 0$ and $t = 1$, has the value $a[4 - \pi/2 + \ln 2]$.

   (b) Find the value of the integral

$$\int \{(y + 3z)\, dx + (2z + x)\, dy + (3x + 2y)\, dz\}$$

taken along the arc of the helix $x = a \cos \theta$, $y = a \sin \theta$, $z = 2a\theta/\pi$ between the points $(a, 0, 0)$ and $(0, a, a)$.   (LU)

19 Show that div $\{f(r)\,\hat{\mathbf{r}}\,\} = \dfrac{1}{r^2} \dfrac{d}{dr} \{r^2 f(r)\}$.

20 Verify that $\mathbf{F} = (3x^2 - 3yz + 2xz)\mathbf{i} + (3y^2 - 3xz + z^2)\mathbf{j} + (3z^2 - 3xy + x^2 + 2yz)\mathbf{k}$
represents a conservative field of force. Find a scalar ϕ such that $\mathbf{F} = \text{grad } \phi$.

Evaluate $\int_C \mathbf{F} \cdot d\mathbf{r}$ where C is any path joining $(-1, 2, 3)$ to $(3, 2, -1)$.

What would be the value of $\oint_C \mathbf{F}.d\mathbf{r}$ once round the circle $x^2 + y^2 = 1$, $z = 0$?

Section 9.5
21 Find the curl of the vectors of Problem 14 of Section 9.3
22 Find the curl of

(i) $\mathbf{v} = x\mathbf{i} + 5xy\mathbf{j}$ (ii) $\mathbf{v} = 3x\mathbf{i} + 2y\mathbf{j} - 4z\mathbf{k}$ (iii) $\mathbf{v} = k\hat{\mathbf{r}}/r^2$.

23 Show that the vector $\mathbf{A} = (4xy - z^3)\mathbf{i} + 2x^2\mathbf{j} - 3xz^2\mathbf{k}$ is irrotational and find a function ϕ such that $\mathbf{A} = \text{grad } \phi$.

24 Show that the force $\mathbf{F}$ defined by
$$\mathbf{F} = (y^2 \cos x + z^3)\mathbf{i} + (2y \sin x + 4)\mathbf{j} + (3xz^2 + 2)\mathbf{k}$$
represents a conservative field of force and find a scalar potential ϕ such that $\mathbf{F} = \nabla\phi$.
Hence, or otherwise, find the work done in moving a particle of unit mass under this field of

force from the point $(0, 1, -1)$ to the point $(\dfrac{\pi}{2}, -1, 2)$.   (EC)

Section 9.6

25 Prove (9.34) to (9.40).

26 Verify (9.34) to (9.40) for $u = yz\mathbf{i}$, $v = x^2y\mathbf{i} + 2xz\mathbf{j} + 6z^2\mathbf{k}$, $\phi = xyz$ and $\psi = x + y + z$.

27 Show that if $\phi = 1/r$, where $r^2 = x^2 + y^2 + z^2$ and $\mathbf{A} = \text{grad } \phi$, then $\text{curl}(\phi\mathbf{A}) = 0$. Find $\text{div}(\phi\mathbf{A})$ in this case.   (LU)

28 Show that $\text{div }(\mathbf{A} \times \mathbf{r}) = 0$ if $\text{curl }\mathbf{A} = 0$, where $\mathbf{r}$ denotes the position vector of any point (x, y, z). Also prove that $\nabla.(r^3\mathbf{r}) = 6r^3$.   (EC)

29 If $\mathbf{A}$ and ϕ are a vector point function and a scalar point function respectively, prove that
$$\text{curl}(\phi\mathbf{A}) = \phi \text{ curl } \mathbf{A} + \text{grad } \phi \wedge \mathbf{A}.$$
Verify the above result when $\mathbf{A} = 2xz^2\mathbf{i} - yz\mathbf{j} + 3xz^3\mathbf{k}$ and $\phi = x^3yz$ at the point $(1, 1, 1)$ (EC)

30 (a) Using Cartesian coordinates, verify that if ϕ is a scalar field and $\mathbf{F}$ is a vector field,
$$\text{div } \phi\mathbf{F} = \phi \text{ div } \mathbf{F} + \text{grad } \phi . \mathbf{F}$$
(b) If $\mathbf{r} = x\mathbf{i} + y\mathbf{j} + z\mathbf{k}$ and $r = |\mathbf{r}|$, show that
$$\text{div } \mathbf{r} = 3 \text{ and grad} f(r) = \frac{1}{r} f'(r)\mathbf{r}$$
Deduce that $\text{div}(\phi r^2\mathbf{r}) = r^2(\mathbf{r} . \text{ grad } \phi + 5\phi)$ (EC)

31 The vector fields $\mathbf{u}$ and $\mathbf{v}$ satisfy the equations
$$\text{curl } \mathbf{u} = c\mathbf{v}, \text{curl } \mathbf{v} = c\mathbf{u}$$
where c is a non-zero constant.
Show that   (i) $\text{div } \mathbf{u} = \text{div } \mathbf{v} = 0$ (ii) $\nabla^2\mathbf{u} + c^2\mathbf{u} = \nabla^2\mathbf{v} + c^2\mathbf{v} = 0$
The following vector identities may be assumed
$$\text{div curl } \mathbf{A} = 0$$
$$\text{curl curl } \mathbf{A} = \text{grad div } \mathbf{A} - \nabla^2\mathbf{A}$$
     (EC)

32 Prove that $\text{div } u\mathbf{v} = u \text{ div } \mathbf{v} + \mathbf{v} . \text{ grad } u$
                  $\text{curl } u\mathbf{v} = u \text{ curl } \mathbf{v} + \text{grad } u \times \mathbf{v}$
where u and $\mathbf{v}$ are arbitrary scalar and vector functions respectively. Hence show that, if $\mathbf{F} = \mathbf{r}/r$, where $\mathbf{r}$ is the position vector of a point and $r = |\mathbf{r}|$ is the distance from the origin, then $\text{div } \mathbf{F} = 2/r$ and $\text{curl } \mathbf{F} = 0$.   (EC)

33 (a) Given the vector $\mathbf{A} = xyz\mathbf{i} + (x^2 + y^2 + z^2)\mathbf{j} - (yz + zx + xy)\mathbf{k}$, verify that
     (i) $\text{div curl } \mathbf{A} = 0$ and (ii) $\text{curl curl } \mathbf{A} = \text{grad div } \mathbf{A} - \nabla^2\mathbf{A}$
     Show further that these identities are true for any vector $\mathbf{A}$.
(b) Two vector fields $\mathbf{u}$ and $\mathbf{v}$ satisfy the equations $\text{curl } \mathbf{u} = c\mathbf{v}$ and $\text{curl } \mathbf{v} = c\mathbf{u}$ where c is a constant. Show that
     (i) $\text{div } \mathbf{u} = \text{div } \mathbf{v} = 0$ (ii) $\mathbf{u}$ and $\mathbf{v}$ satisfy the equation $\nabla^2\mathbf{X} + c^2\mathbf{X} = 0$ (EC)

34 Show, for any arbitrary vectors $\mathbf{u}$ and $\mathbf{v}$, that

$$\text{div } (\mathbf{u} \times \mathbf{v}) = \mathbf{v} \text{ . curl } \mathbf{u} - \mathbf{u}\text{. curl } \mathbf{v}$$

and verify this identity in the particular case

$$\mathbf{u} = (x^2, y^2, z^2), \ \mathbf{v} = (xyz, 0, 0) \tag{EC}$$

35 A vector field **A** is defined by $\mathbf{A} = \mathbf{r} \, f(r)$, where **r** is the position vector
$\mathbf{r} = x\mathbf{i} + y\mathbf{j} + z\mathbf{k}$.
Show that curl **A** = 0
Find the form of $f(r)$ in order that additionally div **A** = 0 $\tag{EC}$

36 (a) Show that $\text{div}(\mathbf{r}/r^3) = 0$ and $\text{curl}(\mathbf{r}/r^3) = \mathbf{0}$

(b) x, y, z are Cartesian coordinates and t represents time. A scalar ϕ is defined such that

$$\phi = \phi(x, z, t) \text{ and } \frac{\partial \phi}{\partial y} = 0$$

Two vectors **E** and **H** are given, in usual notation, by

$$\mathbf{E} = \frac{1}{\varepsilon} \left\{ \frac{\partial \phi}{\partial z} \, \mathbf{i} - \frac{\partial \phi}{\partial x} \, \mathbf{k} \right\}$$

$$\mathbf{H} = -\frac{\partial \phi}{\partial t} \, \mathbf{j}$$

where ε is a constant.
(i) Show that $\qquad \text{div } \mathbf{E} = \text{div } \mathbf{H} = 0$

$$\text{curl } \mathbf{H} = \varepsilon \, \frac{\partial \mathbf{E}}{\partial t}$$

(ii) If also curl $\mathbf{E} = -\mu(\partial \mathbf{H}/\partial t)$, where μ is a constant, show that

$$\frac{\partial^2 \phi}{\partial x^2} + \frac{\partial^2 \phi}{\partial z^2} = \mu\varepsilon \, \frac{\partial^2 \phi}{\partial t^2} \tag{EC}$$

Section 9.7

37 Evaluate the following integrals and sketch the region over which each is taken.

(a) $\displaystyle\int_0^3 \int_1^3 xy \, dy \, dx$

(b) $\displaystyle\int_0^2 \int_{-x}^x (x^2 + y^2) \, dy \, dx$

(c) $\displaystyle\int_{-1}^1 \int_{-1}^x (2xy + x^2y^2) \, dy \, dx$

(d) $\displaystyle\int_{-1}^1 \int_0^{\sqrt{1-y^2}} 2xy^2 \, dx \, dy$

(e) $\displaystyle\int_{-1}^2 \int_{x^2-2}^x (x + y) \, dy \, dx$

38 Convert each integral below into one or more integrals in which the order is reversed. Check by evaluating both forms.

(a) $\displaystyle\int_0^1 \int_{x^2}^x y\,dy\,dx$ (b) $\displaystyle\int_{-2}^2 \int_0^{4-x^2} dy\,dx$ (c) $\displaystyle\int_0^1 \int_{1-x^2}^{\sqrt{1-x^2}} x\,dy\,dx$

39 Find in each case below the volume of the solid bounded above by the given surface and below by the given region of the xy-plane.

(a) $z = e^x \cos y$; $0 \le x \le \ln 3$, $-\dfrac{\pi}{2} \le y \le \dfrac{\pi}{2}$

(b) $z = c\sqrt{1 - \dfrac{x^2}{a^2} - \dfrac{y^2}{b^2}}$; $\dfrac{x^2}{a^2} + \dfrac{y^2}{b^2} \le 1$ (half an ellipsoid)

(c) $z = c\left(1 - \dfrac{x}{a} - \dfrac{y}{b}\right)$; $0 \le x \le a$, $0 \le y \le b\left(1 - \dfrac{x}{a}\right)$ (a tetrahedron)

(d) $z = xy$; the region bounded by $y = x^2$ and $y = x(2 - x)$

(e) $z = 4y^2 - x^2$; the region bounded by $x = 2y$ and $x = y^2$

40 Find the voume of the solid bounded by the given surfaces in each case below
(a) The bounding surfaces are $z = y$, $z = y + 1$, $y = 0$, $y = 1$, $x = 0$, $x = 1$
(b) The bounding surfaces are $z = 2y + 4$, $z = y + 2$, $y^2 = 4 - x$ and $x = 0$
(c) The bounding surfaces are $x^2 + y^2 = 1$, $x = 0$, $y = 0$, $z = y$ and $z = 2$

Section 9.8
41 Sketch the region R and compute its area by means of a double integral for the following:

(a) R is bounded by $y = x^3$ and $y = x^{1/2}$ (b) R is bounded by $y = \sin x$ and $y = \dfrac{2}{\pi} x$

(c) R is determined by the inequalities $x^2 + (y - 1)^2 \le 1$, $y \ge 2 - x^2$

42 Find the mass of the region R corresponding to the given density for the following:
(a) R is the triangle with vertices $(0, 0)$, $(-2, 0)$ and $(-2, 2)$; $\rho = x^2 + y^2$
(b) R is the circular disk $x^2 + y^2 \le 25$; $\rho = x^2 + y^2$

(c) R is bounded by $y = 0$ and $y = \sqrt{4 - x^2}$; $\rho = x^2$

43 Find the moment of inertia of the given mass distribution in the xy-plane, about the given axis for the following:
(a) R bounded by $y = 4 - x^2$ and $y = 0$; $\rho = 2y$; about the y-axis
(b) R the triangle with vertices $(0,0)$, $(a, 0)$ and (b, c): $(a, b, c > 0)$; $\rho = k$, a positive constant; about the x-axis
(c) R bounded by $y = x$ and $y = x^2$; $\rho = xy$; about the z-axis

44 Transform the following integrals into polar coordinates and evaluate them.

(a) $\displaystyle\int_{-1}^{1}\int_{-\sqrt{1-y^2}}^{\sqrt{1-y^2}} e^{-(x^2+y^2)}\, dx\, dy$

(b) $\displaystyle\int_{0}^{\pi/2}\int_{0}^{\sqrt{\frac{1}{4}\pi^2-x^2}} \frac{1+x^2+y^2}{\sqrt{x^2+y^2}}\, dy\, dx$

(c) $\displaystyle\int_{0}^{1}\int_{0}^{\sqrt{1-y^2}} xy\sqrt{x^2+y^2}\, dx\, dy$

45 Evaluate the integrals below.

(a) $\displaystyle\iint_R (x^2+y^2)^{-3/2}\, dA;\quad -3\leq x\leq 0,\ -\sqrt{9-x^2}\leq y\leq 0$

(b) $\displaystyle\iint_R e^{-(x^2+y^2)/2}\, dA;\quad R$ the sector of $x^2+y^2\leq 1$ cut out by $y=|x|$

(c) $\displaystyle\iint_R \sqrt{x^2+y^2}\, dA;\quad R$ bounded by $x^2+y^2=2x$

(d) $\displaystyle\iint_R y\, dA;\quad R$ the first quadrant region bounded by $y=\sqrt{3}x$ and $x^2+y^2=4y$

46 Find the area of the given regions
   (a) The region inside the circle $r=8\cos\theta$ and outside the circle $r=4$
   (b) The region bounded by $y^2=4x$ and $y=x/2$
   (c) The region inside the cardioid $r=1+\cos\theta$ and to the left of the line $4x=3$
   (d) The region bounded on the left by the parabola $y^2=1-2x$ and on the right by
       $x^2+y^2=1$

47 Find the Jacobian of each of the given transformations
   (a) $x=u^2+v^2$ (b) $x=2uv-v^2$ (c) $x=\sin r\cos t$
       $y=u^2-v^2$ $y=v^2-2u$ $y=\cos r\cos t$

48 Show that the area bounded by the isothermal curves $pv=a, pv=b$ and the adiabatic

   curves $pv^\gamma=c, pv^\gamma=d$ is $\dfrac{b-a}{\gamma-1}\log\left[\dfrac{d}{c}\right]$

49 Find the area of the given surface over the given region for the following:
   (a) $z=2x-2y+1;\ 0\leq x\leq 1, 0\leq y\leq 2$
   (b) $z=x+2y+3;\ 1\leq y\leq 4, y\leq x\leq y^2$
   (c) $z=xy;\ x^2+y^2\leq 1, x\geq 0, y\geq 0$
   (d) A cylindrical hole of radius b is cut through the centre of a sphere of radius $a>b$.
       Find the surface area removed from the sphere.

50   $x^2/a^2 + y^2/b^2 = 1$ gives the contour of the base of a right circular cylinder; the height is c. The cylinder is bevelled down so that the height z of any point (x, y) of the base in the first quadrant is given by $z = c(1 - \dfrac{x}{a})(1 - \dfrac{y}{b})$. Express the volume left in this quadrant as a double integral and show that its value is $\dfrac{1}{4} abc (\pi - \dfrac{13}{6})$.   (LU)

51   The length of the side of a square plate ABCD is $2a$, and O is the mid point of AB. If the surface density at any point P on the plate is λOP^2, λ being a constant, express the mass of the plate as a double integral. Show that the distance of the centre of mass of the plate from AB is $1.4a$ and find the moment of inertia of the plate about AB.   (LU)

52   A heavy uniform plate of weight W in the form of a cardioid $r = a(1 + \cos \theta)$ lies on a rough horizontal table (coefficient of friction μ), and is movable about a vertical axis through the pole of coordinates. Show that the couple which must be applied to the plate in order to make it move is $\dfrac{10}{9} \mu a W$.   (LU)

53   The electrical attraction at a point situated at a distance r from an infinite plane due to a surface density of electricity g is given by

$$rg \int\limits_{0}^{\infty} \int\limits_{0}^{2\pi} \frac{x \, d\theta \, dx}{\left(r^2 + x^2\right)^{3/2}}$$

Find the value of this attraction.

54   Let R_{uv} be the rectangular region $1 \le u \le 2, -1 \le v \le 1$. Evaluate

$$\int\limits_{R_{xy}} \int x \, dA_{xy} \text{ when } x = u^2 - v^2, y = 2uv. \text{ Sketch the region } R_{xy}.$$

Section 9.9

55   Evaluate   (a) $\displaystyle\int\limits_{0}^{2} \int\limits_{0}^{x} \int\limits_{0}^{1-x-y} x \, dz \, dy \, dx$ (b) $\displaystyle\int\limits_{0}^{1} \int\limits_{y^2}^{\sqrt{y}} \int\limits_{0}^{y+z} y \, dx \, dz \, dy$

56   Evaluate $\displaystyle\iiint\limits_{R} f(x, y, z) \, dV$ for the given function f and region R.

  (a) $f(x, y, z) = xyz$; R the rectangular solid bounded by the coordinate planes and the planes $x = a, y = b, z = c$; $a, b, c > 0$.

  (b) $f(x, y, z) = x^2$; R the region $x^2 + y^2 + z^2 \le 1$.

  (c) $f(x, y, z) = x^2 + y^2$; R the region bounded by the cylinder $x^2 + z^2 = 4$, and by the planes $y = 0, z = 0$, and $y + z = 2$; $z > 0$.

  (d) $f(x, y, z) = 1$; R the region bounded by $x^2 + y^2 = 1$ and $z = 4 - x^2 - y^2$, $z \ge 0$.

57 Use cylindrical coordinates to evaluate $\iiint\limits_{R} f(x, y, z)\ dV$

(a) $f(x, y, z) = z$; R the region above the cone $z^2 = x^2 + y^2$ and below the plane $z = a, a > 0$.

(b) $f(x, y, z) = e^{(x^2 + y^2)^{3/2}}$; R the region bounded by the planes $z = 0$ and $y = z$ and by the cylinder $x^2 + y^2 = 1, z \geq 0$.

58 Use spherical coordinates to evaluate $\iiint\limits_{R} f(x, y, z)\ dV$

(a) $f(x, y, z) = x^2 + y^2 + z^2$; R the region above the cone $z^2 = x^2 + y^2$ and below the plane $z = a, a > 0$.

(b) $f(x, y, z) = (x^2 + y^2 + z^2)^{-3/2}$; R the region above the plane $z = a(a > 0)$ and below the sphere $x^2 + y^2 + z^2 = 4a^2$.

59 The water face of a dam is vertical and has the form of a trapezium height h and length $2a$ at the top and a at the bottom. The thickness of the dam increases uniformly from zero at the top to a at the bottom. Show that the C.G. of the dam is $5a/16$ from the vertical face and find its distance below the top of the dam. (LU)

Section 9.10
60 Evaluate the integral in Problem 17 using Green's Theorem.

61 Use Green's Theorem to evaluate the integral in Problem 20.

62 Verify Green's theorem for $\oint\limits_{C} \{(1.5x^2 - 4y^2)dx + (2y - 3xy)dy\}$ where C is the

boundary of the region defined by

(a) $y = x^2,\ x = y^2$ (b) $x = 0,\ y = 0,\ x + y = 2$.

Section 9.11
63 Show, using Stokes' theorem, that

$$\iint\limits_{\sigma} \phi\ \text{curl}\ \mathbf{H}.\ d\mathbf{S} = \oint\limits_{\gamma} \phi\mathbf{H}.\ d\mathbf{r} - \iint\limits_{\sigma} (\text{grad}\ \phi \times \mathbf{H})\ .d\mathbf{S}$$

where the open surface σ is bounded by the closed curve γ. (EC)

64 State Stokes' Theorem.

Use the above to evaluate $\iint\limits_{S} \text{curl}\ \mathbf{A}\ .d\mathbf{S}$, where $\mathbf{A} = (x^2 + y - 4)\mathbf{i} + 3xy\mathbf{j} + (2xz +$

$z^2)\mathbf{k}$ and S is the surface of the paraboloid $z = 4 - (x^2 + y^2)$ above the xy-plane. (EC)

65 By using Stokes' theorem, or otherwise, evaluate $\iint_K$ curl $\mathbf{Q} . d\mathbf{S}$ where $\mathbf{Q} = y\mathbf{i} - x\mathbf{j}$ and

K is that part of the surface of the sphere $x^2 + y^2 + z^2 = b^2$ for which $z \geq 0$.   (EC)

66 Use Stokes' Theorem to evaluate $\iint_S$ curl $\mathbf{B} . d\mathbf{S}$

where $\mathbf{S}$ is the surface of the cone $z = \sqrt{x^2 + y^2}$, $0 \leq z \leq 2$ and $\mathbf{B}$ is the vector field
$$\mathbf{B} = (x - y)\mathbf{i} + y^2\mathbf{j} + \sin z\mathbf{k}$$ (EC)

67 Verify Stokes' Theorem $\iint_S$ curl $\mathbf{F} . d\mathbf{S} = \oint_C \mathbf{F} . d\mathbf{r}$

for the vector field $\mathbf{F} = y\mathbf{i} + 3x\mathbf{j} + 2\mathbf{k}$
where $\mathbf{S}$ is the surface of the plane $2x + 3y + 6z = 12$ situated in the first octant and C
is the bounding triangular contour.   (EC)

68 State Stokes' Integral Theorem and verify the theorem for the vector function
$$\mathbf{A} = (2x - y)\mathbf{i} - y^2 z^3 \mathbf{j} - y^3 z^2 \mathbf{k}$$
over the surface S defined by $z = 2, x^2 + y^2 = 4$, its boundary C being the intersection of
the cylinder with the plane $z = 0$.   (EC)

69 State Stokes' theorem and use it to evaluate the line integral

$$\int_C \mathbf{F} . d\mathbf{r}$$

when $\mathbf{F} = (-3y, 3x, 1)$ and C is the circle $x^2 + y^2 = 1$ described anti-clockwise.   (EC)

Section 9.12

70 Verify the Divergence Theorem $\iiint_V$ div $\mathbf{A} \, dV = \iint_S \mathbf{A} . d\mathbf{S}$ where $\mathbf{A} = x\mathbf{i} + y\mathbf{j} + z\mathbf{k}$ and

V is the volume defined by the surfaces, $x = 0, x = 1, y = 0, y = 1, z = 0, z = 1$.
71 Show briefly that div $(\mathbf{u} \wedge \mathbf{v}) = \mathbf{v} . $ curl $\mathbf{u} - \mathbf{u} . $ curl $\mathbf{v}$ and curl $\phi\mathbf{v} = ($grad $\phi) \wedge \mathbf{v} + \phi$ curl $\mathbf{v}$,
where $\mathbf{u}$ and $\mathbf{v}$ are vector point functions and ϕ is a scalar point function.
If $\mathbf{A} = $ curl $\mathbf{B}$ and $\mathbf{B} = $ curl $\mathbf{C}$ show by applying Gauss' divergence theorem to $\mathbf{C} \wedge \mathbf{B}$ that

$$\iiint_V \mathbf{B}^2 \, dV = \iint_S (\mathbf{C} \wedge \mathbf{B}) . d\mathbf{S} + \iiint_V \mathbf{C} . \mathbf{A} \, dV,$$

where the volume V is enclosed by surface S.   (EC)

72 Give a brief proof of the relation div $(\mathbf{a} \wedge \mathbf{b}) = \mathbf{b} . $ curl $\mathbf{a} - \mathbf{a} . $ curl $\mathbf{b}$ where $\mathbf{a}$ and $\mathbf{b}$ are
vector point functions.
In a certain fluid flow problem $\mathbf{v}$ is the fluid velocity and the vorticity Ω and vector
potential $\mathbf{u}$ are defined by the equations $\Omega = (1/2)$ curl $\mathbf{v}$, $\mathbf{v} = $ curl $\mathbf{u}$. Show that

$$\iiint_V v^2 \, dV = \iint_S (\mathbf{u} \wedge \mathbf{v}).d\mathbf{S} + 2 \iiint_V \mathbf{u} \cdot \Omega dV$$

where the closed surface S encloses a volume V of the fluid. (You may assume Gauss' divergence theorem.)   (EC)

73 Verify the Divergence Theorem $\iiint_V \text{div } \mathbf{A} \, dV = \iint_S \mathbf{A} \cdot d\mathbf{S}$

for the vector $\mathbf{A} = x\mathbf{i} + y\mathbf{j} + z\mathbf{k}$ where V is the volume of the solid hemisphere $x^2 + y^2 + z^2 \le a^2, z \ge 0$ and S is its surface (curved upper surface and flat base).

(EC)

74 State Gauss's theorem.
For a certain fluid flow problem the vorticity, Ω, and vector potential, $\mathbf{u}$, are defined in terms of the velocity, $\mathbf{q}$, by

$$\mathbf{q} = \text{curl } \mathbf{u} \qquad \Omega = \frac{1}{2} \text{ curl } \mathbf{q}$$

Show that for a closed surface S bounding a volume V of fluid

$$\iiint_V q^2 dV = \iint_S (\mathbf{u} \times \mathbf{q}).d\mathbf{S} + 2 \iiint_V \mathbf{u}.\Omega \, dV$$

You may assume that $\text{div}(\mathbf{a} \times \mathbf{b}) = \mathbf{b}.\text{curl } \mathbf{a} - \mathbf{a}.\text{curl } \mathbf{b}$.
For a given $\mathbf{q}$, is $\mathbf{u}$ unique? Give reasons.   (EC)

75 Use the Divergence Theorem to show that if S is a closed surface enclosing a volume V,

$$\iint_S r^2 \mathbf{r}.d\mathbf{S} = 5 \iiint_V r^2 dV$$

Deduce the value of the surface integral $\iint_S \mathbf{F} \cdot d\mathbf{S}$ where

$$\mathbf{F} = (x^2 + y^2 + z^2)(x\mathbf{i} + y\mathbf{j} + z\mathbf{k})$$

and S is the surface of the cube bounded by $x = 0, x = 1, y = 0, y = 1, z = 0, z = 1$.

(EC)

10

FUNCTIONS OF A COMPLEX VARIABLE

10.1 Introduction

A large number of problems in engineering can be solved by methods involving complex variables. The theory of analytic functions is important because the results have wide application to problems of heat flow, fluid dynamics, electrostatics and many other areas of engineering.

The real and imaginary parts of an analytic function are each solutions of Laplace's equation and hence two-dimensional potential problems can be dealt with via analytic functions. In particular the technique of conformal mapping allows the solution of a boundary-value problem in potential theory to be effected by transforming a complicated region into a simpler one. Many integrals that arise in practice cannot be solved by real integration techniques and require methods of complex integration for their evaluation. Indeed, the basis of transform calculus is the integration of functions of a complex variable.

10.2 Analytic Functions. The Cauchy-Riemann Equations

Let x and y be real variables; then $z = x + iy$ is called a **complex variable**. Let u and v be two further real variables then $w = u + iv$ is a second complex variable. Suppose that there is a relationship between these complex variables such that to each value of z in a given region of the z-plane there is associated one and only one value of w; then w is said to be a **function** of z, defined on the given region and we write $w = f(z)$.

Examples

(i) $w = z^2 - z$. This function is defined for all values of z. Note that we may write $w = u + iv = (x + iy)^2 - (x + iy)$ and, separating real and imaginary parts, we obtain $u = x^2 - y^2 - x$, $v = 2xy - y$. If we wish to find $f(3 + 2i)$ we merely calculate $(3 + 2i)^2 - (3 + 2i) = 9 - 4 + 12i - 3 - 2i = 2 + 10i$; alternatively, we can proceed via the equations for u and v.

(ii) $w = 3\bar{z} + z$. This function is defined for all z. Note that $w = u + iv = 3(x - iy) + (x + iy)$ so that $u = 4x$ and $v = -2y$. As an example $f(4 - i) = 3(4 + i) + 4 - i = 16 + 2i$.

Representation of a complex function

We are accustomed to drawing a graph of the real-variable function $y = f(x)$ on a plane. However, we are in difficulties when we try to represent a complex function since points on the z-plane will be confused with points on the w-plane. The answer is to keep the planes separate and use *two* Argand diagrams, one for the z-plane and one for the w-plane.

Examples

(i) In Figure 10.1 we show the points $z = 1$, $2 + i$ and $3 + 2i$ being mapped by the function $w = 3z + 2$ into the points $w = 5$, $8 + 3i$, $11 + 6i$ respectively.

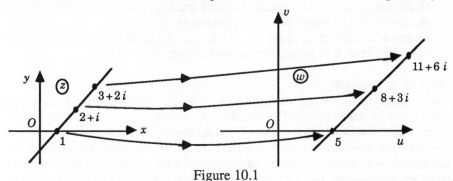

Figure 10.1

Note that the line $y = x - 1$, on which lie all three chosen points in the z–plane, can be shown to map into the line $v = u - 5$ in the w-plane and the images of the z-values all lie on this second line. Since $w = u + iv = 3(x + iy) + 2$ we find that $u = 3x + 2$ and $v = 3y$ so that $y = x - 1$ becomes $v/3 = (u - 2)/3 - 1$ or $v = u - 5$. The mapping $w = 3z + 2$ is an example of a **translation**. See what happens to points on the circle $x^2 + y^2 = 1$. Also investigate the effect on the region $|z| \leq 1$.

(ii) $w = z^2$

Here $u + iv = (x + iy)^2 = x^2 - y^2 + 2ixy$ so that $u = x^2 - y^2$, $v = 2xy$. If we take the line $y = x$ in the z-plane we obtain $u = 0$, a straight line, but if we consider $y = x + 1$, then $u = x^2 - (x + 1)^2 = -2x - 1$, $v = 2x(x + 1)$. Hence eliminating x we obtain $u^2 = 2v + 1$. This is a parabola. Any other straight lines of the form $y = x + k$ ($k \geq 0$) will transform into parabolas, as can be shown. Let us now use polar coordinates, $z = re^{i\theta}$. Then $w = z^2 = r^2 e^{i2\theta}$, i.e. a point (r, θ) in the z-plane produces a point $(r^2, 2\theta)$ in the w-plane; see Figure 10.2.

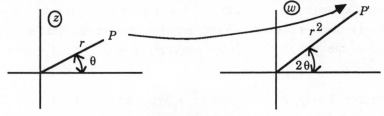

Figure 10.2

Taking r = constant and letting θ vary between 0 and π, then P moves on a semi-circle in the z-plane while in the w-plane, P′ will move round a full circle of radius r^2. We take this matter up again in the next section.

(ii) $w = z^{1/2}$. In this case there are two values of w for each value of $z \neq 0$. We say that the relationship between w and z is **multi-valued**. To be consistent with real variable work we should not regard this relationship as a function. However, most authors will regard a function as associating one *or more* values of w with each value of z. It is customary to speak of single-valued and multi-valued functions and regard a multi-valued function as a collection of **branches**. In this example we may refer to the **principal branch** as giving a value of $z^{1/2}$ in the upper half w-plane and the secondary branch as giving a value in the lower half w-plane.

Limit of a function

We define the limit of $w = f(z)$ as $z \to z_0$ as some number l such that $|f(z) - l|$ can be made as small as we like by making $|z - z_0|$ sufficiently small. For example, the limit of $(z^2 - z)$ as $z \to i$ is, obviously enough, $i^2 - i = -1 - i$. However, we at once encounter a difference from real variables. There, the statement $x \to x_0$ meant a journey along the real axis towards x_0 either from the left or from the right: two directions only. Here the statement $z \to z_0$ must be interpreted as z approaches the point z_0 *along any path whatsoever*. Since $|z - z_0|$ represents the distance between z and z_0 our requirement is merely that this distance decreases to zero.

Example 1

Find $\lim\limits_{z \to 3+i} (z^2 - z)$ along the paths shown in Figure 10.3.

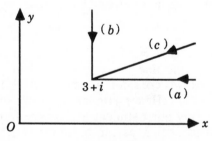

Figure 10.3

(a) Along this path $z = x + i$ and $z^2 - z = (x + i)^2 - (x + i) = x^2 - x - 1 + i(2x - 1)$. As $z \to 3 + i$, $x \searrow 3$ and, assuming that we can add the separate limits of the real and imaginary parts, $z^2 - z \to 3^2 - 3 - 1 + i(6 - 1) = 5 + 5i$.

(b) Here $z = 3 + iy$, $z^2 - z = 6 - y^2 + 5yi$; as $z \to 3 + i$, $y \searrow 1$ and $z^2 - z \to 5 + 5i$.

(c) $z = k(3 + i)$ where k is real; $z^2 - z = k^2(8 + 6i) - k(3 + i)$; as $k \searrow 1$, $z^2 - z \rightarrow 5 + 5i$.

In each case the limit is the same.

Example 2

Find $\lim\limits_{z \rightarrow 0} \dfrac{\bar{z}}{z}$ (a) along the x-axis   (b) along the y-axis.

(a) $z = x, \bar{z} = x$ so that $\bar{z}/z = 1$, which is the limit along the x-axis.

(b) $z = iy, \bar{z} = -iy$ so that $\bar{z}/z = -1$, which is the limit along the y-axis.

Note that, converting to polars, $\bar{z}/z = e^{-i\theta}/e^{i\theta} = e^{-2i\theta}$ and this will give different values along different straight lines into the origin.

Rules for limits of the sums of functions, the difference between functions, the product of functions and the quotient of functions are analogous to those rules for real variables.

Continuity

As you might expect, $f(z)$ is defined to be **continuous** at z_0 if (i) $f(z_0)$ exists and (ii) $\lim\limits_{z \rightarrow z_0} f(z)$ exists and is equal to $f(z_0)$. Again the rules for combining two continuous functions are similar to those for real variables.

Example 1

$$\lim_{z \rightarrow 1+i} \left\{ \frac{(z^2 + 1)(z - i)}{z^2 + z + 1} \right\} = \frac{\lim\limits_{z \rightarrow 1+i} \{z^2 + 1\} . \lim\limits_{z \rightarrow 1+i} \{z - i\}}{\lim\limits_{z \rightarrow 1+i} \{z^2 + z + 1\}}$$

$$= [(1 + i)^2 + 1] [1 + i - i] / [(1 + i)^2 + (1 + i) + 1]$$
$$= (8 + i)/13 = f(1 + i)$$

Example 2

$f(z) = z/(z^2 + 4)$ is continuous everywhere except at $z = \pm 2i$ when $f(z)$ does not exist.

Derivatives of a function

We say that $f(z)$ is **analytic** at z_0 if

$$\lim_{\Delta z \rightarrow 0} \left\{ \frac{f(z_0 + \Delta z) - f(z_0)}{\Delta z} \right\} \qquad (10.1)$$

exists. We denote the limit by $f'(z_0)$ and call it the **derivative** of $f(z)$ at z_0.

Again, note that Δz may tend to zero in *any way whatsoever*. Many simple-looking functions do not possess a derivative at any point; an example follows.

Example

$$f(z) = \bar{z} = x - iy$$

Let $\quad R = \dfrac{f(z_0 + \Delta z) - f(z_0)}{\Delta z} = \dfrac{(x_0 + \Delta x) - i(y_0 + \Delta y) - (x_0 - iy_0)}{\Delta x + i\Delta y}$

$$= \frac{\Delta x - i\Delta y}{\Delta x + i\Delta y}$$

Consider the path parallel to the x-axis so that $\Delta y = 0$
then $R = \Delta x/\Delta x = 1$, which is the limit as $\Delta x \to 0$.
However, the path parallel to the y-axis is defined by $\Delta x = 0$
then $R = -i\Delta y/i\Delta y = -1$, which is the limit as $\Delta y \to 0$.

Clearly $f(z)$ is not analytic at any point.
Where a function $f(z)$ is analytic, the differentiation proceeds as you might expect. For example, if $f(z) = 3z^2 - z, f'(z) = 6z - 1$.

The definition of derivative may seem very restrictive, as it is not sufficient for the limits along directions parallel to the axes to be equal; we must have limits along all possible paths giving the same value. However, most of the common functions of mathematics are analytic at all but a finite number of points. These points are called **singularities**. We shall see that these points have important practical applications; for example, a fluid source or vortex will be represented by singularities of the function which describes the flow pattern.

Cauchy-Riemann conditions

We now seek a set of conditions on the components u and v so that w is analytic. We derive them at an unspecified point z. If $w = f(z)$ is analytic then

$$f'(z) = \lim_{\Delta z \to 0} \left\{ \frac{f(z + \Delta z) - f(z)}{\Delta z} \right\}$$

$$= \lim_{\Delta z \to 0} \left\{ \frac{[u(x + \Delta x, y + \Delta y) + iv(x + \Delta x, y + \Delta y)] - [u(x, y) + iv(x, y)]}{\Delta x + i\Delta y} \right\}$$

The limit exists no matter what path is taken. Consider two particular paths.

(i) We approach z along a path parallel to the x-axis, i.e. $\Delta y = 0$, and we let $\Delta x \to 0$. Then

$$f'(z) = \lim_{\Delta x \to 0} \left\{ \frac{u(x + \Delta x, y) + iv(x + \Delta x, y)}{\Delta x} - \frac{u(x, y) + iv(x, y)}{\Delta x} \right\}$$

$$= \lim_{\Delta x \to 0} \left\{ \frac{u(x + \Delta x, y) - u(x, y)}{\Delta x} + i \frac{v(x + \Delta x, y) - v(x, y)}{\Delta x} \right\}$$

i.e. $$f'(z) = \frac{\partial u}{\partial x} + i \frac{\partial v}{\partial x} = \frac{\partial f}{\partial x} \qquad (10.2)$$

(ii) We now approach z by a path parallel to the y-axis, i.e. $\Delta x = 0$ and we let $\Delta y \to 0$. We leave it to you to show that

$$f'(z) = \frac{1}{i} \frac{\partial u}{\partial y} + \frac{\partial v}{\partial y} = \frac{1}{i} \frac{\partial f}{\partial y} \qquad (10.3)$$

By comparing the real and imaginary parts of (10.2) and (10.3) we find that

$$\frac{\partial u}{\partial x} = \frac{\partial v}{\partial y} \quad \text{and} \quad \frac{\partial u}{\partial y} = -\frac{\partial v}{\partial x} \qquad (10.4)$$

The conditions obtained in (10.4) are known as the **Cauchy-Riemann equations**. We state without proof the converse result: if $u(x, y)$ and $v(x, y)$ and their partial derivatives with respect to x and y are continuous and satisfy (10.4) in a neighbourhood of z_0 then $f(z)$ is analytic at z_0 and $f'(z_0)$ is given by (10.2) or (10.3). (A neighbourhood of z_0 is a region $|z - z_0| < \rho$ for some real number ρ.)

Example 1 $w = f(z) = \bar{z} = x - iy$

Here $u = x$, $v = -y$ so that $\frac{\partial u}{\partial x} = 1, \frac{\partial u}{\partial y} = 0, \frac{\partial v}{\partial x} = 0, \frac{\partial v}{\partial y} = -1$, and therefore the first equation of (10.4) is not satisfied. Hence $f(z)$ is not analytic anywhere.

Example 2 $f(z) = z^2 = x^2 - y^2 + 2ixy$

Here $u = x^2 - y^2$ so that $\frac{\partial u}{\partial x} = 2x$ and $\frac{\partial u}{\partial y} = -2y$. Also $v = 2xy$ so that

$\frac{\partial v}{\partial x} = 2y$ and $\frac{\partial v}{\partial y} = 2x$. Then both equations (10.4) are satisfied for all x and y; hence $f(z)$ is analytic everywhere. From (10.2), $f'(z) = 2x + i2y = 2z$.

Example 3 $f(z) = \frac{1}{z} = (x - iy)/(x^2 + y^2)$

Here $u = \dfrac{x}{x^2 + y^2}$, $v = \dfrac{-y}{x^2 + y^2}$. We leave it to you to show that (10.4) are

both satisfied, except at $x = y = 0$. You should obtain $f'(x) = -1/z^2$ using (10.2) or (10.3). Note that this derivation assumes that not both x and y are zero.

Example 4 $f(z) = z\bar{z} = x^2 + y^2$
We leave it to you to show that $f'(z)$ exists only at $z = 0$.

Connection with Laplace's equation
Suppose $f(z)$ is analytic in some region, then

$$\frac{\partial^2 u}{\partial x \partial y} = \frac{\partial}{\partial x}\left(\frac{\partial u}{\partial y}\right) = \frac{\partial}{\partial x}\left(-\frac{\partial v}{\partial x}\right) = -\frac{\partial^2 v}{\partial x^2}$$

But
$$\frac{\partial^2 u}{\partial y \partial x} = \frac{\partial}{\partial y}\left(\frac{\partial u}{\partial x}\right) = \frac{\partial}{\partial y}\left(\frac{\partial v}{\partial y}\right) = \frac{\partial^2 v}{\partial y^2}$$

Under the conditions demanded of the partial derivatives, we may equate the mixed partial derivatives of u and obtain

$$\frac{\partial^2 v}{\partial x^2} + \frac{\partial^2 v}{\partial y^2} = 0 \qquad\qquad (10.5)$$

In a similar way we find that

$$\frac{\partial^2 u}{\partial x^2} + \frac{\partial^2 u}{\partial y^2} = 0 \qquad\qquad (10.6)$$

We call u and v **conjugate harmonic functions**.

Conversely, if we take two functions u and v satisfying equations (10.6) and (10.5) in some region then $f(z) = u + iv$ is analytic in that region. For example, if $u = xy$, then it is easily seen that u satisfies (10.6). Let us find the conjugate harmonic function $v(x, y)$. We know that $\dfrac{\partial v}{\partial y} = \dfrac{\partial u}{\partial x} = y$ and also that

$\dfrac{\partial v}{\partial x} = -\dfrac{\partial u}{\partial y} = -x$; hence $v = \frac{1}{2}(y^2 - x^2) + c$, where c is a constant. Also we

may form $f(z) = u + iv = -\frac{1}{2}iz^2 + c$ which is analytic everywhere.

We close this section by noting that it can be shown that if $f(z)$ is analytic in a region then it possesses derivatives of *any order* in that region. This emphasises what a strong condition the existence of a first derivative demands.

10.3   Standard Functions of a Complex Variable

We now take a brief look at some properties of the more common functions.

(a)   Exponential function

We define
$$e^z = e^{(x+iy)} = e^x (\cos y + i \sin y) \qquad (10.7)$$
Note that

(i) e^z reduces to e^x when $y = 0$ (ii) e^z is single-valued

(iii) e^z is analytic everywhere and its derivative is e^z (iv) $|e^z| = e^x$

(v) $\arg e^z = y$ (vi) $e^{(z+2k\pi i)} = e^z$ for any integer k

For example, if $w = u + iv = e^z$ then $u = e^x \cos y$, $v = e^x \sin y$ and it is easy to check that the Cauchy-Riemann conditions are satisfied, which proves the analyticity.   Also $dw/dz = \partial u/\partial x + i \, \partial v/\partial x = e^x \cos y + i \, e^x \sin y = e^z$. We leave you to prove properties (i), (ii), (iv), (v) and (vi).

Example

Find all roots of the equation $e^z = -i$.

We have $e^x \cos y = 0$ and $e^x \sin y = -1$ so that $\cos y = 0$ and $y = \pi/2 + k\pi$, k integral.   Then $e^x \sin y = \pm e^x$. Now $e^x \neq -1$ for any x so that $e^x = 1$ and $x = 0$. However, you can easily show (and you should) that k cannot be even, i.e. we cannot have $\sin((\pi/2) + 2k\pi) = -1$ so that the roots we require are given by $z = (-(\pi/2) + 2k\pi)i$, k integer.

(b)   Trigonometric functions

We shall define
$$\cos z = \frac{(e^{iz} + e^{-iz})}{2}, \quad \sin z = \frac{(e^{iz} - e^{-iz})}{2i} \qquad (10.8)$$

which is consistent with our results when z is the real variable x. We may define other trigonometric functions in ways analogous to results for real variables; thus for example, $\tan z = \sin z/\cos z$. You should be able to show easily that $\sin z$ and $\cos z$ are analytic everywhere and that

$$\frac{d}{dz}(\sin z) = \cos z, \quad \frac{d}{dz}(\cos z) = -\sin z \qquad (10.9)$$

All the familiar trigonometric identities hold in the case of complex functions; for example
$$\sin^2 z + \cos^2 z \equiv 1 \qquad (10.10)$$
We can obtain representations for $\sin z$ and $\cos z$ in terms of functions of x and y.   Thus $\sin z = \sin(x + iy) = \sin x \cos (iy) + \cos x \sin (iy)$

$$= \sin x \left[\frac{e^{-y} + e^{+y}}{2} \right] + \cos x \left[\frac{e^{-y} - e^{+y}}{2i} \right]$$

$$= \sin x \cosh y - (1/i) \cos x \sinh y$$
$$= \sin x \cosh y + i \cos x \sinh y \qquad (10.11)$$

Similarly $\cos z = \cos x \cosh y - i \sin x \sinh y \qquad (10.12)$

(c) Hyperbolic functions

We shall define

$$\cosh z = \tfrac{1}{2}(e^z + e^{-z}) \quad \text{and} \quad \sinh z = \tfrac{1}{2}(e^z - e^{-z}) \tag{10.13}$$

Then we can easily deduce that

(i) $\cosh z = \cosh x \cos y + i \sinh x \sin y$ (10.14)
      $\sinh z = \sinh x \cos y + i \cosh x \sin y$ (10.15)

(ii) $\cosh z$ and $\sinh z$ are analytic for all z

(iii) $\dfrac{d}{dz}(\sinh z) = \cosh z, \quad \dfrac{d}{dz}(\cosh z) = \sinh z$ (10.16)

The formulae involving these and other hyperbolic functions extend without difficulty into complex variables.

(d) Logarithmic function

Since the exponential function is one-to-one, it possesses an inverse function which we call log z (or ln z), defined by $e^w = z \ (z \neq 0)$ so that $w = \ln z$. Suppose $w = u + iv$ and $z = re^{i\theta}$, then $e^{(u+iv)} = re^{i\theta}$, so that $e^u = r$ and $e^{iv} = e^{i\theta}$. Then clearly

$$u = \ln r = \ln |z| \tag{10.17}$$

and

$$v = \theta = \arg z \tag{10.18}$$

However, although arg z is usually confined to $-\pi < \theta \le \pi$, it is possible for v to take an infinity of values since $e^{i\theta} = e^{i\theta} \cdot e^{2\pi ki}$ where k is any integer and therefore $v = \theta + 2k\pi i$.

Now we can confine v to the range $-\pi < v \le \pi$, by taking $k = 0$; then the corresponding value of ln z is called the **principal value** of ln z and denoted by Ln z; it is a function. In general, to each value of $z \neq 0$ there are an infinite number of values of ln z each with the same real part.

We can divide up these values into **branches** of range 2π by selecting $k = 0$, $\pm 1, \pm 2, \pm 3, \ldots$ in turn. (This is the customary way of making a **branch cut** for this function.) Notice that each branch is defined on the z-plane without the origin. We shall see later why $z = 0$ is specifically excluded. On each branch ln z is analytic and has derivative $1/z$ except along the negative real axis and at the origin; this is represented schematically in Figure 10.4.

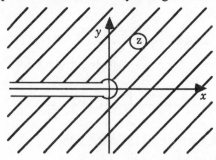

Figure 10.4

The familiar properties of logarithm extend in complex variables to ln z, bearing in mind that with Ln z we may have to adjust the value of argument by a multiple of $2\pi i$ to comply with $-\pi < \arg z \le \pi$.

Examples

(i) $\ln(1+i) = \ln(\sqrt{2}\, e^{i\pi/4}) = \ln \sqrt{2} + i(\pi/4 + 2k\pi)$

$\qquad\qquad = \frac{1}{2}\ln 2 + i(\pi/4 + 2k\pi)$

(ii) $\text{Ln}(1+i) = \frac{1}{2}\ln 2 + i\pi/4$

(iii) $\ln z = 1 - i\pi \Rightarrow z = e^{1-i\pi} = e^1 \cdot e^{-i\pi} = -e$

(e) Powers of a complex number, z^m

If m is a positive integer $w = z^m$ is analytic. Consider though $w = z^2$; as $\arg z$ goes from 0 to π, $\arg w$ goes from 0 to 2π. When $\arg z$ increases from π to 2π, $\arg w$ goes from 2π to 4π.

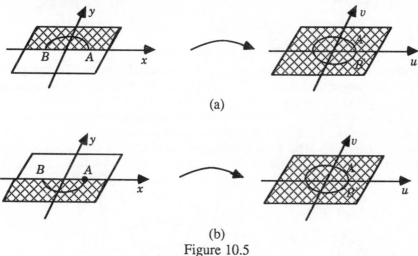

(a)

(b)

Figure 10.5

In Figure 10.5(a) we consider the effect of $w = z^2$ on the upper half z-plane. The semicircle is mapped into the whole-circle in the w-plane and the upper half z-plane is mapped into all the w-plane. In Figure 10.5(b) $\pi \le \arg z < 2\pi$, yet the lower half z-plane is mapped into the entire w-plane. We have represented the mapping $w = z^2$ associated with the **function** $f(z) = z^2$ on two **sheets** in the w-plane.

If m is a non-integer, z^m is multi-valued and it has m branches, all analytic. For example $z^{1/2}$ has two branches on each of which $z^{1/2}$ is single-valued. If $z = re^{i\theta}$, then one branch is found by taking $0 \le \theta < 2\pi$ and the other by taking $2\pi \le \theta < 4\pi$. Other ranges of 2π for θ merely repeat these branches. Since the transition from one branch to another is caused by a rotation of z about the origin $z = 0$, we call the origin a **branch point** of $w = z^{1/2}$. This is one kind of **singularity** of a function. (Note that $z = 0$ is a branch point of $\ln z$.)

(f) Reciprocal function

The function $f(z) = 1/z$ is defined for all $z \neq 0$ and is analytic for all $z \neq 0$ with derivative $(-1)/z^2$. (We can complete the domain of $f(z)$ to include $z = 0$ if we introduce an ideal *'point at infinity'* such that as $z \to 0$, $1/z \to$ the point at infinity.)

10.4   Complex Potential and Conformal Mapping

At the end of Section 10.2 we stated that the real and imaginary parts of an analytic function each satisfied Laplace's equation. Further, the curves $u(x, y)$ = constant and $v(x, y)$ = constant intersect orthogonally.

This follows because along $u(x, y)$ = constant, $\dfrac{\partial u}{\partial x} dx + \dfrac{\partial u}{\partial y} dy = du = 0$

and so $\dfrac{dy}{dx} = -\dfrac{(\partial u/\partial x)}{(\partial u/\partial y)}$, whereas along $v(x, y)$ = constant $\dfrac{dy}{dx} = -\dfrac{(\partial v/\partial x)}{(\partial v/\partial y)}$. The

product of these gradients is $\dfrac{(\partial u/\partial x)}{(\partial u/\partial y)} \cdot \dfrac{(\partial v/\partial x)}{(\partial v/\partial y)} = \dfrac{(\partial u/\partial x)}{(\partial u/\partial y)} \cdot -\dfrac{(\partial u/\partial y)}{(\partial u/\partial x)} = -1$, by the

Cauchy-Riemann conditions. Hence the curves cut at right angles. This property is of great use in potential theory. For example in two-dimensional electrostatic problems if u = constant gives the **equipotential lines** then v = constant are the **electric lines of force**. In Figure 10.6(a) we show some of these two sets of curves for the field due to two oppositely charged particles. In the steady irrotational flow of an ideal fluid in two dimensions if u = constant gives the equipotential lines, v = constant gives the **streamlines** of the flow. In Figure 10.6(b) we show these two sets of curves for flow round a cylinder.

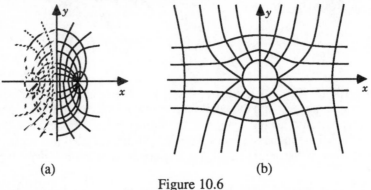

(a)   (b)

Figure 10.6

We call the function $w = u + iv$ the **complex potential** of the field. In Table 10.1 we present some examples of complex potential; σ represents conductivity. In the case of the potential $w = -m \ln z = -m \ln |z| - i\, m \arg z$, the streamlines are $\arg z$ = constant, i.e. lines radiating from the origin, and the equipotential lines are given by $\ln |z|$ = constant, i.e. $|z|$ = constant which are circles centred at the origin.

Other fields that can be studied in this way include
(i)   steady heat flow where $u = $ constant represent isothermals and $v = $ constant represent heat flow lines
(ii)   gravitation where $u = $ constant represent equipotential lines and $v = $ constant represent lines of force.

Table 10.1

Electrostatics / Magnetostatics	Current flow	Hydrodynamics
Uniform field E making angle α with real axis $w = E\,e^{-i\alpha}\,z$	Uniform current J making α with real axis $w = (J/\sigma)\,e^{-i\alpha}\,z$	Uniform stream U making α with real axis $w = U\,e^{-i\alpha}\,z$
Line charge or pole e at origin $w = -2e\ln z$	Electrode with strength J at origin $w = -(J/\sigma)\ln z$	Line source of strength m at origin $w = -m\ln z$
Dipole M $w = -2M/z$		Dipole M $w = -M/z$
Intensity of field $|E| = |dw/dz|$	Magnitude of current $|i| = \sigma|dw/dz|$	Speed of fluid $|q| = |dw/dz|$

Conformal Mapping

Consider the physical situation represented in Figure 10.7(a); a rectangular plate has one edge maintained at $100°$ and the opposite edge at $0°$ whilst the other two edges are insulated. We may assume that flow of heat takes place in the u-v plane until a steady state is reached. It is easy enough to solve Laplace's equation for the steady temperature $T(u, v)$ to obtain

$$T(u, v) = 100v/\pi \qquad (10.19)$$

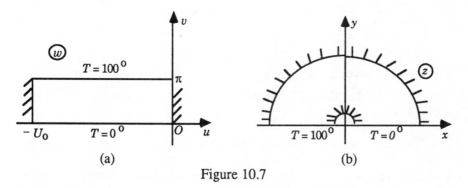

Figure 10.7

Now suppose we wish to find the steady temperature in the semicircular plate of Figure 10.7(b). Consider the relationship $w = \ln z$. Now if $z = re^{i\theta}$, $u = \ln r$ and $v = \theta$. What happens to the rectangle of Figure 10.7(a)? The edge $v = 0$ maps to $\theta = 0$, i.e. the positive real axis in the z-plane; the edge $v = \pi$ maps to $\theta = \pi$, i.e. the negative real axis; the insulated edge $u = 0$ maps to $\ln r = 0$, i.e. $r = 1$ or $|z| = 1$; the insulated edge $u = -u_0$ maps to $\ln r = -u_0$ i.e. $r = e^{-u_0}$, which implies that the further the left-hand edge is to the left the larger u_0 and the smaller r. Hence a semi-infinite rectangle maps to a semicircle with a blob of insulation at the origin. Suppose we now map (10.19). Then

$$T = 100\theta/\pi, \quad 0 \le \theta \le \pi$$

$$= \frac{100}{\pi} \tan^{-1}\left[\frac{y}{x}\right] \tag{10.20}$$

Does this provide the temperature distribution in the plate? It is easy to show by substitution that (10.20) satisfies Laplace's equation. In general, it is also true that the result of transforming a solution of Laplace's equation by an analytic function leads to a function which also satisfies Laplace's equation. Also, since we use the same function to obtain the transformed temperature as to obtain the transformed region, it follows that temperatures on the transformed boundaries are the same as the temperatures on the corresponding original boundaries. We shall state without proof that the rate of change of T perpendicular to a boundary in the u-v plane is proportional to the rate of change of T perpendicular to the corresponding boundary in the x-y plane; hence insulated surfaces map to insulated surfaces.

Now consider a general transformation $w = f(z)$ where $f(z)$ is analytic at $z = z_0$ and $f'(z_0) \ne 0$. See Figures 10.8(a) and (b).

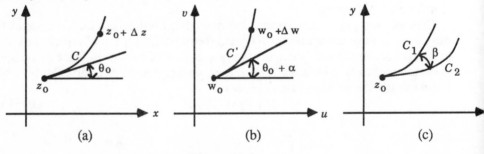

Figure 10.8

As a point moves from z_0 to $z_0 + \Delta z$ along C its image under w moves from w_0 to $w_0 + \Delta w$ along C'. Suppose we characterise the journey along C by a parameter t so that $z = z(t)$; then we may write $w = w(t)$. Now $\dfrac{dw}{dt} = \dfrac{dw}{dz} \cdot \dfrac{dz}{dt}$ and this holds at z_0; dz/dt represents the tangent vector at a point on C and dw/dt represents the tangent vector at the corresponding point on C'. The directions are given by $\theta_0 = \arg(dz/dt)$ and $\theta_0 + \alpha = \arg(dw/dt)$ respectively. Hence $\alpha = \arg(f'(z))$ and has a fixed value at z_0.

If two curves C_1 and C_2 pass through z_0 making an angle β with each other there, see Figure 10.8(c), it follows that the angle β is preserved after mapping by an analytic function $f(z)$.

In our particular problem, since isotherms and lines of heat flow are at right angles, they will remain so after the mapping $w = \ln z$. Hence we have a complete identity between the two physical situations.

In general to solve a potential problem with a complicated geometry, we merely need to find a transformation from a simpler region where the solution can readily be found via an analytic function. Then we transform back the solution. Note that sinks and sources and other **singular points** will map into corresponding singular points.

The property above of preserving the angle, which will be so for any analytic function, makes the mapping **conformal**. We may remark that areas of *small* closed regions around z_0 are scaled by an approximate factor $|f'(z_0)|^2$.

We now present some basic mappings.

(a)	**Translation**	$w = z + b$,	b constant	(10.21a)
(b)	**Magnification**	$w = az$,	a real and constant	(10.21b)
(c)	**Rotation**	$w = iaz$,	a real and constant	(10.21c)
(d)	**Inversion**	$w = 1/z$		(10.21d)

The inversion mapping is worthy of detailed study. We wish to examine the effects on special curves in the z-plane. For this purpose we write the mapping

as $z = 1/w$ and it follows that $x = \dfrac{u}{u^2 + v^2}$, $y = \dfrac{-v}{u^2 + v^2}$. Hence the circle

$x^2 + y^2 = a^2$ is mapped into $\dfrac{u^2}{(u^2 + v^2)^2} + \dfrac{v^2}{(u^2 + v^2)^2} = a^2$ i.e. $\dfrac{1}{a^2} = u^2 + v^2$

which is a circle in the w-plane.

In polar coordinates let $z = re^{i\theta}$ and $w = Re^{i\Theta}$. Then $R = 1/r$ and $\Theta = -\theta$. Try to describe the effects of the mapping via this approach.

Notice that the radius of this circle is the reciprocal of that of the original circle; big circles become small, small circles become big, but the circle $|z| = 1$ does not change. The circle of zero radius, i.e. $z = 0$ is mapped into the point at infinity and vice-versa. These remarks apply to circles whose centres are the origin. It can be shown that other circles may map into circles or straight lines.

Now consider the straight line $\alpha x + \beta y + \gamma = 0$. It is easy to show that this is mapped into the curve $\alpha u - \beta v + \gamma (u^2 + v^2) = 0$. Hence, straight lines through the origin ($\gamma = 0$) are mapped into their reflections in the vertical axis, other straight lines are mapped into circles.

Example

Find the image of the circle $|z - 2i| = 2$ under the inversion mapping.

Three points define the circle; we take $0, 4i, 2 + 2i$ for algebraic simplicity.

The corresponding points in the w-plane are the point at infinity, $-\frac{1}{4}i$ and

$\frac{1}{4}(1-i)$. These define a straight line viz. the line parallel to the real axis through

the point $-\frac{1}{4}i$. Take other points, equally spaced around the given circle and trace the path in the w-plane as we move once round the circle. Note that the point at infinity can appear at $u = -\infty$ and $u = +\infty$.

Application
In electrical circuit theory the **impedance** due to a resistance R is simply R, due to an inductance L is $j\omega L$ ($j^2 = -1$) and due to a capacitance C is $1/j\omega C$. The **admittance** of a circuit is the reciprocal of its total impedance. The impedance Z of a circuit comprising a resistance and an inductance in parallel is

given by $\dfrac{1}{Z} = \dfrac{1}{R} - \dfrac{j}{\omega L}$. The **impedance locus** as L varies between 0 and ∞ is

shown in Figure 10.9(a). The corresponding **admittance locus** is found by the inversion mapping to be that shown in Figure 10.9(b). Such loci are used frequently in the design of electrical circuits.

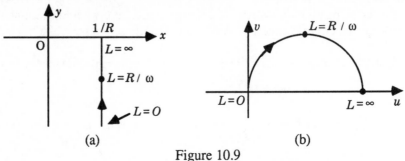

(a)   (b)

Figure 10.9

Example
In the circuit shown in Figure 10.10(a), C is a variable capacitance, R is a fixed resistance and L is a fixed inductance. An alternating voltage of constant frequency $\omega/2\pi$ is applied. We find the impedance and admittance loci as C varies.

The impedance Z is given by $\dfrac{1}{Z} = \dfrac{1}{R + j\omega L} + j\omega C$.

The admittance Y is thus $Y = \dfrac{1}{Z} = \dfrac{R - j\omega L}{R^2 + \omega^2 L^2} + j\omega C$.

Let $Y = u + iv$, so that $u = \dfrac{R}{R^2 + \omega^2 L^2}$ and $v = \dfrac{\omega C\,(R^2 + \omega^2 L^2) - \omega L}{R^2 + \omega^2 L^2}$.

Since R and L are fixed, the Y-locus as ω and C vary is the straight line

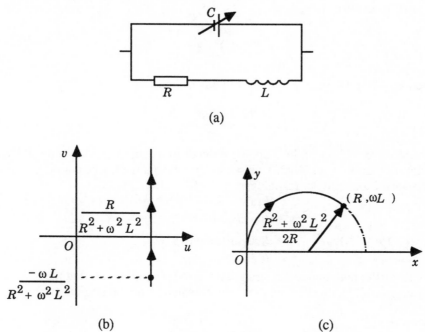

(a)

(b)   (c)

Figure 10.10

$$u = \frac{R}{R^2 + \omega^2 L^2} \quad \text{between} \quad v = \frac{-\omega L}{R^2 + \omega^2 L^2} \quad \text{and} \quad v \to \infty.$$

For the impedance locus we have $\dfrac{1}{Z} = \dfrac{1}{x + iy} = \dfrac{1}{R + j\omega L} + j\omega C$. That is

$$\frac{x - iy}{x^2 + y^2} = \frac{R - j\omega L}{R^2 + \omega^2 L^2} + j\omega C. \quad \text{Hence} \quad \frac{x}{x^2 + y^2} = \frac{R}{R^2 + \omega^2 L^2} \quad \text{giving}$$

$$x^2 + y^2 - \left[\frac{R^2 + \omega^2 L^2}{R} \right] x = 0. \quad \text{This is a circle centre} \quad \left(\frac{R^2 + \omega^2 L^2}{2R}, 0 \right) \text{ and}$$

of radius $\left[\dfrac{R^2 + \omega^2 L^2}{2R} \right]$.

The variation in C is from $C = 0$ to $C = \infty$ and when $C = 0$ we find

$\dfrac{1}{Z} = \dfrac{1}{R + j\omega L}$, i.e. $x = R$, $y = \omega L$. When $C = \infty$ we get $\dfrac{1}{Z} = \infty$ or $Z = 0$,

i.e. $x = 0$, $y = 0$. The loci are shown in Figure 10.10(b) and (c).

Bilinear mapping

$$w = \frac{az + b}{cz + d} \tag{10.22}$$

where $bc - ad \neq 0$.

You should show that we may rewrite this mapping as

$$w = \frac{(bc - ad)/c^2}{z + d/c} + \frac{a}{c} \tag{10.23}$$

provided $c \neq 0$. (If $c = 0$, the mapping reduces to a magnification followed by a translation.) Looking at (10.23) we see that the mapping may be considered as a succession of four simpler mappings:

(i) $z_1 = z + d/c$ (translation)

(ii) $z_2 = 1/z_1$ (inversion)

(iii) $z_3 = (bc - da)z_2/c^2$ (magnification)

(iv) $w = z_3 + a/c$ (translation)

If we restrict our attention to straight lines and circles we can find the result of applying (10.22) by its effect on three points which lie on the original curve.

Example

Consider $w = z/(z - 1)$. We wish to find its effect on the circle $|z| = 1$, the real axis and the imaginary axis. The eight regions of the z-plane thus formed are shown in Figure 10.11(a). We consider the mapping on the five points marked on that diagram together with the point at infinity.

(i) $z = 0$ becomes $w = 0$ (ii) $z = 1$ becomes the point at infinity

(iii) $z = i$ becomes $w = \frac{1}{2} - \frac{1}{2}i$ (iv) $z = -1$ becomes $w = \frac{1}{2}$

(v) $z = -i$ becomes $w = \frac{1}{2} + \frac{1}{2}i$ (vi)   the point at infinity becomes $w = 1$

(For this last result, write $w = 1/(1 - 1/z)$ and let $z \to \infty$.)

The outcomes are shown in Figure 10.11(b)

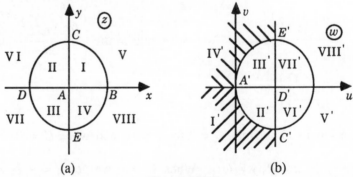

(a)   (b)

Figure 10.11

Note that the circle $|z| = 1$ has become the line $u = \frac{1}{2}$ (any three of results (ii) to (v) will show this.) Furthermore, the real axis, using (iv), (i) and (ii) becomes the real axis; but note where the points moved by considering the mapped images of $z = 0, 1, \infty$ and $z = -\infty$ (still the point at infinity), $-1, 0$. Take $z = 2$ and $z = \frac{1}{2}$ as an aid. Finally the imaginary axis, defined by (v), (i) and (iii), has become the circle $|w - \frac{1}{2}| = \frac{1}{2}$. Examine Figure 10.11 carefully and study the images of the eight regions of the first diagram.

10.5 Further Conformal Mappings

We open this section with a slightly more complicated problem in mapping.

The mapping $w = z + a^2/z$

It is assumed that a is real and positive. Note that at $z = 0$ there is a singularity of w and dw/dz does not exist there. Hence the mapping fails to be conformal at $z = 0$. In addition $dw/dz = 1 - a^2/z^2$ and since this is zero at $z = \pm a$ the mapping is not conformal there either. It is easy to show that

$$u = x\left(1 + \frac{a^2}{x^2 + y^2}\right), \quad v = y\left(1 - \frac{a^2}{x^2 + y^2}\right) \tag{10.24}$$

First of all, if we think of w as a complex potential then the stream function

$$\psi = y\left(1 - \frac{a^2}{x^2 + y^2}\right) = r \sin\theta \left(1 - \frac{a^2}{r^2}\right) \text{ in polar coordinates. Now } \psi = 0 \text{ when}$$

$\sin\theta = 0$ or $r = a$. Thus we could have the flow past the circular cylinder $r = a$ as shown in Figure 10.12. The stagnation points of the flow are where the fluid is at rest, i.e. $u = v = 0$ and these occur on the cylinder where $r = a$ and $y = 0$: the points A and B. In practice the viscosity of a liquid would cause frictional effects to alter markedly the downstream pattern of flow.

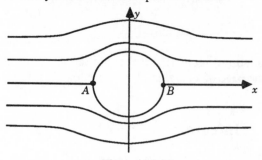

Figure 10.12

In aerofoil theory, the flow at distances from the aerofoil is virtually non-viscous in behaviour and we can calculate drag and lift fairly well via potential theory. Consider our mapping applied to the circle shown in Figure 10.13(a). It passes through $z = -a$ and contains $z = a$ in its interior. It is mapped into a shape depicted in Figure 10.13(b): the so-called **Joukowski aerofoil**. Note that the sharp point or **cusp** at $w = -2a$ is the image of the region near $z = -a$, which was one of the points of non-conformality. Work has been carried out to modify the transformation so as to 'round off' the cusp.

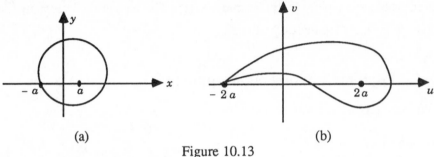

(a)   (b)

Figure 10.13

The mapping $w = z + e^z$

Consider the strip $-\pi \leq y \leq \pi$, $-\infty < x < \infty$ depicted in Figure 10.14(a). The components of w are $u = x + e^x \cos y$ and $v = y + e^x \sin y$. The line $y = \pi$ maps into (u, v) where $u = x - e^x$ and $v = \pi$; now as x varies from $-\infty$ to 0, u varies from $-\infty$ to -1. However, as x varies from 0 to ∞, u retraces its steps from -1 to $-\infty$. If we consider Figure 10.14(a) as representing flow through an open-ended channel then this result implies that a line of flow close to the channel boundary is virtually bent back on itself on reaching the open field. Examine for yourself the images of the lines $y = -\pi$ and $y = 0$.

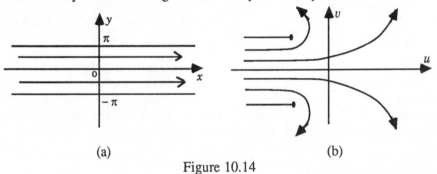

(a)   (b)

Figure 10.14

The pattern depicted in Figure 10.14(b) could represent the equipotentials of the electric field at the edge of a parallel-plate capacitor.

10.6 Complex Integrals

If $f(z)$ is a single-valued, continuous function in some region R then we define the integral of $f(z)$ along a path C in R as

$$\int_C f(z)\, dz = \int_C (u + iv)(dx + i dy)$$

$$= \int_C (u dx - v dy) + i \int_C (v dx + u dy) \qquad (10.25)$$

Note that, whereas real integrals can be interpreted in terms of area, complex integrals are defined in terms of line integrals over paths or curves in the plane.

The line integrals of the right-hand side are evaluated as indicated in Section 9.4. Let us first quote an important result.

$$\left| \int_C f(z)\, dz \right| \le \int_C |f(z)| \,.\, |dz| \le M \int_C ds = ML \qquad (10.26)$$

where M is an upper bound of $|f(z)|$ along C, L is the length of the path C.

Example 1

(i) Find $\displaystyle\int_{C_1} z^2\, dz$ where C_1 is that part of the unit circle from $z = 1$ to $z = i$.

(ii) Find $\displaystyle\oint_C z^2\, dz$ where C is the circumference of the unit circle travelled in

an anti-clockwise sense.

(i) $\displaystyle\int_{C_1} z^2\, dz = \int_{C_1} [(x^2 - y^2)dx - 2xy dy] + i \int_{C_1} [2xy dx + (x^2 - y^2)dy]$

On the unit circle we take $x = \cos\theta$, $y = \sin\theta$ and the path of integration

is from $\theta = 0$ to $\theta = \pi/2$. We leave you to show that $\displaystyle\int_{C_1} f(z)dz = -(1 + i)/3$.

We note that if we merely evaluate $[z^3/3]_1^i$ we again obtain $-(1 + i)/3$; in

other words we carry out an analogue of real integration. Is this result coincidental or is there a deeper underlying result?

(ii) This time the integral is from $\theta = 0$ to $\theta = 2\pi$. Again we leave you to show

that $\displaystyle\oint_C z^2\, dz = 0$. Is this also coincidence?

...ple 2

$\oint_C \dfrac{1}{z} dz$ where C is the unit circle may be evaluated by noting that on the unit

circle $z = \cos\theta + i\sin\theta = e^{i\theta}$ and $dz/d\theta = ie^{i\theta}$. Hence

$$\oint_C \frac{1}{z} dz = \int_0^{2\pi} i\, d\theta = 2\pi i \qquad (10.27)$$

Example 3
In a similar manner we find that for integral n

$$\oint_C \frac{dz}{(z - z_0)^n} = \begin{cases} 0 & n > 1 \\ 2\pi i & n = 1 \end{cases} \qquad (10.28)$$

where C is the circle centre z_0 and radius r, i.e. $|z - z_0| = r$.

This is one of the most important results in integration. We prove the result as
follows. Substituting $z = z_0 + re^{i\theta}$, the integral becomes

$$\oint_C \frac{ire^{i\theta}\, d\theta}{(re^{i\theta})^n} = \int_0^{2\pi} \frac{ie^{(1-n)i\theta}}{r^{n-1}}\, d\theta$$

If $n \neq 1$ we obtain $\dfrac{i}{r^{n-1}} \left[\dfrac{e^{(1-n)i\theta}}{(1-n)i} \right]_0^{2\pi} = \dfrac{[e^{(1-n)2\pi i} - 1]]}{r^{n-1}} \dfrac{1}{(1-n)} = 0$

If $n = 1$ the integral is $\displaystyle\int_0^{2\pi} i\, d\theta = 2\pi i$

Application: Blasius' Theorem
In Figure 10.15(a) we show a section of cylinder in a steady two-dimensional
irrotational flow of an ideal fluid.

The boundary of the cylinder is C. Let X, Y be the components of the force
on the cylinder and let M be the moment about the origin, due to fluid pressure.
In Figure 10.15(b) we depict a small section of the cylinder surface around a
point (x, y). We may write

$$dX = -p\,dy, \qquad dY = p\,dx, \qquad dM = p(x\,dx + y\,dy)$$

Hence $\qquad d(X - iY) = dX - idY = -p(i\,dx + dy) = -ip\,d(x - iy) = -ip\,d\bar{z}$

and $\qquad dM = \mathcal{R}\{p(x + iy)(dx - idy)\} = \mathcal{R}\{pz\,d\bar{z}\}.$

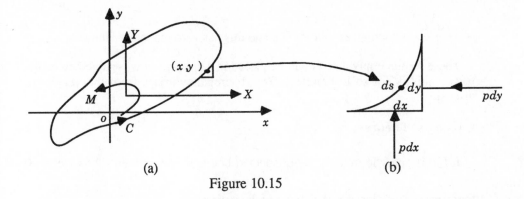

(a)   (b)

Figure 10.15

Now the pressure at any point is $p = p_0 - \frac{1}{2}\rho\, q^2$ where p_0 is the stagnation pressure, ρ is fluid density and the fluid speed q is given by $q^2 = |dw/dz|^2 = \dfrac{dw}{dz} \cdot \dfrac{d\overline{w}}{d\overline{z}}$. Since p_0 is a constant it has no resultant effect on X, Y or M. Then

$$d(X - iY) = +\tfrac{1}{2}i\rho \cdot \frac{dw}{dz} \cdot \frac{d\overline{w}}{d\overline{z}}\,d\overline{z} = \tfrac{1}{2}i\rho\,\frac{dw}{dz}\,d\overline{w} \quad \text{and} \quad dM = \Re\left\{-\tfrac{1}{2}\rho z\,\frac{dw}{dz}\,d\overline{w}\right\}.$$

C is a streamline (on which $\psi = $ constant) and on C, $dw = d\phi$ which is real so that $dw = d\overline{w}$. Hence integrating around C to obtain total effects

$$X - iY = \tfrac{1}{2}i\rho \oint_C \left(\frac{dw}{dz}\right)^2 dz \tag{10.29}$$

$$M = \Re\left\{-\tfrac{1}{2}\rho \oint_C z\left(\frac{dw}{dz}\right)^2 dz\right\} \tag{10.30}$$

These results constitute **Blasius' Theorem.**

Example
A circular cylinder $|z| = a$ is in a uniform stream U. It is known that

$$w = U\left(z + \frac{a^2}{z}\right) \quad \text{so that} \quad \left(\frac{dw}{dz}\right)^2 = U^2\left(1 - \frac{2a^2}{z^2} + \frac{a^4}{z^4}\right). \quad \text{Hence by (10.28)}$$

with $z_0 = 0$ we find from (10.29) that $X = Y = 0$.

Also $z\left(\dfrac{dw}{dz}\right)^2 = U^2\left(1 - \dfrac{2a^2}{z} + \dfrac{a^4}{z^3}\right)$. The only term to contribute to (10.30) is $\dfrac{-2a^2\,U^2}{z}$ which gives $-4\pi a^2\,U^2\,\mathrm{i}$ and this has zero real part. Hence $M = 0$.

These results imply that a cylinder placed in a uniform flow is subject to no net force: this is not so in practice. The discrepancy arises from the neglect of viscosity.

Cauchy's Theorem

If $f(z)$ is analytic in a simply-connected bounded region then $\displaystyle\oint_C f(z)\,\mathrm{d}z = 0$

for every simple closed path C lying in the region.

[By a simply-connected region we mean that any closed curve in the region may be shrunk to a point without any part of it leaving the region; for example, the interior of a square or of a circle.]

We shall follow Cauchy's proof which made a further condition on $f(z)$: that $f'(z)$ was continuous in this region. Goursat later showed this condition to be unnecessary.

Now $\displaystyle\oint_C f(z)\,\mathrm{d}z = \oint_C (u\mathrm{d}x - v\mathrm{d}y) + \mathrm{i}\oint_C (v\mathrm{d}x + u\mathrm{d}y).$

If $f'(z)$ is continuous then $\dfrac{\partial u}{\partial x}, \dfrac{\partial u}{\partial y}, \dfrac{\partial v}{\partial x}$ and $\dfrac{\partial v}{\partial y}$ exist and are continuous so we

may apply Green's theorem in the plane. Then if R be the region bounded by C

$$\oint_C f(z)\,\mathrm{d}z = \iint_R \left(-\frac{\partial v}{\partial x} - \frac{\partial u}{\partial y}\right)\mathrm{d}x\mathrm{d}y + \mathrm{i}\iint_R \left(\frac{\partial u}{\partial x} - \frac{\partial v}{\partial y}\right)\mathrm{d}x\mathrm{d}y$$

Since $f(z)$ is analytic then the Cauchy-Riemann equations are satisfied and both double integrals vanish automatically. Hence the result follows.

Cauchy's Theorem is often claimed to be the most important one in complex variable theory partly because of its consequences, which we shall examine shortly.

Note that the result $\displaystyle\oint_C z^2 = 0$ where C is the unit circle follows immediately

because of the analyticity of z^2. The results of Example 3 for $n \neq 1$ cannot follow from Cauchy's Theorem since the integrand is not analytic at one point, viz. $z = z_0$, of the region contained by C. (Note that $f(z)$ has to be analytic *on* C as well as *inside* it.)

Note finally that $\displaystyle\oint_C \frac{1}{z^2}\,\mathrm{d}z$, where C is the circle $|z| = 2$ and $\dfrac{1}{z^2}$ is defined on

the region $1 < |z| < 2$ to avoid non-analyticity at $z = 0$, cannot be assumed zero since the region is not simply connected. Check this for yourself.

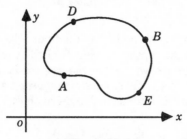

Figure 10.16

By analogy with line integrals and reference to Figure 10.16 in which $f(z)$ is assumed analytic inside and on the contour shown

$$\int\limits_{AEB} f(z)\ dz = \int\limits_{ADB} f(z)\ dz$$

In other words we may choose any path between A and B and $\int f(z)\ dz$ will have the same value and hence depends *only* on the points A and B. This explains the 'coincidence' in Example 1(i). It allows us to use the real variable integration technique.

Example

$$\int\limits_{i}^{1+2i} \cos z\ dz = [\sin z]_{i}^{1+2i} = \sin(1 + 2i) - \sin i \quad \text{(since } \cos z \text{ is analytic)}$$

Consequences of Cauchy's Theorem

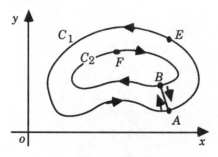

Figure 10.17

Suppose that $f(z)$ is analytic in a region bounded by the closed curves C_1 and C_2 shown in Figure 10.17. Further suppose that we cut this region by a line joining A and B. Then we claim that

$$\oint_{C_1} f(z)\ dz = \oint_{C_2} f(z)\ dz$$

where both curves are travelled round in the anti-clockwise direction.

To prove this consider the closed curve AEABFBA travelled around in the direction indicated by the arrows. *Now the region between C_1 and C_2 has been cut so that no line can cross the cut and be regarded as remaining inside the region.* The cut region is simply-connected and Cauchy's Theorem applies.

Hence
$$\oint_{AEABFBA} f(z)\ dz = 0$$

Now
$$\int_{AB} f(z)\ dz = - \int_{BA} f(z)\ dz$$

Then
$$\oint_{AEABFBA} = \int_{AEA} + \int_{AB} + \int_{BFB} + \int_{BA} = \int_{AEA} + \int_{BFB} = 0$$

Now $\displaystyle\int_{AEA}$ is $\displaystyle\oint_{C_1}$ and $\displaystyle\int_{BFB}$ is $-\displaystyle\oint_{C_2}$ as you can see. The result follows.

In other words to find $\displaystyle\oint_{C_1} f(z)\ dz$ we can replace C_1 by any curve C_2 so long as the region between them contains no singularities of $f(z)$. Quite often a circle is chosen for C_2. (We can also find $\displaystyle\oint_{C_2}$ by replacing C_2 by a suitable C_1.)

Example

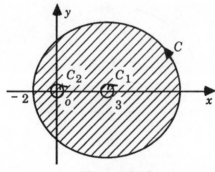

Figure 10.18

Evaluate $\oint\limits_C \dfrac{3}{z(z-3)}\,dz$ where C is the curve $|z-3| = 5$; see Figure 10.18

Note that $f(z) = \dfrac{3}{z(z-3)}$ is not analytic at $z = 0$ and $z = 3$ but is analytic

everywhere else. If we take C_1 as the circle of unit radius centred at $z = 3$ and C_2 as the other unit circle then in the shaded region $f(z)$ is analytic. By an extension of the earlier result, we deduce that

$$\oint\limits_C = \oint\limits_{C_1} + \oint\limits_{C_2}$$

You should prove this by making two suitable cuts in the shaded region. Then

$$\oint\limits_C \frac{3}{z(z-3)}\,dz = \oint\limits_{C_1} \frac{3}{z(z-3)}\,dz + \oint\limits_{C_2} \frac{3}{z(z-3)}\,dz$$

$$= \oint\limits_{C_1} \frac{1}{(z-3)}\,dz - \oint\limits_{C_1} \frac{1}{z}\,dz + \oint\limits_{C_2} \frac{1}{(z-3)}\,dz - \oint\limits_{C_2} \frac{1}{z}\,dz$$

$$= I_1 - I_2 + I_3 - I_4$$

Referring back to (10.28) we can see that $I_1 = 2\pi i$. Similarly from (10.27), $I_4 = 2\pi i$. Since $1/z$ is analytic inside and on C_1, $I_2 = 0$. Since $1/(z - 3)$ is analytic inside and on C_2, $I_3 = 0$. Hence $\oint\limits_C \dfrac{3dz}{z(z-3)} = 2\pi i - 0 + 0 - 2\pi i = 0.$

Cauchy's Integral Formula

If $f(z)$ is analytic within and on the boundary C of a simply-connected region D and if z_0 is any point inside C then

$$\oint\limits_C \frac{f(z)}{(z-z_0)}\,dz = 2\pi i\, f(z_0) \qquad (10.31)$$

This formula is very useful in saving us work in evaluating some integrals.

Example

Find $\oint\limits_C \dfrac{\sin z}{z^2 + 1}\,dz$ where C is the path

(i) $|z - i| = \frac{1}{2}$, (ii) $|z + i| = \frac{1}{2}$, (iii) $|z| = 2$. Refer to Figure 10.19.

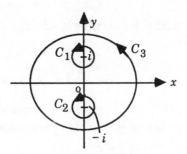

Figure 10.19

For path (i) we write $\dfrac{\sin z}{z^2 + 1} = \dfrac{\sin z/(z + i)}{z - i}$. Since the numerator is analytic

inside and on the path we use (10.31) with $z_0 = i$ to obtain the result

$$2\pi i \left[\frac{\sin z}{z + i}\right]_{z=i} = 2\pi i \frac{\sin i}{2i} = i\pi \sinh 1.$$

Likewise, for path (ii) $\dfrac{\sin z/(z - i)}{z + i}$ is the arrangement which leads us to the result

$2\pi i \sin(-i)/(-2i) = i\pi \sinh 1$.

For path (iii) we use $\displaystyle\oint_{C_3} = \oint_{C_1} + \oint_{C_2}$ to obtain the answer $2\pi i \sinh 1$.

Derivatives of an analytic function

If $f(z)$ is analytic in a simply-connected domain D, then at any interior point z_0 the derivatives of $f(z)$ of all orders exist and are analytic. (Note now the power of the definition of analyticity.) The derivatives at z_0 are given by

$$f^{(n)}(z_0) = \frac{n!}{2\pi i} \oint_C \frac{f(z)}{(z - z_0)^{n+1}} \, dz \tag{10.32}$$

where C is any simple closed curve which encloses z_0 and lies within D.

10.7   Taylor and Laurent Series

Many of the results with which we are familiar from real series extend into complex variables. The idea of **radius of convergence** is extended to a **circle of convergence** $|z - z_0| = \rho$ such that if $|z - z_0| < \rho$ the series under consideration will converge, whereas if $|z - z_0| > \rho$ the series will diverge.

Taylor's Theorem
Let $f(z)$ be analytic at a point z_0. Then there is a power series

$$\sum_{n=0}^{\infty} a_n(\,|z-z_0|)^n, \quad a_n = f^{(n)}(z_0)/n! \tag{10.33}$$

which converges to $f(z_1)$ for every z_1 in every neighbourhood of z_0 on which $f(z)$ is analytic so that

$$f(z_1) = \sum_{n=0}^{\infty} \frac{f^{(n)}(z_0)}{n!}(z_1 - z_0)^n$$

Many of the series can be obtained in ways analogous to those for real series and often we can in effect substitute z for x in series expansions we already know. Note that the concept of radius of convergence now means the distance between z_0 and the nearest singularity (if any). Remember that a singularity (or singular point) is a point at which $f(z)$ is not analytic.

Laurent Series
The trouble with Taylor's series is that the circle of convergence is most often merely a part of the region of analyticity of $f(z)$. For example, the series $\sum z^n$ converges to $f(z) = 1/(1-z)$ only *inside* the circle $|z| = 1$ even though $f(z)$ is analytic everywhere *except* at $z = 1$. The **Laurent series** aims to represent $f(z)$ at as many points as possible.

In effect we can expand around a point of singularity right up to, but not including, the singularity itself. This was not possible for Taylor's series.

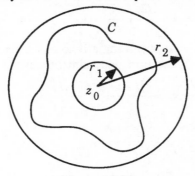

Figure 10.20

In Figure 10.20 we show an **annulus of convergence** $r_1 < |z - z_0| < r_2$ within which the series, which is an extension of Taylor's series, converges. The extension is to allow negative powers of $(z - z_0)$ in addition to the positive powers of Taylor's series.

Laurent's Theorem
If $f(z)$ is analytic throughout a closed annular region D centred at z_0, then at any point z within the annulus the function can be expanded as

$$f(z) = a_0 + a_1(z - z_0) + a_2(z - z_0)^2 + \dots$$
$$+ a_{-1}(z - z_0)^{-1} + a_{-2}(z - z_0)^{-2} + \dots$$

where the coefficients are given by

$$a_n = \frac{1}{2\pi i} \oint_C \frac{f(z)}{(z - z_0)^{n+1}} dz \qquad n = 0, \pm 1, \pm 2 \dots \qquad (10.34)$$

each integral being taken round any simple closed path lying in the annulus and encircling the inner boundary; see Figure 10.20.

Example 1

The function $f(z) = \dfrac{2}{z(z - i)}$ has three distinct series centred at $z_0 = -i$; note the singularities are at $z = 0$ and $z = i$.

(i)　　A Taylor series converging in $0 < |z + i| < 1$
(ii)　　A Laurent series converging in $1 < |z + i| < 2$
(iii)　　A Laurent series converging in $2 < |z + i|$

Sketch the three regions of convergence and mark the positions of the singularities.

Example 2

Find the Laurent series for $f(z) = 1/\{(z - 1)(z + 1)(z - 2)\}$ about the point $z = 1$ in $1 < |z - 1| < 2$.

Now $f(z)$ has singularities at $z = 1, -1,$ and 2. Therefore, in the annulus $1 < |z - 1| < 2$, $f(z)$ is analytic. Make a suitable sketch to verify this.

In partial fractions $f(z) = \dfrac{-1/2}{z - 1} + \dfrac{1/6}{z + 1} + \dfrac{1/3}{z - 2}$. We require an expansion in

powers of $(z - 1)$ which we obtain via a binomial expansion, not via (10.34). We therefore 'force' $(z - 1)$ into each denominator thus

$$f(z) = \frac{-1/2}{z - 1} + \frac{1/6}{(z - 1) + 2} + \frac{1/3}{(z - 1) - 1}$$

$$= \frac{-1/2}{(z - 1)} + \frac{1/6}{2(1 + [z - 1]/2)} + \frac{1/3}{(z - 1)[1 - 1/(z - 1)]}$$

By the Binomial expansion, noting that the second fraction requires $\dfrac{|z - 1|}{2} < 1$

and the third fraction that $\dfrac{1}{|z - 1|} < 1$, i.e. $1 < |z - 1| < 2$,

$$f(z) = \frac{1}{12} - \frac{(z - 1)}{24} + \frac{(z - 1)^2}{48} - \dots - \frac{1/6}{(z - 1)} + \frac{1/3}{(z - 1)^2} + \dots \qquad (10.35)$$

Suppose we now expand in the annulus $0 < |z - 1| < 1$, then the second partial fraction becomes $\dfrac{1/6}{(z-1)+2}$ and the third $\dfrac{-1/3}{1-(z-1)}$ and

$$f(z) = -\frac{1}{4} - \frac{3}{8}(z-1) - \frac{5}{16}(z-1)^2 - \dots + \frac{-1/2}{(z-1)} \qquad (10.36)$$

Finally, suppose we now expand in the annulus $|z-1| > 2$ so that the second

partial fraction is written $\dfrac{1/6}{(z-1)[1+2/(z-1)]}$ and the third $\dfrac{1/3}{(z-1)[1-1/(z-1)]}$

then $\qquad\qquad f(z) = \dfrac{1}{(z-1)^3} - \dfrac{1}{(z-1)^4} + \dfrac{3}{(z-1)^5} - \dots \qquad (10.37)$

Classification of singularities

For the Laurent expansion (10.34) in the annulus $0 < |z - z_0| < \rho$, if the expansion in negative powers of $(z - z_0)$ – the **principal part** of the series – contains an infinite number of terms, then $f(z)$ is said to possess an **essential singularity** at z_0. If the principal part of the series stops short at $a_{-m}(z - z_0)^{-m}$ then $f(z)$ is said to have a **pole of order** m at z_0. A pole of order 1 is called a **simple pole**. Thus the function of the last example has a simple pole at $z = 1$ (as it has at $z = -1$ and at $z = 2$).

Note that $\dfrac{1}{(z^2+1)\,z^2} = \dfrac{1}{(z-i)(z+i)\,z^2}$ has simple poles at $z = i$ and $z = -i$

and a pole of order two at $z = 0$. Also, $\exp\{1/(z-2)\}$ has an essential singularity at $z = 2$.

Consider $f(z) = \dfrac{\sin z}{z} = 1 - \dfrac{z^2}{3!} + \dfrac{z^4}{5!} - \dots$

There are no negative powers of z in the expansion, but the function is not defined at $z = 0$. However, since $\lim\limits_{z \to 0} f(z) = 1$ we *may* define $f(0) = 1$ and overcome any trouble. We say that $z = 0$ is a **removable singularity**.

10.8   The Residue Theorem

Let $f(z)$ be a function which is analytic inside and on a contour C except for a pole of order m at $z = z_0$ which lies within C.

Then, expanding $f(z)$ as a Laurent series in powers of $(z - z_0)$, we evaluate

$\displaystyle\oint_C f(z)\,\mathrm{d}z$ by equating it to $\displaystyle\oint_\Gamma f(z)\,\mathrm{d}z$ where Γ is a circle centre z_0, radius a

lying within C. Then it can be shown that the positive powers of the expansion, being analytic, integrate to zero, all the negative powers of the expansion integrate to zero by (10.28) except in the case of $a_{-1}/(z - z_0)$ which gives the result $2\pi i a_{-1}$, again by (10.28). Since a_{-1} is the only coefficient remaining it is called the **residue** of $f(z)$ at z_0. Hence

$$\oint_C f(z)\ dz = 2\pi i a_{-1} \tag{10.38}$$

Finding the residue

Under the conditions above the evaluation of the integral reduces to finding a_{-1}. Obviously we *could* perform a Laurent expansion but we look for simpler techniques. Suppose $f(z)$ has a simple pole at z_0 then

$$f(z) = \frac{a_{-1}}{(z - z_0)} + a_0 + a_1(z - z_0) + a_2(z - z_0)^2 + \ldots$$

$$(z - z_0)\ f(z) = a_{-1} + a_0(z - z_0) + a_1(z - z_0)^2 + a_2(z - z_0)^3 + \ldots$$

Taking limits as $z \to z_0$

$$\lim_{z \to z_0}\ \{(z - z_0)\ f(z)\} = a_{-1} \tag{10.39}$$

In general, for a pole of order m at z_0,

$$a_{-1} = \frac{1}{(m - 1)!}\ \lim_{z \to z_0}\ \left\{ \frac{d^{m-1}}{dz^{m-1}}\ [(z - z_0)^m\ f(z)] \right\} \tag{10.40}$$

Example 1

Let $f(z) = \dfrac{z - 3}{(z + 2)^3\ (z + 4)}$ and let us find the residue at $z = -4$. Then

$(z + 4)f(z) = (z - 3)/(z + 2)^3$ and, taking the limit as $z \to -4$, we find that $a_{-1} = (-4 - 3)/(-4 + 2)^3 = 7/8$.

Example 2

Let $f(z) = (z^2 + 1)/(z + 1)^3$. Then $f(z)$ has a pole of order 3 at $z = -1$. Hence the residue at $z = -1$ is

$$\frac{1}{2!}\ \lim_{z \to -1}\ \left\{ \frac{d^2}{dz^2}\ [(z + 1)^3\ f(z)] \right\} = \frac{1}{2}\ \lim_{z \to -1}\ \left\{ \frac{d^2}{dz^2}\ (z^2 + 1) \right\} = 1$$

If we seek $\displaystyle\oint_C f(z)\ dz$ where C is $0 < |z + 1| < 2$, the result is $2\pi i.1 = 2\pi i$.

Use of L'Hôpital's Rule

Suppose we wish to find the residue of $f(z) = \dfrac{1}{(z^4 + 1)(z^2 + 4)}$ at $z = \dfrac{1 + i}{\sqrt{2}}$

We require $\lim\limits_{z \to \frac{1+i}{\sqrt{2}}} \left\{ \dfrac{(z - (1 + i)/\sqrt{2})}{(z^4 + 1)(z^2 + 4)} \right\}$

However it is difficult to perform the cancellation needed. If we apply L'Hôpital's rule, the required limit is equal to

$$\lim_{z \to \frac{1+i}{\sqrt{2}}} \left\{ \frac{1}{4z^3(z^2 + 4) + (z^4 + 1)2z} \right\} = \frac{1}{4\left[\dfrac{1+i}{\sqrt{2}}\right]^3 (i + 4)} = \frac{-(5 + 3i)}{68\sqrt{2}}$$

Extension to the Theorem

We seek $\oint\limits_C f(z)\, dz$ where the only singularities of $f(z)$ on and inside C are poles at $z = z_1$, $z = z_2$, ..., $z = z_r$. Then the required answer is $2\pi i \times$ (sum of the residues at the poles inside C).

Example

Find $\oint\limits_C f(z)\, dz$ where C is the unit circle and

$$f(z) = (z^2 + 1)/\{(z - 2)(2z + 1)^2(2z - 1)\}.$$

The poles inside C are at $z = \frac{1}{2}$ and $z = -\frac{1}{2}$. At $z = \frac{1}{2}$, a simple pole, the residue is

$$\lim_{z \to 1/2} \left\{ \frac{(z^2 + 1)}{2(z - 2)(2z + 1)^2} \right\} = \frac{5/4}{2(-3/2) \cdot 4} = -\frac{5}{48}$$

At $z = -\frac{1}{2}$, a pole of order 2, the residue is

$$\lim_{z \to -(1/2)} \left\{ \frac{(z - 2)(2z - 1) \cdot 2z - (z^2 + 1) \cdot (4z - 5)}{4(z - 2)^2 (2z - 1)^2} \right\} = \frac{3}{80}$$

Hence the value of the integral is $2\pi i \left(\dfrac{-5}{48} + \dfrac{3}{80} \right) = -2\pi i/15.$

Expansion method for finding residues

Consider $f(z) = (z^8 + 1)/\{z^3(z^2 + 2)(2z^2 + 1)\}$. This has a pole of order 3 at $z = 0$. To find the residue there it is easiest to expand $f(z)$ in powers of z:

$$f(z) = \frac{z^8 + 1}{2z^3(1 + \frac{1}{2}z^2)(1 + 2z^2)} = \frac{1}{2z^3}(1 + z^8)(1 - \frac{1}{2}z^2 + ...)(1 - 2z^2 + ...)$$

The coefficient of z^2 in the last three factors is $-5/2$ and so the residue is $-5/4$.

10.9 Evaluation of Real Integrals

Several real integrals can be evaluated using contour integration, when other methods fail. We merely list here a few common types of integrals.

Type 1
$$\int_0^{2\pi} f(\cos \theta, \sin \theta)d\theta$$

We let $e^{i\theta} = z$ so that $d\theta = \frac{1}{iz}dz$, $\cos \theta = \frac{1}{2}(z + \frac{1}{z})$, $\sin \theta = \frac{1}{2i}(z - \frac{1}{z})$. Then

$$\int_0^{2\pi} f(\cos \theta, \sin \theta)d\theta \equiv \oint_C g(z)dz$$

where C is the contour $|z| = 1$ and $g(z)$ is the expression obtained by substituting for $d\theta$, $\cos \theta$, and $\sin \theta$ in terms of z. The latter integral is $2\pi i$ times the sum of the residues of $g(z)$ at its poles *inside* $|z| = 1$.

Example
$$I = \int_0^{2\pi} \frac{d\theta}{3 + 2 \cos \theta}$$

Substituting for $d\theta$ and $\cos \theta$ and expressing the limits of integration in terms of z we obtain

$$I = \frac{1}{i} \oint \frac{dz}{z^2 + 3z + 1}$$

The integrand has simple poles at $z = \frac{-3}{2} \pm \sqrt{\frac{9}{4} - 1}$ i.e. at $z = \frac{1}{2}(-3 + \sqrt{5})$ and

$z = \frac{1}{2}(-3 - \sqrt{5})$. Only the first lies inside the contour $|z| = 1$. The residue there is found to be $\dfrac{1}{\sqrt{5}}$ and therefore $\quad I = \dfrac{1}{i} \times 2\pi i \times \dfrac{1}{\sqrt{5}} = \dfrac{2\pi}{\sqrt{5}}$

Type 2 $\quad\displaystyle\int_{-\infty}^{\infty} f(x)\,dx$

where $f(x)$ is a real rational function with a denominator of degree at least 2 greater than the numerator and non-zero for all real x. We write

$$\int_{-\infty}^{\infty} f(x)\,dx = \lim_{R\to\infty} \int_{-R}^{+R} f(x)\,dx$$

and convert this latter integral into a *contour* integral as shown in Figure 10.21.

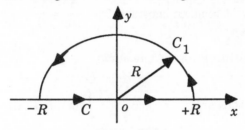

Figure 10.21

The contour C is from $-R$ to $+R$ along the real axis and along the semi-circle C_1. For large enough R the contour will enclose *all* the poles of $f(z)$ in the upper half-plane. Hence $\displaystyle\oint_C f(z)\,dz = 2\pi i \,(\Sigma$ residues in upper half-plane.)

But $\quad\displaystyle\oint_C f(z)\,dz = \int_{-R}^{+R} f(x)\,dx + \int_{C_1} f(z)\,dz$

and it is possible to show that for the functions satisfying the conditions stated above the integral over the semi-circle tends to zero as the radius $R \to \infty$. We have in the limit as $R \to \infty$ that

$$\int_{-\infty}^{\infty} f(x)\,dx = 2\pi i \ (\text{sum of residues of } f(z) \text{ in the upper half-plane})$$

Example

Evaluate $\displaystyle\int_{-\infty}^{\infty} \frac{dx}{(1+x^2)^3}$. Note the integrand satisfies the required conditions.

Now $\displaystyle\frac{1}{(1+z^2)^3} = \frac{1}{(z+i)^3(z-i)^3}$ has *third* order poles at $z = +i$ and $z = -i$.

The pole at $z = +i$ is in the upper half-plane so we evaluate the residue there. By the formula for a residue at a third order pole the residue at $z = +i$ is

$$\frac{1}{2!}\left[\frac{d^2}{dz^2}\left(\frac{1}{(z+i)^3}\right)\right]_{z=i}$$

This gives $\dfrac{3}{16i}$ so that the required integral is $2\pi i \cdot \dfrac{3}{16i} = \dfrac{3\pi}{8}$

Type 3   **Real integrals of the form**

$$\int_{-\infty}^{\infty} f(x)\cos kx\, dx \qquad \text{and} \qquad \int_{-\infty}^{\infty} f(x)\sin kx\, dx$$

where $f(x)$ satisfies conditions similar to those of $f(x)$ in integrals of Type 2.

Take the contour integral $\displaystyle\oint_C f(z)\, e^{ikz}\, dz$ over the same contour as in Type 2 integrals. The real and imaginary parts of the results give the required integrals.

Type 4   **A pole on the real axis.**
   Since the contour of Type 2 would pass through the pole we bend it to by-pass the pole. The contour of Type 2 must be modified by an indentation, formed by a small semi-circle around the pole, which is contracted to the pole. Since we only go 'half-way round the pole' we pick up only πi times the residue there.

Example $\displaystyle I = \int_{-\infty}^{\infty} \frac{(1-\cos 2x)}{x^2}\, dx$

We integrate $f(z) = (1 - e^{2iz})/z^2$ around the contour indicated in Figure 10.22 since there is a double pole at $z = 0$. We let $R \to \infty$ and $r \to 0$.

First note that $\displaystyle\oint_C \to 0$ since it encloses no poles. Also we find $\displaystyle\int_\Gamma \to 0$ by appealing to **Jordan's lemma** which states that if $\phi(z)$ is a rational function,

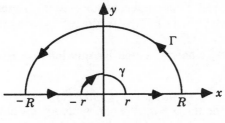

Figure 10.22

i.e. a ratio of two polynomials and if $\psi(z)$ is of the form $e^{imz} g(z)$ where $m > 0$ and $g(z)$ is a rational function then

(i)   if $|z \, \phi(z)| \to 0$ as $R \to \infty$ then $\displaystyle\int_{\Gamma} \phi(z) \, dz \to 0$

(ii)   if $|g(z)| \to 0$ as $R \to \infty$ then $\displaystyle\int_{\Gamma} \psi(z) \, dz \to 0$

Now $\displaystyle\int_{\gamma} \to \pi i$ (residue of $f(z)$ at $z = 0$) $= \pi i \times (-2i) = 2\pi$ and $\displaystyle\int_{\gamma}$ is taken in the

clockwise direction. Also $\displaystyle\int_{-R}^{-r} + \int_{r}^{R} \to I$. Hence in the limit, $0 = \displaystyle\oint_{C} = \int_{-\infty}^{\infty} + 0 - 2\pi.$

Therefore $I = 2\pi$.

10.10   Further Applications of Contour Integration

In this section, we look at two applications of contour integration: inversion of Laplace Transforms and the Nyquist stability criterion for closed loop systems.

Inversion of Laplace Transforms
The Laplace transform of a function $f(t)$, given by

$$F(s) = \int_{0}^{\infty} f(t) \, e^{-st} \, dt \qquad (10.41)$$

has an inverse, viz. $\displaystyle f(t) = \frac{1}{2\pi i} \int_{a - i\infty}^{a + i\infty} F(s) \, e^{st} \, ds \qquad (10.42)$

where the integral is taken along a line parallel to the imaginary axis; s being regarded as complex. The real number a is chosen so that all singularities of $F(s)$ lie to the left of the line of integration parallel to the imaginary axis.

In practice, the integral in (10.42) is evaluated by considering the contour

integral $\oint_C F(s) \cdot e^{st} \, ds$, where C is the **Bromwich contour** in the s-plane

shown in Figure 10.23. Suppose that the only singularities of $F(s)$ are poles which lie to the left of the line $x = a$. We can apply the residue theorem to show that, if the integral around the circular arc tends to zero as $R \to \infty$, then

$$f(t) = \text{the sum of the residues of } e^{st} F(s) \text{ at the poles of } F(s) \qquad (10.43)$$

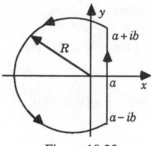

Figure 10.23

Example

Let $q(t)$ represent the charge in a circuit used for charging a capacitor. With the usual notation, the Laplace transform of the charge, viz. $Q(s)$ is given by

$$Q(s) = V_0 \bigg/ \left\{ Rs \left[\frac{1}{RC} + s \right] \right\} \qquad \text{By the inversion formula (10.42) we find}$$

that
$$q(t) = \frac{1}{2\pi i} \int_{a - i\infty}^{a + i\infty} \frac{V_0 \, e^{st}}{Rs(s + 1/RC)} \, ds$$

Here a can have any positive real value (why?). We need the residues at $s = 0$ and $s = -1/RC$. At $s = 0$ we can show that the residue is $V_0 C$. At $s = -1/RC$ we can show that the residue is $-V_0 C e^{-t/RC}$. Hence, by (10.43), $q(t) = V_0 C(1 - e^{-t/RC})$ as is well known.

Note that if $F(s)$ has branch points then the Bromwich contour must be suitably cut. This is beyond our scope in this book.

Stability criteria

In the analysis of physical systems we often want to know merely whether a system is stable, i.e. whether a small transient imposed disturbance produces a response which also dies away. If the Laplace transform of the response is of the form $G(s) = P_1(s)/P_2(s)$ where P_1 and P_2 are polynomials, then by using the residue theorem, we are enabled to state that:

If all the poles of $G(s)$ lie in the left-hand half-plane then the system is stable.

(We have tacitly assumed that the integral of the appropriate function over a semi-circle tends to zero as the radius of the semi-circle tends to infinity.)

Note that the poles of $G(s)$ are the zeros of $P_2(s)$ which may need to be found by numerical methods.

Nyquist stability criterion

First we quote a result from complex variable theory: Let $f(z)$ be analytic inside and on a closed curve C, apart from a finite number, N_p, of poles within C. Further let $f(z)$ have N_z zeros inside C. Let C be mapped into a curve C' by $w = f(z)$. Then the number of times C' encircles the origin is equal to $N_z - N_p$.

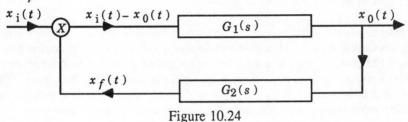

Figure 10.24

A common engineering problem is to make the output $x_0(t)$ of a system follow quickly and closely the input $x_i(t)$. A feedback loop, such as the one shown in Figure 10.24, samples the output, modifies it to $x_f(t)$ and feeds it back to a differential device which produces an error signal $x_i(t) - x_f(t)$. The **transfer function** for a system is the ratio of the Laplace transform of the output to that of the input. Let $G_1(s)$ be the transfer function of the original system and $G_2(s)$ be the transfer function of the feedback loop. Then

$$L\{x_0(t)\} = G_1(s)\,[L\{x_i(t)\} - L\{x_f(t)\}]$$
$$L\{x_f(t)\} = G_2(s)\,L\{x_0(t)\}$$

so that on elimination of $L\{x_f(t)\}$ we obtain

$$L\{x_0(t)\} = \frac{G_1(s)}{[1 + G_1(s)\,.\,G_2(s)]}L\{x_i(t)\}$$

The stability of the feedback system is important. If the system without the feedback loop is stable for $x_i(t)$ then the product $G_1(s)L\{x_i(t)\}$ has no poles in the right-hand half of the s-plane. The stability of the system as a whole depends on the location of the zeros of $[1 + G_1(s)\,.\,G_2(s)]$.

The **Nyquist** criterion, which we state without proof, is that if the locus of $w = G_1(i\omega)\,.\,G_2(i\omega)$ is plotted and does not encircle the point $w = -1$ then the system is stable, otherwise the system is unstable.

Problems

Section 10.2

1   Find the real and imaginary parts $u(x, y)$ and $v(x, y)$ of the functions
(a) z^3 (b) $|z|$ (c) $R(z)$ (d) e^z (e) $1/z$ (f) $z/(z^2 + 1)$ (g) $(2z + 3)/(z + 2)$

2   Plot the z-values given and the corresponding w-values under the function stated. Try to generalise how the z-plane is transformed by the function.
(a) $z = 0, 1 + i, -1, -3 + 2i, -i$; $w = z + 1$
(b) $z = 0, 1, 2, 3, 1 + i, 2 + 2i, i, 2i, 3i$; $w = iz$
(c) $z = i, 1, -i, -1, 1 + i, -1 - i, 2, -4i$; $w = i\bar{z}$

3   Consider $w = z^3$. First express the function in the form $w = u + iv$ where u and v are functions of r and θ. Then show that the region $0 < \arg z < \pi/3$ is mapped onto the upper half-plane $0 < \arg w < \pi$. What region in the z-plane will be mapped into the entire w-plane? Suppose every point of the z-plane is mapped onto the w-plane. Show that apart from $w = 0$, every point on the w-plane is the image under the mapping of three distinct points in the z-plane. Relate the arguments and moduli of these three points.

4   Evaluate the following limits

(a) $\lim\limits_{z \to 1+i} \left\{ \dfrac{z^2 - z + 1 - i}{z^2 - 2z + 2} \right\}$

(b) $\lim\limits_{z \to 2+i} \left\{ \dfrac{1 - z}{1 + z} \right\}$

(c) $\lim\limits_{z \to 2+i} \left\{ \dfrac{z^2 - 2iz}{z^2 + 4} \right\}$

(d) $\lim\limits_{z \to e^{i\pi/4}} \left\{ \dfrac{z^2}{z^4 + z + 1} \right\}$

(e) $\lim\limits_{z \to 1+i} \left\{ \dfrac{z - 1 - i}{z^2 - 2z + 2} \right\}^2$

5   Evaluate $\lim\limits_{z \to 0} \left\{ \dfrac{2xy}{x^2 + y^2} - \dfrac{y^2}{x^2} i \right\}$ along the following paths

(a) the line $y = x$ (b) the line $y = 2x$ (c) the parabola $y = x^2$
What conclusions do you draw?

6   Show that as $z \to 0$ the limit of each of the following functions does not exist

(a) $\dfrac{xy}{x^4 + y^4} + 2xi$

(b) $\dfrac{x^3 y^2}{x^6 + y^4} - \dfrac{x}{y} i$

7   Let $f(z) = (z^2 + 4)/(z - 2i)$ for $z \neq 2i$. Define $f(2i) = 3 + 4i$.
Prove that $\lim\limits_{z \to i} f(z)$ exists and state its value.

8   Where are the following functions discontinuous?

(a) $\dfrac{2z-3}{z^2 + 2z + 2}$ (b) $\dfrac{3z^2 + 4}{z^4 - 16}$ (c) $\dfrac{z^2 + 1}{z^3 + 9}$ for $|z| \le 2$

9 From the definition, find the derivatives of (a) $z^2 + 3z$ (b) z^{-1}
Evaluate these derivatives at $z = -1 + i$.

10 For the following functions find the singular points. Determine the derivatives at other points.
(a) $z/(z + i)$ (b) $(3z - 2)/(z^2 + 2z + 5)$

11 Show that the real and imaginary parts of the following functions satisfy the Cauchy-Riemann equations and state whether the functions are analytic.
(a) $z^2 + 5iz + 3 - i$ (b) $x^2 - iy^2$ (c) $\cos y - i \sin y$
(d) $e^x(\cos y + i \sin y)$

12 The polar form of the Cauchy-Riemann equations is

$$r\frac{\partial u}{\partial r} = \frac{\partial v}{\partial \theta}, \quad r\frac{\partial v}{\partial r} = -\frac{\partial u}{\partial \theta} \quad \text{where } f(z) = u(r, \theta) + iv(r, \theta)$$

Show that these conditions are satisfied by the functions
(a) $\ln r + i\theta$ (b) z^n (c) $|z|$ (d) $|z|^2$

13 Show that the real and imaginary components of each of the following functions are harmonic.
(a) $z^2 + z$ (b) z^4 (c) $1/z$ (d) $e^x(\cos y + i \sin y)$
(e) $z^4 + 5iz + 3 - i$ (f) ze^z (g) $\sin 3z$

14 Show that each of the following functions is harmonic. Then find its conjugate harmonic v and form an analytic function $f(z) = u + iv$.
(a) $u = x$ (b) $u = e^x \cos y$ (c) $u = \ln(x^2 + y^2)$
(d) $xe^x \cos y - ye^x \sin y$ (e) $u = x^2 - y^2 - 2xy - 2x + 3y$
(f) $u = 3x^2y + 2x^2 - y^3 - 2y^2$

15 State the Cauchy-Riemann equations for the function $w = f(z) = u(x, y) + jv(x, y)$, where $z = x + jy$. Show that $u = 4xy - x^3 + 3xy^2$ is harmonic and find the conjugate function v. Hence express $u + jv$ as an analytic function of z. (EC)

16 Determine which of the following functions $u(x, y)$ are harmonic. Find the conjugate harmonic function $v(x, y)$, if it exists, and express $u + iv$ as an analytic function of $z = x + iy$.
(a) $u = 2xy + 3xy^2 - 2y^3$ (b) $u = 2x(1 - y)$ (EC)

Section 10.3
17 Using (10.7), prove properties (i) to (vi) for the exponential function.

18 Prove that

  (i) $e^{z_1/z_2} = e^{z_1 - z_2}$ (ii) $|e^{iz}| = e^{-y}$ (iii) $e^z \neq 0$ for any finite z

19 Find all values z for which

  (i) $e^{3z} = 1$ (ii) $e^{4z} = i$ (iii) $e^z = 1 - i$

20 Find the derivatives of $\sin z$, $\cos z$.

21 Prove that

  (i) $\sin 2z = 2 \sin z \cos z$ (ii) $\cos 2z = \cos^2 z - \sin^2 z$ (iii) $\sin(-z) = -\sin z$
  (iv) $\cos(-z) = \cos z$

22 Split into real and imaginary parts the following: (i) $\cos(-i)$ (ii) $\sin(1 + i)$

23 Prove that   (i) $1 + \tan^2 z = \sec^2 z$ (ii) $\overline{\sin z} = \sin \bar{z}$

24 Find all the values of   (i) $\sin^{-1} 2$ (ii) $\cos^{-1} i$

25 Prove properties (i) to (iii) for hyperbolic functions.

26 Prove that   (i) $\cos (iz) = \cosh z$ (ii) $\sin(iz) = i \sinh z$

27 Find the value of   (i) $2 \sinh(\pi i/3)$ (ii) $\coth(3\pi i/4)$

28 Find the values of $\cosh^{-1} i$.

29 Find the logarithm of   (i) $-i$ (ii) $3 + 4i$ (iii) $2 - i$

30 Find the values of z for which $\ln(z + 1) = \pi i$.

31 Show that $\ln(z - 1) = \frac{1}{2}\ln\{(x - 1)^2 + y^2\} + i \tan^{-1}\{y/(x - 1)\}$.

32 Prove that   (i) $\ln(z_1 z_2) = \ln (z_1) + \ln (z_2)$ (ii) $\ln e^z = z$

33 Find the values of   (i) $(1 + i)^i$ (ii) $(2 - i)^{i+1}$ (iii) 2^{2i}

34 If $w = (z^2 + 1)^{1/2}$ express $\arg w$ in terms of $\arg(z - i)$ and $\arg(z + i)$. Let C be a closed curve around the point i which does not include $-i$. Then let z go once round C anticlockwise and consider the changes in $\arg(z - i)$ and $\arg(z + i)$ to show that the change in $\arg w \neq 0$ and hence that $z = i$ is a branch point for $w = f(z)$. Similarly show that $z = -i$ is also a branch point.

Section 10.4
35 Sketch the following regions in the z-plane

  (i) $|z + 2 - i| \leq 4$ (ii) $1 \leq |z - 2i| \leq 2$ (iii) $|\arg z| < \pi/2$
  (iv) $|z - i| < 2|z + i|$ (v) $R(z^2) < 2$ (vi) $|z + 3| + |z - 3| \leq 10$

36 Find the family of curves orthogonal to
 (i) $x^3y - xy^3 =$ constant (ii) $e^{-x}\cos y + xy =$ constant
 (iii) $r^2\cos 2\theta =$ constant

37 Which of the following functions ϕ could be a possible velocity potential for a fluid flow?
 For those which are, sketch the streamlines and describe the flow.
 (i) $A(x^2 - y^2)$ (ii) $\sin(x + y)$ (iii) $Ux + Vy$

38 For the following complex potentials plot the lines $u =$ constant and $v =$ constant.
 (i) $w = z^2 + 2z$ (ii) $w = z^4$ (iii) $w = A\sin z$

39 Interpret the complex potentials w_1, w_2 and $w_1 + w_2$ in terms of electrostatics, where
 $w_1 = C\ln(z - x_1)$, $w_2 = -C\ln(z - x_2)$ for C a constant.

40 Sketch the equipotential lines of (i) $w = z^2$ (ii) $w = 1/z$

41 Interpret the potential $w = -(iK/2\pi)\ln z$ in terms of fluid flow. Repeat for
 $w = (K/2\pi)\ln z$. K is a positive real constant.

42 Show that $w = \cosh^{-1} z$ may correspond to a flow through an aperture. Sketch the
 streamlines for the flow.

43 Find the images of the curves $\arg z = \pi/6$, $|z| = 2$, $\mathcal{R}(z) = -2$, $\mathcal{J}(z) = 1$ under each of
 the following transformations.
 (i) $w = iz$ (ii) $w = -iz + 2i$ (iii) $w = 2i(z + 1 - i)$

44 Find the image of the parabola $y = x^2$ under the mapping $w = -2z + 2i$.

45 Find the image of the curves (i) to (iv) under the mapping $w = z^2$
 (i) $y = 1 - x$ (ii) $|z| > 2$ (iii) $1 < R(z) < 2$ (iv) $1 < |z| < 2$ and $|\arg z| < \pi/4$

46 Find the image under $w = 1/z$ of the following
 (i) $y = x - 1$ (ii) $x^2 + (y - 1)^2 = 1$ (iii) $x = 0$
 (iv) $(x + 1)^2 + (y - 2)^2 = 4$ (v) $|z + 1| = 1$

47 What points are not altered by $w = 1/z$?

48 Verify that the angle of intersection at each of the points where $|z + 1 + 2i| = 2$ and
 $y = x + 1$ intersect is unaltered by $w = 1/z$.

49 Discuss the mapping $w = 1/(z + 1)$.

50 Find a bilinear transformation which maps the points $z = 0, -i, -1$ into the points
 $w = i, 1, 0$ respectively and find the image under this transformation of the region in the
 z plane represented by $|z| \leq 1$. (EC)

51 Repeat 50 for the z-values and corresponding w-values shown

(i)   0, 1, 2;   4, 10, 16   (ii)   0, 1, ∞;   ∞, 1, 0   (iii)   1, i, –1;   i, –1, –i

(iv) z_1, z_2, z_3;   0, 1, ∞

52   (i)   Show that $w = \dfrac{z - i}{z + i}$ maps the upper half z-plane onto the interior of the unit circle

and indicate the other main features of this transformation.

(ii)   Show that $\phi = \dfrac{1}{\pi} \tan^{-1}\left(\dfrac{y}{x}\right)$, where $0 \le \phi \le 1$, is harmonic in the upper half-plane and

has values 0 and 1 respectively on the positive and negative axes.   (EC)

53   If $z = x + iy$, sketch the region R_1 in the z-plane defined by $|z| \le 1, y \ge 0$

Under the transformation $w = \left[\dfrac{1 + z}{1 - z}\right]^2$

find the image in the w-plane of the points $(0,0)$, $(1,0)$, $(0,1)$ and $(-1,0)$ in the z-plane, and determine the region R_2 in the w-plane which corresponds to the region R_1.   (EC)

54   (a)   If $w = (j - z)/(j + z)$, where $w = u + jv$ and $z = x + jy$, show that the interior of the circle $|z| = 1$ transforms into that half of the w-plane for which $u > 0$.

(b)   If $w = \cosh(\pi z/a)$, a real and positive, find the region in the w-plane which corresponds to the rectangle bounded by the lines $x = 0$, $y = 0$, $x = N > 0$, $y = a$.   (LU)

55   A bilinear transformation is such that the points $z_1 = 0$, $z_2 = 1$, $z_3 = \infty$ in the z-plane are mapped on the w-plane as $w_1 = 1$, $w_2 = e^{j\pi/4}$, $w_3 = j$. Obtain this transformation and find the curve in the w-plane corresponding to the line $y = 0$. Into what area in the w-plane is the half plane $y > 0$ mapped?   (LU)

56   If $w = \dfrac{z - 1}{z + 1}$ where $w = u + iv$ and $z = x + iy$, express u and v in terms of x and

y.   Show that the straight line $u = k$ (where k is a real constant) in the w-plane transforms into a circle through the point -1 in the z-plane. Sketch the circles given by $k = 0$ and $k = 1/2$.

57   Find the image under $w = e^z$ of the following

(i) $x = -2$, $-\pi/2 < y < \pi$ (ii) $x = 1$, $0 \le y \le \pi$

58   Find the image under $w = \ln z$ of the following   (i) $|z| = 9$ (ii) $\arg z = \pi/4$

59   Find the image under $w = \sin z$ of

(i) $y \ge 0$, $-\pi/2 < x < \pi/2$ (ii) $0 < x < \pi/2$, $y \ge 0$

60   If $w = \sin(\pi z/a)$, where a is real and positive and $z = x + iy$, find the region in the w-plane which corresponds to the interior of the square bounded by the lines $x = a/2$, $x = -a/2$, $y = 0$ and $y = a$ in the z-plane.   (EC)

Section 10.5

61 Write down in complex exponential form the equation of the circle $|z| = r$. Hence determine how the circle $|z| = r$, where $r > 1$, maps from the z-plane on to the w-plane under the transformation $w = z + z^{-1}$.   (EC)

62 Show that w describes an ellipse as z describes the circle $|z| = d$, where $d > a$, under the

transformation $w = \dfrac{1}{2}\left[z + \dfrac{a^2}{z} \right]$.   (EC)

63 If $w = f(z)$, where $w = u + jv$ and $z = x + jy$, obtain the Cauchy-Riemann equations connecting the derivatives of u and v with respect to x and y, and deduce that u satisfies Laplace's equation $\partial^2 u/\partial x^2 + \partial^2 u/\partial y^2 = 0$.

In the transformation to parabolic coordinates defined by $z = \zeta^2$, where $\zeta = \xi + j\eta$, show that the Laplacian $(\partial^2\phi/\partial x^2 + \partial^2\phi/\partial y^2)$ becomes

$$\frac{1}{4\rho^2}\left(\frac{\partial^2\phi}{\partial\xi^2} + \frac{\partial^2\phi}{\partial\eta^2} \right)$$

where $\rho^2 = \xi^2 + \eta^2$. Sketch the coordinate lines ξ = constant, η = constant.   (LU)

64 (a) Show, that if a and b are real, the transformation $\dfrac{z-a}{z+a} = j\, e^{jw\pi/b}$, where

$w = u + jv$ and $z = x + jy$, transforms the infinite strip bounded by the lines $u = 0$, $u = b$ into the interior of the circle $|z| = a$.

(b) If $w = \coth(c/z)$ where c is real show that $v = 0$ when $y = 0$ and when

$\left| z - \dfrac{jc}{n\pi} \right| = \dfrac{c}{n\pi}$ where n is an integer.   (LU)

Section 10.6

65 Evaluate the following integrals along the contours shown

(a) $\displaystyle\int_{(0,1)}^{(2,5)} \{(3x + y)\mathrm{d}x + (2y - x)\mathrm{d}y\}$ along the straight line joining end points.

(b) $\displaystyle\oint_C |z|^2\, \mathrm{d}z$ around the square with vertices $(0,0)$, $(1,0)$, $(1,1)$, $(0,1)$.

(c) $\displaystyle\int_C (z^2 + 3z)\mathrm{d}z$ along the circle $|z| = 2$ from $(2,0)$ to $(0,2)$ anti-clockwise.

(d) $\displaystyle\oint_C \frac{\mathrm{d}z}{z - 2}$ (i) around $|z - 2| = 4$, (ii) around $|z - 1| = 5$, (iii) around the square with vertices $\pm 3 \pm 3i$.

66 Verify Cauchy's theorem for the functions below with C being the square with vertices $\pm 1 \pm i$ (a) $3z^2 - iz - 4$ (b) $10 \sin 2z$

67 Evaluate $\int_C f(z) \, dz$ for functions $f(z)$ and paths C specified below.

(a) $f(z) = z^3 - 1$; $|z - 1| = 1$ (b) $f(z) = z/(z^2 - 1)$; $|z - \pi i| = 1$

(c) $f(z) = z^2 + z + 1$; from $z = -i$ to $z = 2i$ (d) $f(z) = 1/(z^2 - 2z)$; $|z| = 1$

(e) $f(z) = 2i/(z^2 + 1)$; $|z - 1| = 6$ (f) $\sin z/(2z - \pi)$; $|z| = 1$

(g) $\sin z/(2z - \pi)$; $|z| = 2$

68 Repeat Problem 67 for the following

(a) $f(z) = e^z/(z - 2)$; $|z| = 3$ (b) $f(z) = \sin 3z/(z + \frac{1}{2}\pi)$; $|z| = 5$

(c) $f(z) = e^{3z}/(z - \pi i)$; $|z - 1| = 4$ (d) $\sin^6 z/(z - \pi/6)$; $|z| = 1$

(e) $f(z) = 1/\{(z + i)z^4\}$; $|z - i| = 1.5$ (f) $f(z) = 3/\{z^2(z + i)^2\}$; $|z| = 5$

(g) $f(z) = 1/(z^4 - 1)$; $|z - 1| = 5$

69 Using (10.32) evaluate $\displaystyle\oint_C \frac{e^{2z}}{(z + 1)^4} \, dz$ where C is $|z| = 3$.

Hint: take $z_0 = -1$ and $n = 3$.

70 (i) Evaluate $\displaystyle\oint_C \frac{z + 4}{z^2 + 2z + 5} \, dz$

(a) if C is the circle $|z| = 1$ (b) if C is the circle $|z + 1 - j| = 2$

(ii) By putting $z = e^{j\theta}$ and integrating around the unit circle, show that

$$\int_0^{2\pi} \frac{d\theta}{25 - 16 \cos^2 \theta} = \frac{2\pi}{15} \tag{EC}$$

71 State Cauchy's integral formula for an analytic function and use it to integrate the function $(z^2 + 1)/(z^2 - 1)$ along a circle of unit radius with centre at (a) $z = 1$, (b) $z = -1$.
Use the formula to show that if an analytic function takes known values $u + iv$ on the circle $|z| = R$, its value at an interior point z_0 is

$$u_0 + iv_0 = \frac{1}{2\pi} \int_0^{2\pi} \frac{(u + iv) z \, d\theta}{z - z_0}, \quad z = R \, e^{i\theta}$$

Prove also that, if z_0 is on the real axis at $z = r$,

$$u_0 = \frac{R}{2\pi} \int_0^{2\pi} \frac{Ru - ru \cos \theta + rv \sin \theta}{R^2 + r^2 - 2Rr \cos \theta} d\theta \tag{LU}$$

72 Evaluate $\displaystyle\int_C \frac{(z+2)dz}{4z^2+4z+5}$ where C is the circle $|z + \frac{1}{2} - \frac{1}{2}i| = 1$.   (LU)

73 State Cauchy's Integral formula for the nth derivative. Hence evaluate

(a) $\displaystyle\int_C \frac{dz}{(z-1)^3(z-3)(z-4)}$, where C is the circle $|z-1| = 1$

(b) $\displaystyle\int_C \frac{\cos^4 z}{z-\pi/6}dz$, where C is the circle $|z| = 1$.   (EC)

Section   10.7

74 Find the Taylor series for each of the following functions with centre of expansion c.
(a) e^z, $c = i$ (b) $1/(z + 2i)$, $c = 0$ (c) $1/(z^2 - 1)$, $c = 0$

75 Find the series expansion of each of the functions in the given region
(a) $1/\{(z + 2)(z - 1)\}$; $1 < |z - 2| < 4$ (b) as (a) in $4 < |z - 2|$
(c) $\cos(z - 1)/z^3$; $0 < |z| < \infty$ (d) $\sinh z/z^2$; $0 < |z| < \infty$

76 Find all the expansions centred at the origin for $f(z) = 1/\{z^2(z - 1)(z - 2)\}$.

77 Expand $\displaystyle\frac{1}{z} + \frac{1}{z-1} + \frac{1}{z-i}$ in $0 < |z| < 1$.

78 Expand (a) $1/z(z - 1)$ (b) $(z - 1)/[z^3(z - 2)]$ in all Laurent series possible about the origin.

79 Expand $1/\{(z + 1)(z + 3)\}$ in Laurent series valid for
(a) $1 < |z| < 3$ (b) $|z| > 3$ (c) $0 < |z + 1| < 2$ (d) $|z| < 1$

80 Classify the singularities of the following functions
(a) $(z^2 + 1)/z$ (b) $\cos(1/z)$ (c) $1/\{(z - 1)(z - 2)^2\}$
(d) $(\cos z - 1)/z^2$ (e) $(z^2 - 3z + 2)/(z - 2)$ (f) $(e^{2z} - 1)/z^4$
(g) $(z^2 - 1)/(z^2 + 1)$ (h) $\tan^2 z$

Section 10.8

81 Find the residue of each of the following functions at each of its singularities.
(a) $(e^{2z} - 1)/z$ (b) $(z^2 + 1)/(z - 1)$ (c) $\sinh z/z^2$
(d) $(z^2 - 1)/\{(z - 2)(z + 1)(z - \pi)\}$ (e) $\tan z/z^3$ (f) $ze^z/(z^4 - z^2)$
(g) $(z - 2)/\{z(z - 1)\}$ (h) $(z - 2)/\{z^2(z - 1)^2\}$ (i) $(1 - \cos 2z)/z^3$
(j) $(z + 2)/(z^2 + 9)$ (k) $z^2/(z^4 + 16)$ (l) $e^{iz}/(z^2 + 4)$

82 Find the integral of the following functions round the contour $|z| = 1$
(a) $z/(2z - 1)$ (b) $(2z^3 + 4)/(4z - \pi)$ (c) $(z^2 + 1)/(z^2 - 2z)$

    (d) $\cosec z$ (e) $(1 - e^z)^{-1}$

83   Find the integral of the following functions round the given contour

    (a) $(z^2 + 1)e^z / \{(z + i)(z - 1)^3\}$, $|z| = 0.5$ (b) $e^{1/z}$, $|z| = 6$

    (c) $z/(z^4 - 1)$, $|z| = 4$

Section 10.9

84   By contour integration, show that $\displaystyle\int_{-\infty}^{\infty} \frac{dx}{(x^2 + 1)(x^2 + 4)^2} = \frac{5\pi}{144}$ (EC)

85   Using suitable contours and Cauchy's Residue theorem, evaluate

$$\int_{0}^{2\pi} \frac{d\theta}{5 - 3\sin\theta} \quad \text{and} \quad \int_{-\infty}^{\infty} \frac{x^2\,dx}{(x^2 + 1)^2} \qquad \text{(EC)}$$

86   By using suitable contours show that

    (i) $\displaystyle\int_{0}^{2\pi} \frac{d\theta}{5 + 4\cos\theta} = \frac{2\pi}{3}$ (ii) $\displaystyle\int_{0}^{\infty} \frac{dx}{x^6 + 1} = \frac{\pi}{3}$ (EC)

87   Use contour integration to show that

$$\int_{-\infty}^{\infty} \frac{\cos ax}{(x^2 + b^2)(x^2 + c^2)}\,dx = \frac{\pi}{(b^2 - c^2)} \left\{ \frac{1}{c}\exp(-ac) - \frac{1}{b}\exp(-ab) \right\}$$

    Deduce the values of

$$\int_{-\infty}^{\infty} \frac{\sin ax}{(x^2 + b^2)(x^2 + c^2)}\,dx \qquad \int_{-\infty}^{\infty} \frac{1}{(x^2 + b^2)(x^2 + c^2)}\,dx \qquad \int_{-\infty}^{\infty} \frac{1}{(x^2 + b^2)^2}\,dx \qquad \text{(EC)}$$

88   Use the complex integral calculus to show

    (a) $\displaystyle\int_{0}^{2\pi} \frac{\sin^2\theta}{5 + 4\cos\theta}\,d\theta = \frac{\pi}{4}$ (b) $\displaystyle\int_{0}^{\infty} \frac{x^2\,dx}{x^4 + a^4} = \frac{\pi\sqrt{2}}{4a}$, $a > 0$

    (c) $\displaystyle\int_{0}^{2\pi} \frac{\cos 2\theta\,d\theta}{5 - 4\cos\theta} = \frac{\pi}{6}$ (d) $\displaystyle\int_{0}^{\infty} \frac{\cos x\,dx}{(1 + x^2)^2} = \frac{\pi}{2e}$ (LU)

89   (i)   Evaluate $\displaystyle\int \frac{z^2\,dz}{(z - 1)(z + 2)^2}$, taken round the circle $|z| = 3$.

(ii) By integrating round a suitable semi-circle, prove that $\displaystyle\int_0^\infty \frac{dx}{x^4 + 1} = \frac{\pi}{2\sqrt{2}}$

(iii) Show that the function $e^{iz}/(z^2 + \pi^2)$ has only one singularity, a simple pole, above the real axis and find the residue at this pole. By integrating the function round a

suitable semi-circle prove that $\displaystyle\int_0^\infty \frac{\cos x}{x^2 + \pi^2}\,dx = \frac{1}{2}e^{-\pi}.$ (LU)

90 (a) Evaluate the contour integral $\displaystyle\oint_C \frac{\exp z}{(z^2 + \pi^2)^2}\,dz$ where C is the circle $|z| = 4$

(b) Use contour integration to find the value of $\displaystyle\int_0^{2\pi} \frac{d\theta}{(5 - 3\sin\theta)^2}$ (EC)

91 (i) A contour C consists of the part of the real axis between $-R$ and $+R$ together with the semi-circle $|z| = R$, $0 \le \arg z \le \pi$. By integrating $e^{imz}/(1 + z^4)$, where $m > 0$, round C and letting $R \to \infty$, show that

$$\int_0^\infty \frac{\cos mx}{1 + x^4}\,dx = \frac{\pi}{2\sqrt{2}}e^{-k}(\cos k + \sin k), \text{ where } k = m/\sqrt{2}$$

(i) By integrating e^{iz}/z round the same contour as in (i) above indented at the origin

show that $\displaystyle\int_0^\infty \frac{\sin x}{x}\,dx = \frac{\pi}{2}$.   (LU)

92 (i) By writing $2\cos\theta = z + \dfrac{1}{z}$ evaluate $\displaystyle\int_0^{2\pi} \frac{d\theta}{1 - 2x\cos\theta + x^2}$ in the two cases $x < 1$

and $x > 1$.

(ii) Evaluate $\displaystyle\int_{-\infty}^\infty \frac{\cos 2x}{x^2 + a^2}\,dx, (a > 0)$, by contour integration.   (LU)

Section 10.10

93 The Laplace inverse of the transformed function $F(s)$ is given by the contour integral

$$f(t) = \frac{1}{2\pi i}\oint_C F(z)\,e^{zt}\,dz$$

where C is a suitably chosen closed contour which encloses the poles of $F(z)$.
Use this result to show that the inverses of $(3s + 2)/(s^2 + 4)$ and $1/[s(s + 1)^2]$ are

$(3 \cos 2t + \sin 2t)$ and $1 - (t + 1)e^{-t}$ respectively.   (EC)

94 Invert the following transforms using the method of residues

(a) $\dfrac{1}{(s + 1)(s - 2)^2}$ (b) $\dfrac{s}{(s + 1)^2 (s - 1)^2}$ (c) $\dfrac{1}{(s^2 + 1)^2}$

95 If $f(z)$ is regular on and inside a simple closed contour C, show, by consideration of the

integral $\displaystyle\oint_C \dfrac{f'(z)}{f(z)}\,dz$, that the change in $\arg\{f(z)\}$ as the point z moves once round C is

$2\pi N$ where N is the number of zeros of $f(z)$ inside C.
Show that the equation $z^4 + z + 1 = 0$ has one root inside each of the four quadrants of
the z-plane. Show also that each of these roots lies outside the circle $|z| = 2/3$.   (LU)

96 Show that the equation $2z^4 + 8z + 7 = 0$ has no real or purely imaginary roots. Show
that there is only one root which lies in the first quadrant.   (LU)

11

STATISTICAL METHODS

11.1 Introduction

For samples of size 30 or more, it is a reasonable approximation to say that sample means, sample proportions, differences between sample means and certain other variables are normally distributed. The Central Limit Theorem guarantees that, in the limit as the sample size tends to infinity, the variables mentioned above *are* normally distributed.

The problems fall into two categories: those concerned with estimating a **population parameter** from the observed **sample statistic** (e.g. population mean μ from sample mean $\bar{x}$) and those concerned with testing the validity of a hypothesis by examining a suitable sample statistic (e.g. testing the hypothesis that a machine produces articles of a given average length by studying the mean of a sample). In the former case we quote a **confidence interval** for the parameter, an interval about the sample value in which we are reasonably sure that the value of the population parameter lies: the less the risk of being wrong that we are prepared to accept, the wider the interval quoted. In the latter case we state in advance a risk of wrong judgement and base a decision on whether the probability of the observed sample statistic exceeds a value determined by a risk. We have to accept that the confidence interval may not include the true value of the population parameter or that the decision we took may be wrong: in either case our results should be accompanied by a statement of the risk of being wrong. With regard to the test of a hypothesis the errors are categorised as follows. If we reject the hypothesis when it is, in fact, correct we commit a **Type I error;** conversely, if we accept the hypothesis when it is false we commit a **Type II error.** The probability of committing a Type I error is called the **level of significance** for the test employed, or the α-**risk;** the probability of committing a Type II error is called a β-**risk.**

11.2 Tests of Hypotheses

We develop a strategy for testing statistical hypotheses based on a simple example.

Wire cables are compounded from separate wires: the cables are produced with a mean breaking load of 30kN and the standard deviation of these loads is 0.4kN. The manufacturing process is modified slightly at an increased cost, and to examine the effect on the end product a sample of 100 cables produced by the modified process is tested. The average breaking load for this sample is found to be 30.08kN. What evidence is there that the modification to the process has caused a real change in the mean breaking load of the cables produced?

We begin by noting that the discrepancy between the sample mean of 30.08kN and the supposed population mean of 30kN can be due (i) to sampling errors or (ii) to the possibility that the population mean is not 30kN or (iii) to both causes. We assume that the population mean *is* in fact 30kN and that, consequently, the sole reason for the discrepancy is sampling error. Under this assumption – the **null hypothesis** H_0 – we calculate the probability of a discrepancy at least as large as the one observed. Any conclusion we reach is subject to a risk of being wrong and we must first decide on a level of risk. We shall arbitrarily choose a risk of 5%. This means that were the experiment (of taking a sample of size 100 and measuring the sample mean) to be repeated a *large* number of times and were the population mean 30kN then in about 5% of experiments the sample mean would be 30.08kN or more (it should be noted that were the population mean less than 30kN, the chances of a sample mean being 30.08kN or more will be smaller than 5%). For a population mean much smaller than 30kN the chances of a sample mean > 30.08kN become remote.

From the Central Limit Theorem, we know that sample means are approximately normally distributed and therefore we proceed to calculate an acceptance region; see Figure 11.1 and refer to Table A.I, page 470.

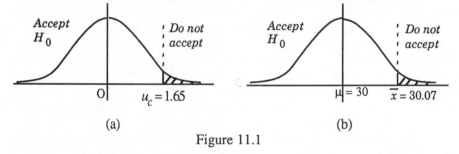

(a)   (b)

Figure 11.1

The critical value u_c is found from Table A.I to be 1.65 (2 d.p.). In order to find the corresponding value for $\bar{x}$ we use the formula

$$u_c = \frac{\bar{x} - \mu}{\sigma/\sqrt{n}} \quad \text{which becomes} \quad 1.65 = \frac{\bar{x} - 30}{0.4/10} \quad \text{and this yields}$$

$$\bar{x} = 30 + 0.04 \times 1.65 = 30.066 = 30.07\text{kN} \quad (2 \text{ d.p.})$$

Our decision rule is: accept H_0 if the observed $\bar{x} < 30.07$, otherwise do not accept H_0. Note that the α-risk has now decreased because we have rounded up $\bar{x}$ to 2 d.p. The observed value of $\bar{x}$ lies outside the acceptance region and therefore we conclude that we should not accept the hypothesis that the modified process still produces cables of mean breaking load 30kN. We may then say that the modification has caused a real change in the mean breaking load of the cables and that the risk of our judgement being wrong is somewhat smaller than 5%.

Before we proceed further we examine the effect of reducing the risk of wrong judgement to 1%. From Table A.I the critical value u_c is now 2.33 and then

$\bar{x} = 30 + 0.04 \times 2.33 = 30.0932 = 30.09$ (2 d.p.). The new decision rule is:

accept H_0 if observed $\bar{x} < 30.09$, otherwise do not accept H_0. From Figure 11.2 we see that the observed value 30.08 now lies within the acceptance region. This time we state that there is not sufficient evidence that the modification has caused a change in mean breaking load.

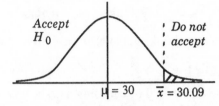

Figure 11.2

In the first case we could commit an α-risk or Type I error and the probability of that error is 5%. In the second case we could commit a Type II error or β-risk: we cannot yet state the probability of this happening.

If the population mean were less than 30kN then the probability of a Type I error would be smaller: see Figure 11.3.

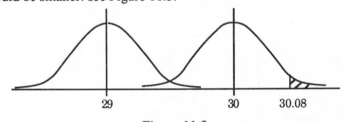

Figure 11.3

Given the decision rule, the α-risk is greatest when $\mu = 30$kN. To fix a value for β we must state a clear alternative hypothesis. As things stand we have a null hypothesis H_0: $\mu = 30$kN and an alternative hypothesis H_1: $\mu = 30.1$ kN; this might correspond to the situation where a mean breaking load of more than 30.1kN would justify the more expensive modified process.

We will assume that we have decided upon an α-risk of 5%. This means that we will accept H_0: $\mu = 30$kN if the observed $\bar{x}$ is less than 30.07kN. If we accept H_0 there is a possibility that the alternative hypothesis H_1: $\mu = 30.1$kN is true and we should have committed a Type II error; what is the probability β, of this happening? In other words, what is the probability that we could get a sample mean of less than 30.07 when $\mu = 30.1$? See Figure 11.4. We find a critical value

$$u_c = \frac{30.07 - 30.1}{0.4/10} = \frac{-0.03}{0.04} = -0.75$$

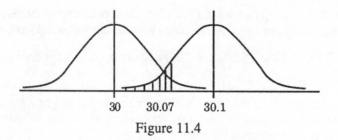

| 30 | 30.07 | 30.1 |

Figure 11.4

By reference to Table A.I we find that the β-risk is 0.2266.

It should be noted that the β-risk is greatest if $\bar{x} = 30.07$; smaller values of $\bar{x}$ will provide a smaller value of β. See whether you can reason why.

We now examine the effect on β of choosing α to be 1%. We remember that this gave a critical $\bar{x}$ of 30.09. The corresponding value of u_c is found to be

$$u_c = \frac{30.09 - 30.1}{0.04} = -0.25 \text{ and from Table A.I we find that } \beta = 0.4013; \text{ see}$$

Figure 11.5. This means we run a risk of more than 40% of accepting the null hypothesis

$$H_o : \mu = 30.0\text{kN}.$$

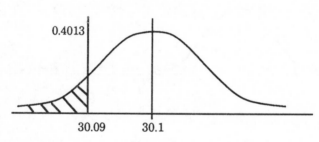

Figure 11.5

In decreasing the risk of a Type I error we have increased the risk of a Type II error. In general, for a fixed sample size, decreasing one of these errors leads to an increase in the other. The only way to reduce both α- and β- risks simultaneously is to increase the sample size. For example, suppose the sample size n were 200 and that we have the decision rule: accept H_0 if $\bar{x} < 30.07$, otherwise do not accept H_0. Then, to find the α-risk we calculate the value

$$u_c = \frac{30.07 - 30}{0.4/\sqrt{200}} = 2.475 \text{ and this corresponds to an α-risk of 0.0066. The β-}$$

risk is found via $u_c = \dfrac{30.07 - 30.1}{0.4/\sqrt{200}} = -1.06$ and this gives $\beta = 0.1446$.

However, a sample of size 200 may be too costly and we should perhaps consider a better decision rule. Presumably we are concerned to minimise the risk of passing the modification unless the mean breaking load is at least as high as 30.1kN. Given any two of the parameters n, α and β, the third is automatically fixed. Suppose then that we wish to make β at most 0.05 and retain n as 100 then what will the decision rule become and what will α be? For $\beta = 0.05$ we

shall find $-1.65 = \dfrac{\bar{x} - 30.1}{0.04}$ and therefore $\bar{x} = 30.1 - 0.04 \times 1.65 = 30.03$

(2 d.p.). The decision rule is now: accept H_0 if $\bar{x} < 30.03$, otherwise do not

accept H_0. Then to determine the α-risk we calculate $u_c = \dfrac{30.03 - 30}{0.04} = 0.75$

and discover that $\alpha = 0.2266$. This may seem too high a risk and it serves to emphasise the point that we need to decide on our priorities before choosing a course of action.

We might ask what is the effect on β of varying H_1. We assume that the decision rule hinges on the critical $\bar{x}$ value of 30.07 and that $n = 100$. It should be clear that the probability of accepting H_0 when the true population mean is μ

is the area under the normal curve to the left of $u_c = \dfrac{30.07 - \mu}{0.4/10}$. Let us represent

this area by $f(\mu)$; then we can plot a graph of $f(\mu)$ against μ. This is the **Operating Characteristic Curve** for the test criterion, see Figure 11.6(a).

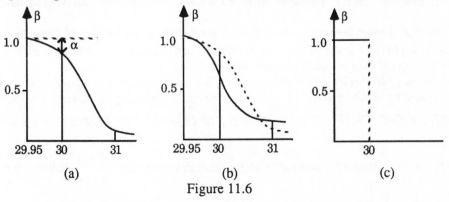

Figure 11.6

In Figure 11.6(b) we compare the operating characteristic curves for sample sizes 100 and 200. Notice how the larger sample size has improved the test in that it better discriminates between $\mu = 30$ and any neighbouring value of μ. The more sharply the operating characteristic curve falls from a high chance of acceptance to a low chance the better is the discrimination. The extreme situation would be that we accept H_0 for $\mu \leq 30$ and do not accept H_0 for $\mu > 30$. The 'ideal' operating characteristic curve is shown in Figure 11.6(c).

A related concept is that of the **power** of a test, which is the ability of the test to discriminate between alternatives. The power function is $(1 - \beta)$ and you should construct its graph for the problem under consideration.

General Strategy

There is a general strategy that we have adopted in testing the null hypothesis and we summarise the steps involved.

1. *The Null Hypothesis and its alternative are stated*
2. *The probability distribution model is selected*
3. *The α-risk is chosen*
4   *Determine the acceptance and rejection regions*
5. *Perform the calculations*
6. *Draw the statistical inference*
7. *Consider further action*

11.3   Mean of a Small Sample: The *t* Test

In many experiments it is not posssible or not economic to take large sized samples: also, we may be unaware of the population variance. Under these circumstances the normal model will be inappropriate even though the parent

population *is* normally distributed. The statistic $(\bar{x} - \mu)/(\hat{\sigma}/\sqrt{n}\,)$, where $\hat{\sigma}^2$ is

the estimate of the variance provided by the sample, will not be normally

distributed since not only does $\bar{x}$ vary from sample to sample but so does $\hat{\sigma}$. In addition to the sample mean, we need therefore to consider two further variables: the sample variance *and* the sample size.

A suitable model is provided by the *t* **distribution** (sometimes referred to as Student's *t* distribution). It relies on the parent population having a normal distribution $N(\mu, \sigma^2)$. Then if the mean of a random sample of size n is $\bar{x}$ and

the sample standard deviation is s, the random variable $t = \dfrac{\bar{x} - \mu}{s/\sqrt{n}}$ follows the t

distribution with $v = (n - 1)$ degrees of freedom.

Instead of one table for the normal distribution $N(0, 1^2)$, we now need a table for each value of v (starting with $v = 1$). In fact, the values provided by t -

tables for $v > 30$ are extremely close to the corresponding values from $N(0, 1^2)$ and this ties in with our claim that sample means for sample sizes of about 30 or more can be assumed to be approximately normally distributed. Figure 11.7 shows graphs of the t distribution for $v = 1$, 5 and ∞. The case $v \to \infty$ is the normal distribution, though had we also drawn the graph of the case $v = 30$ we should not have been able to distinguish between them. Note that the curves are symmetrical about the vertical axis. The expectation of the distribution is 0 and its

variance is $\dfrac{v}{v - 2}$ (for $v > 2$).

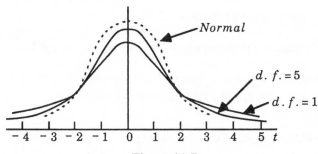

Figure 11.7

Table A.II on page 471 shows selected percentage points of the t distribution. Note that the interpretation is similar to that for the normal distribution. For example, the probability that with 4 degrees of freedom $t > 2.13$ is 5%, that $t < -2.13$ is (by symmetry) 5% and that $|t| > 2.78$ is 5%. Try Problems 1 and 2 to ensure that you are familiar with the use of this table.

Confidence Intervals

Consider the following example.

A random sample of 5 steel beams produced by a mill was tested for ultimate compressive strength. The results were, in MN/m^2

301500   329600   318700   307200   322400

We require to find the 95% confidence interval for the mean ultimate compressive strengths of steel beams produced at the mill.

Calculations produce $\bar{x} = 315880$ $s^2 = 130210 \times 10^3$

Hence the estimated standard error of the mean $\dfrac{s}{\sqrt{5}} = 5103$.

From Table A.II we see that for 4 degrees of freedom there is a 5% chance that $|t| > 2.78$ and hence a 95% chance that $|t| \leq 2.78$. Therefore the 95% confidence interval for the mean ultimate compressive strength is

315880 $\pm$ 2.78 $\times$ 5103   i.e.   315880 $\pm$ 14186

It is now probably reasonable to round off these results to give

315900 $\pm$ 14200   and hence the interval is   301700 to 330100

Show that the 99% confidence interval is 292400 to 339400.

Tests of Hypothesis

A machine is set to pack 10 kg of cement in a bag. A sample of 9 bags is taken and for this sample the mean weight of cement in a bag is found to be 9.5 kg and the sample standard deviation is 0.5 kg. Test the hypothesis that the machine is in proper working order using a risk of wrong rejection of (a) 5% (b) 1%.

(a) We follow the strategy laid down earlier.

   1. The null hypothesis is $H_0 : \mu = 10$ and the alternative hypothesis is $H_1 : \mu \neq 10$.

   2. The risk of wrong rejection is 5%.

   3. The t distribution is chosen (why?) with 8 degrees of freedom.

   4. From Table A.II we find that we should reject H_0 if $|t| > 2.314$. We use a two-tailed test because of the nature of H_1.

   5. From the sample of size 9 we find $t = \dfrac{\bar{x} - \mu}{s/\sqrt{9}} = \dfrac{3(9.5 - 10)}{0.5} = -3$

   6. We therefore do not accept H_0, with a risk of wrong judgement of less than 5%.

(b) The only changes are

   2. The risk of wrong rejection is 1%.

   4. From Table A.II we find that we should reject H_0 if $|t| > 3.36$.

   6. We therefore do not reject it.

11.4   Test of Sample Variance: The χ^2 Distribution

Often, articles need to be manufactured with precision; if components are to interlock it is important that certain parameters, for example diameter, do not vary much from a quoted mean. It is often necessary to check the variance of a sample against a quoted population variance to see whether the machine is now producing articles with greater variability.

The sample variance, $s^2 = \dfrac{1}{n-1} \displaystyle\sum_{i=1}^{n} (x_i - \bar{x})^2$, is an unbiased estimator of the population variance σ^2.

It follows that $\dfrac{\sum (x_i - \bar{x})^2}{\sigma^2}$, i.e. $\dfrac{(n-1)s^2}{\sigma^2}$ is an unbiased estimator of

$(n-1)$. In fact, if the sample is drawn from a normal population $N(\mu, \sigma^2)$ the value of $(n-1)s^2/\sigma^2$ is a random variable which follows the chi-square distribution having $v = (n-1)$ degrees of freedom. Chi-square is written χ^2.

Note, as for the t test, the assumption of a normal population. If the population is 'nearly' normal, the χ^2 distribution is 'nearly' followed. Different values of v give rise to different shapes of χ^2 distribution. In Figure 11.8 we graph the χ^2 distribution for three values of v. Notice that the curves are markedly skew. In the special case $v = 2$ the curve is different from those with other values of v since it possesses no local maximum.

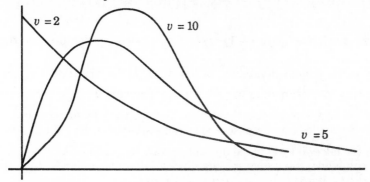

Figure 11.8

Table A.III on page 472 gives selected percentage points for various χ^2 distributions. For example, the probability that, with $v = 1$, $\chi^2 > 6.635$ is 1%, that $\chi^2 > 3.841$ is 5% and with $v = 4$ the probability that $\chi^2 < 0.2971$ is 1%. You should try Problem 22 to gain familiarity with the table.

Degrees of Freedom
It should be pointed out at this stage that the number of degrees of freedom v for a statistic is not always one less than the number of items n in the sample chosen. In general, the number of degrees of freedom for a statistic is given by $v = n - \gamma$ where γ is the number of population parameters that must be estimated from a sample. For the one-sample t test we needed only to estimate μ, for χ^2 only σ^2; in both these cases, then, $\gamma = 1$. We shall see when we come to analyse regression lines that $\gamma = 2$ there since we use the sample of observations to estimate the slope and the intercept of the line.

Confidence Intervals
A sample of 10 batteries were tested for useful life. These lives, in months, were recorded to 1 d.p. and are as follows: 15.0, 14.2, 15.9, 14.8, 14.9, 15.1,

16.0, 14.8, 15.1, 15.3. We now seek a 95% confidence interval for the standard deviation of the lives of batteries produced by the process under consideration.

From the sample readings we find $s^2 = 0.281$. There are 9 degrees of freedom. In Figure 11.9 we see that the critical χ^2 values are those corresponding to percentage points 0.025 and 0.975; these, from Table A.III are found to be 2.70 and 19.0 respectively.

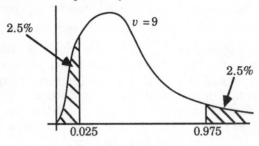

Figure 11.9

Now σ^2 is estimated by $(n-1)s^2/\chi^2$. Corresponding, therefore, to the value

$\chi^2 = 2.70$ is the estimated value $\hat{\sigma}^2$ of $9 \times 0.281/2.70$, i.e. 0.937 (3 d.p.). In the same way, corresponding to $\chi^2 = 19.0$ is the value $9 \times 0.281/19.0 = 0.133$. Therefore the 95% interval for σ^2 is $0.133 < \sigma^2 < 0.937$. Notice that this interval is *not* symmetrical about the observed sample value of 0.281. We can therefore obtain a 95% interval for σ by taking square roots to obtain $0.365 < \sigma < 0.968$.

If we now widen the limits to obtain a 99% confidence interval we find that $0.107 < \sigma^2 < 1.458$ and hence $0.327 < \sigma < 1.207$.

Test of Hypothesis

A manufacturer claims that certain machine parts he produces will have mean diameter 4cm with standard deviation 0.01 mm. The diameters of 5 parts are measured and found to be (in mm): 39.98, 40.01, 39.96, 40.03, 40.025. Is the claim on the standard deviation valid?

We test whether the sample standard deviation is significantly higher than 0.01 mm (or, equivalently, whether the sample variance is significantly higher than 0.0001 (mm)^2).

1.   The null hypothesis is $H_0 : \sigma^2 = 0.0001$ and the alternative hypothesis is $H_1 : \sigma^2 > 0.0001$.
2.   We select the χ^2 distribution with 4 degrees of freedom.
3.   The α-risk is chosen to be 5%.
4.   The acceptance region is found from Table A.III to be $\chi^2 < 9.488$.

5. $\chi^2 = \dfrac{(n-1)s^2}{\sigma^2} = \dfrac{4 \times (0.0301)^2}{0.0001} = \dfrac{0.003625}{0.0001} = 36.25.$

6.   We conclude, with much less than 5% risk, that the variability in diameter is greater than was claimed.

Note The χ^2 distribution can also be used to study inferences about frequencies; we do not pursue these matters here.

11.5   Sample Variances: The F Test

We have not yet compared the means of two samples and this problem is tackled in the next section. We require first to decide whether the variances of the two samples s_1^2 and s_2^2 can be regarded as equal: if so we are able to 'pool' the variances so that we get a total estimate of the variance. This decision can be made by reference to the so-called F test.

F distribution

If samples of two independent random variables have sample variances s_1^2 and s_2^2 based on v_1 and v_2 degrees of freedom respectively, where the corresponding population variances are σ_1^2 and σ_2^2, then the ratio $F = \dfrac{s_1^2 \sigma_2^2}{s_2^2 \sigma_1^2}$

follows an F distribution with v_1 and v_2 degrees of freedom.

In many cases we are concerned with working under a null hypothesis that σ_1^2 and σ_2^2 are equal and then we test the ratio s_1^2/s_2^2 against the F distribution. In this book we always arrange that $s_1^2 \geq s_2^2$ so that the calculated ratio ≥ 1.

If s_1^2 and s_2^2 are the sample variances of two independent random variables, of sizes n_1 and n_2 respectively, both taken from normal populations with the same variance then the ratio s_1^2/s_2^2 follows an F distribution with $v_1 = n_1 - 1$ and

$v_2 = n_2 - 1$ degrees of freedom; this is written $F_{v_2}^{v_1}$

Table A.IV on page 473 shows selected percentage points of several F distributions.

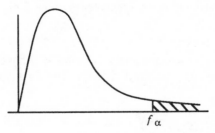

Figure 11.10

The tabulated value, f_α, is the value of F above which there is an area of α

(i.e. the probability of obtaining a value of $F > f_\alpha$ is α) see Figure 11.10 ; of course, we must first specify v_1 and v_2. Sometimes we write $f_\alpha(v_1, v_2)$, e.g. $f_{0.05}(2,3)$, to emphasise all three parameters. Table A.IV is used as follows.

1.   Select the column corresponding to v_1.
2.   Select the row corresponding to v_2.
3.   Observe that there are 3 entries corresponding to v_1 and v_2. Of these the top one will be relevant to a 5% risk, the middle one to a 2.5 % risk and the bottom one to a 1% risk.

Examples
(i)   We seek $f_{0.05}(2,3)$. The three numbers in the table are 9.55, 16.0 and 30.8 and of these we require 9.55.
(ii)   We seek $f_{0.01}(8,60)$. The three numbers are 2.10, 2.41 and 2.82: of these we require 2.82.

Notice that with 8 and 60 degrees of freedom respectively we are saying that the probability of an F score > 2.82 is 0.01 or 1%. Observe also that for a fixed value of α, the values of F decrease towards 1 as v_1 and v_2 get larger. To gain familiarity with the table you should try Problem 1.

Where the particular combination of v_1 and v_2 is not present you must either use interpolation or consult more comprehensive tables.

When testing null hypotheses $\sigma_1^2 = \sigma_2^2$ we shall have the alternative hypothesis $\sigma_1^2 > \sigma_2^2$ or $\sigma_1^2 < \sigma_2^2$ or $\sigma_1^2 \neq \sigma_2^2$. In the second case we can express the alternative hypothesis as $\sigma_2^2 > \sigma_1^2$ and calculate s_2^2/s_1^2. In the third case we note that we work with a risk of $\alpha/2$, and use a two-tailed test.

Properties of the F distribution

(i) $v_2 \rightarrow \infty; \quad F(v_1, \infty) = \chi_1^2/v_1$

(Hence the χ^2 distribution is a special case of the F distribution.)

(ii) $v_1 = 1; \quad F(1, v_2) = t^2(v_2)$
(In some books the variance analysis tables will have a column headed 'F or t^2'. This implies that the t distribution is a special case of the F distribution.)

(iii) $f_\alpha(v_1, v_2) = 1/f_{1-\alpha}(v_2, v_1)$

For example, $f_{0.99}(2,3) = \dfrac{1}{f_{0.01}(3,2)} \cong \dfrac{1}{99.2} \cong 0.01$

Example 1

Observations of water hardness were taken from the output of two separate water treatments. The first sample comprised 11 observations and the second comprised 15 observations. The results were recorded in parts per million of calcium carbonate and it was found that the variance of the first sample was (in coded form) 94.6 and the variance of the second sample was 36.8. Is there a significant difference in the variances?

Although $s_1^2 > s_2^2$ it does not follow that $\sigma_1^2 > \sigma_2^2$, however, since $s_1^2 \gg s_2^2$ we are led to suspect that σ_1^2 may be greater than σ_2^2. It is now a question of how much greater than 1 the ratio s_1^2/s_2^2 can be.

1. The Null Hypothesis, H_0 is that $\sigma_1^2 = \sigma_2^2$ and the alternative hypothesis, H_1 is that $\sigma_1^2 > \sigma_2^2$. We adhere to our convention that the larger variance is divided by the smaller and so the variance ratio is always ≥ 1.
2. The α-risk is chosen to be 5%.
3. We choose an F distribution with 10 and 14 degrees of freedom respectively.
4. The acceptance and rejection regions are shown in Figure 11.11. Note that the upper limit of the acceptance region is given by $f_{0.05}(10,14)$. The upper level is found to be 2.60. Therefore we cannot accept H_0 at the 5% level of significance if $F > 2.60$.
5. We calculate $F = 94.6/36.8 = 2.57$ (3 s.f.).
6. Since this lies in the acceptance region we cannot reject H_0.

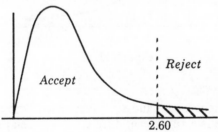

Figure 11.11

Example 2

The effectiveness of diluting a poisonous substance was determined by taking 6 samples from the container after 5 minutes of mixing and a further 6 samples after 5 more minutes. The following results were obtained for the percentage of substance in each sample.

after 5 minutes:	21.3	21.6	19.3	20.8	21.2	21.8
after 10 minutes:	19.6	20.6	20.8	21.8	22.0	21.2

We require to determine whether there is a significant increase in the homogeneity of the dilution.

1. The null hypothesis, H_0, is that there is no change in the homogeneity i.e.

$\sigma_1^2 = \sigma_2^2$. The alternative hypothesis, H_1, is that there is an *increase* in the homogeneity, i.e. $\sigma_1^2 > \sigma_2^2$.

2.   We work with an α-risk of 0.05.

3.   We work with an F distribution with 5 and 5 degrees of freedom and use a one-tailed test.

4.   We find that $f_{0.05}(5, 5) = 5.05$; notice that we have concentrated the 5% at one end: see Figure 11.12.

5.   We calculate s_1^2 (relating to 5 minutes) as 0.812 and s_2^2 (relating to 10 minutes) as 0.768. This gives $F = 0.812/0.768 = 1.06$.

6.   Clearly we must accept H_0 and state that at the 5% level of significance the extra 5 minutes mixing has had no marked effect on the homogeneity of the mixing.

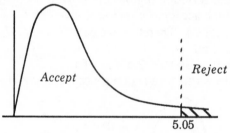

Figure 11.12

Pooling of Variances

If two sample variances, s_1^2 and s_2^2 when tested with an F distribution do not suggest that there is a significant difference between them we may wish to obtain an overall (or pooled) estimate of the common variance. We take a weighted average of the sample variance, where the weights are the respective degrees of freedom. The larger sample (the amalgamation of the two original samples) gives a better estimate of the variance. The pooled estimate is, then

$$\frac{(v_1 s_1^2 + v_2 s_2^2)}{v_1 + v_2}$$ based on $v = n_1 + n_2 - 2 = v_1 + v_2$ degrees of freedom.

Example

In Example 1 above we may pool the variances and obtain an estimate of

$$\frac{10 \times 94.6 + 14 \times 36.8}{10 + 14} = \frac{1461.2}{24} = 60.9$$

11.6 Comparison of Sample Means

When population variances are known we may compare sample means using the normal distribution. We concentrate in this section on situations where one or both of the following conditions obtain
(i) the population variance is unknown   (ii) the sample sizes are small.

Independent Samples

Consider the situation where we wish to compare the outputs from two machines or two factories where it is not necessarily true that the numbers in each sample are equal. Even if the numbers in each sample are the same, the conditions under which the samples were taken may vary and we are entitled to assume that the samples are independent.

We **assume** that the variances of the populations from which the samples were taken are, in fact, *equal*. We further **assume** that the two populations are *normally* distributed. For reference we state that the two populations will be distributed as $N(\mu_1, \sigma_1^2)$ and $N(\mu_2, \sigma_2^2)$, that samples are drawn of size n_1

with sample mean $\bar{x}_1$ and sample variance s_1^2 from the first population and of

size n_2 with sample mean $\bar{x}_2$ and sample variance s_2^2 from the second.

We assume that $\mu_1 = \mu_2$ and then calculate

$$\frac{(\bar{x}_1 - \bar{x}_2) - (\mu_1 - \mu_2)}{s.e.} = \frac{(\bar{x}_1 - \bar{x}_2)}{s.e.}$$ where s.e. (the standard error of the

difference in the means) is equal to $\sqrt{\dfrac{\sigma_1^2}{n_1} + \dfrac{\sigma_2^2}{n_2}}$. However, we have assumed

that $\sigma_1^2 = \sigma_2^2 = \sigma^2$ and hence we take the standard error to be

$\sqrt{\dfrac{\sigma^2}{n_1} + \dfrac{\sigma^2}{n_2}} = \sigma\sqrt{\dfrac{1}{n_1} + \dfrac{1}{n_2}}$. Then $\dfrac{(\bar{x}_1 - \bar{x}_2)}{s.e.}$ will follow the distribution

$N(0,1)$. For each sample the number of degrees of freedom is one less than the number of items in the sample since the determination of the mean reduces the degrees of freedom by 1; hence the total number of degrees of freedom in the system is $(n_1 - 1) + (n_2 - 1)$. How valid is the assumption that the variances are equal? We shall first have to check the validity by means of an F test.

Example 1

Two makes of steel cable are tested for strength. The breaking loads for samples of each make were measured and the results recorded below in kN. Is there a significant difference in the mean breaking loads? Work with a 5% risk of wrong judgement.

Make I	30.38	30.25	30.34	30.27		
Make II	30.37	30.36	30.35	30.30	30.32	30.39

For ease in following the argument we perform some calculations first.

For Make I $v_1 = 3$, $\bar{x}_1 = 30.31$ and $s_1^2 = \dfrac{110 \times 10^{-4}}{3}$

For Make II $v_2 = 5$, $\bar{x}_2 = 30.35$ and $s_2^2 = \dfrac{56 \times 10^{-4}}{5}$

(a)   *Are the sample variances significantly different?*

1.   We set up the null hypothesis that the population variances are equal; the alternative hypothesis is that they are not equal.
2.   We work with an α-risk of 5%.
3.   We use an F distribution with 3 and 5 degrees of freedom.
4.   We find that $f_{0.05}(3,5) = 5.41$. The rejection region is $F > 5.41$.

5.   We calculate $F = \dfrac{s_1^{\,2}}{s_2^{\,2}} = \dfrac{110 \times 10^{-4}}{3} \times \dfrac{5}{56 \times 10^{-4}} = 3.27$ (3 s.f.)

6.   Hence we cannot conclude (with a 5% risk) that the variances are not significantly different, though we prefer a value of F closer to 1.

(b)   *Pooling of variances*

We now estimate σ^2 as $\dfrac{3 \times \left(\dfrac{110}{3} \times 10^{-4}\right) + 5\left(\dfrac{56}{5} \times 10^{-4}\right)}{3 + 5} = 2.075 \times 10^{-3}$

(c)   *Are the sample means significantly different?*

1.   The null hypothesis is that $\mu_1 = \mu_2$. The alternative hypothesis is that $\mu_1 \neq \mu_2$.
2.   The α-risk is 5%.
3.   We use a t distribution with 8 degrees of freedom since we estimate σ from the sample data and employ a two-tailed test.
4.   The critical values of t are ± 2.75 and hence the rejection region is $|t| > 2.75$.

5.   We calculate $t = \dfrac{(30.35 - 30.31) - 0}{\sqrt{2.075 \times 10^{-3}\left[\dfrac{1}{4} + \dfrac{1}{6}\right]}} = 1.36$ (3 s.f.)

6.   Since the value calculated lies in the acceptance region, we conclude that the mean breaking loads are not significantly different at the 5% level.

Example 2

A standard cell, the voltage of which was known to be 1.10 volts was used to test the accuracy of two voltmeters A and B. Nine independent readings of the voltages from each meter give the following values.

Meter	Variate	Measured Voltages
A	x_1	1.12 1.16 1.15 1.11 1.10 1.13 1.16 1.14 1.10
B	x_2	1.12 1.09 1.04 1.09 1.09 1.05 1.06 1.03 1.08

Using a maximum risk of 5% of finding spurious differences, determine whether there is a significant difference in the mean readings of the voltmeters. Give a 95% confidence interval for the difference in mean readings.

First we perform some calculations.

$$\bar{x}_1 = 1.13, \quad \bar{x}_2 = 1.07$$

$$s_1^2 = \text{var}(x_1) = 5.75 \times 10^{-4}; \quad s_2^2 = \text{var}(x_2) = 8.44 \times 10^{-4}$$

(a) *Are the sample variances significantly different?*

We list merely the differences from the previous examples. The α-risk is 1% and this together with an F distribution with 8 and 8 degrees of freedom produces an acceptance region of $F < 3.44$. In fact,

$$F = \frac{s_2^2}{s_1^2} = \frac{8.44 \times 10^{-4}}{5.75 \times 10^{-4}} = 1.47 \text{ and this lies in the range appropriate to}$$

an α-risk of 5%. Hence the sample variances are not significantly different.

(b) *Pooling of variances*

$$\text{Estimate } \sigma^2 \text{ as } \frac{8 \times \left(8.44 \times 10^{-4}\right) + 8 \times \left(5.75 \times 10^{-4}\right)}{8 + 8} = 7.10 \times 10^{-4} \quad \text{(3 s.f.)}$$

(c) *Are the means significantly different?*

With a null hypothesis that $\mu_1 = \mu_2$ and an α-risk of 5%, we employ a t distribution with 16 degrees of freedom and operate a two-tailed test. Table A.II yields the critical value of 2.12. We calculate

$$t = \frac{(1.13 - 1.07) - 0}{\sqrt{7.10 \times 10^{-4} \left[\frac{1}{9} + \frac{1}{9}\right]}} = 4.78$$

We reject the null hypothesis and conclude that the means are significantly different.

(d)   *Confidence Interval for difference in means*

Now that we have established a significant difference in the means we determine a 95% confidence interval for the difference in means. The observed difference was 0.06, hence the confidence interval required will be

$$0.06 \pm 2.12 \text{ s.e.}, \quad \text{i.e. } 0.06 \pm 2.12 \times \sqrt{7.10 \times 10^{-4}\left[\frac{1}{9}+\frac{1}{9}\right]}$$

i.e. 0.0600 ± 0.0266 (4 d.p.)

This gives a 95% confidence interval of (0.0334, 0.0866) which would seem to be quite large.

Note: If one end of the interval were positive and the other end were negative we should expect the corresponding null hypothesis $\mu_1 = \mu_2$ to have been accepted at an α-risk of 5%. If on applying the F test, we are forced to reject the null hypothesis $\sigma_1^2 = \sigma_2^2$ then the procedures above will be inapplicable.

Paired samples

Suppose we are testing whether a particular factor in some experimental situation is causing the mean of a parameter to be significantly different from that value which was expected. How can we be sure that no other factor may be contributing to the difference being so large? One solution, which is extensively used, is to set up a controlled experiment in which for each item in the test sample there is one item in the control sample which is subject to the same conditions.

Alternatively we may wish to compare the effects of two processes on several items which are paired off (in the sense of possessing identical or similar characteristics or in the sense of being subjected to identical conditions). We can then examine the different effects of the processes and be reasonably sure that we have eliminated or at least reduced the effects of other experimental factors. Previously, with independent samples we had ignored these other effects.

We then work with a set of differences and the null hypothesis will be that the mean of such differences is zero.

Example

Ten mixes of concrete were sampled after 2 and 4 minutes of mixing. The samples were tested for porosity. The results, in suitable units, are shown below. Is there evidence that time in mixing affects the porosity of the concrete?

Mix	1	2	3	4	5	6	7	8	9	10
After 2 mins	2.0	3.1	2.5	2.3	3.0	2.7	2.1	2.4	2.8	2.2
After 4 mins	2.2	2.7	2.8	2.4	2.6	2.5	2.4	2.6	2.9	2.5

We shall first effect a solution by the method of independent samples.

Let the variate x_1 be the porosity after 2 minutes and the variate x_2 the

porosity after 4 minutes. Calculations yield $\bar{x}_1 = 2.51$; $\bar{x}_2 = 2.56$; $s_1^2 = 0.143$;

$s_2^2 = 0.043$.

A cursory glance at the ratio s_1^2/s_2^2 shows that we may pool the variances to estimate σ^2 as $\dfrac{(9 \times 0.143) + (9 \times 0.043)}{18} = 0.093$

1. The null hypothesis is that $\mu_1 = \mu_2$ and the alternative hypothesis is that $\mu_1 \neq \mu_2$.
2. We work with an α-risk of 5%.
3. We use a t distribution with 18 degrees of freedom; a two-tailed test.
4. The acceptance region is $|t| < 2.10$.
5. We calculate $t = \dfrac{(2.56 - 2.51) - 0}{\sqrt{0.093 \left[\dfrac{1}{10} + \dfrac{1}{10}\right]}} = \dfrac{0.05}{\sqrt{0.0186}} = 0.367$
6. We conclude that mixing time does not affect porosity.

If however, we proceed via the paired sample approach then we first form a set of differences, being careful to subtract consistently the first reading in each pair from the second.

Mix	1	2	3	4	5	6	7	8	9	10
Differences (4 min – 2 min)	0.2	–0.4	0.3	0.1	–0.4	–0.2	0.3	0.2	0.1	0.3

From this set of 10 readings, we calculate

$$\bar{d} = 0.05, \quad \mathrm{var}(d) \equiv s_d^2 = 0.0783$$

1. We set up a null hypothesis, that the expected difference $\mu_d = 0$. The alternative hypothesis is that $\mu_d \neq 0$.
2. We work with an α - risk of 5%.
3. We use a t distribution with 9 degrees of freedom (we have effectively reduced our problem to a study of 10 numbers) and apply a two-tailed test.
4. The acceptance region is found to be $|t| < 2.26$.
5. We calculate $t = \dfrac{0.05 - 0}{\sqrt{\dfrac{0.0783}{10}}} = 0.565$
6. Hence the main difference is not significant at the 5% level and we have not sufficient evidence that the time in mixing affects porosity.

It can be seen that although both methods yield the same conclusion the values of t obtained are different. What then would be the interpretation if the t-values had been such that in one test the t-value lay in the acceptance region but in the other test the t-value did not. What must be borne in mind is that although we may not be able to detect a significant difference by one of our tests, there may be an underlying real difference in population means. Consequently, any indication of such a difference must be heeded and the fact that one test does give a significant result is sufficient. Note too, that the pairing approach does eliminate factors other than the true difference in effects of the two treatments and we would expect it to be more sensitive. Whether the unpaired test will also give a significant result depends on the amount to which the other factors obscure this essential difference in means.

Tests of Hypothesis or Confidence Intervals?

It may be argued that a confidence interval may be more meaningful under some circumstances than the test of a hypothesis. A confidence interval can provide a measure of reliability on a sample estimate, whereas a test of hypothesis will only yield rejection or non-rejection. We have seen that the two concepts are linked. That is, for example, a 95% confidence interval on the mean of a sample may include the proposed mean which forms the basis of the null hypothesis. However, since the confidence interval also gives an index of reliability, it is often preferred.

Yet tests of hypothesis do have a role to play: in statistical quality control where the output from a process is continually monitored. It should be borne in mind that economic considerations must also be taken into account and these may modify a decision reached on purely statistical grounds.

11.7   Introduction to Analysis of Variance

Three brands of lubricating oil were tested on three cars of the same make. Each car was test-driven over the same routes on the same days and carried the same load. At the end of the test period, measurements were made of the maximum wear in each cylinder. The results are shown in Table 11.1; the wear is recorded in hundredths of a millimetre.

Table 11.1

Lubricant	Cylinder wear			
A	22	21	25	18
B	16	19	21	18
C	10	17	12	15

We have taken readings for each of the four cylinders in the cars, and, if we take the sample mean reading for each car, we may assume that their difference reflects the difference in the performance of the lubricating oil. At first glance it would seem that oil A is least effective and that C is most effective, but it remains to be seen whether the differences are significant.

We can attempt a simple-minded analysis of the data. The row means are, respectively $21\frac{1}{2}$, $18\frac{1}{2}$, $13\frac{1}{2}$. To eliminate the effects of differences between lubricants we shall subtract $3\frac{1}{2}$ from each observation in the first row, $\frac{1}{2}$ from the observations in the second row and add $4\frac{1}{2}$ to those in the third row; this will make all row means equal to 18. The results will then be as in Table 11.2.

Table 11.2

					Means
A	$18\frac{1}{2}$	$17\frac{1}{2}$	$21\frac{1}{2}$	$14\frac{1}{2}$	18
B	$15\frac{1}{2}$	$18\frac{1}{2}$	$20\frac{1}{2}$	$17\frac{1}{2}$	18
C	$14\frac{1}{2}$	$21\frac{1}{2}$	$16\frac{1}{2}$	$19\frac{1}{2}$	18

What then are the remaining variations due to? We attribute them to *experimental error*; this could be pure observational error or its combination with unmentioned factors. We can see that there is still a variation *within* each sample against which must be weighed the variation *between sample means*.

Assumptions for model of one-factor analysis of variance
1. The observations are formed into k rows which are effectively random samples from normal distributions.
2. The normal populations have the same variance, viz. σ^2.

We have not specified that the sample sizes are equal. The analysis can easily be modified for unequal sizes, but we shall work with k samples each of size s.
Table 11.3 illustrates the general case.

Table 11.3

Sample	Observations	Totals	Means
1	$x_{11}, x_{12}, ..., x_{1s}$	T_1	$\bar{x}_1$
2	$x_{21}, x_{22}, ..., x_{2s}$	T_2	$\bar{x}_2$
.	. . .	.	.
k	$x_{k1}, x_{k2}, ..., x_{ks}$	T_k	$\bar{x}_k$

The mean of sample i is $\bar{x}_i = T_i/s$. The total number of observations, $N = ks$

The grand total $T = \displaystyle\sum_{i=1}^{k} T_i$; the overall mean $\bar{x} = T/ks = T/N$

We have effectively assumed that the ith sample comes from a population $N(\mu_i, \sigma^2)$.

Partition of total sums of squares
It can be shown that

$$\sum_{i=1}^{k} \sum_{j=1}^{s} (x_{ij} - \bar{x})^2 = \sum_{i=1}^{k} s(\bar{x}_i - \bar{x})^2 + \sum_{i=1}^{k} \sum_{j=1}^{s} (x_{ij} - \bar{x}_i)^2 \qquad (11.1)$$

or   Total sum of squares $= \dfrac{\text{Between samples}}{\text{sum of squares}} + \dfrac{\text{Within samples}}{\text{sum of squares}}$

This result shows that we can split the total variation of the observations about the overall mean into a component representing the variation between sample means and a component representing the variation within each sample. The first term on the right-hand side of (11.1) represents the departures of sample means from the overall mean. The presence of the factor s indicates the size of each sample and suggests that one of the summations has been done.

Estimates of σ^2
(i)   Since the k samples came from populations with assumed equal variances σ^2, we are entitled to pool the variances of each of the k samples to obtain a better estimate of σ^2.

The variance of sample i is $s^2_i = \dfrac{1}{(s-1)} \sum_{j=1}^{s} (x_{ij} - \bar{x}_i)^2$.

Since we are assuming that the sample sizes are equal, the pooled estimate of σ^2 is

$$s^2_p = \frac{1}{k} \sum_{i=1}^{k} s^2_i = \frac{1}{k(s-1)} \sum_{i=1}^{k} \sum_{j=1}^{s} (x_{ij} - \bar{x}_i)^2. \qquad (11.2)$$

based on $k(s-1)$ degrees of freedom. This is equal to

$$\frac{1}{k(s-1)} \times \left(\begin{matrix} \text{within samples} \\ \text{sum of squares} \end{matrix} \right).$$

(ii)   We now set up a null hypothesis that the k samples come from populations with equal means, i.e.

$$H_0: \mu_1 = \mu_2 = \dots = \mu_k = \mu \qquad (11.3)$$

The alternative hypothesis H_1 is that at least one of the population means is not equal to the others.

Under the null hypothesis we can make a second and independent estimate of σ^2. We may now regard the original N observations as comprised of k samples from the same population $N(\mu, \sigma^2)$. In this case we can consider the k sample means as k observations from the same population. We know that the variance of the means of samples of size s from a population with variance σ^2 is the standard error of sample means, viz. σ^2/s.

If we consider the sample means, $\bar{x}_1$, $\bar{x}_2$, ..., $\bar{x}_k$, the variance of this set of k

observations is $\dfrac{1}{(k-1)} \displaystyle\sum_{i=1}^{k} (\bar{x}_i - \bar{x})^2$. If this is an estimate of $\dfrac{\sigma^2}{s}$ it follows that

we obtain a second estimate of σ^2 as

$$s^2_m = \frac{s}{(k-1)} \sum_{i=1}^{k} (\bar{x}_i - \bar{x})^2 \tag{11.4}$$

based on $(k-1)$ degrees of freedom.

This is equal to $\dfrac{1}{(k-1)} \times$ (between samples sum of squares).

Notice that since $sk - 1 \equiv (k-1) + k(s-1)$ we have partitioned the total degrees of freedom in the system as well as the total sum of squares. We illustrate the partition of degrees of freedom by Figure 11.13. The cross-shaded area represents the 'unexplained' part of sum of squares. We shall examine whether the ratio s^2_m/s^2_p is significantly larger than 1 by means of an F test. We can hence construct a theoretical variance analysis table. (see Table 11.4)

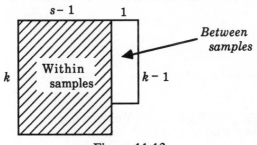

Figure 11.13

Simplification of Calculations
The following simplifications ease the calculations.

(i)   Total sum of squares: $\displaystyle\sum_{i=1}^{k} \sum_{j=1}^{s} (x_{ij} - \bar{x})^2 = \sum_{i=1}^{k} \sum_{j=1}^{s} x^2_{ij} - \frac{T^2}{N}$ (11.5a)

(ii)   Between samples sum of squares:

$$s \sum_{i=1}^{k} (\bar{x}_i - \bar{x})^2 = \frac{1}{s} \sum_{i=1}^{k} T_i^2 - \frac{T^2}{N} \tag{11.5b}$$

The term T^2/N is known as the **correction factor**.

Table 11.4

Source of variation	Components of sum of squares	Degrees of Freedom	Components of Variance	F ratio
Between samples	$s \displaystyle\sum_{i=1}^{k} (\bar{x}_i - \bar{x})^2$	$k - 1$	$\dfrac{s}{(k-1)} \displaystyle\sum_{i=1}^{k} (\bar{x}_i - \bar{x})^2$ $= s_m^2$	
				$\dfrac{s_m^2}{s_p^2}$
Within samples	$\displaystyle\sum_{i=1}^{k}\sum_{j=1}^{s} (x_{ij} - \bar{x}_i)^2$	$k(s-1)$	$\dfrac{1}{k(s-1)} \displaystyle\sum_{i=1}^{k}\sum_{j=1}^{s} (x_{ij} - \bar{x}_i)^2$ $= s_p^2$	
Total	$\displaystyle\sum_{i=1}^{k}\sum_{j=1}^{s} (x_{ij} - \bar{x})^2$	$ks - 1$		

(iii) $\dfrac{\text{Within samples}}{\text{sum of squares}} = \dfrac{\text{Total}}{\text{sum of squares}} - \dfrac{\text{Between samples}}{\text{sum of squares}}$ (11.5c)

We may now present a strategy of calculation using these simplifications for a new analysis of variance table. The overall strategy is as follows.

(i)   Calculate $T_i, T_i^2, T, T^2/N;\quad \displaystyle\sum_{i=1}^{k} T_i^2.$

(ii) Calculate $\displaystyle\sum_{i=1}^{k}\sum_{j=1}^{s} x_{ij}^2.$

(iii) Calculate total sum of squares via (11.5a), between samples sum of squares via (11.5b) and within samples sum of squares via (11.5c).

(iv) Complete ANOVA (analysis of variance) table.

(v)   Carry out decision-making strategy.

Example

Let us now apply the technique to the example on page 436.

We see that $s = 4$, $k = 3$ and hence that $N = ks = 12$. First we re-present the data and provide an extended table (Table 11.5).

Table 11.5

i	x_{ij}				T_i	T_i^2
1	22	21	25	18	86	7396
2	16	19	21	18	74	5476
3	10	17	12	15	54	2916
			Total		214	15788

(i) The correction factor $T^2/N = (214)^2/12 = 3816.\dot{3}$.

(ii) $\displaystyle\sum_{i=1}^{k} \sum_{j=1}^{s} x_{ij}^2 = 22^2 + 21^2 + ... + 15^2 = 4014$

(iii) Total sum of squares $= \displaystyle\sum_{i=1}^{k} \sum_{j=1}^{s} x_{ij}^2 - \frac{T^2}{N} = 4014 - 3816.\dot{3} = 197.\dot{6}$

Between samples sum of squares $= \dfrac{1}{s} \displaystyle\sum_{i=1}^{k} T_i^2 - \frac{T^2}{N} = \left(\frac{1}{4} \times 15788\right) - 3816.\dot{3} = 130.\dot{6}$

Within samples sum of squares $= 197.\dot{6} - 130.\dot{6} = 67$.

(iv) The variance analysis table is as follows.

Source of variation	Components of sum of squares	Degrees of freedom	Components of variance	F ratio	Critical value of F
Between samples	130.$\dot{6}$	2	65.3		
				8.78	1%
					$F_9^2 = 8.02$
Within samples	67	9	7.$\dot{4}$		
Total	197.$\dot{6}$	11			

(v) The decision-making strategy is
   1. $H_0 : \mu_1 = \mu_2 = \mu_3$.

2. We work with an α-risk of 1%.
3. The appropriate F distribution has 2 and 9 degrees of freedom.
4. The rejection region is $s_m^2/s_p^2 > 8.02$
5. The calculated F value is 8.78
6 There is a significant difference between sample means, and hence we conclude with slightly less than 1% risk of wrong judgement that there is difference in effectiveness of lubricating oils.

Note: The method of analysis of variance can be applied to two or more factors.

11.8 Introduction to Simple Linear Regression

In the next three sections we shall be concerned mainly with studying the relationship between two variables which appear from a cursory glance at the data to be related. We first tackle these problems from a *regression* approach, that is, we regard one of the variables as the dependent variable and the others as independent variables; the second line of attack is to study correlation effects which show to what degree the variables are related. We shall see that correlation is a less powerful technique than regression; however, there is a connection between the two ideas. We now look at an example which will be used to illustrate the main ideas of *simple regression* and *correlation*.

Example
The following data gives the percentage of sand and soil at different depths.

depth (cm) x	0	15	30	45	60	75	90	105	120
% sand $\quad y$	75.6	58.0	59.3	57.5	52.5	54.2	35.8	41.9	32.6

We shall seek to answer the following questions.
(i)   Is there a significant relationship between depth of measurement and percentage of sand?
(ii) Is there a simple relationship between depth and percentage of sand which will enable us to predict percentage of sand given the depth?
(iii) Could we find a simple relationship to estimate depth, having measured the percentage of sand in a soil sample?
(iv) In particular, is a linear relationship a good model?
(v)   If we are able to obtain a linear relationship, then how reliable will its predictions be? Can we perhaps place confidence intervals on the predictions?

Note that this situation is typical of many laboratory experiments. Often a straight line is fitted either to the observed data or data derived from the observations in such a way as to make a straight line reasonable. For example, if we have a theoretical model equation $pv^\gamma = c$ where c and γ are constants and both pairs of readings for p and v are experimentally observed then we can rearrange the model equation as

$$\ln p + \gamma \ln v = \ln c \quad \text{or} \quad P + \gamma V = D \quad \text{or} \quad P = D - \gamma V$$

which suggests a linear relationship between $P(= \ln p)$ and $V(= \ln v)$. Usually we would fit a straight line to the derived data and then estimate D and γ from this line. By a simple calculation we can deduce c, the original constant. Regression analysis puts this procedure on a rigorous footing by deriving the equation of the straight line algebraically (hence removing subjective fitting of the straight line) and also by measuring the goodness of fit of the line, placing confidence limits on the line slope and intercept and placing a confidence interval on any prediction from the line.

In this example, a straight line fit is attempted in the hope that it is a reasonable fit and that it will provide a reasonable prediction model.

Provided the observations are relatively few in number, it is worthwhile plotting each observation (x_i, y_i) on a graph with Cartesian axes. Once again, we emphasise that our observations are but a sample of all possible observations. The observations from the above example are plotted in Figure 11.14. The points show a general downwards trend from left to right, but it is by no means clear that the fitting of a straight line will give reasonable predictions. In Figure 11.15 we have drawn in a straight line which qualitatively seems to fit the plotted data.

It has the special feature that it passes through the point with coordinates $(\bar{x}, \bar{y})$; the reason for this will become apparent later. We shall fit the line

$$y = \alpha + \beta x \tag{11.6}$$

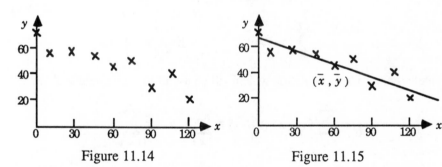

Figure 11.14   Figure 11.15

The following points are to be noted

(i)   For each observed value of x_i it may be that the observed value y_i does not equal the value of y obtained when substituting $x = x_i$ in (11.6) We write

$$y_i = \alpha + \beta x_i + \xi_i, \text{ where } i = 1, ..., n. \tag{11.7}$$

Equation (11.7) is the **simple regression model**. The term ξ_i is an error term and the ξ_i are independent random variables having mean 0 and variance σ^2.

(ii)   We have to obtain estimates of α and β from the observations.

(iii)   If $\beta = 0$ there is no regression model and we must test whether $\hat{\beta} = 0$ is a possible estimate.

(iv)   We *assume* that the values x_i are not subject to error but that y_i alone is subject to the errors in the model.

(v)   The model itself may not be an ideal relationship for the true population of x and y, but even if it were, the samples of observations may not fit any such model exactly because of observational errors in y. In other words, we will obtain an *estimated regression line*

$$\hat{y}_i = \hat{\alpha} + \hat{\beta} x_i$$

Note that the 'hats' on α, β and y_i indicate that they are estimates.

We shall first estimate the parameters α and β and then place confidence intervals on them.

From (11.7) we have an expression for the error at each observed point

$$\xi_i = y_i - \alpha - \beta x_i \tag{11.8}$$

The method of least squares minimises

$$\sum_{i=1}^{n} \xi_i^2 = \sum_{i=1}^{n} (y_i - \alpha - \beta x_i)^2 = S(\alpha, \beta)$$

We solve the equations

$$\frac{\partial S}{\partial \alpha} = 0 = \frac{\partial S}{\partial \beta}$$

These lead to the *normal equations*

$$\left. \begin{array}{l} \displaystyle\sum_{i=1}^{n} y_i = n\alpha + \beta \sum_{i=1}^{n} x_i \\[3mm] \text{and } \displaystyle\sum_{i=1}^{n} x_i y_i = \alpha \sum_{i=1}^{n} x_i + \beta \sum_{i=1}^{n} x_i^2 \end{array} \right\} \tag{11.9}$$

We shall now introduce a notation which will be useful in subsequent analysis.

Let

$$\left. \begin{array}{l} \displaystyle S_{xy} = \sum_{i=1}^{n} (x_i - \bar{x})(y_i - \bar{y}) \equiv \sum_{i=1}^{n} x_i y_i - \bar{x} \sum_{i=1}^{n} y_i \\[4mm] \equiv \displaystyle\sum_{i=1}^{n} x_i y_i - \bar{y} \sum_{i=1}^{n} x_i \\[4mm] \text{and } \displaystyle S_{xx} = \sum_{i=1}^{n} (x_i - \bar{x})^2 \equiv \sum_{i=1}^{n} x_i^2 - \bar{x} \sum_{i=1}^{n} x_i \\[4mm] \text{and } \displaystyle S_{yy} = \sum_{i=1}^{n} (y_i - \bar{y})^2 \equiv \sum_{i=1}^{n} y_i^2 - \bar{y} \sum_{i=1}^{n} y_i \end{array} \right\} \tag{11.10}$$

Eliminating α from equations (11.9) we find that the estimate of β is

$$\hat{\beta} = \frac{n\sum_{i=1}^{n} x_i y_i - (\sum_{i=1}^{n} x_i)(\sum_{i=1}^{n} y_i)}{n\sum_{i=1}^{n} x_i^2 - (\sum_{i=1}^{n} x_i)^2} = \frac{S_{xy}}{S_{xx}} \qquad (11.11)$$

From the first equation (11.9), division by n and rearrangement produces

$$\hat{\alpha} = \bar{y} - \hat{\beta}\bar{x} \qquad (11.12)$$

Equations (11.11) and (11.12) produce least squares estimates of the parameters

α and β. The estimated regression is $\hat{y} = \hat{\alpha} + \hat{\beta}x$

From (11.12) we obtain this equation in the form

$$\hat{y} - \bar{y} = \hat{\beta}(x - \bar{x}) \qquad (11.13)$$

and we see immediately that the point with coordinates $(\bar{x}, \bar{y})$ lies on the line as hinted at earlier.

Example

For our example we shall calculate α and β. We must first calculate S_{xx} and S_{xy}. We shall also calculate S_{yy} since this will be required in future analysis. We obtain these values by first calculating

$$S_{xx} = 13500 \qquad S_{yy} = 1422.36 \qquad S_{xy} = -4059$$

Hence, from (11.11), $\hat{\beta} = -4059/13500 = -0.301$ (3 s.f.)

and from (11.12), $\hat{\alpha} = 51.93 + 0.301 \times 60 = 69.99$ (2 d.p.)

This gives the required regression line as $y = 69.99 - 0.301x$

Alternatively, we could have used (11.13) to obtain $y - 51.93 = -0.301(x - 60)$.

Confidence intervals for α and β

We quote the 95% confidence intervals for α and β as

$$\hat{\alpha} \pm t_{0.025}\, s \sqrt{\frac{1}{n} + \frac{\bar{x}^2}{S_{xx}}} \qquad (11.14a)$$

and

$$\hat{\beta} \pm t_{0.025}\, \frac{s}{\sqrt{S_{xx}}} \qquad (11.14b)$$

where

$$s^2 = \frac{1}{n-2} \sum_{i=1}^{n} (y_i - \hat{y}_i)^2.$$

Example

We return to our example to first calculate s^2. Although we shall see later how this can be obtained by a back-door approach, we shall, for the moment have to calculate each ξ_i directly. Calculations (find $\hat{y}_i$ first from the equation

$\hat{y}_i = \hat{\alpha} + \hat{\beta} x_i$) produce $s^2 = \dfrac{201.98}{7} = 28.85$ and hence $s = 5.372$ (3 d.p.).

A 95% confidence interval is first found by taking from Table A.II the value of $t_{0.025}$ with 7 degrees of freedom to be 2.365.

For α the 95% confidence interval is $69.99 \pm 2.365 \times 5.372 \sqrt{\dfrac{1}{9} + \dfrac{3600}{13500}}$

i.e. 69.99 ± 7.81 (3 s.f.) i.e. $(62.18, 77.80)$

For β the 95% confidence interval is $-0.301 \pm 2.365 \times \dfrac{5.372}{\sqrt{13500}}$

i.e. -0.301 ± 0.109 (3 s.f.) i.e. $(-0.410, -0.192)$

Now, since the confidence interval for β does not include the possibility $\beta = 0$ we assume that we can reject (at the 5% level) the hypothesis $\beta = 0$ and this suggests that the regression equation is a significant one.

Then the 95% confidence interval for the mean value is (for a given value of x, x_0 say)

$$\hat{\alpha} + \hat{\beta} x_0 \pm t_{0.025}\, s \sqrt{\frac{1}{n} + \frac{(x_0 - \bar{x})^2}{S_{xx}}} \qquad (11.15)$$

The 95% confidence interval for a single observation y corresponding to the value x_0 is given by

$$\hat{\alpha} + \hat{\beta} x_0 \pm t_{0.025}\, s \sqrt{\frac{1}{n} + \frac{(x_0 - \bar{x})^2}{S_{xx}} + 1} \qquad (11.16)$$

In each case, the t value is from the distribution with $(n - 2)$ degrees of freedom.

For the current example, the 95% confidence interval for the mean of the y-values corresponding to $x = 80$ is

$$69.99 - 0.301 \times 80 \pm 2.365 \times 5.372 \sqrt{\frac{1}{9} + \frac{(80 - 60)^2}{13500}}$$

i.e. 45.91 ± 4.77 or $(41.14, 50.68)$

This interval seems quite large: if we examine equation (11.15) we can see that the further x_0 is from $\bar{x}$ then the larger the quantity $(x_0 - \bar{x})^2$ and hence the wider the confidence interval.

The 95% confidence interval for a single observation is, from (11.16)

$$69.99 - 0.301 \times 80 \pm 2.365 \times 5.372 \sqrt{\frac{1}{9} + \frac{(80 - 60)^2}{13500} + 1}$$

i.e. 45.91 ± 13.57 or $(32.34, 59.48)$

This interval is clearly so large as to be of little comfort with regard to the precision of an estimate.

Finally, we find the 95% confidence interval for a single observation corresponding to $x = 180$. Using equation (11.16) we obtain

$$69.99 - 0.301 \times 180 \pm 2.365 \times 5.372 \sqrt{\frac{1}{9} + \frac{(180 - 60)^2}{13500} + 1}$$

i.e. 15.81 ± 18.75 or $(-2.94, 34.56)$

We have clearly got a useless result here. This pinpoints the dangers of *extrapolation*. In the first place, the value of $(x_0 - \bar{x})^2$ is so large that the confidence interval is very wide indeed. In the second place, we cannot be sure that a linear model is applicable outside the range of observations; see Figure 11.16. We should therefore be very wary of extrapolation and avoid it wherever possible.

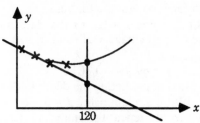

Figure 11.16

Further conclusions

It is worth pursuing this example a little further. First we shall see what is the minimum width of confidence interval by taking $x_0 = 60$. Then equation (11.15) gives the 95% confidence interval for the mean of the y-values as

$$69.99 - 0.301 \times 60 \pm 2.365 \times 5.372 \sqrt{\frac{1}{9} + 0} \quad (\bar{x} \text{ being } 60)$$

i.e. 51.93 ± 4.23 or $(47.70, 56.16)$

Even this interval is quite wide – about 16% of the middle value. Second, we remark that for a single observation even at $x_0 = \bar{x}$ and even if $n \to \infty$, the confidence interval is still not zero.

The 95% confidence interval has a limiting minimum value of

$$69.99 - 0.301 \times 60 \pm 2.365 \times 5.372 \sqrt{0 + 0 + 1}$$

i.e. 51.93 ± 12.70 or $(39.23, 64.63)$

This width is due to the inherent variability in the data.

Finally, we illustrate diagrammatically the effect on the estimates of the mean y–value and on the single observation of varying certain parameters. Figure 11.17(a) shows the effect on the interval estimate of the mean value y assuming that only the slope β is subject to error; Figure 11.17(b) shows the effect if only the intercept α is subject to error; Figure 11.17(c) shows the combined effects on the mean value and on a single observation.

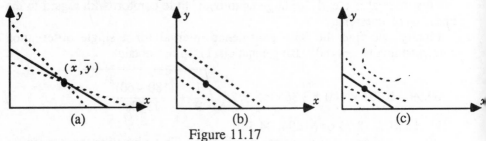

| (a) | (b) | (c) |

Figure 11.17

It should be noted that the effects increase as we move away from $\bar{x}$.

11.9 Further Features

What we have done so far is to assume that only the y-values are subject to error and to fit a straight line $y = \alpha + \beta x$ to the data using a least squares criterion. In our example on percentage y of sand in soil at a given depth x we obtained the estimated regression line for y on x:
$$y = 69.99 - 0.301x \qquad (11.17)$$
Suppose we were interested in predicting the depth knowing the percentage of sand. We should then be fitting a relationship $x = a + by$ (the regression line of x on y) to the data and we would repeat the analysis interchanging the roles of y and x.

Thus $\hat{b} = \dfrac{S_{xy}}{S_{yy}} = \dfrac{-4059}{1422.36} = -2.85$ (3 s.f.)

and $\hat{a} = \bar{x} - \hat{b}\,\bar{y} = 60 + 2.85 \times 51.93 = 207.9$

Therefore the regression equation of x on y is
$$x = 207.9 - 2.85y \qquad (11.18a)$$
Now if we re-express this equation with y as the left-hand side we obtain

$$y = \frac{207.9}{2.85} - \frac{x}{2.85} \text{ i.e. } y = 72.95 - 0.351x \qquad (11.18b)$$

This is not the same as equation (11.17). However, the lines given by equations (11.17) and (11.18a) pass through the point with coordinates $(\bar{x}\ ,\ \bar{y})$; see Figure 11.18 (which is not to scale).

If the data points lay exactly on a straight line we would expect the regression line of y on x and the regression line of x on y to coincide as the line on which the data points lay. The worse the fit of either regression line the greater the difference in their slopes. Notice that since both lines pass through $(\bar{x}, \bar{y})$ we could transfer the origin of axes to this point. This would introduce new variables $(x_i - \bar{x})$ and $(y_i - \bar{y})$ which explains the presence of these terms in S_{xx}, S_{xy} and S_{yy}.

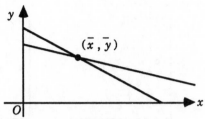

Figure 11.18

What fitting a regression line does not tell us is the degree of linear relationship between y and x, i.e. how closely they are related linearly. We shall seek to measure this closeness (and hence the goodness of fit of either regression line). This technique will measure the degree of correlation between x and y; neither variable is assumed to be independent as is the assumption with regression. The sample correlation coefficient, r, is given by

$$r^2 = \frac{(S_{xy})^2}{S_{xx} S_{yy}} \tag{11.19}$$

It can be shown that $r^2 \le 1$; since, clearly, $r^2 \ge 0$ it follows that $-1 \le r \le +1$

In our sand and soil example, $S_{xx} = 13500$, $S_{yy} = 1422.36$, $S_{xy} = -4059$ and hence $r^2 = 0.858$ (3 d.p.).

Interpretation of r^2

For the regression line $y = \hat{\alpha} + \hat{\beta}x$ we defined its slope $\hat{\beta} = \dfrac{S_{xy}}{S_{xx}}$, whilst for

the regression line $x = a + by$ we defined $\hat{b} = \dfrac{S_{xy}}{S_{yy}}$.

If we re-express the second line as $y = -\dfrac{\hat{a}}{\hat{b}} + \dfrac{1}{\hat{b}} x$ its slope is $\dfrac{1}{\hat{b}} = \dfrac{S_{yy}}{S_{xy}}$.

The ratio of these two slopes is $\hat{\beta} / (\dfrac{1}{\hat{b}}) = \dfrac{S_{xy}}{S_{xx}} \bigg/ \dfrac{S_{yy}}{S_{xy}} = \dfrac{(S_{xy})^2}{S_{xx} S_{yy}} = r^2$

If the two lines had coincided then $r^2 = 1$ and we have perfect linear correlation. If the slope of either regression line is positive we take r positive and in the case of $r^2 = 1$ we see that $r = 1$ gives perfect positive linear correlation; similarly $r = -1$ gives perfect negative linear correlation. It should be clear that the closer r^2 is to 1 then the better the fit of the regression line of y on x (and of the regression line of x on y) and the more closely the variables x and y are correlated.

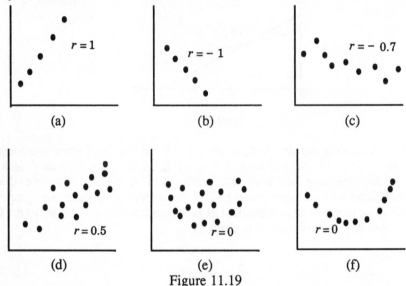

Figure 11.19

In Figure 11.19 we see schematic representations of various values of r. In (a) we have a case of perfect positive correlation whilst in (b) we have perfect negative correlation. In (c) and (d) we see the lowering of $|r|$ as the scatter increases. The last two diagrams have the same value of r but represent different situations altogether; whereas (e) represents random scatter, (f) has definite indications of a relationship between x and y which, because it is not linear, produces a zero r score. As long as we accept the reasonableness of a linear model, then the higher the value of $|r|$ the better we are pleased.

It is not possible to give definite guidelines as to what values of r^2 correspond to good fits. In some scientific experiments a value of r^2 of 0.96 would perhaps be considered only as fair; on the other hand in social science studies an r^2 value of 0.70 is considered very good. It all depends on the data and how it can be collected.

We have already mentioned that there are several sources of variability associated with a simple linear regression model. We shall now endeavour to test the goodness of fit of the model to the data by comparing these sources of variability which we shall measure by means of variances.

Notice that we assumed that the statistically independent errors ξ_i in our model were distributed with mean 0 and variance σ^2; from the model equation (11.7) it follows that the distribution of each of the independent random variables y_i has mean $\alpha + \beta x_i$ and variance σ^2.

We can show that

$$\sum_{i=1}^{n} \hat{\xi}_i^2 = \sum_{i=1}^{n} (y_i - \bar{y})^2 - \sum_{i=1}^{n} \hat{\beta}^2 (x_i - \bar{x})^2 = \sum_{i=1}^{n} (y_i - \bar{y})^2 - \sum_{i=1}^{n} (\hat{y}_i - \bar{y})^2$$

where $\hat{y}_i$ is the estimated mean value of y_i from the regression equation [see equation (11.13)].

Hence we may rearrange the relationship to give

$$S_{yy} = \sum_{i=1}^{n} (y_i - \bar{y})^2 = \sum_{i=1}^{n} (\hat{y}_i - \bar{y})^2 + \sum_{i=1}^{n} \hat{\xi}_i^2 \qquad (11.20)$$

If we refer to Figure 11.20, we find that $(y_i - \bar{y})$ is the difference between the observation y_i and the mean of observations $\bar{y}$ and that $(\hat{y}_i - \bar{y})$ is the difference between the estimated value $\hat{y}_i$ and the mean $\bar{y}$. Hence we may write equation (11.20) in words:

$$\begin{pmatrix} \text{Total sum of squares due to} \\ \text{observed values } y_i \end{pmatrix} = \begin{pmatrix} \text{Sum of squares due to} \\ \text{estimated values } \hat{y}_i \end{pmatrix} + \begin{pmatrix} \text{Sum of squares} \\ \text{due to errors} \end{pmatrix} \quad (11.21)$$

We have therefore partitioned the total sum of squared deviations of the original observations about the mean into
(a)   a component due to values estimated by the regression equation and
(b)   a component due to errors about the regression line.

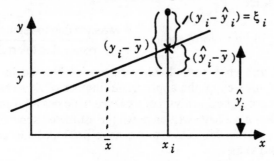

Figure 11.20

Now $(\hat{y}_i - \bar{y})^2 = \hat{\beta}^2 (x_i - \bar{x})^2$ and hence the component

$$\sum_{i=1}^{n} (\hat{y}_i - \bar{y})^2 = \hat{\beta}^2 \sum_{i=1}^{n} (x_i - \bar{x})^2 = \hat{\beta}^2 S_{xx} = \frac{(S_{xy})^2}{S_{xx}} = \frac{(S_{xy})^2 S_{yy}}{S_{yy} S_{xx}}$$

Also $r^2 = \dfrac{(S_{xy})^2}{S_{xx} S_{yy}}$ from equation (11.19).

Therefore

$$\sum_{i=1}^{n} (\hat{y}_i - \bar{y})^2 = r^2 S_{yy} \tag{11.22a}$$

Hence

$$\sum_{i=1}^{n} \hat{\xi}_i^{\,2} = (1 - r^2) S_{yy} \tag{11.22b}$$

Second interpretation of r^2

The quantity $r^2 S_{yy}$ may be regarded as that part of the total sum of squares which is 'explained' by the regression line. The quantity $(1 - r^2)\, S_{yy}$ is the sum of squares due to errors about the regression line. Hence r^2 measures the proportion of the total sum of squares explained by the regression equation. We see that the closer r^2 is to 1, the less the sum of squares due to errors (all the sum of squares being explained by the regression line) and the line fits perfectly, since

if $\displaystyle\sum_{i=1}^{n} \hat{\xi}_i^{\,2} = 0$ it follows that *all* the individual errors $\hat{\xi}_i$ are zero.

Analysis of variance

The technique will be to form the ratio $\left(\dfrac{\text{variance due to regression}}{\text{variance due to errors}} \right)$ and to examine its value with an F test. Each variance is obtained by dividing the appropriate sum of squares by the appropriate number of degrees of freedom. We now refer to the theoretical analysis of variance table below, Table 11.6.

In the second column we have written down the estimated sums of squares that we derived earlier; in the third column we have produced the corresponding theoretically expected values.

In column 4 we meet the appropriate degrees of freedom. With n sample items there is a total of $(n-1)$ degrees of freedom available in the system.

Since 2 data points are sufficient to determine any line it follows that $(n-2)$ points are competing to share the total error about the regression line. This leaves one degree of freedom to be associated with the sum of squares due to the regression.

Alternatively, we notice that the number of degrees of freedom is the coefficient of σ^2 in the appropriate expected sum of squares. We will then obtain unbiased estimates of the variance.

In columns 5 and 6 we produce the variance components by dividing a sum of squares by its associated number of degrees of freedom.

Table 11.6

Source of variability	Components of sums of squares		Degrees of freedom	Components of variance		F
	estimated	expected		estimated	expected	
Due to regression line	$r^2 S_{yy}$	$\sigma^2 + \beta^2 S_{xx}$	1	$r^2 S_{yy}$	$\sigma^2 + \beta^2 S_{xx}$	$F^1_{(n-2)}$
About regression line	$(1 - r^2) S_{yy}$	$(n - 2)\sigma^2$	$(n - 2)$	$\dfrac{(1 - r^2) S_{yy}}{(n - 2)}$	σ^2	$\dfrac{r^2 (n - 2)}{1 - r^2}$
Total	S_{yy}	$(n - 1)\sigma^2 + \beta^2 S_{xx}$	$(n - 1)$			

Finally, we divide the two variances to obtain the ratio $r^2 (n - 2)/(1 - r^2)$.

For the sandy soil example, we had $S_{yy} = 1422.36$, $n = 9$ and $r^2 = 0.858$. The variance analysis table is as below.

Source of variability	Component of sum of squares	Degrees of freedom	Component of variance	F	Critical values of F
Due to regression line	1220.38	1	1220.38		5% 5.59
				42.2	
Errors about line	201.98	7	28.91		1% 12.25
Total	1422.36	8			

The calculated F value is again much greater than the 1% value and we therefore conclude, with a much less than 1% risk, that the regression is a significant one.

Error variance

It is customary to speak of the second entry in the component of variance column of the VA table, $(1 - r^2) S_{yy}/(n - 2)$, as the **error variance** associated with the linear regression model. Hence we might quote the estimated regression equation together with the error variance. For the sandy soil example the equation is $\hat{y} = 69.99 - 0.301x$ with error variance 28.91.

Review example

We now collect together all the techniques we have developed so far and apply them to the example below.

Age of machine in years, x_i	1	1.5	2	3	3	3.5	4	4	5	5	6	7
Cost of previous year's maintenance (£100's), y_i	22	20	22	21	23	25	24	27	25	27	26	29

The steps we shall perform are as follows

(i)   Calculate $\bar{x}$, $\bar{y}$, S_{xx}, S_{xy}, S_{yy}

(ii)   Calculate the percentage overall fit; $100r^2$.

(iii)   Validate the regression equation by a variance analysis.

(iv)   Estimate the regression equation.

(v)   Place confidence intervals on $\hat{\alpha}$ and $\hat{\beta}$.

(vi)   Place confidence intervals on the estimated cost $\hat{y}$ and on an observed cost y_i for a given value x_0.

(vii)   Draw appropriate conclusions.

We assume a regression line of the form $y = \alpha + \beta x$

(i)   The calculations give $\bar{x} = 3.75$, $\bar{y} = 24.25$; $S_{xx} = 35.75$, $S_{xy} = 47.25$, $S_{yy} = 82.25$

(ii)   Since $r^2 = \dfrac{(S_{xy})^2}{S_{xx} S_{yy}} = \dfrac{(47.25)^2}{35.75 \times 82.25} = 0.7593$ (4 s.f.),

the percentage overall fit is 75.93%

(iii)   We now form the variance analysis table on the next page: note that the component of sum of squares due to regression is $r^2 S_{yy} = 62.45$ and that due to errors is $(1 - r^2) S_{yy} = 82.25 - 62.45 = 19.80$.

Hence we conclude, with considerably less than 1% risk of being wrong, that the regression effect is significant.

(iv)   To find the regression equation we first calculate

$$\hat{\beta} = \frac{S_{xy}}{S_{xx}} = \frac{47.25}{35.75} = 1.322$$

Then we use the equation for the estimated regression line in the form

$$\hat{y} - \bar{y} = \hat{\beta}(x - \bar{x})$$

i.e. $\hat{y} - 24.25 = 1.322(x - 3.75)$

We obtain the estimated regression equation as $\hat{y} = 1.32x + 19.29$ with error variance 1.98.

Source of variability	Component of sum of squares	Degrees of freedom	Component of variance	F	Critical values of F
Due to regression line	62.45	1	62.45		5% F^1_{10} = 4.96
				31.54	
Errors about line	19.80	10	1.98		1% F^1_{10} = 10.04
Total	82.25	11			

(v) To estimate the 95% confidence intervals we note that there are 10 degrees of freedom, and the appropriate t – score is found to be 2.228.

We also require $s = \sqrt{1.98} = 1.407$. Hence the 95% confidence interval for β is

$$\hat{\beta} \pm 2.228 \; \frac{s}{\sqrt{S_{xx}}} \quad \text{i.e. } 1.322 \pm 2.228 \times \frac{1.407}{5.979} \quad \text{i.e. } 1.322 \pm 0.524$$

or (0.798, 1.846).

The 95% confidence interval for α is

$$\hat{\alpha} \pm 2.228 \, s \sqrt{\frac{1}{n} + \frac{\bar{x}^2}{S_{xx}}} \quad \text{i.e. } 19.29 \pm 2.228 \times 1.407 \sqrt{\frac{1}{12} + \frac{(3.75)^2}{35.75}}$$

i.e. 19.29 ± 2.16 or (17.13, 21.45).

(vi) The 95% confidence interval for the expected value $\hat{y}$ corresponding to the

value x_0 is $\hat{\alpha} + \hat{\beta} x_0 \pm 2.228s \sqrt{\dfrac{1}{n} + \dfrac{(x_0 - \bar{x})^2}{S_{xx}}}$ and that for a single observation is $\hat{\alpha} + \hat{\beta} x_0 \pm 2.228s \sqrt{\dfrac{1}{n} + \dfrac{(x_0 - \bar{x})^2}{S_{xx}} + 1}$

(vii) Our conclusions are that, with much less than a 1% error of wrong judgement, there is a significant increase in the annual cost of maintaining the machines with age of the machine. The relationship is represented by

the regression equation $\hat{y} = 1.32x + 19.29$ with error variance 1.98, $\hat{y}$ is

the estimated annual maintenance cost at age x years. In the sample of machines examined whose ages ranged between 1 and 7 years, 75.93% of the variability observed in annual maintenance cost could be attributed to age

effects. Since the 95% confidence interval for α does not include the value zero we can conclude (with less than 5% risk) that the expected cost of the first year's maintenance of a machine of the type studied is not zero. (Perhaps this may represent some basic fixed annual charge.)

The marginal increase in annual maintenance cost is estimated as 1.32 $(\hat{\beta})$.

11.10 Concluding Remarks

(i)   One question with reference to the review example remains to be answered: what effect, if any, do the repeated observations have on the variance analysis table? Is there a possibility of subdividing further the sum of squares due to erros about the regression line? Suppose, in general that there are k distinct values of x: $x_1, x_2, ..., x_k$ out of the n observations. Let n_1 be the number of occasions on which x_1 occurs, n_2 the number of occasions on which x_2 occurs etc. (any of these values n_i can be 1). Then, of course, $n = n_1 + n_2 + ... + n_k$. We denote by y_{i1}, y_{i2} etc. those observed values of y which correspond to the value x_i and

denote by $\bar{y}_i$ the average of these values y_{i1}, y_{i2} etc. Now we can partition the error sum of squares into two halves: that part which is due to the variability of y_{i1}, y_{i2} etc. which correspond to each given x_i and a second part which really represents lack of fit of the model. In other words, the first part reflects experimental or observational error, whilst the second part is the error due to the simplicity of a linear model. Perhaps we can apply this reasoning to the example.

First we will compute the component of sum of squares due solely to experimental error. We note that this has $(n - k)$ associated degrees of freedom (k different values of x_i). Then we subtract this from the error sum of squares and produce a variance analysis table which contains three sub-components.

Applying this procedure, we had three instances of repeated measurements in 12 observations i.e. $k = 9$. For $x_4 = 3$ we had $y_{41} = 21$, $y_{42} = 23$

and therefore $\bar{y}_4 = 22$. Similarly, for $x_6 = 4$ we had $y_{61} = 27$, $y_{62} = 24$ and $\bar{y}_6 = 25.5$; for $x_7 = 5$ we had $y_{71} = 25$, $y_{72} = 27$ and $\bar{y}_7 = 26$. If we form the quantity

$$\sum_{i=1}^{n_1} (y_{1i} - \bar{y}_1)^2 + \sum_{i=1}^{n_2} (y_{2i} - \bar{y}_2)^2 + \ldots + \sum_{i=1}^{n_k} (y_{ki} - \bar{y}_k)^2$$

we notice that, since most of the observations are not replicated, all but three of these terms will vanish and we shall be left with

$$\sum_{i=1}^{2} (y_{4i} - \bar{y}_4)^2 + \sum_{i=1}^{2} (y_{6i} - \bar{y}_6)^2 + \sum_{i=1}^{2} (y_{7i} - \bar{y}_7)^2$$

$$= (21 - 22)^2 + (23 - 22)^2 + (27 - 25.5)^2 + (24 - 25.5)^2 + (25 - 26)^2 + $$
$$(27 - 26)^2$$

$$= 1 + 1 + 2.25 + 2.25 + 1 + 1 = 8.5$$

This sum of squares is associated with $n - k = 12 - 9$ i.e. 3 degrees of freedom. If we refer back to page 455, we find that the sum of squares due to errors about the line was 19.80 with 10 associated degrees of freedom. Since replication of observations has accounted for 8.5 of this total and has 3 associated degrees of freedom, it follows that we still have to account for $19.80 - 8.5 = 11.30$ of the sum of squares and $10 - 3 = 7$ degrees of freedom; this variability is attributable to lack of fit of the line to the data. The extended variance analysis table follows.

Source of variability	Component of sum of squares	Degrees of freedom	Component of variance	F	Critical values of F
Due to regression line	62.45	1	62.45	22.07	5% $F_7^1 = 5.59$ 1% $F_7^1 = 12.25$
Errors about line:					
Lack of fit	11.3	7	1.61	0.57	5% $F_3^7 = 8.85$
Experimental error	8.5	3	2.83		1% $F_3^7 = 27.7$
Total	82.25	11			

We can now see that the regression is still clearly significant at the 1% level $(22.07 > 12.25)$. Furthermore, the contribution of the lack of fit is not significant at even the 5% level (0.57 << 8.85). This latter result implies that we need not consider a model containing higher order powers of x, since a linear model seems to fit the data quite well. Note that in this example the presence of replications has shifted the emphasis onto the experimental error and the value 2.83 is now the unbiased estimate of σ^2. It might well be argued that it is not worth partitioning the error sum of squares unless all observations are replicated, certainly only when the majority are; this is a conclusion we should support as only 3 terms contributed to the value 2.83 and the estimate is unlikely to be a good one.

(ii)   We ought to say a little more about the connection between regression and correlation. Correlation states the degree to which two normal variates are related, whereas regression (and it need not be linear regression) tells us how the variables are related. Regression answers more interesting questions than does correlation and we have, in fact, relegated correlation to the role of an aid to regression analysis. It must be borne in mind that a high value of the correlation coefficient r *does not necessarily indicate a cause and effect relationship*. If two variables are highly correlated it suggests further investigation of the situation. It would be foolish to assume that a high correlation coefficient between output of steel and production of sulphuric acid implied that the one had a marked effect on the other; the truth is more likely that both are influenced by a third factor – growing demand for industrial products. Yet even a linear regression analysis might give a value of $\hat\beta$ significantly different from zero. More investigation is clearly called for before any definite conclusions can be reached.

Problems

Section 11.2

1   In a chemical plant the acid content of the effluent is monitored. From 400 measurements the acid content in g/litre of effluent had mean 0.142 and standard deviation 0.008. Is the result consistent with a mean acid content of 0.1413 g/litre obtained from previous tests? State any assumptions made.

2   Two observers took separate sets of pressure readings for the same orifice, 100 readings were taken by each and the results were (in mm of mercury).
     First observer: mean reading 3.571,   Standard deviation 0.203
     Second observer: mean reading 3.637,   Standard deviation 0.203
     Assuming that each set of readings can be regarded as a random sample of the variations in pressure, determine whether the difference in mean readings is significant at the 5% level.

3   The distribution of daily-output from a particular machine has mean 500 units and standard deviation 50 units. With a new mode of operation a sample of 10 daily runs gave a mean of 550 units. Show that the new method has made a significant change in the daily rate and quote 95% confidence limits for the new mean daily rate.

4 Past experience indicates that wire rods of a particular type have a breaking load which is distributed with standard deviation 60 N. In testing large batches of rods, what sample size should be used to obtain a mean breaking load within 8 N of the mean for the batch with probability 0.99?

5 It is considered satisfactory to know the mean of a measurement within 0.1 of its true value with a probability of 0.95. A batch was measured and found to have standard deviation 0.53. What sample size should we use to obtain a satisfactory estimate of the mean of each batch?

6 A firm manufactures fibres with breaking strengths having 400 N and standard deviation 40 N. A new process is claimed to improve the mean breaking strength. What should be the decision rule for rejecting the old process at 1% risk if it is decided to test 100 fibres? What is the probability of accepting the old process if the new process has increased the mean breaking strength to 420 N without altering the standard deviation? Draw an operating characteristic curve and a power curve for values of the new mean of 390, 400, 410, ..., 450.

7 Draw an operating characteristic curve and a power curve for Problem 3.

8 A coin is thrown 100 times. With a 5% risk calculate the probability of accepting the coin as being fair ($p = 0.5$) when $p = 0.1, 0.2,....,0.9$. Draw the operating characteristic curve and the power curve.

Section 11.3

9 Using Table A.II, calculate the probabilities that

(a) $|t| > 3.01$, $v = 13$	(b) $|t| > 2.08$, $v = 21$	(c) $|t| > 1.76$, $v = 14$
(d) $|t| \leq 2.84$, $v = 7$	(e) $|t| \leq 4.77$, $v = 5$	(f) $t > 7.45$, $v = 3$
(g) $t < -2.39$, $v = 24$	(h) $t \leq 11.24$, $v = 8$	

10 Find 5% critical values of t for $v = 15, 28, 35, 250$.

11 A machine is designed to produce bolts of diameter 0.14 cm. To find whether the machine is in working order a sample of 10 bolts is taken. Recorded thicknesses are 0.1325, 0.14, 0.135, 0.1425, 0.135, 0.1375, 0.1375, 0.1275, 0.14, 0.1325. With a level of significance of 5%, determine whether the machine is in working order.

12 Paper is produced with a supposed mean thickness of 0.004 cm. 9 sample measurements are taken and their mean is 0.0037 cm with a standard deviation of 0.0003 cm. With a 5% risk, decide whether the output differs significantly from the supposed standard.

13 A random sample of 10 items has a mean of 224 and a standard deviation of 13.2. Could the sample have come from a population with a mean of 208?

14 A machine produces components to a nominal length of 4.800 cm. It is reset every morning. The first hour's production was sampled and the following lengths measured. 4.802, 4.801, 4.804, 4.801. 4.798, 4.802, 4.800, 4.801, 4.801, 4.802, 4.803, 4.801, 4.799, 4.800, 4.801, 4.805. Is there evidence of an incorrect setting?

15 Measurements of the diameters of a random sample of 200 bearings made by a firm during one week showed a mean of 1.224 cm and a standard deviation of 0.042 cm. Find 95% and 99% confidence limits for the mean diameter of all the bearings.

16 Boilers at a factory used a mean weight of fuel per day of 30.10 units. A new design of boiler was tested over 10 days and the following weights of fuel were recorded: 30.15, 29.84, 28.32, 27.50, 30.32, 29.41, 29.57, 28.54, 32.43, 29.82. Is there evidence of an economy in fuel with the new design?

17 A sample of 10 items gave the following results: 33.5, 34.8, 34.0, 33.4, 34.6, 35.1, 33.6, 34.2, 34.2, 34.6. Find 95% and 99% confidence intervals for the population mean. State any assumptions made.

18 A sample of 5 items has masses 6.7, 5.9, 6.4, 5.8 and 6.2 kg. What are the 95% confidence limits for the mean of the population?

19 Ten measurements of a physical constant gave a mean of 123.77 units and the sum of their squared deviations from this value was 0.9261 squared units. Find 95% confidence limits for the true value of the constant.

20 A random sample of size 12 had a mean of 14.3 and variance of 2.1. Test at the 5% level of significance the claim that the population mean is 15.0 as opposed to the alternative hypothesis that the mean is less than 15.0

21 For a type of paint the drying time (in hours) when it was tried out on 9 test specimens was
$$4.0, \quad 3.7, \quad 3.8, \quad 4.5, \quad 5.0, \quad 4.3, \quad 3.6, \quad 4.1, \quad 3.0$$
   (i)   Calculate the mean drying time for the 9 specimens.
   (ii)   Calculate an unbiased estimate of the variance of the population of all drying times for this paint.
   (iii) Find 95% confidence limits for the population mean if it is known from past experience that the standard deviation of drying times for this type of paint is 0.6 hours.
   (iv)   Find 95% confidence limits for the mean if no previous information about drying times is available.   (EC)

Section   11.4
22 (i)   Find the values of χ^2 which are exceeded with a probability of 5% for $v = 12, 7, 3,$ 80, 200.
    (ii)   Repeat the above for a probability of 1%.
    (iii) Repeat for a probability of 50%.
    (iv)   What are the approximate probabilities that χ^2 will be greater than 20 for $v = 12,$ 30?
    (v)   What is the probability that (a) $\chi^2 \leq 26.30$, (b) $\chi^2 \leq 23.54$, (c) $7.962 \leq \chi^2 \leq 15.34$ with $v = 16$?

23 The variance of a group of 8 observations is 3.52. What are the 95% confidence limits for the population variance? State any assumptions made.

24 A sample of 20 observations gave a standard deviation 3.72. Is this compatible with the hypothesis that the sample is from a normal distribution with variance 4.35?

25 The following data are from a normal population. With a 5% risk, test the claim that $\sigma = 2.5$ as opposed to the alternative claim that $\sigma \neq 2.5$: 20.2, 40.2, 37.6, 31.2, 26.0, 36.6, 26.4, 21.6, 32.0.

26 Repeat 25 for the alternative claim that $\sigma > 2.5$.

27 A company manufactures components with a certain life. A random sample of 25 components has a standard deviation of 140 hours. What are the 95% confidence limits for the standard deviation of all components?

28 The standard deviation of wires manufactured by a firm is 533.8 N. A new process is tested and the standard deviation of a sample of 10 wires is 667.2 N. Is the increase in variability significant?

29 Find a 95% confidence interval for the variance of the normal population from which a random sample of 6 items gave $\Sigma(x_i - \bar{x})^2 = 40$.

30 A sample of size 20 gave an estimate for σ^2 of 16.4. Find a 99% confidence interval for the population standard deviation.

31 Find a 99% confidence interval for the variance of the population from which the following sample was taken : 7.4, 12.5, 9.7, 9.0, 10.6, 11.3.

Section 11.5

32 Find
   (a) $f_{0.05}(4, 6), f_{0.05}(10, 7), f_{0.05}(12, 32), f_{0.05}(12, 34), f_{0.05}(12, 35), f_{0.05}(11, 35)$
   (b) $f_{0.025}(3, 3), f_{0.025}(10, 120)$ (c) $f_{0.01}(10, 8), f_{0.01}(8, 10)$
   (d) $f_{0.99}(3, 10), f_{0.95}(3, 10)$

33 A sample of 8 observations has variance 2.59 and a sample of 16 has variance 4.92. Can these samples be reasonably regarded as coming from populations with the same variance?

34 For each of the cases below we show the sample size and sample variance for two samples. Find whether they can be reasonably regarded as coming from the same normal population and if so obtain a pooled estimation of the population. Work with 5% risk.
   (a)   size 4, variance 0.537; size 16, variance 0.155
   (b)   10, 760/9; 8, 587.5/7   (c)   16, 6.2; 21, 3.9
   (d)   21, 42.9; 13, 146.0

35 Two voltmeters are tested by comparing ten independent readings from each with the voltage of a standard cell (whose voltage is known to be 2.10 volts). The results are shown below. Is there evidence that one meter is more consistent than the other?

| First meter | 2.11 | 2.15 | 2.14 | 2.10 | 2.09 | 2.11 | 2.12 | 2.15 | 2.13 | 2.14 |
| Second meter | 2.12 | 2.06 | 2.02 | 2.08 | 2.11 | 2.05 | 2.06 | 2.03 | 2.05 | 2.08 |

36 The following data for two samples is given. Assuming the observations from each sample to be normally distributed test the hypothesis that the samples could have come from the same population.

First sample $n = 11$, $\Sigma x_i = 66$ $\Sigma x_i^2 = 402.4$

Second sample $n = 6$, $\Sigma x_i = 37.6$ $\Sigma x_i^2 = 195.36$

37 The data for two samples of a test for water hardness are shown below. The hypothesis is that there is no significant difference in the variances. Is it justified?

First sample $n = 9$, $\Sigma x_i = 252$ $\Sigma x_i^2 = 7275$

Second sample $n = 14$, $\Sigma x_i = 434$ $\Sigma x_i^2 = 13550$

38 A machine was modified to improve its precision. 14 measurements of a quantity were made before modifications with results:
2.20, 2.36, 2.22, 2.39, 2.41, 2.25, 2.38, 2.19, 2.57, 2.52, 2.49, 2.33, 2.70, 2.30.
After modification 10 measurements of the same quantity were made with results:
2.22, 2.27, 2.50, 2.52, 2.33, 2.35, 2.40, 2.44, 2.51, 2.32.
Is there evidence that modification has improved the precision of the machine?

39 The outputs of two machines were measured for length. From machine A the variance of 21 measurements was 42.9 and from machine B the variance of 13 measurements was 146.0. Is it likely that the variances of the two outputs are equal?

Section 11.6

40 Tests were made on the output of two machines. The number of hours of useful life of the items from the two machines are tabulated below. Is there a significant difference in mean useful life?

   A   300, 310, 290, 300, 290, 300, 280, 300, 300, 310
   B   290, 300, 310, 290, 280, 290, 300, 290

41 In a determination of the percentage content of a metallic element in an alloy produced by two manufacturers the results were as follows. Are the differences significant?

   A   19.2, 15.8, 18.7, 16.3, 12.4
   B   13.2, 11.9, 15.2, 11.7, 12.9

42 Six operators made tests on each of 2 machines. The times taken to perform the tasks are shown below. Compare the results if
(a) the difference between operators is ignored
(b) the difference between operators is eliminated.

	I	II	III	IV	V	VI
A	8.2	7.1	7.4	7.2	7.5	7.0
B	7.5	7.4	7.6	7.0	7.7	6.9

43 The lifetimes of two kinds of abrasive stones are compared. For 10 stones of the first kind the average is 29 units with standard deviation 3 units; for 12 stones of the second kind the average is 33 units with a standard deviation of 2 units. Is the average of the second kind significantly higher than the first?

44 Two makes of component, one plastic and one metal, were tested by finding how many operations they could perform before surface imperfections rendered them inoperable. 4 plastic components gave results 2120, 2300, 2245, 2095, whilst 4 metal components gave results 3350, 3270, 3100 and 3360. Is there sufficient evidence that the metal components last 1000 operations more than plastic ones? Use a 1% level of significance.

45 Two balances A and B are tested for comparability. 8 items were weighed on each balance. Do the results suggest that there is a significant discrepancy between the balances?

A	1.003	2.537	2.538	1.528	4.533	4.518	4.025	5.032
B	1.000	2.535	2.535	1.530	4.531	4.516	4.026	5.032

46 The tensile strengths of two alloys were compared by taking samples of 10 specimens of each alloy. The first alloy has mean strength 15700 with standard deviation 2983; the second alloy has mean strength 13550 with standard deviation 2032. Test the hypothesis that the mean strengths of the two alloys do not differ; work with a 1% risk.

47 Nine pieces of fibre are treated with two chemicals A and B, one half of each piece is treated with A and the other half with B. From the results for breaking strengths given, can you conclude that there is a significant difference between the mean breaking strengths of those parts treated with A and those treated with B?

	I	II	III	IV	V	VI	VII	VIII	IX
A	331	381	384	367	365	403	409	346	378
B	355	376	365	367	411	368	395	336	373

48 Pressures at two stations A and B are recorded at random times and the results in millibars are shown below.

A	1007	1001	1003	1032	1036	1020	996
B	997	989	992	1004	984	970	

Test the hypothesis that the mean pressure at A is the same as at B. Is there a more suitable way of testing this hypothesis?

49 The percentages of a substance found in six random samples taken from a large quantity of metal were   4.21   4.57   4.60   4.55   4.32   4.60
Find 95% confidence limits for the mean percentage of the substance in the metal. Subsequently a further seven random samples were taken from a similar quantity of a second metal giving percentages   4.40   4.60   4.80   4.85   4.70   4.45   4.55
Use the variance ratio test to compare the variability of the two metals and the t test to compare the mean percentages of the substance in the metals. Comment on the results of the tests.   (LU)

50 (a)   Tests on 8 randomly selected spark plugs made by one manufacturer gave the following hours of acceptable service
x (hours)   192 199 211 195 176 187 203 189
Obtain 95% confidence limits for the mean length of service of the spark plugs made by this manufacturer.

(b)   Similar tests on a random sample of 10 spark plugs made by a rival manufacturer gave the results:

      y (hours)   206   214   198   204   199   203   184   210   207   215

      (i)   Using an F-test show that there is no significant difference at the 5% level between the variances of populations from which the samples were taken.

      (ii)   Find the "pooled sum of squares" estimate of the variance of the populations.

      (iii)   Test at both the 5% and 1% levels whether there is a significant difference between the mean values of the two populations from which the samples were taken.   (EC)

Section 11.7

51   Five mixes of the same alloy were made and for each mix, five determinations of density are made. Is there any evidence, on the basis of the results below, that certain mixes have a higher mean density than others?

Mix			Density (g.ml^{-1})		
A	5.6	5.5	5.7	5.1	5.2
B	5.3	5.5	5.4	5.2	5.4
C	5.5	5.3	5.4	5.4	5.2
D	5.5	5.4	5.0	5.3	5.8
E	5.7	5.4	5.6	5.5	5.4

52   For measuring strains in members, thin wire gauges are fixed to the surface of the strained member with an adhesive. The fixing of the adhesive requires that it should be cured at a high temperature. In order to analyse the effect of temperature on the sensitivity of the gauge, five different curing temperatures were used each on six gauges. The gauges were applied to suitable surfaces and these surfaces were subjected to known strains. The gauge response was expressed as a percentage of the known applied strain. From the results given does it seem that changes in curing temperature affect the ability of the gauges?

Curing Temperature °C	Percentage gauge response					
80	76	79	68	74	81	78
90	95	82	90	100	92	87
105	103	91	98	90	93	69
120	77	78	65	69	80	69
130	70	56	69	68	61	60

53   From the following measurements made in a precipitator determine whether the flow rate affects the exit dust loading.

Total flow (ft^3/h)	Exit dust loading (grains per cubic yard in flue gas)		
200	1.2	1.0	1.6
300	2.0	1.8	2.5
400	2.4	3.0	3.5
500	3.1	3.8	4.4

                                                                          (EC)

54   The resistance of wire from four sources was tested by taking a sample of five from each source. From the data below is there a significant difference between the resistances of

wires from the four sources?

Source	Resistance (Ω)				
A	15.4	15.0	14.4	14.6	15.8
B	17.2	17.2	17.6	17.6	18.0
C	18.8	18.8	18.4	19.0	19.2
D	11.6	11.0	11.4	11.6	11.0

Section 11.8

For each of the sets of data in Problems 55 to 64, estimate the parameters of the regression line of y on x, $y = \alpha + \beta x$ and find 95% and 99% confidence intervals for the parameters α and β.

55	x	100	200	300	400	500						
	y	4.1	8.0	12.6	16.3	19.4						

56	x	1	2	3	4	5	6					
	y	15	35	41	63	77	84					

57	x	0	10	20	30	40	50					
	y	52	60	64	73	76	81					

58	x	26.8	25.4	28.9	23.6	27.7	23.9	24.7	28.1	26.9	27.4	22.6	25.6
	y	26.5	27.3	24.2	27.1	23.6	25.9	26.3	22.5	21.7	21.4	25.8	24.9

59	x	1.0	1.1	1.2	1.3	1.4	1.5	1.6	1.7	1.8	1.9	2.0	
	y	8.1	7.8	8.5	9.8	9.5	8.9	8.6	10.2	9.3	9.2	10.5	

60	x	65	63	67	64	68	62	70	66	68	67	69	71
	y	68	66	68	65	69	66	68	65	71	67	68	70

61	x	0.01	0.02	0.03	0.04	0.05	0.06	0.07					
	y	1.1	1.3	1.3	1.6	1.5	1.9	1.8					

62	x	56	42	72	36	63	47	55	49	38	42	68	50
	y	147	125	160	118	149	128	150	145	115	140	152	150

63	x	43	35	51	47	46	62	32	36	41	39	53	48
	y	12	8	14	9	11	16	7	9	12	10	13	11

64	x	43	35	51	47	46	62						
	y	12	8	14	9	11	16						

65 Write a computer program to carry out the tasks required from Problems 55 to 64.

66 An experiment was conducted to evaluate a new method for determining the carbon content of clays used in the manufacture of sewer pipes. The carbon contents of 37 clays were determined both by the new method, represented by y, and by a standard method known to be very accurate, represented by x. Estimate the regression equation which assesses the results y obtained by the new method in terms of the true carbon contents, x, using the

crude sums of squares and products given below.

$$n = 37, \quad \sum_i^n x_i = 74, \quad \sum_i^n y_i = 111, \quad \sum_i^n x_i^2 = 348, \quad \sum_i^n y_i^2 = 958, \quad \sum_i^n x_i y_i = 522$$

67 (i) Given a set of observations $(x_1, y_1), (x_2, y_2), ..., (x_n, y_n)$, derive the Normal equations which are used for the least-squares estimation of the parameters a, b in the linear relation $y = a + bx$ connecting the variates x and y.

(ii) Tensile tests on a steel specimen yielded the following results:

Tensile force	x	1	2	3	4	5	6
Elongation	y	15	35	41	63	77	84

Assuming the regression of y on x to be linear, estimate the parameters of the regression line and determine 95% confidence limits for its slope. (EC)

68 The following data is believed to follow a relationship of the form $y = a + b/x$. Obtain the regression equation

x	2	3	4	5	6	7	8	9
y	14.3	14.8	15.1	15.3	15.5	15.7	15.8	16.0

69 Fit a regression curve $y = ax + bx^2$ to the data

x	1	2	3	4	5
y	2.2	5.7	9.4	14.7	20.4

70 In a study of machine maintenance costs in a factory, data on 12 machines was collected, consisting of, for each machine, its age (x) in months and its maintenance cost rate (y) in pence per hour.

From this data, calculations gave:

$$\bar{x} = 24, \quad \bar{y} = 18, \quad \Sigma(x - \bar{x})^2 = 1950, \quad \Sigma(y - \bar{y})^2 = 472, \quad \Sigma(x - \bar{x})(y - \bar{y}) = 780.$$

The regression of cost rate on age can be assumed to be linear.

On this basis, obtain an estimate of the mean maintenance cost rate of machines which are 29 months old and find the 95% confidence limits for the maintenance cost rate of one such machine.

71 Find 95% confidence intervals for particular values of y predicted by the regression equations of Problems 55 to 64 of Section 11.8. The corresponding x-values are indicated below; ((i) refers to problem 55 (ii) to problem 56 and so on)

(i) $x = 250$ (ii) $x = 5.2$ (iii) $x = 15$ (iv) $x = 25$ (v) $x = 1.43$
(vi) $x = 63.8$ (vii) $x = 0.058$ (viii) $x = 60$ (ix) $x = 50$ (x) $x = 50$

72 The following data is given

x	1.0	1.1	1.2	1.3	1.4	1.5	1.6	1.7	1.8	1.9	2.0
y	5.6	5.3	6.0	7.3	7.1	6.4	6.1	7.7	6.8	6.7	8.0

Estimate the linear regression line $y = \alpha + \beta x$ and estimate the value of y when $x = 1.75$. Find 95% and 99% confidence intervals for α and β. Sketch the regression lines and the 90% confidence limits for the mean value of y for $x = 1.6$ and for the value of y when $x = 1.6$.

Section 11.9

73 Evaluate the coefficient r^2 for each of the Problems 55 to 64.

74 Find r^2 for the data below. Comment.

x	4	5	9	14	18	22	24
y	1.6	2.2	1.1	1.6	0.7	0.3	1.7

75 Repeat Problem 74 for the data

x	70	92	80	74	65	83
y	70	80	59	83	74	86

76 The extension y cm of a spring under a load of x newtons is measured for eight values of x and the results are given in the following table:

x	1	2	3	4	5	6	7	8
y	1.9	3.8	5.1	6.0	7.0	7.8	8.4	8.8

It is thought that there is a relation between x and y of the form $y = ax + bx^2$. Show that the data indicate a significant correlation between y/x and x^2, and use this correlation to estimate values of a and b. (LU)

77 For each of Problems 55 to 64, calculate the percentage overall fit $(100r^2)$ and validate (or otherwise) the regression equation by a variance analysis.

78 The following data was recorded from an experiment.

x	0	5	10	15	20	25
y	22	36	70	92	125	143

Find the regression equation $y = \alpha + \beta x$; estimate the value of y for $x = 17$; test the hypothesis that $\alpha = 6$ as opposed to the alternative $\alpha \neq 6$, using a 1% level of significance; decide whether a linear model is appropriate.

APPENDIX

DIFFERENTIATION UNDER THE INTEGRAL SIGN

Suppose $f(x, y)$ and its partial derivative

$$\frac{\partial f}{\partial y}$$

are both continuous functions of x and y, and let

$$\int_a^b f(x, y)\mathrm{d}x = F(y).$$

Then
$$\frac{\mathrm{d}F}{\mathrm{d}y} = \int_a^b \frac{\partial f}{\partial y}\,\mathrm{d}x \qquad\qquad (A.1)$$

This result can apply to infinite integrals also.

As a consequence of (A.1) we can derive two useful definite integrals:

$$\int_0^\infty \frac{\sin x}{x}\mathrm{d}x = \frac{\pi}{2} \qquad\qquad (A.2)$$

and
$$\int_0^\infty \frac{\sin \alpha x}{x}\mathrm{d}x = \frac{\pi}{2}\,\mathrm{sgn}(\alpha), \quad \alpha \neq 0 \qquad\qquad (A.3)$$

where $\mathrm{sgn}(\alpha) = \begin{cases} 1, & \alpha > 0 \\ 0, & \alpha = 0 \\ -1, & \alpha < 0 \end{cases}$

VARIABLE LIMITS OF INTEGRATION

If, now, a and b are functions of α and $F(\alpha) = \int_a^b f(x, \alpha)dx$ then, under

suitable conditions,

$$\frac{dF}{d\alpha} = \int_a^b \frac{\partial f}{\partial \alpha} dx + f(b, \alpha)\frac{db}{d\alpha} - f(a, \alpha)\frac{da}{d\alpha} \qquad (A.4)$$

INTEGRATION UNDER THE INTEGRAL SIGN

It can be shown that
$$\int_0^\infty e^{-x^2} \cos 2\alpha x \, dx = \frac{\sqrt{\pi}}{2} e^{-\alpha^2} \qquad (A.5)$$

By integrating both sides with respect to α it follows that

$$\int_0^\infty e^{-x^2} \frac{\sin 2\alpha x}{x} dx = \frac{\pi}{2} \operatorname{erf} \alpha \qquad (A.6)$$

where
$$\operatorname{erf} \alpha = \int_0^\alpha e^{-u^2} du.$$

Table A.I.   The Normal Probability Integral

$\dfrac{x-\mu}{\sigma}$	0	1	2	3	4	5	6	7	8	9
0	0000	0040	0080	0120	0160	0199	0239	0279	0319	0359
0.1	0398	0438	0478	0517	0557	0596	0636	0675	0714	0753
0.2	0793	0832	0871	0909	0948	0987	1026	1064	1103	1141
0.3	1179	1217	1255	1293	1331	1368	1406	1443	1480	1517
0.4	1555	1591	1628	1664	1700	1736	1772	1808	1844	1879
0.5	1915	1950	1985	2019	2054	2088	2123	2157	2190	2224
0.6	2257	2291	2324	2357	2389	2422	2454	2486	2517	2549
0.7	2580	2611	2642	2673	2703	2734	2764	2794	2822	2852
0.8	2881	2910	2939	2967	2995	3023	3051	3078	3106	3133
0.9	3159	3186	3212	3238	3264	3289	3315	3340	3365	3389
1.0	3413	3438	3461	3485	3508	3531	3554	3577	3599	3621
1.1	3643	3665	3686	3708	3729	3749	3770	3790	3810	3830
1.2	3849	3869	3888	3907	3925	3944	3962	3980	3997	4015
1.3	4032	4049	4066	4082	4099	4115	4131	4147	4162	4177
1.4	4192	4207	4222	4236	4251	4265	4279	4292	4306	4319
1.5	4332	4345	4357	4370	4382	4394	4406	4418	4429	4441
1.6	4452	4463	4474	4484	4495	4505	4515	4525	4535	4545
1.7	4554	4564	4573	4582	4591	4599	4608	4616	4625	4633
1.8	4641	4649	4656	4664	4671	4678	4686	4693	4699	4706
1.9	4713	4719	4726	4732	4738	4744	4750	4756	4761	4767
2.0	4772	4778	4783	4788	4793	4798	4803	4808	4812	4817
2.1	4821	4826	4830	4834	4838	4842	4846	4850	4854	4857
2.2	4861	4865	4868	4871	4875	4878	4881	4884	4887	4890
2.3	4893	4896	4898	4901	4904	4906	4909	4911	4913	4916
2.4	4918	4920	4922	4925	4927	4929	4931	4932	4934	4936
2.5	4938	4940	4941	4943	4945	4946	4948	4949	4951	4952
2.6	4953	4955	4956	4957	4959	4960	4961	4962	4963	4964
2.7	4965	4966	4967	4968	4969	4970	4971	4972	4973	4974
2.8	4974	4975	4976	4977	4977	4978	4979	4979	4980	4981
2.9	4981	4982	4982	4983	4984	4984	4985	4985	4986	4986

3.0	3.1	3.2	3.3	3.4	3.5	3.6	3.7	3.8	3.9
4987	4990	4993	4995	4997	4998	4998	4999	4999	4999

Table A.II. Percentage points of the t distribution

$P\%$ v	50	25	10	5	2.5	1	0.5	0.1
1	1.00	2.41	6.31	12.7	25.5	63.7	127.	637.
2	.816	1.60	2.92	4.30	6.21	9.92	14.1	31.6
3	.765	1.42	2.35	3.18	4.18	5.84	7.45	12.9
4	.741	1.34	2.13	2.78	3.50	4.60	5.60	8.61
5	.727	1.30	2.01	2.57	3.16	4.03	4.77	6.86
6	.718	1.27	1.94	2.45	2.97	3.71	4.32	5.96
7	.711	1.25	1.89	2.36	2.84	3.50	4.03	5.40
8	.706	1.24	1.86	2.31	2.75	3.36	3.83	5.04
9	.703	1.23	1.83	2.26	2.68	3.25	3.69	4.78
10	.700	1.22	1.81	2.23	2.63	3.17	3.58	4.59
11	.698	1.21	1.80	2.20	2.59	3.11	3.50	4.44
12	.695	1.21	1.78	2.18	2.56	3.05	3.43	4.32
13	.694	1.20	1.77	2.16	2.53	3.01	3.37	4.22
14	.692	1.20	1.76	2.14	2.51	2.98	3.33	4.14
15	.691	1.20	1.75	2.13	2.49	2.95	3.29	4.07
16	.690	1.19	1.75	2.12	2.47	2.92	3.25	4.01
17	.689	1.19	1.74	2.11	2.46	2.90	3.22	3.96
18	.688	1.19	1.73	2.10	2.44	2.88	3.20	3.92
19	.688	1.19	1.73	2.09	2.43	2.86	3.17	3.88
20	.687	1.18	1.72	2.09	2.42	2.85	3.15	3.85
21	.686	1.18	1.72	2.08	2.41	2.83	3.14	3.82
22	.686	1.18	1.72	2.07	2.41	2.82	3.12	3.79
23	.685	1.18	1.71	2.07	2.40	2.81	3.10	3.77
24	.685	1.18	1.71	2.06	2.39	2.80	3.09	3.74
25	.684	1.18	1.71	2.06	2.38	2.79	3.08	3.72
26	.684	1.18	1.71	2.06	2.38	2.78	3.07	3.71
27	.684	1.18	1.70	2.05	2.37	2.77	3.06	3.69
28	.683	1.17	1.70	2.05	2.37	2.76	3.05	3.67
29	.683	1.17	1.70	2.05	2.36	2.76	3.04	3.66
30	.683	1.17	1.70	2.04	2.36	2.75	3.03	3.65
40	.681	1.17	1.68	2.02	2.33	2.70	2.97	3.55
60	.679	1.16	1.67	2.00	2.30	2.66	2.91	3.46
∞	.674	1.15	1.64	1.96	2.24	2.58	2.81	3.29

Table A.III.   Percentage points of the χ^2 distribution

ν \ P	0.990	0.950	0.900	0.500	0.100	0.050	0.010
1	1571×10^{-7}	3932×10^{-6}	0.01579	0.4549	2.705	3.841	6.635
2	0.0201	0.1026	0.2107	1.386	4.605	5.991	9.210
3	0.1148	0.3518	0.5844	2.366	6.251	7.815	11.34
4	0.2971	0.711	1.064	3.357	7.779	9.488	13.28
5	0.5543	1.145	1.610	4.351	9.236	11.07	15.09
6	0.8721	1.635	2.204	5.348	10.64	12.59	16.81
7	1.239	2.167	2.833	6.346	12.02	14.07	18.48
8	1.646	2.733	3.490	7.344	13.36	15.51	20.09
9	2.088	3.325	4.168	8.343	14.68	16.92	21.67
10	2.558	3.940	4.865	9.342	15.99	18.31	23.21
11	3.053	4.575	5.578	10.34	17.28	19.68	24.73
12	3.571	5.226	6.304	11.34	18.55	21.03	26.22
13	4.107	5.892	7.042	12.34	19.81	22.36	27.69
14	4.660	6.571	7.790	13.34	21.06	23.68	29.14
15	5.229	7.261	8.547	14.34	22.31	25.00	30.58
16	5.812	7.962	9.312	15.34	23.54	26.30	32.00
17	6.408	8.672	10.09	16.34	24.77	27.59	33.41
18	7.015	9.390	10.86	17.34	25.99	28.87	34.81
19	7.633	10.12	11.65	18.34	27.20	30.14	36.19
20	8.260	10.85	12.44	19.34	28.41	31.41	37.57
21	8.897	11.59	13.24	20.34	29.62	32.67	38.93
22	9.542	12.34	14.04	21.34	30.81	33.92	40.29
23	10.20	13.09	14.85	22.34	32.01	35.17	41.64
24	10.86	13.85	15.66	23.34	33.20	36.42	42.98
25	11.52	14.61	16.47	24.34	34.38	37.65	44.31
26	12.20	15.38	17.29	25.34	35.56	38.89	45.64
27	12.88	16.15	18.11	26.34	36.74	40.11	46.96
28	13.56	16.93	18.94	27.34	37.92	41.34	48.28
29	14.26	17.71	19.77	28.34	39.09	42.56	49.59
30	14.95	18.49	20.60	29.34	40.26	43.77	50.89
40	22.16	26.51	29.05	39.34	51.81	55.76	63.69
50	29.71	34.76	37.69	49.33	63.17	67.50	76.15
60	37.48	43.19	46.46	59.33	74.40	79.08	88.38
70	45.44	51.74	55.33	69.33	85.53	90.53	100.4
80	53.54	60.39	64.28	79.33	96.58	101.9	112.3
90	61.75	69.13	73.29	89.33	107.6	113.1	124.1
100	70.06	77.93	82.36	99.33	118.5	124.3	135.8

Table A.IV. F distribution – 5% 2.5% and 1% points

v_2 \ v_1	1	2	3	4	5	6	7	8	10	12	24	∞
1	161	200	216	225	230	234	237	239	242	244	249	254
	648	800	864	900	922	937	948	957	969	977	997	1018
	4052	5000	5403	5625	5764	5859	5928	5981	6056	6106	6235	6366
2	18.5	19.0	19.2	19.2	19.3	19.3	19.4	19.4	19.4	19.4	19.5	19.5
	38.5	39.0	39.2	39.2	39.3	39.3	39.4	39.4	39.4	39.4	39.5	39.5
	98.5	99.0	99.2	99.2	99.3	99.3	99.4	99.4	99.4	99.4	99.5	99.5
3	10.13	9.55	9.28	9.12	9.01	8.94	8.89	8.85	8.79	8.74	8.64	8.53
	17.4	16.0	15.4	15.1	14.9	14.7	14.6	14.5	14.4	14.3	14.1	13.9
	34.1	30.8	29.5	28.7	28.2	27.9	27.7	27.5	27.2	27.1	26.6	26.1
4	7.71	6.94	6.59	6.39	6.26	6.16	6.09	6.04	5.96	5.91	5.77	5.63
	12.22	10.65	9.98	9.60	9.36	9.20	9.07	8.98	8.84	8.75	8.51	8.26
	21.2	18.0	16.7	16.0	15.5	15.2	15.0	14.8	14.5	14.4	13.9	13.5
5	6.61	5.79	5.41	5.19	5.05	4.95	4.88	4.82	4.74	4.68	4.53	4.36
	10.01	8.43	7.76	7.39	7.15	6.98	6.85	6.76	6.62	6.52	6.28	6.02
	16.26	13.27	12.06	11.39	10.97	10.67	10.46	10.29	10.05	9.89	9.47	9.02
6	5.99	5.14	4.75	4.53	4.39	4.28	4.21	4.15	4.06	4.00	3.84	3.67
	8.81	7.26	6.60	6.23	5.99	5.82	5.70	5.60	5.46	5.37	5.12	4.85
	13.74	10.92	9.78	9.15	8.75	8.47	8.26	8.10	7.87	7.72	7.31	6.88
7	5.59	4.74	4.35	4.12	3.97	3.87	3.79	3.73	3.64	3.57	3.41	3.23
	8.07	6.54	5.89	5.52	5.29	5.12	4.99	4.90	4.76	4.67	4.42	4.14
	12.25	9.55	8.45	7.85	7.46	7.19	6.99	6.84	6.62	6.47	6.07	5.65
8	5.32	4.46	4.07	3.84	3.69	3.58	3.50	3.44	3.35	3.28	3.12	2.93
	7.57	6.06	5.42	5.05	4.82	4.65	4.53	4.43	4.30	4.20	3.95	3.67
	11.26	8.65	7.59	7.01	6.63	6.37	6.18	6.03	5.81	5.67	5.28	4.86
9	5.12	4.26	3.86	3.63	3.48	3.37	3.29	3.23	3.14	3.07	2.90	2.71
	7.21	5.71	5.08	4.72	4.48	4.32	4.20	4.10	3.96	3.87	3.61	3.33
	10.56	8.02	6.99	6.42	6.06	5.80	5.61	5.47	5.26	5.11	4.73	4.31
10	4.96	4.10	3.71	3.48	3.33	3.22	3.14	3.07	2.98	2.91	2.74	2.54
	6.94	5.46	4.83	4.47	4.24	4.07	3.95	3.85	3.72	3.62	3.37	3.08
	10.04	7.56	6.55	5.99	5.64	5.39	5.20	5.06	4.85	4.71	4.33	3.91
11	4.84	3.98	3.59	3.36	3.20	3.09	3.01	2.95	2.85	2.79	2.61	2.40
	6.72	5.26	4.63	4.28	4.04	3.88	3.76	3.66	3.53	3.43	3.17	2.88
	9.65	7.21	6.22	5.67	5.32	5.07	4.89	4.74	4.54	4.40	4.02	3.60
12	4.75	3.89	3.49	3.26	3.11	3.00	2.91	2.85	2.75	2.69	2.51	2.30
	6.55	5.10	4.47	4.12	3.89	3.73	3.61	3.51	3.37	3.28	3.02	2.72
	9.33	6.93	5.95	5.41	5.06	4.82	4.64	4.50	4.30	4.16	3.78	3.36

4.60	3.74	3.34	3.11	2.96	2.85	2.76	2.70	2.60	2.53	2.35	2.13
6.30	4.86	4.24	3.89	3.66	3.50	3.38	3.29	3.15	3.05	2.79	2.49
8.86	6.51	5.56	5.04	4.70	4.46	4.28	4.14	3.94	3.80	3.43	3.00
4.49	3.63	3.24	3.01	2.85	2.74	2.66	2.59	2.49	2.42	2.24	2.01
6.12	4.69	4.08	3.73	3.50	3.34	3.22	3.12	2.99	2.89	2.63	2.32
8.53	6.23	5.29	4.77	4.44	4.20	4.03	3.89	3.69	3.55	3.18	2.75
4.41	3.55	3.16	2.93	2.77	2.66	2.58	2.51	2.41	2.34	2.15	1.92
5.98	4.56	3.95	3.61	3.38	3.22	3.10	3.01	2.87	2.77	2.50	2.19
8.29	6.01	5.09	4.58	4.25	4.01	3.84	3.71	3.51	3.37	3.00	2.57
4.35	3.49	3.10	2.87	2.71	2.60	2.51	2.45	2.35	2.28	2.08	1.84
5.87	4.46	3.86	3.51	3.29	3.13	3.01	2.91	2.77	2.68	2.41	2.09
8.10	5.85	4.94	4.43	4.10	3.87	3.70	3.56	3.37	3.23	2.86	2.42
4.26	3.40	3.01	2.78	2.62	2.51	2.42	2.36	2.25	2.18	1.98	1.73
5.72	4.32	3.72	3.38	3.15	2.99	2.87	2.78	2.64	2.54	2.27	1.94
7.82	5.61	4.72	4.22	3.90	3.67	3.50	3.36	3.17	3.03	2.66	2.21
4.20	3.34	2.95	2.71	2.56	2.45	2.36	2.29	2.19	2.12	1.91	1.65
5.61	4.22	3.63	3.29	3.06	2.90	2.78	2.69	2.55	2.45	2.17	1.83
7.64	5.45	4.57	4.07	3.75	3.53	3.36	3.23	3.03	2.90	2.52	2.06
4.15	3.29	2.90	2.67	2.51	2.40	2.31	2.24	2.14	2.07	1.86	1.59
5.53	4.15	3.56	3.22	3.00	2.84	2.72	2.62	2.48	2.38	2.10	1.75
7.50	5.34	4.46	3.97	3.65	3.43	3.26	3.13	2.93	2.80	2.42	1.96
4.11	3.26	2.87	2.63	2.48	2.36	2.28	2.21	2.11	2.03	1.82	1.55
5.47	4.09	3.51	3.17	2.94	2.79	2.66	2.57	2.43	2.33	2.05	1.69
7.40	5.25	4.38	3.89	3.58	3.35	3.18	3.05	2.86	2.72	2.35	1.87
4.08	3.23	2.84	2.61	2.45	2.34	2.25	2.18	2.08	2.00	1.79	1.51
5.42	4.05	3.46	3.13	2.90	2.74	2.62	2.53	2.39	2.29	2.01	1.64
7.31	5.18	4.31	3.83	3.51	3.29	3.12	2.99	2.80	2.66	2.29	1.80
4.00	3.15	2.76	2.53	2.37	2.25	2.17	2.10	1.99	1.92	1.70	1.39
5.29	3.93	3.34	3.01	2.79	2.63	2.51	2.41	2.27	2.17	1.88	1.48
7.08	4.98	4.13	3.65	3.34	3.12	2.95	2.82	2.63	2.50	2.12	1.60
3.92	3.07	2.68	2.45	2.29	2.18	2.09	2.02	1.91	1.83	1.61	1.25
5.15	3.80	3.23	2.89	2.67	2.52	2.39	2.30	2.16	2.05	1.76	1.31
6.85	4.79	3.95	3.48	3.17	2.96	2.79	2.66	2.47	2.34	1.95	1.38
3.84	3.00	2.60	2.37	2.21	2.10	2.01	1.94	1.83	1.75	1.52	1.00
5.02	3.69	3.12	2.79	2.57	2.41	2.29	2.19	2.05	1.94	1.64	1.00
6.63	4.61	3.78	3.32	3.02	2.80	2.64	2.51	2.32	2.18	1.79	1.00

Row labels (second column of each three-row block): 14, 16, 18, 20, 24, 28, 32, 36, 40, 60, 120, ∞

For each value of $F_{v_2}^{v_1}$, the upper number is the 5% point, the middle one is the 2.5% point and the lowest one is the 1% point.

ANSWERS

Chapter 1
Section 1.2
1 Any two of the first set; any pair of the second set except for the last pair.
3 Two, in each case.   4 (i), (iii) and (v) are subspaces.
5 (i) $(-2\lambda, -2\lambda, \lambda)$; (ii) $(\mu, 0, -\mu)$, (iii) $(3\nu, -2\nu, \nu)$

6 (i) $\left\{ \left[\dfrac{2}{\sqrt{14}}, \dfrac{1}{\sqrt{14}}, \dfrac{3}{\sqrt{14}} \right], \left[\dfrac{4}{\sqrt{42}}, \dfrac{5}{\sqrt{42}}, \dfrac{1}{\sqrt{42}} \right], \left[\dfrac{1}{\sqrt{3}}, \dfrac{1}{\sqrt{3}}, -\dfrac{1}{\sqrt{3}} \right] \right\}$

   (ii) $\left\{ \left[\dfrac{\sqrt{2}}{2}, \dfrac{-\sqrt{2}}{2}, 0 \right], \left[\dfrac{\sqrt{2}}{6}, \dfrac{\sqrt{2}}{6}, \dfrac{-2\sqrt{2}}{3} \right], \left[\dfrac{-2}{3}, \dfrac{-2}{3}, \dfrac{-1}{3} \right] \right\}$

   (iii) $\left\{ \left[\dfrac{\sqrt{2}}{2}, 0, \dfrac{\sqrt{2}}{2} \right], (0, 1, 0), \left[\dfrac{\sqrt{2}}{2}, 0, \dfrac{-\sqrt{2}}{2} \right] \right\}$

   (iv) $\left\{ \left[\dfrac{2\sqrt{5}}{5}, \dfrac{-\sqrt{5}}{5}, 0 \right], \left[\dfrac{\sqrt{5}}{5}, \dfrac{2\sqrt{5}}{5}, 0 \right], (0, 0, 1) \right\}$

7 (i) $\left\{ \dfrac{1}{\sqrt{3}}(1, 1, -1), \dfrac{1}{\sqrt{2}}(1, 0, 1), \dfrac{1}{\sqrt{6}}(2, -1, 1) \right\}$

   (ii) $\left\{ \dfrac{1}{\sqrt{51}}(7, -1, -1), \dfrac{1}{\sqrt{2}}(0, 1, -1), \dfrac{7}{\sqrt{102}} \left[\dfrac{2}{7}, 1, 1 \right] \right\}$

8 $\dfrac{1}{\sqrt{2}}(1, 0, 1, 0), \dfrac{1}{\sqrt{6}}(1, 2, -1, 0), \dfrac{1}{\sqrt{21}}(-2, 2, 2, 3), \dfrac{1}{\sqrt{7}}(-1, 1, 1, 2)$

9 $\left\{ \dfrac{1}{\sqrt{2}}(0, 1, 1, 0), \dfrac{1}{3}(0, 2, -2, -1), \dfrac{1}{9}(-3, -2, 2, -8) \right\}$

Section 1.3
10 $\begin{bmatrix} 1 & 3 & 2 \\ 2 & 1 & 1 \\ 3 & 2 & 3 \end{bmatrix}$ 11 $(6, 4, 8), (8, 9, 19), (14, 13, 27)$

12 (i), (iii) and (iv); $(1, 2, -2), (-1, 3, 1)$ and $(1, 1, 1)$
13 $(1, -1, 1)$; $(2, 1, -1)$; $(2, -1, 0)$ and $(3, 0, -1)$
14 (i) 2, 1 (ii) 2, 1 (iii) 2, 2 (iv) 2, 1 (v) 2, 1 (vi) 2, 2
15 $(-x_2, -x_1)$ and (x_2, x_1) 16 Space with basis $(1, 1, 1, 0)$ and $(2, 1, 0, 1)$

17 (i) $\left[4 - \dfrac{7}{3}a, \dfrac{4}{3}a - 1, a \right]$ (ii) $(a, -a, -a)$

18 (i)   2, {(1, 0, 1), (0, 1, 1)}   (ii) 1, {(1, −1, 1)}   (iii) 1, {(2, 3, 5)}
     (iv) 0, no basis   (v)   0, no basis   (vi) 2, {(1, −1, 0), (1, 0, −1)}

19 (i)   2, {(1, 0, 2), (3, −2, 0)}, (0, 0, −1), $\alpha \begin{bmatrix} 1 \\ 0 \\ 2 \end{bmatrix} + \beta \begin{bmatrix} 3 \\ -2 \\ 0 \end{bmatrix} + \begin{bmatrix} 0 \\ 0 \\ -1 \end{bmatrix}$

   (ii) 1, {(1, 0, −2)}, (0, 0, 1), $\begin{bmatrix} 0 \\ 0 \\ 1 \end{bmatrix} + \alpha \begin{bmatrix} 1 \\ 0 \\ -2 \end{bmatrix}$

   (iii) 1, {(5, −2, 13)}, (1, 0, 3), $\begin{bmatrix} 1 \\ 0 \\ 3 \end{bmatrix} + \alpha \begin{bmatrix} 5 \\ -2 \\ 13 \end{bmatrix}$ (iv) 0, no basis, gen. sol. is $\begin{bmatrix} 3 \\ 2 \\ 0 \end{bmatrix}$

Section 1.4
20 2   **21** 2, 2, 2, 2, 3

Section 1.5

30 $L = \begin{bmatrix} 2 & 0 & 0 \\ 1 & -3 & 0 \\ -1 & 0 & 7/2 \end{bmatrix}$, $U = \begin{bmatrix} 1 & -1 & 3/2 \\ 0 & 1 & -11/6 \\ 0 & 0 & 1 \end{bmatrix}$, Det = −21

31 (i)   1, −2, 3   (ii) −6/125, 47/125, 99/125   (iii) 77/18, 32/9, 28/9, −7/3
     (iv) 4, 2, −3, 3   (v)   5, 16, −6

32 1, 1, 1, $A^{-1} = \dfrac{1}{64} \begin{bmatrix} 39 & 14 & -22 \\ 14 & -36 & 20 \\ -22 & 20 & -4 \end{bmatrix}$

33 (a) $L = \begin{bmatrix} 1 & 0 & 0 & 0 \\ 0 & 1 & 0 & 0 \\ 1 & 1 & -2 & 0 \\ 2 & 0 & 0 & -2 \end{bmatrix}$, $U = \begin{bmatrix} 1 & 1 & 2 & 1 \\ 0 & 1 & 1 & 1 \\ 0 & 0 & 1 & 0.5 \\ 0 & 0 & 0 & 1 \end{bmatrix}$

   (b) −1.5, 2.5, 0.5, −1   (c) $A^{-1} = \dfrac{1}{4} \begin{bmatrix} 4 & -6 & 2 & -1 \\ -4 & 2 & 2 & 1 \\ 0 & 2 & -2 & 1 \\ 4 & 0 & 0 & -2 \end{bmatrix}$

34 $A^{-1} = -\dfrac{1}{64} \begin{bmatrix} -39 & -14 & 22 \\ -14 & 36 & -20 \\ 22 & -20 & 4 \end{bmatrix}$

35 (a)   7/16, 1/2, 1/2, 1/2   **36** $x^{(2)} = (2.052, 4.791, 8.147)$

Section 1.6

38 (i) $\begin{bmatrix} 23 & 20 \\ 55 & 48 \\ 87 & 76 \end{bmatrix}$ (ii) $\begin{bmatrix} 47 & 50 \\ 54 & 62 \\ 65 & 70 \\ 72 & 82 \end{bmatrix}$ (iii) $\begin{bmatrix} 10 \\ 28 \\ 46 \end{bmatrix}$ (iv) $\begin{bmatrix} ax + by \\ cx + dy \end{bmatrix}$

(v) $\begin{bmatrix} a^2 + h^2 + g^2 & ah + hb + gf & ag + hf + gc \\ ha + bh + fg & h^2 + b^2 + f^2 & hg + bf + fc \\ ga + fh + cg & gh + fb + cf & g^2 + f^2 + c^2 \end{bmatrix}$

(vi) $\begin{bmatrix} A_1 & A_2 \\ A_3 & A_4 \end{bmatrix}$ (vii) $\begin{bmatrix} A_1 & 0 \\ A_3 & A_4 - A_3 A_1^{-1} A_2 \end{bmatrix}$

41 (i) $\begin{bmatrix} -23 & 29 & -64/5 & -18/5 \\ 10 & -12 & 26/5 & 7/5 \\ 1 & -2 & 6/5 & 2/5 \\ 2 & -2 & 3/5 & 1/5 \end{bmatrix}$ (i) and (iii) are inverses of each other

(iv) $\dfrac{1}{18}\begin{bmatrix} 2 & 5 & -7 & 1 \\ 5 & -1 & 5 & -2 \\ -7 & 5 & 11 & 10 \\ 1 & -2 & 10 & 5 \end{bmatrix}$ (v) $\begin{bmatrix} \cos\alpha & -\sin\alpha & -\cos(\alpha - \beta + \gamma) & \sin(\alpha - \beta + \gamma) \\ \sin\alpha & \cos\alpha & -\sin(\alpha - \beta + \gamma) & -\cos(\alpha - \beta + \gamma) \\ 0 & 0 & \cos\gamma & -\sin\gamma \\ 0 & 0 & \sin\gamma & \cos\gamma \end{bmatrix}$

(vi) $\dfrac{1}{13}\begin{bmatrix} 7 & -6 & -3 & 4 \\ -6 & 7 & 10 & -9 \\ 5 & -8 & -4 & 14 \\ -4 & 9 & -2 & -6 \end{bmatrix}$

Chapter 2
Section 2.2
1 (i) 2(−1, 1), 3(−2, 1) (ii) 1(−1, 1, 2), 2(0, 1, 2), 3(0, 0, 1)
(iii) 1(−1, 0, 1), 1(−2, 1, 0) are possible solutions, 5(1, 1, 1)
(iv) 1(1, −1, 0); 2(2, −1, −2), 3(1, −1, −2) (v) 0(1, −1, 0); 1(0, 0, 1); 4(1, 1, 0)
(vi) 1(1, 0, −1) & (0, 1, −1); 3(1, 1, 0) (vii) −1(0, 1, −1); i(1+i, 1, 1); −i(1−i, 1, 1)
(viii) 2(2, −1, 0); 0(4, −1, 0); 1(4, 0, −1) (ix) −1(1, 0, 1); 2(1, 3, 1); 1(3, 2, 1)
(x) 1(1, 0, −1, 0) & (1, −1, 0, 0); 2(−2, 4, 1, 2); 3(0, 3, 1, 2)

2 (ii) $\begin{bmatrix} -3 & 0 & 3 \\ -2 & 1 & 2 \\ -6 & 0 & 6 \end{bmatrix}$ **3** 1(−1, 1, 1); −2(11, 1, −14); 3(1, 1, 1)

Section 2.3
4 (i) −2(5, −4); 7(1, 1) (ii) 3+2i(1, −i); 3 − 2i(1, i)
(iii) 4(2, 3, −8); −2(0, 9, −8); 7(0, 0, 1) (iv) −9(0, 1, 11); −7(4, 1, 15); 7(4, 1,1)
5 0(1, −1, 0), 1(0, 0 1), 4(1, 1, 0). Real, symmetric matrix
6 0(1, 1, −1), (3 + √3)(1, 1 + √3, 2 + √3), (3 − √3)(1, 1 − √3, 2 − √3)
7 2(1, 0, −1), (2 + √2)(1, √2, 1), (2 − √2)(1, −√2, 1)
8 −2, 2(1, 1, −1), 4 **9** (2, 1, 2) $\begin{bmatrix} 3 & -2 & 4 \\ -2 & -2 & 6 \\ 4 & 6 & -1 \end{bmatrix}$
11 (0, 0, 1), (1, −1, 0), (1, 0, −1) are possible answers

14 (i) $\begin{bmatrix} 7 & 10 \\ 5 & 7 \end{bmatrix}$, $\begin{bmatrix} 17 & 24 \\ 12 & 17 \end{bmatrix}$, $\begin{bmatrix} -1 & 2 \\ 1 & -1 \end{bmatrix}$

(ii) $\begin{bmatrix} 42 & 31 & 29 \\ 45 & 39 & 31 \\ 53 & 45 & 42 \end{bmatrix}$, $\begin{bmatrix} 193 & 160 & 144 \\ 224 & 177 & 160 \\ 272 & 224 & 193 \end{bmatrix}$, $\dfrac{1}{11}\begin{bmatrix} -2 & 5 & -1 \\ -1 & -3 & 5 \\ 7 & -1 & -2 \end{bmatrix}$

(iii) $\begin{bmatrix} 15 & 19 & -26 \\ 14 & 46 & -52 \\ 14 & 19 & -25 \end{bmatrix}$, $\begin{bmatrix} 31 & 65 & -80 \\ 30 & 146 & -160 \\ 30 & 65 & -79 \end{bmatrix}$, $\dfrac{1}{6}\begin{bmatrix} 0 & -1 & 4 \\ -6 & 1 & 8 \\ -6 & -1 & 10 \end{bmatrix}$

15 $\omega^2 = 2, 3, 6$ with corresponding eigenvectors $(0, 1, -1)$, $(1, -1, -1)$, $(2, 1, 1)$

16 (i) $2\begin{bmatrix} -1 & -2 \\ 1 & 2 \end{bmatrix} + 3\begin{bmatrix} 2 & 2 \\ -1 & -1 \end{bmatrix}$

(iv) $\dfrac{1}{2}\begin{bmatrix} 0 & -2 & 1 \\ 0 & 2 & -1 \\ 0 & 0 & 0 \end{bmatrix} + 2\begin{bmatrix} 2 & 2 & 0 \\ -1 & -1 & 0 \\ -2 & -2 & 0 \end{bmatrix} + 3\dfrac{1}{2}\begin{bmatrix} -2 & -2 & -1 \\ 2 & 2 & 1 \\ 4 & 4 & 2 \end{bmatrix}$

(v) $1\begin{bmatrix} 0 & 0 & 0 \\ 0 & 0 & 0 \\ 0 & 0 & 1 \end{bmatrix} + 4\dfrac{1}{2}\begin{bmatrix} 1 & 1 & 0 \\ 1 & 1 & 0 \\ 0 & 0 & 0 \end{bmatrix}$

Section 2.4
17 (i) $2x_1^2 + 2x_2^2 - 5x_3^2 - 6x_1 x_2 + 2x_1 x_3 + 8x_2 x_3$

(ii) $3x_1^2 + 4x_2^2 + 6x_3^2 + 4x_1 x_3 - 10x_2 x_3$

(iii) $x_1^2 + 2x_3^2 - 4x_1 x_2 + 8x_1 x_3 + 6x_2 x_3$

18 (i) $\begin{bmatrix} 2 & -3 & 0 \\ -3 & 0 & 0 \\ 0 & 0 & 1 \end{bmatrix}$ (ii) $\begin{bmatrix} 3 & -4 & 5/2 \\ -4 & 4 & 0 \\ 5/2 & 0 & 0 \end{bmatrix}$ (iii) $\begin{bmatrix} 1 & 2 & 3 \\ 2 & -2 & -4 \\ 3 & -4 & -3 \end{bmatrix}$

19 (i) $y_1^2 + y_2^2 + y_3^2$ (ii) $y_1^2 - y_2^2 + y_3^2$ (iii) $4y_1^2 - 16y_2^2 + 16y_3^2$

(iv) $y_1^2 + 128y_2^2 - 128y_3^2$ (v) $4y_1^2 - 16y_3^2 + 12y_4^2$ (vi) $y_1^2 - 8y_2^2$

20 $4, 6, H = \begin{bmatrix} 0 & 1 & 1 \\ 1 & 1 & 0 \\ -1 & 0 & -1 \end{bmatrix}$ **21** $P = \begin{bmatrix} 1 & 1 \\ 1 & -1 \end{bmatrix}$, $B^{-1} = \dfrac{1}{7}\begin{bmatrix} 4 & 3 \\ 3 & 4 \end{bmatrix}$

22 $3(0, 1, -1)$, $3(1, 1, 1)$, $-3(-2, 1, 1)$ are possible answers, indefinite.

24 $H = \begin{bmatrix} 0 & 1 & 2 \\ 1 & -1 & 1 \\ -1 & -1 & 1 \end{bmatrix}$ **25** $P = \begin{bmatrix} 2 & 1 & 2 \\ -2 & 2 & 1 \\ 1 & 2 & -2 \end{bmatrix}$ **26** Positive definite

Section 2.5
27 $2.436EI/L^2$ [0.076, 0.293, 0.617, 1]

28 $0[0, 0, 1, 0, 1, 1]$ is smallest, $-24.208[0, 0.526, 0, -0.526, -1, 1]$ is largest

29 $0.6176\sqrt{\delta_{11}}$ [1, 0.5325, 0.1573], $3.755\sqrt{\delta_{11}}$ [-0.7502, 1, 0.9223],

$9.689\sqrt{\delta_{11}}$ [−0.3210, 1, −0.8971]

Section 2.6
31 7.873[0.225, 0.549, 1]   **32** 12.289[0.274, 1, 0.871]
33 [1, −1.492, 1.376, −0.784]   **34** 12.59[0.77, 1, 0.32]

35 19.29[20, 13, 9]   **36** $\dfrac{1}{18}\begin{bmatrix} 2 & 2 & 4 \\ 2 & 5 & -2 \\ 4 & -2 & 2 \end{bmatrix}$

Section 2.7
39 6.425[0.731, 0.233, 1]
40 (i)   2[1, 0, −1], $(2 + \sqrt{2})$[1, $−\sqrt{2}$, 1], $(2 − \sqrt{2})$[1, $\sqrt{2}$, 1]
    (ii)   26.305[0.070, 0.233, 0.530, 1],   2.203[−0.828, −1, −0.612, 0.615],
        0.454[1, −0.207, −0.676, 0.337],   0.038[0.427, −1, 0.822, −0.233]
41 5.851[0.106, 1, 0.663]   **42** 19.286[1, 0.668, 0.450]
43 1[1, 1, −1], 2.38742[1, −0.38742, 0.07504], 0.27924[0.58112, 1, 0.86037]
44 19.2[1, 0.7, 0.4], −7.2[−1, 0.9, 0.9]   **45** 12.265[1, −1.485, 1.322]

Chapter 3
Section 3.1
1 (a)   5(3, 2)   (b)   7(3, 2); 9, any positive coordinates on $x + 3y = 9$; 8(4, 0)
   (c)   138(30, 26)   (d)   6(0, 6) is max, unbounded solution   (e)   no solution
   (f)   max. 23(4, 15), min 10

Section 3.2
4 22(4, 1)
5 (a)   53/3(17/3, 2/3, 0)   (b)   23/3(5/3, 4/3, 0)   (c)   40(0, 12, 2)
   (d)   15(3, 0, 0)   (e)   244(24, 24, 28)
6 7X, 4Y, 2Z   **7** (a)   10(0, 0, 5/3, 0)   (b)   6.5 at (2.5, 0, 2, 0)
8 Lead, 34 tons; tin, 6 tons; B, 60 tons   **9** 0A, 600B, 0C

Section 3.3
11 (i) $f(x)$ is not differentiable at its minimum   (ii) $f'(x)$ never zero for $x \geq 2$

Section 3.4
14 (b) $\tau = \dfrac{1}{2}(1 + \sqrt{5})$

16 (a)   0.54 at $x = −1$ or 1   (b)   0.416 at $x = 2$ (c)   0.94492 at $x = 0.104042$
    (d)   −2.7290 at $x = 1.32472$ (e)   0 at $x = 1$

Section 3.5
20 −1.205 at (1, 1.35)   **21** −108 at (8, 6)
22 (a)   0 at (0, 0)   (b)   0.2222 at (1.1111, 1.1111)   (c)   0 at (1, 1)   (d)   0 at (1, 1)
    (e)   0 at (3, 0.148)   (f)   four minimum values of 0 at (±1, 0), (0, ±1)

Section 3.6

24 (i) minimum $z = x^2 + xy + y^2$ (ii) minimum $z = (x + y)^2$

26 $\pm 52/\sqrt{104}$ at $(\pm 10/\sqrt{104}, \pm 1/\sqrt{104})$ **27** minimum at $(0, 0)$

28 Saddle $f = 0$ at $(0, 0)$, minimum $f = -108a^3$ at $(6a, 18a)$

30 48.3 at $(1, -1, 1)$ **31** $\left[\dfrac{-1}{\sqrt{14}}, \dfrac{-3}{\sqrt{14}}, \dfrac{-2}{\sqrt{14}}\right]$

35 $[0.2265, 0.3727, 0.8999]$ **36** (i) $4\pi k^3/3$, (ii) e^{-1} **37** (i) $6\sqrt{6}$

40 $r = 5.419, h = 10.838$ **41** $\dfrac{17}{23}$ at $\left[\dfrac{10}{23}, \dfrac{1}{23}, -\dfrac{1}{23}, \dfrac{17}{23}\right]$

43 Hexagon of side 10 and length $10\sqrt{3}$

Section 3.7

44 $(2i + j)/\sqrt{5}$, $(4i - 5j)/\sqrt{41}$, $(-i + 5j)/\sqrt{26}$ **45** 10.0 at point $(0, 0)$

47 $f_{min} = \frac{1}{2}$ at $(1\frac{1}{2}, \frac{1}{2})$ **48** $\{-58, 12, 152\}/362 \cong \{-0.160, 0.033, 0.420\}$ **49** 2.50

51 (a) $f = 0$ at $(5, 5)$ (b) 8.024 **52** $x^{(3)} = x^{(4)} = [2.37, 1.84]$

Chapter 4

Section 4.2

2 $x(0.2) = 0.1627$
3 $y(0.2) = 1.248$, $y(0.4) = 1.670$; by Taylor $y(0.2) = 1.253$, $y(0.4) = 1.675$
4 (i) $y(1) = 12.4451$ (ii) $y(1) = 1.41421$ (iii) $y(1) = 2.37797$
  (iv) $y(1) = 0.25720$

Section 4.3

6 $y(-0.1) = 0.895$, $y(0.1) = 1.094$, $y(0.2) = 1.176$, $y(0.3) = 1.241$
7 1.0100, 1.0405, 1.0923, 1.1668 **8** 0.684 **9** 1.015
10 0.6536 **11** 0.30934 **12** 0.851, 0.780

Section 4.4

13 (a) $3^n[A \cos \dfrac{n\pi}{3} + B \sin \dfrac{n\pi}{3}] + 3^{n-2} + \dfrac{13}{49}(7n + 1)$

14 (a) $A(-7)^n + B + 3^n/20$ (b) $\{1 + (-7)^{n+1}\}/\{1 + (-7)^n\} - 2$

15 (a) $(An + B)2^n + 3^n$ (b) $x_n = 4^n + 2 + n^2/2 - n/2$, $y_n = 2.4^n - 2 - n$

16 (a) $(An + B)3^n + 2^n + \dfrac{1}{4}(n + 1)$ (b) $n(A \, 4^n + B)$

17 (a) $13^{n/2}[A \cos 0.588n + B \sin 0.588n] + 2^n/5$
  (b) $(An + B)3^n + (n + 4)2^n + n^2 \, 3^n/18$

Section 4.5

20 $y = x + 1$, $y = x + 1 + 0.1e^x$, $y = x + 1 - 0.1e^x$, $y(5) = 20.84$ or -8.84

23 See problem 21   **24** Analytical solution $y = \frac{1}{2}x$

Section 4.8

28 $y(0.5) = 1.32$, $y(1) = 1.41$

29 $y'_1 = y_2$, $y'_2 = y_3$, $y'_3 = -\frac{1}{2}y_1 y_3$, $y_1(1) = 0.2$, $y_2(1) = 0.397$, $y_3(1) = 0.387$

31 As an example $y(0.5) = -0.241$

33 First iteration gives 0.0704, 0.1213, 0.1392, 0.1213, 0.0704

34 0.1002, 0.2013, 0.3046   **35** Analytical solution is $y = \sin x$

36 For example $y(0.5) \cong 0.30$ **37** For example $y(0.5) = 0.363$

38 For example $y(0.5) = -0.0694$

Chapter 5

Section 5.1

1   (a) $A[1 + 4x + 8x^2/3 + ...] + Bx^{3/4}[1 + x + 4x^2/15 + ...]$

      (b) $Ax[1 + x^2/10 + x^4/360 + ...] + B\sqrt{x}[1 + x^2/6 + x^4/168 + ...]$

      (c) $A[1 + 2x/3 + x^2/3] + Bx^4[1 + 2x^5 + 3x^6 + ...]$

      (d) $A[1 - x/3 + 5x^2/42 - ...] + Bx^{1/4}[1 - 2x/5 + 2x^2/15 - ...]$

      (e) $A\sqrt{x}[1 - x + x^2/2 - ...] + Bx[1 - 2x/3 + 4x^2/15 - ...]$

      (f) $Ax^{-1}[1 - x^2/2! + x^4/4! - ...] + B[1 - x^2/3! + x^4/5! - ...]$

      (g) $A[1 + x + 2x^2 + 14x^3/3 - ...] + Bx^{7/3}[1 + 12x/5 + 396x^2/65 + ...]$

      (h) $y = au + bv$ where $u = A[1 - 2x + 3x^2/2 - 2x^3/3 + ...]$,

           $v = u \ln x + [3x - 13x^2/4 + 31x^3/18 + ...]$

3 $y = A(1 - x + x^2/6 - x^3/90 + ...) + B\sqrt{x}[1 - x/3 + x^2/30 - x^3/630 + ...]$; $A = 1, B = 0$

4 $A\left[1 + \dfrac{2x}{3} + \dfrac{x^2}{6}\right] + Bx^4\left[1 + \dfrac{2!\,x}{5.1} + \dfrac{3!\,x^2}{6.5.2.1} + \dfrac{4!\,x^3}{7.6.5.3.2.1} + ...\right]$

6   (i) $x = 0, x = 1$ are singular points, $x = 1$ is regular

7 $A\left[1 + x + \dfrac{x^2}{(2!)^2} + \dfrac{x^3}{(3!)^2} + ...\right] = u(x),\ Bu\,(x)\ln x - 2B\left[\dfrac{x}{(1!)^3} + \dfrac{3x^2}{(2!)^3} + \dfrac{11x^3}{(3!)^3} + ...\right]$

8   (a) $a_0 x^{1-\lambda}\left[1 - \left(\dfrac{2+\lambda}{2-\lambda}\right)x + \left(\dfrac{1+\lambda}{3-\lambda}\right)\left(\dfrac{2+\lambda}{2-\lambda}\right)x^2 + ...\right]$

      (b) $ku(x)\ln x + kx^{-1}\left[5x - \dfrac{37x^2}{2} + \dfrac{53x^3}{6} + ...\right]$

9 $A[1 - nx^2/2! + n(n-2)x^4/4! - ...] + Bx[1 - (n-1)x^2/3! + (n-1)(n-3)x^4/5 - ...]$,

   $y = \exp(-x^2/4)[1 - 2x^2 + x^4/3]$

10 $T = A[1 + \dfrac{14000}{3}x^{3/2} + 14{,}000{,}000x^{5/2} + ...] + B[x + 3750x^2 + \dfrac{2800}{3}x^{5/2} + ...]$, yes

Section 5.3

12 $1.66\,\Gamma(1.6)$, $1.6\,\Gamma(1.6)$, $1.9644 \times 10^8\,\Gamma(1.6)$, $-6.6672 \times 10^{-17}\,\Gamma(1.6)$ **13** 5.24

14 3.50   **15** $\sqrt{\pi}\,a^{(n/2+1)}\,\Gamma\left(\dfrac{1}{n}\right) \Big/ 2n\,\Gamma\left(\dfrac{3}{2}+\dfrac{1}{n}\right)$ **16** $\dfrac{1}{5}\left[\Gamma\left(\dfrac{1}{3}\right)\right]^2$

Section 5.4

21 (i) $(384/x^4 - 72/x^2 + 1)J_1(x) + (12/x - 192/x^3)J_0(x)$

(ii) $-\sqrt{\dfrac{2}{\pi x}}\left[\dfrac{\cos x}{x} + \sin x\right]$

(iii) $2x\,J_3(2x) + 2x^2\left[\left\{1 - \dfrac{3}{x^2}\right\}J_2(2x) + \dfrac{3}{2x}J_1(2x)\right]$, $J_0(x^2) - 2x^2\,J_1(x^2)$,

$\left[1 - \dfrac{4}{x^2}\right]J_1(x) + \dfrac{2}{x}J_0(x)$

22 $\dfrac{1}{2}J_0(x)(\sin x + \cos x)$ $\dfrac{1}{2}J_0(x)(\sin x - \cos x)$ $-\dfrac{1}{2}J_0(x)(\sin x + \cos x)$

$\dfrac{1}{2}J_0(x)(\cos x - \sin x)$ $\dfrac{x}{2}J_0(x)(\sin x + \cos x) - \dfrac{1}{2}J_0(x)\sin x$

$\dfrac{x}{2}J_0(x)(\sin x - \cos x) + \dfrac{1}{2}J_0(x)\cos x$

$\dfrac{x}{2}J_0(x)(\sin x + \cos x) - \dfrac{1}{2}J_0(x)\sin x - xJ_1(x)\cos x$

$xJ_1(x)\sin x - \dfrac{x}{2}J_0(x)(\sin x - \cos x) - \dfrac{1}{2}J_0(x)\cos x$ $-xJ_1(x)$

$x^2\,J_1(x) + xJ_0(x) - \int J_0(x)dx$ $x^3\,J_1(x) - 2x^2\,J_2(x)$ $-J_0(x)$,

$-xJ_0(x) + \int J_0(x)dx$

26 0.9385, 0.2423, 0.0307

Section 5.6

29 (i) $AJ_0(x) + BY_0(x)$ (ii) $AJ_3(x) + BJ_{-3}(x)$ (iii) $x(AJ_1(x) + BY_1(x))$

(iv) $AJ_3(3x) + BJ_{-3}(3x)$ (v) $\dfrac{1}{x}[AJ_1(x) + BY_1(x)]$

(vi) $AJ_0\left[\dfrac{3\sqrt{x}}{2}\right] + BY_0\left[\dfrac{3\sqrt{x}}{2}\right]$ (vii) $AJ_{n/k}(ke^x) + BJ_{-n/k}(ke^x)$

33 From flanges $-\dfrac{1}{2}\pi k(\partial T/\partial r)_{r=a}$. From rim $-2\pi k(\partial T/\partial r)_{r=b}$,

$T = T_1 + AI_0(\lambda r) + BK_0(\lambda r)$, $\lambda = \sqrt{2h/k}$ and $AI_0(\lambda a) + BK_0(\lambda a) = T_0 - T_1$,

$$AI_1(\lambda b) - BK_1(\lambda b) = -\tfrac{1}{2}\lambda[AI_0(\lambda b) + BK_0(\lambda b)]$$

Section 5.7

34 $P_4(x) = \dfrac{5.7}{2.4}x^4 - 2 \cdot \dfrac{3.5}{2.4}x^2 + \dfrac{1.3}{2.4}$ $\qquad P_5(x) = \dfrac{7.9}{2.4}x^5 - \dfrac{2.5.7}{2.4}x^3 + \dfrac{3.5}{2.4}x$

$P_6(x) = \dfrac{7.9.11}{2.4.6}x^6 - \dfrac{3.5.7.9}{2.4.6}x^4 + \dfrac{3.3.5.7}{2.4.6}x^2 - \dfrac{1.3.5}{2.4.6}$

36 $P_3(x) = \dfrac{3.5}{2.3}x^3 - \dfrac{3.1.3}{2.3}x$, $P_4(x)$ as in problem 34

37 $1 = P_0(x)$, $x = P_1(x)$, $x^2 = \dfrac{1}{3}P_0(x) + \dfrac{2}{3}P_2(x)$, $x^3 = \dfrac{2}{5}P_3(x) + \dfrac{3}{5}P_1(x)$,

$x^4 = \dfrac{8}{35}P_4(x) + \dfrac{4}{7}P_2(x) + \dfrac{7}{5}P_0(x)$

41 $T = \dfrac{1}{2}(T_1 + T_2) + \dfrac{3}{4}(T_1 - T_2)\left(\dfrac{r}{a}\right)P_1(\cos\theta) + \dfrac{7}{16}(T_2 - T_1)\left(\dfrac{r}{a}\right)^3 P_3(\cos\theta) + \ldots$

Section 5.8

45 $T = 2A \displaystyle\sum_{\alpha} J_0\left(\dfrac{\alpha r}{a}\right)\sinh\left(\dfrac{\alpha Z}{a}\right) \bigg/ \alpha J_1(\alpha)\sinh\left(\dfrac{\alpha l}{a}\right)$

46 $z = J_n\left(\dfrac{\omega r}{c}\right)\cos n\theta \cos(\omega t - \varepsilon)$, $n = 0, 1, 2, \ldots$ and $J_n\left(\dfrac{\omega a}{c}\right) = 0$

47 $F(r) = AJ_n\left[\dfrac{\omega r}{c}\right] + BY_n\left[\dfrac{\omega r}{c}\right]$

Chapter 6
Section 6.3

1 (a) $\dfrac{3}{2} - \dfrac{2}{\pi}\left[\sin x + \dfrac{1}{3}\sin 3x + \dfrac{1}{5}\sin 5x + \ldots\right]$

(b) $\dfrac{\pi}{4} - \dfrac{2}{\pi}\left[\cos x + \dfrac{1}{3^2}\cos 3x + \dfrac{1}{5^2}\cos 5x + \ldots\right] + \left[\sin x - \dfrac{1}{2}\sin 2x + \dfrac{1}{3}\sin 3x + \ldots\right]$

(c) $\dfrac{\pi^2}{3} - 4\left\{\dfrac{\cos x}{1^2} - \dfrac{\cos 2x}{2^2} + \dfrac{\cos 3x}{3^2} - \ldots\right\}$

(d) $\dfrac{2}{\pi}\left[1 - \displaystyle\sum_{n=1}^{\infty}\dfrac{\cos 2nx}{(4n^2 - 1)}\right]$ (e) $\dfrac{2}{\pi}\left[\sin x + \dfrac{1}{3}\sin 3x + \ldots\right]$

(f) $\dfrac{1}{4}(\pi + 2) - \dfrac{2}{\pi}\sum\limits_{n=1}^{\infty}\dfrac{\cos(2n-1)x}{(2n-1)^2} - \dfrac{1}{\pi}\sum\limits_{n=1}^{\infty}\dfrac{(\pi - 1 + (-1)^n)(-1)^n}{n}\sin nx$

(g) $\dfrac{2}{\pi}\left[\cos x - \dfrac{1}{3}\cos 3x + \ldots + \sin 2x + \dfrac{1}{3}\sin 6x + \dfrac{1}{5}\sin 10x + \ldots\right]$

2    $\dfrac{\pi}{4} - \dfrac{1}{2}$ 3 $\dfrac{4}{\pi\sqrt{2}}\left[\sin x + \dfrac{\sin 3x}{3^2} - \dfrac{\sin 5x}{5^2} - \ldots\right]$

5    $\dfrac{2}{\pi}\left[(\pi^2 - 4)\sin x - \dfrac{\pi^2}{2}\sin 2x + \left[\dfrac{\pi^2}{3} - \dfrac{4^2}{3^3}\right]\sin 3x + \ldots\right]$

Section 6.4
6   (a) even   (b) neither   (c) neither   (d) odd   (e) even   (f) neither

7    $\dfrac{1}{\pi} + \dfrac{1}{2}\cos x - \dfrac{2}{\pi}\sum\limits_{n=1}^{\infty}\dfrac{(-1)^n\cos 2nx}{(4n^2 - 1)}$

8   (a) $\dfrac{8}{\pi}(\sin x + \dfrac{1}{3}\sin 3x + \ldots);\; 2$

     (b) $2\left[\left\{\dfrac{\pi}{1} - \dfrac{4}{1^3\,\pi}\right\}\sin x - \dfrac{\pi}{2}\sin 2x + \left\{\dfrac{\pi}{3} - \dfrac{4}{3^3\,\pi}\right\}\sin 3x - \ldots\right];$

            $\dfrac{\pi^3}{3} - 4(\cos x - \dfrac{1}{4}\cos 2x + \dfrac{1}{9}\cos 2x - \ldots)$

     (c) $\dfrac{2}{\pi}\left[(\pi + 2)\sin x - \dfrac{\pi}{2}\sin 2x + \dfrac{1}{3}(\pi + 2)\sin 3x - \dfrac{\pi}{4}\sin 4x + \ldots\right]$

9    $\dfrac{4}{\pi}\left[\sin x + \dfrac{2}{3}\sin 2x + \dfrac{1}{3}\sin 3x + \dfrac{4}{15}\sin 4x + \dfrac{1}{5}\sin 5x + \ldots\right]$

10   $\dfrac{\pi}{4} - \dfrac{1}{2}$ 11 $2\left[\sin x - \dfrac{1}{2}\sin 2x + \dfrac{1}{3}\sin 3x - \ldots\right]$

12   $\dfrac{\pi}{9} + \dfrac{2}{\pi}\left[\left\{\dfrac{\pi}{2\sqrt{3}} - 1\right\}\cos x + \dfrac{\pi}{4\sqrt{3}}\cos 2x - \dfrac{4}{9}\cos 2x + \ldots\right]$

Section 6.5
13   (a) $1 + \dfrac{2}{\pi}\sum\limits_{n=1}^{\infty}\dfrac{1}{n}[1 - 2(-1)^n]\sin n\pi x$

(b) $\displaystyle\sum_{n=1}^{\infty} b_n \sin \frac{n\pi x}{2}$ where $b_n = \dfrac{-2}{n\pi}$ for n even and $\dfrac{-2}{n\pi}\left[1 + \dfrac{1}{n\pi}(-1)^{(n-1)/2}\right]$ for n odd

(c) $\displaystyle\frac{\sqrt{3}}{\pi}\left[\frac{\sin x}{1^2} + \frac{\sin 2x}{2^2} - \frac{\sin 4x}{4^2} - \frac{\sin 5x}{5^2} + \ldots\right]$

(d) $\displaystyle\frac{8a}{\pi^2}\left[\sin \frac{2\pi x}{l} - \frac{1}{3^2}\sin \frac{6\pi x}{l} + \frac{1}{5^2}\sin \frac{10\pi x}{l} - \ldots\right]$

(e) $\displaystyle\frac{2al^2}{\pi^2 b(l-b)}\sum_{n=1}^{\infty}\frac{1}{n^2}\sin \frac{n\pi b}{l}\sin \frac{n\pi x}{l}$

14 (a) $\displaystyle\frac{3a}{\pi}\left[\sin 2x + \frac{1}{2}\sin 4x + \frac{1}{4}\sin 8x + \frac{1}{5}\sin 10x + \frac{1}{7}\sin 14x + \ldots\right]$

(b) $\displaystyle\frac{8\sqrt{2}\,a}{\pi^2}\left[\cos \frac{\pi x}{4b} - \frac{1}{3^2}\cos \frac{3\pi x}{4b} - \frac{1}{5^2}\cos \frac{5\pi x}{4b} + \ldots\right]$

15 (a) $\displaystyle\frac{16}{\pi}\sum_{n=1}^{\infty}\frac{(1-\cos n\pi)}{n}\sin \frac{n\pi x}{2}$ (b) $\displaystyle 2 - \frac{8}{\pi^2}\sum_{n=1}^{\infty}\frac{(1-\cos n\pi)}{n^2}\cos \frac{n\pi x}{4}$

(c) $\displaystyle 20 - \frac{40}{\pi}\sum_{n=1}^{\infty}\frac{1}{n}\sin \frac{n\pi x}{5}$

(d) $\displaystyle\frac{3}{2} + \sum_{n=1}^{\infty}\left[\frac{6(\cos n\pi - 1)}{n^2\pi^2}\cos \frac{n\pi x}{3} - \frac{6\cos n\pi}{n\pi}\sin \frac{n\pi x}{3}\right]$

Section 6.6

20 $-1.69 - 16.36 \cos x + 3.27 \cos 2x + 0.49 \cos 3x + 23.37 \sin x + 2.73 \sin 2x - 2.03 \sin 3x$

21 Coefficients in same order
    (a) 1.92, -1.02, 0.92; 2.83, 0.03, -0.63
    (b) 1.63, -1.19, -0.22, -0.07; 1.00, -0.17, -0.07
    (c) 1.13, 0.75, -0.77, -0.40; 1.03, -0.09, -0.47

22 (a) 1.21, 1.00, -0.01, 0.23, 0.25; 2.73, -0.21, 0.27, -0.12
    (b) $a_i = 0$; 3.95, -1.27, -0.76, 0.69, 0

Chapter 7
Section 7.3

3 (a) hyperbolic   (b) hyperbolic   (c) elliptic   (d) elliptic
   (e) hyperbolic   (f) hyperbolic   (g) parabolic

Section 7.4

5 $\theta = e^{-\pi y}\sin \pi x + e^{-2\pi y}\sin 2\pi x$ 6 $V = 4\cosh 15x \sin 5t$

9 $A_n = \begin{cases} 0, & n \text{ even} \\[2mm] \dfrac{4a(-1)^{(n-1)/2}}{n^2 \pi^2 \sinh(n\pi b/a)}, & n \text{ odd} \end{cases}$

Section 7.5

10 (a) $\theta = -200 \sum_1^\infty \dfrac{(-1)^n}{n\pi} \sin \dfrac{n\pi x}{50} e^{(-n^2 \pi^2 Kt)/50}$

     (b) $\theta = 100 - 200 \sum_1^\infty \dfrac{(2(-1)^n - 1)}{n\pi} \sin \dfrac{n\pi x}{50} e^{(-n^2 \pi^2 Kt)/50}$

     (c) $\theta = x - 100 \sum_1^\infty \dfrac{(-1)^n}{n\pi} \sin \dfrac{n\pi x}{50} e^{(-n^2 \pi^2 Kt)/50}$

     (d) $\theta = 25 + x - 50 \sum_1^\infty \dfrac{((-1)^n + 1)}{n\pi} \sin \dfrac{n\pi x}{50} e^{(-n^2 \pi^2 Kt)/50}$

     (e) $\theta = 50 + 2x + 100 \sum \dfrac{((-1)^n + 1)}{n\pi} \sin \dfrac{n\pi x}{50} e^{(-n^2 \pi^2 Kt)/50}$

11 $u = \dfrac{4}{\pi}\left(e^{-kt} \sin x - \dfrac{1}{9} e^{-9kt} \sin 3x + \dfrac{1}{25} e^{-25kt} \sin 5x - \ldots\right)$

12 $u = \sum \beta_n \exp\left[-\left[\dfrac{n^2 \pi^2 a^2}{L^2} + h\right]t\right] \sinh \dfrac{n\pi x}{L}, \quad \beta_n = \dfrac{2}{L} \int_0^L f(x) \sin \dfrac{n\pi x}{L} dx$

Particular solution $\beta_n = 8L^2/(n^3 \pi^3)$, n odd

13 $u = u_0 + \dfrac{8e^{-kt}}{\pi^3} \sum \dfrac{1}{(2s-1)^3} \sin(2s-1)\pi x \cdot e^{-\{(2s-1)h\pi\}^2 t}$

     $u = u_0 + e^{-kt} \sum B_r \sin r\pi x \cdot e^{-(r\pi h)^2 t}$

15 $\bar{U} = 1 - \dfrac{8}{\pi^2} \sum_{1,3,5,\ldots}^\infty \dfrac{1}{n^2} \exp(-(n^2 \pi^2/4H^2)c_v t)$ **22** (ii) $F = 3e^{-2y} \sin 3x$

23 (i) $y = A \sin 2\pi x \cos 2\pi ct$ (ii) $y = \sum_1^\infty \dfrac{12A}{n^3 \pi^3}(-1)^{n+1} \sin n\pi x \cos n\pi ct$

     (iii) $y = \dfrac{2}{\pi^2}\left\{\dfrac{(\sqrt{2}-1)}{1^2} \sin \pi x \cos \pi ct + \dfrac{2 \sin 2\pi x \cos 2\pi ct}{2^2} + \dfrac{(\sqrt{2}+1)}{3^2} \sin 3\pi x \cos 3\pi ct\right.$

$$-\frac{(\sqrt{2}+1)}{5^2}\sin 5\pi x \cos 5\pi ct - \frac{2}{6^2}\sin 6\pi x \cos 6\pi ct - ...\Bigg\}$$

(iv) $y = \sum_{1}^{\infty} \frac{-4A}{n^3 \pi^3}\left\{1 + 5(-1)^n + \frac{12}{n^2\pi^2}(1 - (-1)^n)\right\}\sin n\pi x \cos n\pi ct$

24 $y = \Sigma(\alpha \cosh ax + \beta \sinh ax + \gamma \cos ax + \delta \sin ax)\sin(ca^2 t + \varepsilon),\ c^2 = EI/\rho A$

25 $2\pi/\omega$ where $\rho A\omega^2 = EI\pi^4/l^4$

29 $a^2 = \frac{1}{2}[(RG - LC\,\omega^2) + K],\ \ b^2 = \frac{1}{2}[-(RG - LC\,\omega^2) + K],$

$$K = \sqrt{(RG - LC\,\omega^2)^2 + (RC + GL)^2\,\omega^2}$$

30 $f(\theta) = A_1 \cos n\theta + B_1 \sin n\theta,\ V = \frac{1}{r}\sin 2\theta$

31 (i) $V = a^2 - r^2 \cos 2\theta$ \qquad (ii) $f = Cx^2\,e^{k(x^3/3-t)},\ f = x^2\,e^{(x^3/3-1/3-t)}$

Section 7.6

32 $u(10, 0.4) = 10$ (nearest integer), $u(10, 0.8) = 14$ (nearest integer)

33 $r = 0.1,\ u(10, 0.4) = 9$ (nearest integer), $u(10, 0.5) = 10$ (nearest integer)

$r = 0.4,\ u(10, 0.4) = 12$ (nearest integer), $u(10, 0.8) = 14$ (nearest integer)

$r = 0.8,\ u(10, 0.8) = 24$ (nearest integer)

$r = 1.0,\ u(10, 1) = 30$ (nearest integer)

34 $u(12, 0.4) \cong 110,\ \ u(12, 0.8) \cong 90,\ \ u(10, 0.4) \cong 16,\ \ u(10, 0.8) \cong 29$

35 $u(10, 0.4) \cong 6,\ \ u(10, 0.8) \cong 10$ \qquad **36** $u(10, 0.4) \cong 0,\ \ u(10, 0.8) \cong 1$

37 $h = 2$ gives \quad (32) $r = 0.8,\ u(10, 3.2) \cong 30,\ u(10, 16) \cong 75$

$r = 2,\ u(10, 8) \cong 55,\ u(10, 16) \cong 72$

$r = 5,\ u(10, 20) \cong 99,\ u(10, 40) \cong 123$

(36) $r = 0.8,\ u(10, 3.2) \cong 10,\ u(10, 16) \cong 29$

$r = 2,\ u(10, 8) \cong 24,\ u(10, 16) \cong 34$

Section 7.7

38 Temperatures at A, B, C, D are 24, 23, 25, 27 to nearest integer

39 Values at A, B, C, D are 0.7, 1.4, 0.4, 1.0

40 Values at A, C, C, D are 64, 71, 37, 43 to nearest integer

41 Values at A, B, C, D are 15, 17, 15, 13 to nearest integer

42 $h = 1$ gives for example $V(1, 2) = 27, V(2, 2) = 8, V(3, 3) = -10$ all to nearest integer

43 $h = 1$ gives $\theta(0, 0) \cong 30, \theta(1, 0) = \theta(-1, 0) \cong 20, \theta(1, 1) = \theta(-1, 1) \cong -32, \theta(1, -1) = \theta(-1, -1) \cong 82, \theta(0, 1) \cong -28, \theta(0, -1) \cong 108$

44 $h = 1$ gives with origin at centre of pipe section, $\theta(0, 3) = 133, \theta(-2, 3) = \theta(2, 3) = 114, \theta(3, 3) = \theta(-3, 3) = 122, \theta(4, 4) = \theta(-4, 4) = 95, \theta(-3, 0) = \theta(3, 0) = 128$

Section 7.8

46 Elliptic, parabolic or hyperbolic for

(a) $x > 0, y > 0$ and $x < 0, y < 0$; $x = 0, y = 0$; $x > 0, y < 0$ and $x < 0, y > 0$

(b) no region \qquad all x and y \qquad no region

(c) $x^2 + y^2 < 1$ \qquad $x^2 + y^2 = 1$ \qquad $x^2 + y^2 > 1$

(d) $x^2 < 4y$ $x^2 = 4y$ $x^2 > 4y$

48 (i) $\xi = 3x + y$, $\eta = 2x + y$, $\dfrac{\partial^2 u}{\partial \xi \, \partial \eta} = 0$

     (ii) $\xi = 4x + y$, $\eta = -2x + y$, $2\dfrac{\partial^2 u}{\partial \xi \, \partial \eta} = 2\dfrac{\partial u}{\partial \xi} - \dfrac{\partial u}{\partial \eta}$

     (iii) $\xi = 2x + y$, $\eta = y$, $\dfrac{\partial^2 u}{\partial y^2} = 0$ (iv) $\xi = y - x$, $\eta = y$, $\dfrac{\partial^2 u}{\partial \eta^2} = \dfrac{\partial u}{\partial \xi} - 9u$

49 $a = 1, b = 2, z = f(x + y) + g(x + 2y)$
50   See answer to Problem 13 of Section 7.5   **51** $-0.59, -1.90, -1.12$

Chapter 8
Section 8.2

1 $s/(s + 1)(s^2 + 2s + 5)$, $(2s^5 - 4s^3 - 6s)/(s^2 + 1)^4$, $k/(s^2 + 4k^2)$,

     $2\omega(s + 2)/[(s + 2)^2 + \omega^2]^2$, $(s + 4)/(s^2 + 8s + 12)$, $(6s^2 - 29s + 36)/(s - 2)^3$,

     $\frac{1}{2}\ln[(s + 1)/(s - 1)]$

2 $N_1(t) = \dfrac{a}{\lambda_1}(1 - e^{-\lambda_1 t})$, $N_2(t) = \dfrac{a}{(\lambda_2 + b)} - \dfrac{a\,e^{-\lambda_1 t}}{(\lambda_2 - \lambda_1 - b)} + \dfrac{a\lambda_1 e^{-(\lambda_2 + b)t}}{(\lambda_2 + b)(\lambda_2 - \lambda_1 + b)}$

     After shut down $N_1(t) = \dfrac{a}{\lambda_1} e^{-\lambda_1 t}$, $N_2(t) = \dfrac{a}{(\lambda_2 + b)}\left[e^{-\lambda_2 t} + \dfrac{(\lambda_2 + b)}{(\lambda_2 - \lambda_1)}(e^{-\lambda_1 t} - e^{-\lambda_1 t}) \right]$

4 $y = \cos 2t + 3U(t - 2)\left[\frac{1}{4} - \frac{1}{4}\cos 2(t - 2)\right]$

5   (i) $f(t) = n + 1$ for $n\pi < t < (n + 1)\pi$
     (ii) $f(t) = \cos t, \ 0 < t < \pi, = 0, \ \pi < t < 2\pi$, solution $f(t) = $ (i) $-$ (ii)
6 $be^{-sc}/(s^2 + b^2)(1 - e^{-s\pi/b})$
9 $te^{-4t}(e^4 - 5)/16 - e^{-4t}(e^4 + 3)/32, \ 0 < t < 1$,
     $-5te^{-4t}/16 - 3e^{-4t}/32 - t/16 + 3/32, \ t > 1$

10   (i) e^{-bs}/s (ii) e^{-bs}/s^2 (iii) $\dfrac{x^2}{2} + \dfrac{1}{6a}(x - a/2)^3 \, U(x - a/2)$

11 $53\omega l^4/30720EI$

12   If $n = 1/\sqrt{LC}$ then $CE_0\left(t - \dfrac{1}{n}\sin nt\right)/T, \ 0 < t < T$ and

     $CE_0\left[T \cos n(t - T) + \dfrac{1}{n}\sin n(t - T) - \dfrac{1}{n}\sin nt\right]/T, \ t > T$

Section 8.3

13   (a)   (i) $\dfrac{1}{s^2}$ (ii) $\dfrac{1}{(s + 1)^2}$; $y = 2t\,e^{-t} + 2e^{-t} + t - 2$

(b) $\quad F(s) = \dfrac{2(s^2 + 25)}{s^2(s^2 + 20)}$, $\quad f(t) = \dfrac{5}{2}t - \dfrac{1}{2\sqrt{20}}\sin\sqrt{20}\,t$

15 (a) $\quad (e^{-t} - \cos t + \sin t)/2$ (b) $\quad (e^{2t} - e^{-t} - 3te^{-t})/9$

16 (b) $\quad 0.5t^2 - 2t + 3 - (t + 3)e^{-t}$ (c) $\quad (5t - 2)e^{-t} + t^2 - 4t + 6$

Section 8.5

20 (i) $\quad \dfrac{\sin \lambda b}{2\lambda a}$ $(\lambda \ne 0)$, $\quad \dfrac{b}{a}$ $(\lambda = 0)$ (ii) $\quad \dfrac{\lambda}{\pi^2 + \lambda^2}$, $\dfrac{\pi}{2}e^{-\pi m}$

21 (i) $\quad 4\lambda/(1 + \lambda^2)^2$ (ii) $\quad [(2\lambda^2 - 1)\sin 2\lambda + 2\lambda \cos 2\lambda]/\lambda^3$

22 (i) $\quad \sin \lambda\pi/(1 - \lambda^2)$ (ii) $\quad 1/(1 + \lambda^2)$

23 $\quad \dfrac{1}{\lambda}\sin \lambda a$, $f(x) = \begin{cases} 0, & x < 0 \\ x, & 0 < x < a \\ 0, & x > a \end{cases}$

Section 8.7

25 (i) $\quad 2\left[\dfrac{\sin 2\omega}{\omega}\right]e^{-2j\omega}$ (ii) $\quad 2\dfrac{\sin \omega a}{\omega} + 2\dfrac{\sin 2\omega a}{\omega}e^{-2j\omega}$ (iii) $\quad 2\dfrac{\sin(\omega + 1)a}{(\omega + 1)}$

26 (i) $\quad 8\pi\, e^{-j\pi/2} \displaystyle\sum_{n=-\infty}^{\infty} \dfrac{(1 - \cos n\pi)}{n\pi}\, \delta(\omega - n\pi)$

     (ii) $\quad 2[\text{Si}(\omega + 1) - \text{Si}(\omega - 1)]$ where $\text{Si}(x) = \displaystyle\int_0^x \dfrac{\sin \tau}{\tau}d\tau$

27 (i) $\quad \dfrac{A}{a + j\omega}$ (ii) $\quad \dfrac{2Aa}{a^2 + \omega^2}$ (iii) $\quad \dfrac{A}{a - j\omega}$

28 (i) $\quad 2p_3(t + 1)$ (ii) $\quad 3e^{(1+t)}\, U(-1 - t)$ (iii) $\quad \dfrac{1}{\sqrt{\pi}}e^{-t^2/4}$

     (iv) $\quad \dfrac{2}{\pi t}\sin at \cos \omega_1 t$ (v) $\quad 2p_4(t - 3)$

30 $\quad (1 - e^{-t})\, U(t)$

Section 8.8

31 (i) $\quad \dfrac{4z}{4z - 1}$ (ii) $\quad \dfrac{a(1 + 2a) - 2z}{z(z - a)}$

32 (i) $\quad -\dfrac{1}{3} + \dfrac{4}{3} \cdot 4^k$ (ii) $\quad (k + 4).2^k$

33   (i) $\dfrac{z^2 - z \cos 3T}{z^2 - 2z \cos 3T + 1}$ (ii) $\dfrac{2^T z \sin 3T}{z^2 - 2z \cdot 2^T \cos 3T + 2^{2T}}$ (iii) $\dfrac{z}{z - 2}$

     (iv) $\dfrac{z^2 e^{-2T} \sin 2T}{z^2 - 2e^{-2T} z \cos 2T + e^{-4T}}$

35   (i) $\dfrac{z}{z + 1}$ (ii) $\dfrac{2z}{2z - 1}$ (iii) $\dfrac{z^3 + 4z^2 + z}{(z - 1)^4}$ (iv) $\exp(1/z)$

     (v) $\dfrac{z\,(z - \cosh 2)}{z^2 - 2z \cosh 2 + 1}$ (vi) $\dfrac{ze^3}{(ze^3 - 1)^2}$ (vii) $2T2\,\dfrac{z(z + 1)}{(z - 1)^3}$

36 $\dfrac{z^2}{z^2 - 1}$

37   (i) $8 - 8(0.5)^k - 6k(0.5)^k$ (ii) $\dfrac{8}{9}(-2)^k + \dfrac{k}{3} + \dfrac{5}{9}$

    (iii) $\delta(k) + \delta(k - 1) + 2^{k+1}$ (iv) ke^{-k}
    (v) $0.59408k(0.2)^k + 0.36(0.2)^k$

38   (i) $y(k) = 2 \cdot 2^k - 2 - k$ (ii) $y(k) = 21 \times 3^k - 3$
    (iii) $y(k) = 12 \times 2^k + 50 \times 5^k$ (iv) $y(k) = (6 + 3^k).3^k$

39 $1.5 - 0.5(1/3)^k$; $2, k = 0$ and $3.5(1/3)^k + 1.5, k > 0$

40   (a)   (i) and (ii) $3^{k+1} - 2^{k+1}$ (iii) $2 - (0.5)^k$

    (b) $y(0) = 1,\ y(k) = (0.5)^k(2^{k+1} - 1)\ k = 1, 2, 3$
       and $y(k) = 15 \times (0.5)^k,\ k = 4, 5, \ldots$

    (c) $y(0) = 1,\ y(k) = ((0.3)^k + (-1)^k\,0.3)/3,\ k = 1, 2, \ldots$

Chapter 9
Section 9.1
1   (i) Parallel planes   (ii) hyperboloids
2   Directions of (i) x axis   (ii) x axis $(x \neq \pm 1)$ (iii) $-2\mathbf{i} \pm 4x^2\mathbf{j}$ (iv) all
3 $-e^{-t}\mathbf{i} - 12 \sin 3t\mathbf{j} + 2 \cos t\mathbf{k}$, $e^{-t}\mathbf{i} - 36 \cos 3t\mathbf{j} - 2 \sin t\mathbf{k}$
4 $(4xy - 4x^3)\mathbf{i} + y \cos x\mathbf{j}$, $2x^2\mathbf{i} + \sin x\mathbf{j} - \sin y\mathbf{k}$
6 $m = -n(1 \pm i)$, $\mathbf{r} = e^{-nt}(\mathbf{A} \cos nt + \mathbf{B} \sin nt)$, $\mathbf{r} = \mathbf{v}/n\ e^{-nt} \sin nt$

Section 9.2

7   (a) $\mathbf{i} + \mathbf{j} + \mathbf{k}$, $\sqrt{3}$ (b) $2(x\mathbf{i} + y\mathbf{j} - z\mathbf{k})$, $2\sqrt{x^2 + y^2 + z^2}$

    (c) $2(z^4 - xy)\mathbf{i} - x^2\mathbf{j} + 8xz^3\mathbf{k}$, $\sqrt{4(z^4 - xy)^2 + x^4 + 64x^2 z^6}$

    (d) $-3x^2\,y\mathbf{i} - x^3\mathbf{j} + 2\mathbf{k}$, $\sqrt{9x^4 y^2 + x^6 + 4}$

8 (i) 0   (ii) $28/\sqrt{114}$ (iii) $-3/\sqrt{6}$

9 (i) $(x^2 + y^2)/2$ (ii) $(3x^2 - 2y^2 + z^2)/2$

10 (i) $(2\mathbf{i} + \mathbf{j} + \mathbf{k})/\sqrt{6}$ (ii) $(\mathbf{i} + \mathbf{j})/\sqrt{2}$

11 Sphere $x^2 + y^2 + z^2 = 49$, $\pm(6\mathbf{i} + 2\mathbf{j} + 3\mathbf{k})/7$, $\dfrac{-4}{2401}(6\mathbf{i} + 2\mathbf{j} + 3\mathbf{k})$

13 $\dfrac{m}{4\pi\varepsilon \cdot 7^5}(36\mathbf{i} - 37\mathbf{j} + 18\mathbf{k})$

Section 9.3

14 (i) 6   (ii) $2x + 2y + 2z$ (iii) $y + z + x$ (iv) $\dfrac{3}{2}\left[\dfrac{1}{r^{3/2}}\right]$

Section 9.4

16 (i) $4a^2 \tan^{-1}\dfrac{b}{a} + 4b^2 \tan^{-1}\dfrac{a}{b}$ (ii) $3k^2 \pi/2$ (iii) $c^2 \pi/4$

17 $1 - 2/\pi - \pi/4$

18 (b) $2a^2$ **20** $\phi = x^3 + y^3 + z^3 - 3xyz + x^2z + z^2y$, -28, 0

Section 9.5

21 (i) 0   (ii) 0   (iii) $-y\mathbf{i} - z\mathbf{j} - x\mathbf{k}$ (iv) 0

22 (i) $5y\mathbf{k}$ (ii) 0   (iii) 0

23 $2x^2 y - xz^3$

24 $\phi = y^2 \cos x + xz^3 + 4y + 2z + C$, $4\pi - 1$

Section 9.6

27 r^{-4} **35** const.$/r^3$

Section 9.7

37 (a) 18   (b) 32/3   (c) 2/9   (d) 4/15   (e) 119/20

38 (a) 1/15   (b) 32/3   (c) 1/12

39 (a) 4   (b) $\dfrac{2}{3}\pi abc$ (c) $\dfrac{1}{6}abc$ (e) 1/10   (e) 64/35

40 (a) 1   (b) 64/3   (c) $\pi/2 - 1/3$

Section 9.8

41 (a) 5/12   (b) $2 - \pi/2$ (c) $4/3 + \pi/2$

42 (a) 16/3   (b) $625\pi/2$ (c) 2π

43 (a) 2048/105   (b) $kac^3/12$ (c) 3/80

44 (a) $\pi(1 - 1/e)$ (b) $\pi^2(\pi^2 + 12)/48$ (c) 1/10

45 (a) $\pi\sqrt{3}$ (b) $\pi(\tan\frac{1}{2} - \tan\frac{1}{4})$ (c) 32/9   (d) $8\pi/3 - 3\sqrt{3}$

46 (a) $16\pi/3 + 8\sqrt{3}$ (b) 64/3   (c) $\pi - 9\sqrt{3}/16$ (d) $\pi/2 - 2/3$

47 (a) $-8uv$ (b) $4(v^2 - v - u)$ (c) $-\sin t \cos t$

49 (a) 6   (b) $\sqrt{6}$ (c) $\pi(2\sqrt{2}-1)/6$ (d) $4\pi a[a-\sqrt{a^2-b^2}\,]$

51 M. of I. $= 656\lambda a^6/45$ **53** $2\pi\gamma$ **54** 48

Section 9.9
55 (a) $-10/3$ (b) 71/420
56 (a) $a^2 b^2 c^2/8$ (b) $4\pi/15$ (c) $40\pi/3 - 256/9$ (d) $7\pi/2$
57 $\pi a^4/4$ (b) $2(e-1)/3$ **58** (a) $3\pi a^5/10$ (b) $\pi(2\ln 2 - 1)$
59 $3h/8$

Section 9.11
64 -4π **65** $-2\pi b^2$ **66** 4π **69** 6π

Section 9.12
75 5

Chapter 10
Section 10.2
1 (a) $u = x^3 - 3xy^2$, $v = 3x^2 y - y^3$ (b) $(x^2 + y^2)^{1/2}$, 0   (c) x, 0
   (d) $e^x \cos y$, $e^x \sin y$ (e) $x/(x^2 + y^2)$, $-y/(x^2 + y^2)$
   (f) $x(x^2 + y^2 + 1)/[(x^2 - y^2 + 1)^2 + 4x^2 y^2]$, $y(1 - x^2 - y^2)/[(x^2 - y^2 + 1)^2 + 4x^2 y^2]$
   (g) $2(x^2 - y^2) + 7x + 6$, $y(4x + 7)$
2 (a) Translation by 1 unit   (b) Rotation by $\pi/2$
   (c) Reflection in the x-axis and rotation by $\pi/2$
3 (a) $r^3 \cos 3\theta + ir^3 \sin 3\theta$ (c) $z = 0$ and $\alpha \le \arg z < \alpha + 2\pi/3$
   (d) Equal moduli; arguments differ by $2\pi/3$

4 (a) $1 - \frac{1}{2} i$ (b) $-(2 + i)/5$ (c) $5(7 - 4i)/66$ (d) $(1 + i)/\sqrt{2}$ (e) $\frac{1}{4}$
5 (a) $1 - i$ (b) $4/5 - 4i$ (c) 0. Limit as $z \to 0$ does not exist
7 $3i$, no, yes
8 (a) $-1 \pm i$ (b) $\pm 2 \pm 2i$ (c) nowhere in that region
9 (a) $2z + 3$ (b) $-z^{-2}$
10 (a) $-i$, $i/(z + i)^2$ (b) $-1 \pm 2i$, $(10 + 4z - 3z^2)/(z^2 + 2z + 5)^2$
11 (a) analytic   (b) not   (c) not   (d) analytic
14 (a) y, z (b) $e^x \sin y$, e^z (c) $2 \tan^{-1}(y/x)$, $\ln z^2$
   (d) $ye^x \cos y + xe^x \sin y$, ze^z
   (e) $x^2 - y^2 + 2xy - 3x - 2y$, $(1 + i)z^2 - (2 + 3i)z$
   (f) $4xy - x^3 + 3xy^2$, $2z^2 - iz^3$
15 $v = 2y^2 - 2x^2 - 3x^2 y + y^3$, $f(z) = -z^3 - 2iz^2$
16 (a) not harmonic   (b) harmonic, $v = x^2 - y^2 + 2y$, $f(z) = 2z + iz^2$

Section 10.3

19 (i) $2k\pi\, i/3$ (ii) $\frac{1}{8}\pi i + \frac{1}{2} k\pi i$ (iii) $\frac{1}{2}\ln 2 + (2k\pi - \frac{\pi}{4})i$

20 $\cos z, -\sin z$

22 (i) $\cosh 1$ (ii) $\sinh 1 \cosh 1 + i \cos 1 \sinh 1$

24 (i) $\pm i \ln(2 + \sqrt{3}) + \pi/2 + 2k\pi$

     (ii) $-i \ln (\sqrt{2} + 1) + \pi/2 + 2k\pi, -i \ln(\sqrt{2} - 1) + 3\pi/2 + 2k\pi$

27 (i) $i\sqrt{3}$ (ii) i

28 $\ln(\sqrt{2} + 1) + \pi\, i/2 + 2k\pi i, \ \ln(\sqrt{2} - 1) + 3\pi i/2 + 2k\pi i$

29 (i) $(2k\pi - \pi/2)i$ (ii) $\ln 5 + (\tan^{-1} 4/3 + 2k\pi)i$ (iii) $\frac{1}{2}\ln 5 + (\tan^{-1}(-\frac{1}{2}) + 2k\pi)i$

30 -2

33 (i) $e^{-\pi/4 + 2k\pi} \{\cos(\frac{1}{2}\ln 2) + i \sin(\frac{1}{2}\ln 2)\}$

     (ii) $\sqrt{5}\, e^{(\tan^{-1} \frac{1}{2} - 2k\pi)} [\operatorname{cis}(\ln \sqrt{5} - \tan^{-1}\frac{1}{2} + 2k\pi)]$

     (iii) $e^{-4k\pi} \operatorname{cis}(2 \ln 2); \ \operatorname{cis} \theta \equiv \cos \theta + i \sin \theta$

34 $\arg w = \frac{1}{2}\arg(z - i) + \frac{1}{2}\arg(z + i)$

Section 10.4

35 (i) Circular region, centre $(-2 + i)$, radius 4   (ii) Annulus, centre 2i, radii 1 and 2

     (iii) Right half plane   (iv) Circle centre $- 5i/3$, radius 4/3

     (v) Region between branches of hyperbola $x^2 - y^2 = 2$

     (vi) Region inside ellipse with foci $(\pm 3, 0)$

36 (i) $6x^2 y^2 - x^4 - y^4 = \text{const.}$ (ii) $y^2 - x^2 - 2e^{-x} \sin y = \text{const.}$

     (iii) $r^2 \sin 2\theta = \text{const.}$

37 (i) Yes, streamlines are rectangular hyperbolas   (ii) No

     (iii) Yes, straight line flow

39 w_1 – line charge at point x_1, w_2 – opposite line charge at point x_2, $w_1 + w_2$ is combination of these giving circular lines of force

41 Vortex of strength K at the origin; sink at the origin

43

	$\arg z = \pi/6$	$\lvert z \rvert = 2$	$R(z) -2$	$I(z) = 1$
(i)	$v + u\sqrt{3} = 0$	$u^2 + v^2 = 4$	$v = -2$	$u = -1$
(ii)	$v + u\sqrt{3} = 2$	$u^2 + (v - 2)^2 = 4$	$v = 4$	$u = -1$
(iii)	$v + u\sqrt{3} = 2(1 + \sqrt{3})$	$\left[\dfrac{u}{2} - 1\right]^2 + \left[\dfrac{v}{2} - 1\right]^2 = 4$	$v = -2$	$u = 0$

44 $u^2 = 4\left[1 - \dfrac{v}{2}\right]$

45 $u^2 + 2v = 1$ (ii) $\lvert w \rvert > 4$ (iii) Region between $v^2 = 4(1 - u)$ and $v^2 = 16(4 - u)$

     (iv) $1 < \lvert z \rvert < 4$ and $\operatorname{larg} w\lvert < \pi/2$

46 (i) $u^2 + v^2 - u - v = 0$ (ii) $v = -1/2$ (iii) $u = 0$

     (iv) $u^2 + v^2 + 2u + 4v + 1 = 0$ (v) $u = -1/2$

47 $(0, 0)$ **48** Translation + inversion   **49** $w = (10z - 12)/(z - 1)$

50 $-i(z + 1)/(z - 1)$

51 (i) $w = 6z + 4$ (ii) $w = \dfrac{1}{z}$ (iii) $w = iz$ (iv) $w = \dfrac{(z_2 - z_3)(z - z_1)}{(z_2 - z_1)(z - z_3)}$

53 $(1, 0), \infty, (-1, 0), (0, 0),$ Upper half plane

54 (b) Region bonded by ellipse $u^2/\cosh^2 \dfrac{N\pi}{a} + v^2/\sinh^2 \dfrac{N\pi}{a} = 1$ and real axis

55 $w = \left[iz + \dfrac{1}{\sqrt{2}}(1 + i) \right] \Big/ \left[z + \dfrac{1}{\sqrt{2}}(1 + i) \right], \ u^2 + v^2 = 1$

56 $u = (x^2 + y^2 - 1)/[(x + 1)^2 + y^2], \ v = 2y/[(x + 1)^2 + y^2]$

57 (i) major arc of circle $u^2 + v^2 = \dfrac{1}{e^4}$ between $\left[0, \ -\dfrac{1}{e^2} \right]$ and $\left[-\dfrac{1}{e^2}, \ 0 \right]$

   (ii) semicircle $u^2 + v^2 = e^2$, upper half

58 (i) straight line $u = 3 \ln 3$ (ii) straight line $v = \pi/4$

59 (i) $v \geq 0, -\infty < u < \infty$ (ii) $v \geq 0, u > 0$

60 $v > 0, \ u^2/\cosh^2 \pi + v^2/\sinh^2 \pi = 1$

Section 10.5

61 $z = re^{i\theta}$, ellipse $u^2/(r + \dfrac{1}{r})^2 + v^2/(r - \dfrac{1}{r})^2 = 1$

Section 10.6

65 (a) 32 (b) $-1 + i$ (c) $-44/3 - 8i/3$ (d) (i) $2\pi i$ (ii) $2\pi i$ (iii) $2\pi i$

67 (a) 0 (b) 0 (c) $-3/2$ (d) $-\pi i$ (e) 0 (f) 0 (g) πi

68 (a) $2\pi i e^2$ (b) $2\pi i$ (c) $-2\pi i$ (d) $\pi i/32$ (e) $-2\pi i$ (f) 0 (g) 0

69 $8\pi i/3e^2$ **70** (i) (a) 0   (b) $\pi(3 + 2i)/2$

71 (a) $2\pi i$ (b) $-2\pi i$ **72** $\pi(3/2 + i)$

73 (a) $19\pi i/108$ (b) $9\pi i/8$

Section 10.7

74 (a) $e^i \displaystyle\sum_{n=0}^{\infty} \dfrac{(z - i)^n}{n!}$ (b) $\displaystyle\sum_{n=0}^{\infty} \dfrac{(-1)^n}{(2i)(n + 1)} z^n$ (c) $-\displaystyle\sum_{n=0}^{\infty} z^{2n}$

75 (a) $\dfrac{1}{3} \displaystyle\sum_{n=0}^{\infty} (-1)^n \left[\dfrac{1}{(z - 2)^{n+1}} - \dfrac{(z - 2)^n}{4^{n+1}} \right]$ (b) $\dfrac{1}{3} \displaystyle\sum_{n=0}^{\infty} (-1)^n \dfrac{1 - 4^n}{(z - 2)^{n+1}}$

   (c) $\displaystyle\sum_{n=0}^{\infty} \dfrac{(-1)^n}{(2n)!}(z - 1)^{2n-1}$ (d) $\displaystyle\sum_{n=0}^{\infty} \dfrac{z^{2n-1}}{(2n + 1)!}$

76 $\dfrac{1}{2z^2} + \dfrac{3}{4z} + \displaystyle\sum_{n=0}^{\infty} \left[1 - \dfrac{1}{2^{n+3}} \right] z^n$ $|z| < 1$

$$\frac{1}{2z^2} + \frac{3}{4z} + \sum_{n=0}^{\infty} \frac{1}{z^{n+1}} - \sum_{n=0}^{\infty} \frac{1}{2^{n+3}} z^n \qquad 1 < |z| < 2$$

$$\frac{1}{2z^2} + \frac{3}{4z} - \sum_{n=0}^{\infty} \frac{1}{z^{n+1}} - \sum_{n=0}^{\infty} \frac{2^{n-2}}{z^{n+1}} \qquad 2 < |z| < \infty$$

77 $\dfrac{1}{z} + \displaystyle\sum_{n=0}^{\infty} \left[\dfrac{1}{i^{n-1}} - 1 \right] z^n$

78 (a) $-\dfrac{1}{z} - 1 - z - z^2, \ 0 < |z| < 1; \ z^{-2} + z^{-3} + z^{-4} + \dots \ |z| > 1$

(b) $-\dfrac{1}{4} z^3 - \dfrac{1}{2} z^4 - \dfrac{11}{16} z^5 - \dfrac{13}{16} z^6 - \dots, \ |z| < 1$

$1 + z + z^2 + \dfrac{3}{4} z^3 + \dots + \dfrac{1}{z} + \dfrac{1}{z^2} - \dfrac{1}{z^3} + \dots, \ 1 < |z| < 2$

$1 + 5/z + 49/z^3 + 129/z^4, \ |z| > 2$

79 (a) $-\dfrac{1}{2z^4} + \dfrac{1}{2z^3} - \dfrac{1}{2z^2} + \dfrac{1}{2z} - \dfrac{1}{6} + \dfrac{z}{18} - \dfrac{z^2}{56} + \dfrac{z^3}{162}$ (b) $\dfrac{1}{z^2} - \dfrac{4}{z^3} + \dfrac{13}{z^4} - \dfrac{40}{z^5}$

(c) $\dfrac{1}{2(z+1)} - \dfrac{1}{4} + \dfrac{1}{8}(z+1) - \dfrac{1}{16}(z+1)^2 + \dots$ (d) $\dfrac{1}{3} - \dfrac{4}{9}z + \dfrac{13}{27}z^2 - \dfrac{40}{81}z^3$

80 (a) Pole of order 1 at $z = 0$ (b) Essential at $z = 0$ (c) Pole of order 1 at $z = 1$, Pole of order 2 at $z = 2$ (d) Removable singularity at $z = 0, f(0) = -1/2$
(e) Removable singularity at $z = 2, f(2) = 1$ (f) Pole or order 3 at $z = 0$
(g) Poles of order 1 at $z \pm i$ (h) Pole of order 2 at $z = (2k + 1)\pi/2$

Section 10.8
81 (a) 0 (b) 2 (c) 1 (d) $1/(2 - \pi), 0, (\pi - 1)/(\pi - 2)$ (e) 0

(f) $0, -1; \ -1, e^{-1}/2; \ 1, e/2$ (g) $0, 2; \ 1, -1$ (h) $0, -3; \ 1, 3$ (i) $-\dfrac{1}{16} + \dfrac{1}{24} i$

(j) $\dfrac{3 - 2i}{6}; -3i, \dfrac{3 + 2i}{6}$ (k) $2e^{\pm i\pi/4}, \dfrac{(1 \mp i)}{8\sqrt{2}}; 2e^{\pm 3i\pi/4}, \dfrac{(-1 \mp i)}{8\sqrt{2}}$ (l) $\pm 2i, \mp ie^2/4$

82 (a) $\pi i/2$ (b) $(2 + \pi^3/64)\pi i$ (c) $-\pi i$ (d) $2\pi i$ (e) $-2\pi i$
83 (a) 0 (b) $2\pi i$ (c) 0

Section 10.9
85 $\pi/2, \pi/2$ **87** $0, \pi/[bc(b + c)], \pi/(2b^3)$
89 (i) $2\pi i$ **90** $i/\pi, 5\pi/32$
92 (i) $2\pi/(1 - x^2), 0 < x < 1, 2\pi/(x^2 - 1), x > 1$ (ii) $\pi e^{-2a}/a$

Section 10.10
94 (a) $[e^{-t} + e^{2t}(3t - 1)]/9$ (b) $[e^{-t} + e^t(2t - 1)]/4$ (c) $(\sin t - t \cos t)/2$

Chapter 11
Section 11.2
1   Yes   2   No   3   (519, 581)   4   375   5   108
6   Reject old process if mean breaking strength of sample exceeds 409.4 kN; 0.004

Section 11.3
9   (a) 1%   (b) 5%   (c) 10%   (d) 97.5%   (e) 99.5%
    (f) 0.25%   (g) 1.25%   (h) 87.5%
10 2.13, 2.05, 2.03, 1.96   11 No; highly significant   12 $t = -3$; significant
13 No   14 $t = 3.014$; yes   15 1.224 ± 0.006 cm; 1.024 ± 0.008 cm
16 No; $t = 1.203$ 17 34.2 ± 0.4, 34.2 ± 0.6 18 6.2 ± 0.46
19 123.77 ± 0.23 20 Cannot reject claim
21 (i) 4.0   (ii) 0.33   (iii) (3.61, 4.39)   (iv) (3.56, 4.44)

Section 11.4
22 (i)   21.03, 14.07, 7.815, 101.9, 232.9   (ii) 26.22, 18.48, 11.34, 112.3, 246.6
    (iii) 11.34, 6.346, 2.366, 79.33, 200   (iv) 8%, 91%
    (v) (a) 95%   (b) 90%   (c) 45%
23 (1.54, 14.6)   24 No, $\chi^2 = 60.44$ 25 $\chi^2 = 64.96$, reject H_0
26 Reject first claim   27 $111.6 < \sigma < 198.8$ 28 No
29 $3.125 \leq \sigma^2 \leq 48.135$ 30 (2.84, 6.75)   31 (0.962, 39.1)

Section 11.5
32 (a) 4.53, 3.64, 2.07, 2.05, 2.04, 2.08   (b) 15.4, 2.16
    (c) 5.81, 5.06   (d) 0.037, 0.153
33 Yes, $F = 1.90$
34 (a) No   (b) Yes, $\hat{\sigma}^2 = 84.2$ (c) Yes, $\hat{\sigma}^2 = 4.9$ (d) No
35 No, $F = 2.289$ 36 Not significant at 5% level
37 Significant at 5% level, not at 1% level   38 No, $F = 2.04$ 39 No

Section 11.6
40 $t = 0.976$, No   41 Doubtful, $t = 3.22$
42 Not significant in either case   43 Yes   44 No
45 No significant difference between balances
46 No significant difference in means   47 No
48 Reject; take readings at same times
49 (i) 95% confidence between variances (almost too small an F value to be true) and no
        significant difference between means
50 (a) 194 ± 8.91 (b) (i) 95.625   (iii) significant at 5% level, but not at 1% level.

Section 11.7
51 No   52 Yes
53 Reject hypothesis that mean dust loadings are the same for different flow rates
54 Yes

Section 11.8

55	95% C.I. for α:	0.41 ± 1.58	95% C.I. for β:	0.039 ± 0.005
	99% C.I. for α:	0.41 ± 2.89	99% C.I. for β:	0.039 ± 0.008
56	95% C.I. for α:	3.20 ± 11.22	95% C.I. for β:	14.09 ± 2.88
	99% C.I. for α:	3.20 ± 18.58	99% C.I. for β:	14.09 ± 4.78
57	95% C.I. for α:	53.24 ± 3.25	95% C.I. for β:	0.58 ± 0.11
	99% C.I. for α:	53.24 ± 5.39	99% C.I. for β:	0.58 ± 0.18
58	95% C.I. for α:	42.60 ± 14.49	95% C.I. for β:	-0.69 ± 0.56
	99% C.I. for α:	42.60 ± 20.61	99% C.I. for β:	-0.69 ± 0.79
59	95% C.I. for α:	6.41 ± 2.09	95% C.I. for β:	1.81 ± 1.36
	99% C.I. for α:	6.41 ± 2.17	99% C.I. for β:	1.81 ± 1.42
60	95% C.I. for α:	35.82 ± 22.68	95% C.I. for β:	0.48 ± 0.34
	99% C.I. for α:	35.82 ± 32.25	99% C.I. for β:	0.48 ± 0.48
61	95% C.I. for α:	1.00 ± 0.24	95% C.I. for β:	12.50 ± 5.42
	99% C.I. for α:	1.00 ± 0.38	99% C.I. for β:	12.50 ± 8.51
62	95% C.I. for α:	83.70 ± 23.62	95% C.I. for β:	1.09 ± 0.45
	99% C.I. for α:	83.70 ± 33.59	99% C.I. for β:	1.09 ± 0.64
63	95% C.I. for α:	-0.86 ± 4.57	95% C.I. for β:	0.27 ± 0.10
	99% C.I. for α:	-0.86 ± 6.49	99% C.I. for β:	0.27 ± 0.14
64	95% C.I. for α:	-2.20 ± 11.00	95% C.I. for β:	0.29 ± 0.23
	99% C.I. for α:	-2.20 ± 18.20	99% C.I. for β:	0.29 ± 0.38

66 $y = 1.5x$ **67** $3.2, 14.1$ **68** $y = 16.27 - 4.21/x$

69 $y = 0.43x - 0.01x^2$ **70** 20p/hour, 95% C.I., 20 ± 9.461

71
(i)	10.14 ± 1.66	(ii)	76.45 ± 13.90	(iii)	61.90 ± 4.96
(iv)	25.43 ± 3.84	(v)	9.00 ± 1.50	(vi)	66.22 ± 3.40
(vii)	1.73 ± 0.32	(viii)	149.20 ± 18.40	(ix)	12.49 ± 3.04
(x)	12.45 ± 1.97				

72 $3.94, 1.80$

95% C.I. For α, β 3.94 ± 2.13, 1.8 ± 1.39

99% C.I. For α, β 3.94 ± 3.05, 1.8 ± 2.00

$y(1.6) = 6.82 \pm 1.24$ [90% C.I.]

Mean value of $y(1.6) = 6.82 \pm 0.73$ [90% C.I.]

Section 11.9

73 $0.996, 0.98, 0.98, 0.43, 0.50, 0.49, 0.87, 0.75, 0.78, 0.76$

74 0.28 **75** 0.058 **76** $\hat{a} = 1.83$, $\hat{b} = -1.3 \times 10^{-2}$, $r^2 = 0.91$

77 $F_3^1 = 680.6$, $F_4^1 = 184.3$, $F_4^1 = 223.8$, $F_{10}^1 = 7.54$, $F_9^1 = 9.00$, $F_{10}^1 = 9.75$, $F_5^1 = 35.0$,

$F_{10}^1 = 29.46$, $F_{10}^1 = 34.65$, $F_4^1 = 12.67$

78 $17.48, 5.11$; 104.3. Reject hypothesis that $\alpha = 6$ at 5% level

INDEX